Yakov Berkovich and Zvonimir Janko

Groups of Prime Power Order

De Gruyter Expositions in Mathematics

Volume 62

Yakov Berkovich and Zvonimir Janko

Groups of Prime Power Order

Volume 5

DE GRUYTER

Mathematics Subject Classification 2010
20-02, 20D15, 20E07

Authors
Prof. Yakov G. Berkovich
18251 Afula
Israel
berkov@math.haifa.ac.il

Prof. Dr. Zvonimir Janko
Ruprecht-Karls-Universität Heidelberg
Mathematisches Institut
Im Neuenheimer Feld 288
69120 Heidelberg
Germany
janko@mathi.uni-heidelberg.de

ISBN 978-3-11-029534-4
e-ISBN (PDF) 978-3-11-029535-1
e-ISBN (EPUB) 978-3-11-038904-3
Set-ISBN 978-3-11-029536-8
ISSN 0938-6572

Library of Congress Cataloging-in-Publication Data
A CIP catalog record for this book has been applied for at the Library of Congress.

Bibliographic information published by the Deutsche Nationalbibliothek
The Deutsche Nationalbibliothek lists this publication in the Deutsche Nationalbibliografie; detailed bibliographic data are available on the Internet at http://dnb.dnb.de.

Typesetting: le-tex publishing services GmbH, Leipzig
Printing and binding: CPI Books GmbH, Leck
♾ Printed on acid-free paper
Printed in Germany

www.degruyter.com

Contents

List of definitions and notations

Set theory

- $|M|$ is the cardinality of a set M (if G is a finite group, then $|G|$ is called its order).
- $x \in M$ ($x \notin M$) means that x is (is not) an element of a set M. $N \subseteq M$ ($N \nsubseteq M$) means that N is (is not) a subset of the set M; moreover, if $M \neq N \subseteq M$, we write $N \subset M$.
- $\emptyset$ is the empty set.
- N is called a nontrivial subset of M, if $N \neq \emptyset$ and $N \subset M$. If $N \subset M$, we say that N is a proper subset of M.
- $M \cap N$ is the intersection and $M \cup N$ is the union of sets M and N. If M, N are sets, then $N - M = \{x \in N \mid x \notin M\}$ is the difference of N and M.

Number theory and general algebra

- p is always a prime number.
- π is a set of primes; π' is the set of all primes not contained in π.
- m, n, k, r, s are, as a rule, natural numbers.
- If $\pi(m)$ is the set of prime divisors of m, then m is a π-number if $\pi(m) \subseteq \pi$.
- n_p is the p-part of n, and n_π is the π-part of n.
- $\mathrm{GCD}(m, n)$ is the greatest common divisor of m and n.
- $\mathrm{LCM}(m, n)$ is the least common multiple of m and n.
- $m \mid n$ should be read as: m divides n.
- $\mathrm{GF}(p^m)$ is the finite field containing p^m elements.
- F^* is the multiplicative group of a field F.
- $\mathcal{L}(G)$ is the lattice of subgroups of a group G.
- $\mathcal{L}_N(G)$ is the lattice of normal subgroups of a group G.
- If $n = p_1^{\alpha_1}, \dots, p_k^{\alpha_k}$ is the standard prime decomposition of n, then $\lambda(n) = \sum_{i=1}^{k} \alpha_i$.

Groups

We consider only finite groups which are denoted, with a pair exceptions, by upper case Latin letters.

- If G is a group, then $\pi(G) = \pi(|G|)$.
- G is a p-group if $|G|$ is a power of p; G is a π-group if $\pi(G) \subseteq \pi$.
- G is, as a rule, a finite p-group.
- $H \leq G$ means that H is a subgroup of G.

- $H < G$ means that $H \leq G$ and $H \neq G$ (in that case H is called a *proper* subgroup of G). $\{1\}$ denotes the group containing only one element.
- H is a nontrivial subgroup of G if $\{1\} < H < G$.
- H is a maximal subgroup of G if $G > \{1\}$, $H < G$ and it follows from $H \leq M < G$ that $H = M$.
- If H is a proper normal subgroup of G, then we write $H \triangleleft G$. Expressions "normal subgroup of G" and "G-invariant subgroup" are synonyms.
- A normal subgroup H of G is nontrivial provided $G > H > \{1\}$.
- H is a minimal normal subgroup of G if (a) H normal in G; (b) $H > \{1\}$; (c) $N \triangleleft G$ and $N < H$ implies $N = \{1\}$. Thus, the group $\{1\}$ has no minimal normal subgroup.
- A group G is *metabelian* if G/A is abelian for some abelian $A \triangleleft G$. Groups of class 2 are metabelian but the converse is not true.
- A p-group G is said to be *Dedekindian* if all its subgroups are normal.
- A group G is said to be *minimal nonabelian* if it is nonabelian but all its proper subgroups are abelian.
- $\mathcal{A}_1(G)$ is the set of all minimal nonabelian subgroups of a p-group G.
- A group H is said to be *capable* if there exists a group G such that $G/\mathrm{Z}(G) \cong H$.
- A p-group G is said to be *metahamiltonian* if all its minimal nonabelian (so all nonabelian) subgroups are normal.
- $H \leq G$ is quasinormal if it is permutable with all subgroups of G.
- A p-group is said to be *modular* if all its subgroups are quasinormal.
- H is a maximal normal subgroup of G if $H < G$ and G/H is simple.
- The subgroup generated by all minimal normal subgroups of G is called the *socle* of G and denoted by $\mathrm{Sc}(G)$. We put, by definition, $\mathrm{Sc}(\{1\}) = \{1\}$.
- $\mathrm{N}_G(M) = \{x \in G \mid x^{-1}Mx = M\}$ is the normalizer of a subset M in G.
- $\mathrm{C}_G(x)$ is the centralizer of an element x in $G : \mathrm{C}_G(x) = \{z \in G \mid zx = xz\}$.
- (G) is the number of nonidentity cyclic subgroups of G.
- $\mathrm{C}_G(M) = \bigcap_{x \in M} \mathrm{C}_G(x)$ is the centralizer of a subset M in G.
- If $A \leq B$ and A, B are normal in G, then $\mathrm{C}_G(B/A) = H$, where $H/A = \mathrm{C}_{G/A}(B/A)$.
- $H < G$ is a TI-subgroup if $T \cap T^x = \{1\}$ for all $x \in G - \mathrm{N}_G(H)$.
- $A \text{ wr } B$ is the wreath product of the "passive" group A and the transitive permutation group B (in what follows we assume that B is a regular permutation group, as a rule, a p-group); B is called the active factor of the wreath product). Then the order of that group is $|A|^{|B|} \cdot |B|$.
- $\mathrm{Aut}(G)$ is the group of automorphisms of G (the automorphism group of G)
- $\mathrm{Inn}(G)$ is the group of all inner automorphisms of G.
- $\mathrm{Out}(G) = \mathrm{Aut}(G)/\mathrm{Inn}(G)$ is the outer automorphism group of G.
- $\mathcal{N}(G)$ is the norm of G, the intersection of normalizers of all subgroups of G.
- If $a, b \in G$, then $a^b = b^{-1}ab$ is a conjugate to a in G.
- $a \in G$ is real if it conjugate with a^{-1}.
- An element $x \in G$ inverts a subgroup $H \leq G$ if $h^x = h^{-1}$ for all $h \in H$.

- If $M \subseteq G$, then $\langle M \rangle = \langle x \mid x \in M \rangle$ is the subgroup of G generated by M.
- $M^x = x^{-1}Mx = \{y^x \mid y \in M\}$ for $x \in G$ and $M \subseteq G$.
- $[x, y] = x^{-1}y^{-1}xy = x^{-1}x^y$ is the *commutator* of elements x, y of G. If $M, N \subseteq G$ then $[M, N] = \langle [x, y] \mid x \in M, y \in N \rangle$ is a subgroup of G.
- $o(x)$ is the order of an element x of G.
- An element $x \in G$ is a π-element if $\pi(o(x)) \subseteq \pi$.
- G is a π-group, if $\pi(G) \subseteq \pi$. Obviously, G is a π-group if and only if all of its elements are π-elements.
- G' is the subgroup generated by all commutators $[x, y]$, $x, y \in G$ (i.e., $G' = [G, G]$), $G^{(2)} = [G', G'] = G'' = (G')'$, $G^{(3)} = [G'', G''] = (G'')'$ and so on. G' is called the *commutator* (or *derived*) subgroup of G.
- $\mathrm{Z}(G) = \bigcap_{x \in G} \mathrm{C}_G(x)$ is the center of G.
- $\mathrm{Z}_i(G)$ is the ith member of the upper central series of G; in particular, $\mathrm{Z}_0(G) = \{1\}$, $\mathrm{Z}_1(G) = \mathrm{Z}(G)$.
- $\mathrm{K}_i(G)$ is the ith member of the lower central series of G; in particular, $\mathrm{K}_2(G) = G'$. We have $\mathrm{K}_i(G) = [G, \dots, G]$ ($i \geq 1$ times). We set $\mathrm{K}_1(G) = G$.
- If G is nonabelian, then $\eta(G)/\mathrm{K}_3(G) = \mathrm{Z}(G/\mathrm{K}_3(G))$.
- $\mathcal{M}(G) = \langle x \in G \mid \mathrm{C}_G(x) = \mathrm{C}_G(x^p) \rangle$ is the Mann subgroup of a p-group G.
- $\mathrm{Syl}_p(G)$ is the set of p-Sylow subgroups of an arbitrary finite group G.
- S_n is the *symmetric* group of degree n.
- A_n is the *alternating* group of degree n.
- Σ_{p^n} is a Sylow p-subgroup of S_{p^n}.
- $\mathcal{H}_{2,p}$ is a nonabelian metacyclic p-group of order p^4 and exponent p^2.
- $\mathrm{GL}(n, F)$ is the set of all nonsingular $n \times n$ matrices with entries in a field F, the n-dimensional *general linear* group over F, $\mathrm{SL}(n, F) = \{A \in \mathrm{GL}(n, F) \mid \det(A) = 1 \in F\}$, the n-dimensional *special linear* group over F.
- If $H \leq G$, then $H_G = \bigcap_{x \in G} x^{-1}Hx$ is the *core* of the subgroup H in G and H^G, the intersection of all normal subgroups of G containing H, is the *normal closure* or *normal hull* of H in G. Obviously, H_G is normal in G.
- If G is a p-group, then $p^{b(x)} = |G : \mathrm{C}_G(x)|$; $b(x)$ is said to be the *breadth* of $x \in G$, where G is a p-group; $b(G) = \max\{b(x) \mid x \in G\}$ is the *breadth* of G.
- If $H \leq G$ and $|G : \mathrm{N}_G(H)| = p^{\mathrm{sb}(H)}$, when $\mathrm{sb}(H)$ is said to be the *subgroup breadth* of H. Next, $\mathrm{sb}(G) = \max\{\mathrm{sb}(H) \mid H \leq G\}$.
- $\Phi(G)$ is the Frattini subgroup of G (= the intersection of all maximal subgroups of G), $\Phi(\{1\}) = \{1\}$, $p^{\mathrm{d}(G)} = |G : \Phi(G)|$.
- $\Gamma_i = \{H < G \mid \Phi(G) \leq H, |G : H| = p^i\}$, $i = 1, \dots, \mathrm{d}(G)$, where $G > \{1\}$.
- If $H < G$, then $\Gamma_1(H)$ is the set of all maximal subgroups of H.
- $\exp(G)$ is the exponent of G (the least common multiple of the orders of elements of G). If G is a p-group, then $\exp(G) = \max\{o(x) \mid x \in G\}$.
- $k(G)$ is the number of conjugacy classes of G (= G-classes), the class number of G.
- K_x is the G-class containing an element x (sometimes we also write $ccl_G(x)$).

- C_m is the cyclic group of order m.
- $A \times B$ is the direct product of groups A and B.
- $G^m = G \times \cdots G$ (m times) is the direct product of m copies of a group G.
- $A * B$ is a central product of groups A and B, i.e., $A * B = AB$ with $[A, B] = \{1\}$. In particular, the direct product $A \times B$ is a central product of groups A and B.
- $E_{p^m} = C_p^m$ is the elementary abelian group of order p^m. G is an elementary abelian p-group if and only if it is a p-group $> \{1\}$ and G coincides with its socle. Next, $\{1\}$ is elementary abelian for each prime p, by definition.
- A group G is said to be *homocyclic* if it is a direct product of isomorphic cyclic subgroups (obviously, elementary abelian p-groups are homocyclic).
- $ES(m, p)$ is an *extraspecial* group of order p^{1+2m} (a p-group G is said to be extraspecial if $G' = \Phi(G) = Z(G)$ is of order p). Note that for each positive integer m, there are exactly two nonisomorphic extraspecial groups of order p^{2m+1}.
- $S(p^3)$ is a nonabelian group of order p^3 and exponent $p > 2$.
- A *special* p-group is a nonabelian p-group G such that $G' = \Phi(G) = Z(G)$ is elementary abelian. Direct products of extraspecial p-groups are special.
- D_{2m} is the *dihedral* group of order $2m$, $m > 2$. Some authors consider E_{2^2} as the dihedral group D_4.
- Q_{2^m} is the *generalized quaternion* group of order $2^m \geq 2^3$.
- SD_{2^m} is the *semidihedral group* of order $2^m \geq 2^4$.
- M_{p^m} is a nonabelian p-group containing exactly p cyclic subgroups of index p (see Theorem 1.2).
- $cl(G)$ is the *nilpotence class* of a p-group G.
- $dl(G)$ is the *derived length* of a p-group G.
- $CL(G)$ is the set of all G-classes.
- A p-group of *maximal class* is a nonabelian group G of order p^m with $cl(G) = m - 1$.
- A p-group is s-self-dual if every its subgroup is isomorphic to a quotient group.
- A p-group is q-self-dual if every its quotient group is isomorphic to a subgroup.
- $GL(n, F)$ ($SL(n, p)$) the general (special) linear group of degree n over the field F.
- $\Omega_m(G) = \langle x \in G \mid o(x) \leq p^m \rangle$, $\Omega_m^*(G) = \langle x \in G \mid o(x) = p^m \rangle$ and $\mho_m(G) = \langle x^{p^m} \mid x \in G \rangle$.
- A p-group G is said to be *regular*, if for any $x, y \in G$ there exists $z \in \langle x, y \rangle'$ such that $(xy)^p = x^p y^p z^p$.
- A p-group is *absolutely regular* if $|G/\mho_1(G)| < p^p$. Absolutely regular p-groups are regular.
- A p-group is *thin* if it is either absolutely regular or of maximal class.
- $G = A \cdot B$ is a *semidirect product* with *kernel* B and *complement* A.
- A group G is an extension of a normal subgroup N by a group H if $G/N \cong H$. A group G splits over N if $G = H \cdot N$ with $H \leq G$ and $H \cap N = \{1\}$ (in that case, G is a semidirect product of H and N with kernel N).

- $H^\# = H - \{e_H\}$, where e_H is the identity element of the group H. If $M \subseteq G$, then $M^\# = M - \{e_G\}$.
- An automorphism α of G is *regular* (= *fixed-point-free*) if it induces a regular permutation on $G^\#$ (a permutation is said to be *regular* if it has no fixed points).
- An *involution* is an element of order 2 in a group.
- A group G is said to *metacyclic* if it contains a normal cyclic subgroup C such that G/C is cyclic.
- A group G is said to be *minimal nonmetacyclic* if it is nonmetacyclic but all its proper subgroups are metacyclic.
- A subgroup A of a group G is said to be *soft*, if $\mathrm{C}_G(A) = A$ and $|\mathrm{N}_G(A) : A| = p$.
- A *section* of a group G is an epimorphic image of some subgroup of G.
- If $F = \mathrm{GF}(p^n)$, then we usually write $\mathrm{GL}(m, p^n), \mathrm{SL}(m, p^n), \dots$ instead of $\mathrm{GL}(m, F), \mathrm{SL}(m, F), \dots$.
- $\mathrm{c}_n(G)$ is the number of cyclic subgroups of order p^n in a p-group G.
- $\mathrm{c}(G)$ is the number of non-identity cyclic subgroups of G.
- $\mathrm{s}_n(G)$ is the number of subgroups of order p^n in a p-group G.
- $v_n(G)$ is the number of normal subgroups of order p^n in a p-group G.
- $e_n(G)$ is the number of subgroups of order p^n and exponent p in G.
- $M_p(m, n) = \langle a, b \mid o(a) = p^m,\ o(b) = p^n,\ a^b = a^{1+p^{m-1}} \rangle$ is the metacyclic minimal nonabelian group of order p^{m+n}.
- $M_p(m, n.1) = \langle a, b \mid o(a) = p^m,\ o(b) = p^n,\ [a, b] = c,\ [a, c] = [b, c] = c^p = 1 \rangle$ is the nonmetacyclic minimal nonabelian group of order p^{m+n+1}.
- An $\mathcal{A}_n$-group is a p-group G all of whose subgroups of index p^n are abelian but G contains a nonabelian subgroup of index p^{n-1}. In particular, $\mathcal{A}_1$-group is a minimal nonabelian p-group for some p.
- $\alpha_n(G)$ is the number of $\mathcal{A}_n$-subgroups in a p-group G.
- $\mathcal{D}_n$-group is a 2-group all of whose subgroups of index p^n are Dedekindian, containing a non-Dedekindian subgroup of index p^{n-1} and which is not an $\mathcal{A}_n$-group.
- $\mathcal{MA}(G)$ is the set of minimal nonabelian subgroups of a p-group G.
- $\mathcal{MA}_k(G) = \{H \in \mathcal{MA}(G) \mid \Omega_k(H) = H\}$.
- $\mathrm{D}_k(G) = \langle \mathcal{MA}_k(G) \rangle = \langle H \mid H \in \mathcal{MA}_k(G) \rangle$.
- $L_n = |\{x \in G \mid x^n = 1\}|$.
- If G is a metacyclic p-group and $w(G) = \max\{i \mid |\Omega_i(G)| = p^{2^i}\}$, then $R(G) = \Omega_{w(G)}(G)$. In that case, $G/R(G)$ is either cyclic or a 2-group of maximal class.

Characters and representations

- $\mathrm{Irr}(G)$ is the set of all *irreducible* characters of G over complex numbers.
- A character of degree 1 is said to be *linear*.
- $\mathrm{Lin}(G)$ is the set of all *linear* characters of G (obviously, $\mathrm{Lin}(G) \subseteq \mathrm{Irr}(G)$).

- $\mathrm{Irr}_1(G) = \mathrm{Irr}(G) - \mathrm{Lin}(G)$ is the set of all *nonlinear* irreducible characters of G; $\mathrm{n}(G) = |\mathrm{Irr}_1(G)|$.
- $\chi(1)$ is the *degree* of a character χ of G.
- χ_H is the *restriction* of a character χ of G to $H \leq G$.
- χ^G is the character of G induced from the character χ of some subgroup of G.
- $\bar{\chi}$ is a character of G defined as follows: $\bar{\chi}(x) = \overline{\chi(x)}$ (here $\bar{w}$ is the complex conjugate of a complex number w).
- $\mathrm{X}(G)$ is a character table of G.
- $\mathrm{Irr}(\chi)$ is the set of irreducible constituents of a character χ of G.
- 1_G is the principal character of G.
- $\mathrm{Irr}^{\#}(G) = \mathrm{Irr}(G) - \{1_G\}$.
- If χ is a character of G, then $\ker(\chi) = \{x \in G \mid \chi(x) = \chi(1)\}$ is the *kernel* of a character χ.
- $\mathrm{Z}(\chi) = \{x \in G \mid |\chi(x)| = \chi(1)\}$ is the *quasikernel* of χ.
- If N is normal in G, then $\mathrm{Irr}(G \mid N) = \{\chi \in \mathrm{Irr}(G) \mid N \not\leq \ker(\chi)\}$.
- $\langle \chi, \tau \rangle = |G|^{-1} \sum_{x \in G} \chi(x)\tau(x^{-1})$ is the *inner product* of characters χ and τ of G.
- $\mathrm{I}_G(\phi) = \langle x \in G \mid \phi^x = \phi \rangle$ is the *inertia subgroup* of $\phi \in \mathrm{Irr}(H)$ in G, where $H \triangleleft G$.
- 1_G is the *principal character* of G ($1_G(x) = 1$ for all $x \in G$).
- $\mathrm{M}(G)$ is the *Schur multiplier* of G.
- $\mathrm{cd}(G) = \{\chi(1) \mid \chi \in \mathrm{Irr}(G)\}$.
- $\mathrm{mc}(G) = \mathrm{k}(G)/|G|$ is the *measure of commutativity* of G.
- $\mathrm{T}(G) = \sum_{\chi \in \mathrm{Irr}(G)} \chi(1)$.
- $\mathrm{f}(G) = \mathrm{T}(G)/|G|$.

Preface

This is the fifth volume of the series devoted to elementary parts of finite p-group theory. The material presented here has appeared in a book form for the first time.

Below we list some new characterizations of certain classes of p-groups and results presented in this volume:

(1) classification of non-Dedekindian groups in which the normal closures of nonnormal cyclic subgroups are nonabelian (abelian),
(2) (i) p-groups with all subgroups isomorphic to quotient groups (s-self-dual groups), (ii) minimal non-s-self p-groups,
(3) p-groups all of whose maximal subgroups, except one, are s-self-dual,
(4) nonabelian p-groups all of whose subgroups are q-self-dual,
(5) p-groups with all subgroups isomorphic to quotient groups (q-self-dual groups),
(6) minimal non-q-self-dual 2-groups,
(7) p-groups with absolutely regular normalizer of some subgroup,
(8) p-groups all of whose $\mathcal{A}_2$-subgroups are metacyclic,
(9) p-groups all of whose $\mathcal{A}_1$-subgroups are pairwise nonisomorphic,
(10) metacyclic groups of exponent p^e with a normal cyclic subgroup of order p^e,
(11) another proof of Baer's theorem about 2-groups with nonabelian norm,
(12) properties of metahamiltonian p-groups,
(13) p-groups all of whose nonnormal subgroups are elementary abelian,
(14) large elementary abelian subgroups of a p-group of maximal class,
(15) p-groups all of whose nonabelian two-generator subgroups are minimal nonabelian,
(16) p-groups all of whose three pairwise noncommuting elements generate a p-group of maximal class,
(17) § 190 p-groups containing a subgroup of maximal class and index p,
(18) groups G such that $H < G$ and $|H|^2 < |G|$ imply that $H \triangleleft G$,
(19) nonabelian p-groups in which any nonabelian subgroups contains its centralizer,
(21) groups in which the normal closure of any cyclic subgroup is abelian,
(22) nonabelian p-groups with an abelian subgroup of index p covered by minimal nonabelian subgroups,
(23) non-Dedekindian p-groups which are not generated by their nonnormal subgroups,
(24) certain groups cannot be normal subgroups of capable p-groups,
(25) nonabelian p-groups, $p > 2$, of exponent p^e all of whose cyclic subgroups of order p^e are normal,
(26) non-Dedekindian p-groups in which the normal closures of nonnormal cyclic subgroups have cyclic centers,

(27) groups all of whose minimal nonabelian (maximal abelian) subgroups are isolated, p-groups saturated by isolated subgroups,
(28) p-groups whose the proper Hughes subgroup have Frattini subgroup of order p,
(29) computation of orders of some 2-groups all of whose subgroups of given index are Dedekindian,
(30) an estimate of the number of minimal nonabelian subgroups in a group with a given quotient group G/A, where A is maximal abelian normal subgroup of G,
(31) the number of epimorphic images of maximal class and order in a 2-group,
(32) the number of cyclic subgroups of given order in a metacyclic p-group,
(33) the estimate of the number of minimal nonabelian subgroups in a p-group,
(34) p-groups G all of whose nonnormal subgroups contain G' in its normal closure, etc.

For further information, see Contents.

The problems from section "Research problems and themes," Parts V, are posed, if it is not stated otherwise, by the first author. That section contains more than 570 items some of which are solved by the second author. Many problems from the lists in previous three volumes were solved mainly by the second author and mathematicians from Shanxi Normal University.

There are, in the text, many hundreds exercises most of which are solved (these exercises, if it is not stated otherwise, written by the first author).

Sections and appendices written by the authors, are listed in the Author Index. Avinoam Mann wrote Appendix 65, Qinhai Zhang wrote § 205. In § 204 we have used papers of Robert van der Waall. Some sections were discussed with Isaacs and Mann. The problems were discussed with Mann and, of course, with the second author. M. Roitman (Haifa University) helped in whole project. We are indebted to all named mathematicians.

We added in the Bibliography of volume III those items that are cited in this volume and omitted from the so-obtained Bibliography those items that are not cited here.

We are grateful to the publishing house of Walter de Gruyter and all its workers supporting and promoting the publication of this and four previous volumes.

§ 190 On p-groups containing a subgroup of maximal class and index p

If a p-group G contains an absolutely regular subgroup H of index p and order $> p^p$, then it contains at most one normal subgroup of order p^p and exponent p (Theorems 12.1 (b) and 9.6 (c)). In this section, we consider the case when H, an irregular subgroup of maximal class, has index p in a p-group G. Such p-groups have appeared many times in our book (see, for example, Theorem 12.12). It is natural to treat the structure of such groups in detail. If G is such a p-group, it is either of maximal class or $\mathrm{d}(G) = 3$ (Theorem 12.12 (a)). If G is not of maximal class, it possesses a normal subgroup of order p^p and exponent p (Theorem 12.1 (a)). We show that then our G, provided $|G| \geq p^{p+3}$, has exactly one normal subgroup of order p^p and exponent p. Note that p^{p+3} is the best possible value here (indeed, the group $G = \Sigma_{p^2} \times C$ of order p^{p+2}, where $\Sigma_{p^2} \in \mathrm{Syl}_p(\mathrm{S}_{p^2})$, $|C| = p$ and S_{p^2} is the symmetric group of degree p^2, has two distinct normal subgroups of order p^p and exponent p). We also consider the case when $|G| = p^{p+2}$ and $e_p(G) = 1$, where $e_n(G)$ is the number of subgroups of order p^n and exponent p in G.

First, we prove the following.

Lemma 190.1. *Suppose that G is a p-group of order $> p^{p+3}$. If G contains a subgroup H of maximal class and index p, then one of the following holds:*
(a) *G is of maximal class.*
(b) *G contains only one normal subgroup of order p^p and exponent p.*

Proof. By Theorem 9.5, H is irregular, and so is G. Assume that G is not of maximal class. By Theorem 12.1 (a), G contains a normal subgroup R of order p^p and exponent p. Since $R \nleq H$ (Theorem 9.6 (c)), one obtains $HR = G$ so that (below we use Exercise 9.1 (b))

$$|H \cap R| = p^{p-1} \Rightarrow H \cap R = \Omega_1(\Phi(H)) \Rightarrow \mathrm{d}(H/(H \cap R))) = 2$$

since $H/(H \cap R)$ is of maximal class and order $\geq p^4$. Then, in view of $G/R \cong H/(H \cap R)$ and $H \cap R \leq \Phi(H) \leq \Phi(G)$, one has $G/(H \cap R) = H/(H \cap R) \times R/(H \cap R)$ so that (see Theorem 12.12 (a))

$$\mathrm{d}(G) = \mathrm{d}(H/(H \cap R)) + \mathrm{d}(R/(H \cap R) = 2 + 1 = 3\,.$$

(The last equality also follows from Theorem 12.12 (a).)

Assume, by way of contradiction, that G contains a G-invariant subgroup $R_1 \neq R$ of order p^p and exponent p. Write $D = RR_1$. As above, $|H \cap D_1| = p^{p-1}$. As H has only one normal subgroup of order p^{p-1} (Exercise 9.1 (b)), the subgroup $R_1 \cap H = \Omega_1(\Phi(H)) = R \cap H = R \cap R_1$ has order p^{p-1}; then, by the product formula, $D \triangleleft G$ is of order p^{p+1}. Next, $\Omega_1(D) = RR_1 = D$. Assume that D is regular. Then $\exp(D) = p$ (Theorem 7.2 (b)) so that $H \cap D \lhd H$ is of order p^p and exponent p, contrary to Theorem 9.6 (c). Thus, D is irregular of order p^{p+1} so that it is of maximal class (Theorem 7.1 (b)).

Let $T/D \lhd G/D$ be of order p. As $R < T$ and $|T| = p^{p+2} > p^{p+1}$, the subgroup T is not of maximal class (Theorem 9.6 (b)). Next, $T \nleq H$ since $D \nleq H$. The intersection $K = H \cap T$ is an H-invariant maximal subgroup of T so that $|K| = p^{p+1}$, by the product formula. As $|H : K| \geq (p^{p+3})/(p^{p+1}) = p^2$, it follows that $K \leq \Phi(H)$, and therefore K is absolutely regular (Theorem 9.6 (e)). Application of Theorem 12.1 (b) to the pair $K < T$ shows that $|\Omega_1(T)| = p^p$, a contradiction since $\Omega_1(T) \geq D = \Omega_1(D)$ and $|D| = p^{p+1}$. Thus, R is the unique normal subgroup of G of order p^p and exponent p. □

As the following theorem shows, one can assume, in Lemma 190.1, that $|G| \geq p^{p+3}$ instead of $|G| > p^{p+3}$.

Theorem 190.2. *Let H be a subgroup of maximal class and index p in a p-group G, $|G| = p^m \geq p^{p+3}$. If G is not of maximal class, then G has exactly one normal subgroup, say R, of order p^p and exponent p. There are in the set Γ_1 exactly p^2 subgroups of maximal class (all these subgroups do not contain R).*

Proof. In view of Lemma 190.1, one may assume that $m = p + 3$; then $|H| = p^{p+2} > p^{p+1}$ so H is irregular (Theorem 9.5). Assume that in G there are two distinct normal subgroups R and S of order p^p and exponent p. By Theorem 9.6 (c), $R, S \nleq H$. Write $D = RS$; then $D \lhd G$. As in the proof of Lemma 190.1, D is of maximal class and order p^{p+1} and $\exp(D \cap H) = p^2$. Let $D < T \lhd G$, where $|T : D| = p$; then $|T| = p^{p+2}$ so $T(\in \Gamma_1)$ is not of maximal class since $R < D < T$ (Theorem 9.6 (c) again).

Let $R_1, \dots, R_{p+1}$ be pairwise distinct normal subgroups of order p^p and exponent p is G (they exist, by Theorem 13.5 and assumption). Let H_1 be the fundamental subgroup of H (see Theorem 9.6). Write $T_i = H_1R_i$, $i = 1, \dots, p + 1$; H_1 is maximal in T_i and T_i is not of maximal class for all i (Theorem 9.6 (c)). Then $R_i = \Omega_1(T_i)$ for $i = 1, \dots, p + 1$ (Theorem 12.1 (b)) and hence $T_1, \dots, T_{p+1} \in \Gamma_1$ are pairwise distinct. By Theorem 12.12 (b), $\Gamma_1 = \{M_1, \dots, M_{p^2}, T_1, \dots, T_{p+1}\}$. where $M_1, \dots, M_{p^2}$ are pairwise distinct subgroups of maximal class and $T_1, \dots, T_{p+1}$ are not of maximal class. As $T \in \Gamma_1$ (see the previous paragraph) is not of maximal class and $|\Omega_1(T)| \geq |D| > p^p = |\Omega_1(T_i)|$ for all i (see the first paragraph of the proof), it follows that $T \notin \{M_1, \dots, M_{p^2}, T_1, \dots, T_{p+1}\}(= \Gamma_1)$, a final contradiction. □

Proposition 190.3. *Let a p-group G of order p^{p+2} be not of maximal class. Suppose that G contains a subgroup H of maximal class and index p. If G has a unique normal subgroup, say R, of order p^p and exponent p, then $|\Omega_1(H)| \in \{p^p, p^{p-1}\}$.*

Proof. By Theorem 12.1 (a), G has a normal subgroup R of order p^p and exponent p, and, by hypothesis, R is a unique subgroup satisfying this condition. It follows that $e_p(H) \in \{0, 1, p\}$. If $e_p(H) = 0$, then $|\Omega_1(H)| = p^{p-1}$. If $R < H$, then $e_p(H) = 1$ since all subgroups of order p^p and exponent p are G-invariant. It remains to consider that case when $e_p(H) = p$. Let $R_1 \neq R$ be a subgroup of order p^p and exponent p in H. Then, by hypothesis, $\mathrm{N}_G(R_1) = H$. In that case, G is of maximal class (Remark 10.5), a contradiction. □

Remark 1 (= Remark 10.5). Let us prove that if H is a proper subgroup of a p-group G and $\mathrm{N}_G(H)$ is of maximal class so is G. If $|\mathrm{N}_G(H) : H| = p$, then $\mathrm{N}_G(H)$ is the unique subgroup of G containing H as a subgroup of index p, and so G is of maximal class, by Exercise 10.10. Now let $|\mathrm{N}_G(H) : H| > p$. Then, by Exercise 9.1 (b), H is characteristic in $\mathrm{N}_G(H)$, and this implies that $G = \mathrm{N}_G(H)$, so that G is of maximal class, by hypothesis. (This is an alternate proof of Remark 10.5.)

If H is a proper normal subgroup of maximal class and order $> p^{p+1}$ of a p-group G and G/H is either cyclic or generalized quaternion, then G has at most one normal subgroup of order p^p and exponent p. Let $R < G$ be a normal subgroup of order p^p and exponent p (Theorem 12.1 (a)). Then $M = HR$ has order $p|H|$, by the product formula (indeed, $|HR : H| = p$ since G/H has no subgroup $\cong \mathrm{E}_{p^2}$). Note that $M/H = \Omega_1(G/H)$. Now the result follows from Theorem 190.2 applied to M.

Suppose that a 2-group G of order $\geq 2^5$ possesses a metacyclic subgroup H of index 2. If E, E_1 be two distinct G-invariant subgroups $\cong \mathrm{E}_8$, then $F = EE_1 = D \times C$, where $D \cong \mathrm{D}_8$ and $|C| = 2$. Indeed, $\Omega_1(H) = E \cap E_1 \cong \mathrm{E}_4$ and $E \cap E_1 \leq Z(F)$ so that $|F| = 16$, by the product formula. Clearly, F is nonabelian (otherwise, $F \cong \mathrm{E}_{16}$, and consideration of $H \cap F$ yields a contradiction). As $\Omega_1(F) = F$ is of order 2^4, F is not minimal nonabelian (Lemma 65.1). Let $K < F$ be minimal nonabelian. Then $F = K(E \cap E_1) = K \times C$, where $C < E \cap E_1$ is of order 2, $C \not\leq K$. Now, $K \cong \mathrm{D}_8$ since K is nonabelian and has a subgroup $K \cap E \cong \mathrm{E}_4$.

§ 191 *p*-groups *G* all of whose nonnormal subgroups contain *G′* in its normal closure

In Theorem 62.1, all non-Dedekindian p-groups G have been determined (up to isomorphism) with the property that the normal closure of each nonnormal subgroup is a maximal subgroup of G. In particular, this normal closure contains the derived subgroup G'.

Therefore, it is natural to try to classify non-Dedekindian p-groups G such that the normal closure of each nonnormal subgroup contains G' (Problem 148 (iii)). A non-Dedekindian p-group G is called a Cl-group if the normal closure H^G of each nonnormal subgroup H of G contains the derived subgroup G'. For such groups G, we show here that G' is abelian (Theorem 191.1) and if $p > 2$, then G' is an elementary abelian, the class of G is at most 3 and $\mho_1(G) \leq Z(G)$ (Theorem 191.5).

Theorem 191.1. *Let G be a Cl-group. Then G' is contained in each maximal normal abelian subgroup of G. In particular, G' is abelian and so G is metabelian.*

Proof. Let A be a maximal normal abelian subgroup of G and assume that $G' \nleq A$. If X is any subgroup of A, then $X^G \leq A$ and so (by our assumption), $X^G = X$. It follows that each subgroup of A is normal in G.

Let Z be any cyclic subgroup in A and note that $Z \trianglelefteq G$. But Aut(Z) is abelian and so

$$A \leq C_G(Z) \trianglelefteq G \text{ and } G/C_G(Z) \text{ is abelian so that } G' \leq C_G(Z)\,.$$

Hence G' centralizes each cyclic subgroup in A and so G' centralizes A. On the other hand, $C_G(A) = A$ and therefore $G' \leq A$, a contradiction. We have proved that $G' \leq A$ and so G is metabelian. □

Proposition 191.2. *Let G be a Cl-group and let A be a maximal normal abelian subgroup of G. If $\Omega_1(A) \nleq Z(G)$, then G' is elementary abelian.*

Proof. Let $a \in \Omega_1(A) - Z(G)$. Then $\langle a \rangle$ is not normal in G. Hence

$$G' \leq \langle a \rangle^G \leq \Omega_1(A)\,,$$

and so G' is elementary abelian. □

Proposition 191.3. *Let G be a Cl-group with G' elementary abelian. Then* cl$(G) \leq 3$ *and $\mho_1(G) \leq Z(G)$ in case $p > 2$ and $\mho_2(G) \leq Z(G)$ in case $p = 2$.*

Proof. Suppose that cl$(G) > 2$. Let $G_0 < G'$ be a G-invariant subgroup such that $|G' : G_0| = p$. Then each element in G_0 is central in G. Indeed, let $g_0 \in G_0 - Z(G)$. Then $o(g_0) = p$ and $\langle g_0 \rangle$ is not normal in G and $\langle g_0 \rangle^G \leq G_0$, a contradiction. We have proved that $G_0 \leq Z(G)$ and so cl$(G) = 3$. Hence, in any case, cl$(G) \leq 3$.

Assume $p > 2$. By Lemma 188.2, for any $x, y \in G$,

$$[x, y^p] = [x, y]^p [x, y, y]^{\binom{p}{2}} = 1 \text{ and so } \mho_1(G) \le Z(G)\,.$$

Suppose $p = 2$. By Lemma 188.2, for any $x, y \in G$,

$$[x, y^4] = [x, y]^4 [x, y, y]^{\binom{4}{2}} = 1 \text{ and so } \mho_2(G) \le Z(G)\,. \qquad \square$$

Proposition 191.4. *Let G be a Cl-group. Then each factor-group G/N, $N \trianglelefteq G$, is also a Cl-group or G/N is Dedekindian.*

Proof. If G/N is Dedekindian, then we are done. Let X/N be a nonnormal subgroup in G/N. Then X is not normal in G and so $G' \le X^G$. But then (using the bar convention) $\bar{X}^{\bar{G}} = X^G/N$ and $\bar{G}' = (G'N)/N$ so that $\bar{G}' \le \bar{X}^{\bar{G}}$ and we are done. $\square$

Theorem 191.5. *Let G be a Cl-group with $p > 2$ or $p = 2$ and G is of class 2. Then G' is elementary abelian and so* $\mathrm{cl}(G) \le 3$ *and* $\mho_1(G) \le Z(G)$.

Proof. Let G be a minimal counter-example to the assertion that G' is elementary abelian. If G does not contain a normal subgroup of type (p, p), then Lemma 1.4 gives that G is either cyclic or a 2-group of maximal class. But G is a Cl-group and so G is non-Dedekindian and this implies that G is noncyclic and in case $p = 2$, G is of class 2 and therefore $G \cong D_8$ or Q_8. But D_8 has a normal four-subgroup and Q_8 is Dedekindian. Hence in any case G has a normal subgroup of type (p, p). Let A be a maximal normal abelian subgroup of G containing a normal subgroup of type (p, p). By Theorem 191.1, we have $G' \le A$. If $\Omega_1(A) \nleq Z(G)$, then Proposition 191.2 gives that G' is elementary abelian, a contradiction.

We have $\Omega_1(A) \le Z(G)$ and so $\exp(A) > p$. Assume for a moment that $\Omega_1(A) \nleq G'$. Let M be a subgroup of order p in $\Omega_1(A)$ such that $M \cap G' = \{1\}$. But then we obtain $(G/M)' \cong G'$ and $(G/M)'$ is elementary abelian, a contradiction. Hence we have $\Omega_1(A) \le \Omega_1(G')$ and so $\Omega_1(A) = \Omega_1(G')$. If $G' = \Omega_1(A)$, then we have a contradiction to the fact that G is a minimal counter-example. Hence, we have $G' > \Omega_1(A)$ so that $\exp(G') > p$.

Suppose that $|\Omega_1(A)| > p$ and let $a \in G' - \Omega_1(A)$ so that $1 \ne a^p \in \Omega_1(A)$. Let $b \in \Omega_1(A) - \langle a^p \rangle$. But then $(G/\langle b \rangle)' = G'/\langle b \rangle$ is not an elementary abelian, a contradiction. Hence, $|\Omega_1(A)| = p$ and so A is cyclic, a final contradiction.

We have proved that G' is an elementary abelian and so, by Proposition 191.3 in case $p > 2$, G is of class ≤ 3 and $\mho_1(G) \le Z(G)$. If $p = 2$, then G is of class 2 and so for any $x, y \in G$, $[x^2, y] = [x, y]^2 = 1$ and so we get again $\mho_1(G) \le Z(G)$. Our theorem is proved. $\square$

Remark 1. By Theorem 62.1, Theorem 191.5 does not hold for $p = 2$ and G of arbitrary class and so Problem 148 (iii) remains open for $p = 2$.

Problem 1. Study the p-groups G all of whose minimal nonabelian subgroups contain G'.

Problem 2. Study the p-groups G such that $H \cap G' = H'$ for any nonabelian $H < G$.

Problem 3. Study the irregular p-groups G all of whose maximal regular subgroups contain G'.

Problem 4. Study the non-Dedekindian p-groups G such that $G' \leq C^G$ for any nonnormal cyclic $C < G$.

§ 192 *p*-groups with all subgroups isomorphic to quotient groups

A group G is s-self-dual if every subgroup of G is isomorphic to a quotient of G. It is q-self-dual if every quotient of G is isomorphic to a subgroup of G. It is self-dual if it is both s-self-dual and q-self-dual. It is known that abelian p-groups are self-dual. Also, $S(p^3)$, a nonabelian group of order p^3 and exponent p is self-dual. The dihedral group D_8 is q-self-dual but not s-self-dual.

It is well known that finite abelian groups are self-dual (see [Fuch, Exercise 22 (b), p. 53]). Here we determine s-self-dual p-groups (Problem 706 (i)) in Theorem 192.1 and in Corollary 192.2 self-dual p-groups are completely determined, where we use to a certain extent a paper of Ying [Yin1]. Through many years the results of [Yin1] were repeated in [ADZ]. However, a classification of q-self-dual p-groups (Problem 706 (ii)) is still an open problem.

Theorem 192.1 ([Yin1]). *A nonabelian p-group G is s-self-dual if and only if G is isomorphic to one of the following groups:*

(a) $\mathrm{M}(n,n)\times A$, $n\ge 2$, *where* $\mathrm{M}(n,n)=\langle a,b\mid a^{p^n}=b^{p^n}=1,\ a^b=a^{1+p^{n-1}}\rangle$ *is a metacyclic minimal nonabelian group of order* p^{2n} *and exponent* p^n *and A is abelian of exponent* $<p^n$.

(b) $S(p^3)\times B$, *where* $S(p^3)$, $p>2$, *is the nonabelian group of order* p^3 *and exponent p and B is elementary abelian.*

Corollary 192.2 (Ying). *A nonabelian p-group G is self-dual if and only if it is isomorphic to* $S(p^3)\times B$ *from Theorem 192.1 (b).*

We shall prove first a series of lemmas about s-self-dual p-groups.

Lemma 192.3. *If G is an s-self-dual p-group and* $H\le G$, *then* $\mathrm{d}(H)\le\mathrm{d}(G)$.

Proof. We have for some $N\trianglelefteq G$, $G/N\cong H$ and so $\mathrm{d}(H)\le\mathrm{d}(G)$. □

Lemma 192.4. *A p-group G is s-self-dual if and only if each maximal subgroup of G is isomorphic to a homomorphic image of G.*

Proof. If G is s-self-dual, then its every maximal subgroup is isomorphic to a homomorphic image of G.

Let us prove the reverse assertion. Suppose that each maximal subgroup of G is isomorphic to a homomorphic image of G. Let $H<G$ and we use induction on $|G:H|$ to prove that H is isomorphic to a homomorphic image of G. If $|G:H|=p$, then, by the assumption, H is a homomorphic image of G. Suppose that $|G:H|q>p$ and let $K>H$ be a subgroup of G such that $|K:H|=p$; then $|G:H|<|G:K|$. By induction, there is $N\trianglelefteq G$ such that $G/N\cong K$. Then G/N, being isomorphic to K, contains a maximal subgroup T/N such that $T/N\cong H$. By the assumption, since $T\in\Gamma_1$, there is $P\trianglelefteq G$

such that $G/P \cong T$. Since H is a homomorphic image of $T \cong G/P$, there is $M \trianglelefteq G$ such that $P \leq M \leq G$ with $G/M \cong H$, and we are done. □

Lemma 192.5. *Let $G = H \times A$ be a p-group, where $H \cong \mathrm{M}(n, n)$, $n \geq 2$, is defined in the statement of Theorem 192.1 and A is abelian of exponent $< p^n$. Then G is s-self-dual.*

Proof. We have $H = \langle a, b \mid a^{p^n} = b^{p^n} = 1,\ a^b = a^{1+p^{n-1}}\rangle$, and so

$$|H| = p^{2n}, \quad n \geq 2, \quad H' = \langle a^{p^{n-1}}\rangle \cong \mathrm{C}_p, \quad H' \leq \mathrm{Z}(G),$$
$$\exp(H) = p^n, \quad \Omega_1(H) = \langle a^{p^{n-1}}\rangle \times \langle b^{p^{n-1}}\rangle$$
$$\text{and} \quad \mho_1(H) = \Phi(H) = \mathrm{Z}(H) = \langle a^p\rangle \times \langle b^p\rangle .$$

Each maximal subgroup V of H is abelian of type (p^n, p^{n-1}) and so $V \cong H/H'$. By Lemma 192.4, H is s-self-dual.

Now we show that G is s-self-dual. By Lemma 192.4, it suffices to show that each maximal subgroup U of G is isomorphic to a homomorphic image of G. If $A \leq U$, then $U = A \times (H \cap U)$. But H is s-self-dual and so there is a normal subgroup N of H such that $H/N \cong H \cap U$. Then $G/N \cong (H/N) \times A \cong U$.

If $A \nleq U$, then there is $g \in A - U$ such that $G = \langle g\rangle U$ because U is a maximal subgroup of G. In that case, $a = g^i u$ and $b = g^j v$ for some integers i, j, and elements $u, v \in U$. Furthermore, the facts that $\exp(A) < p^n = o(a) = o(b)$ and $g \in A \leq \mathrm{Z}(G)$ give the following equalities:

$$o(u) = o(v) = p^n, \quad u^{p^{n-1}} = a^{p^{n-1}}, \quad v^{p^{n-1}} = b^{p^{n-1}}, \text{ and } [u, v] = [a, b] = a^{p^{n-1}} = u^{p^{n-1}}.$$

It follows that $\langle u^{p^{n-1}}\rangle \cap \langle v^{p^{n-1}}\rangle \leq \langle a\rangle \cap \langle b\rangle = \{1\}$, and so $\langle u\rangle \cap \langle v\rangle = \{1\}$. As a consequence, $|\langle u, v\rangle| = |H|$ and $\langle u, v\rangle \cong H$.

Notice that $G = \langle A, a, b\rangle = \langle A, g^i u, g^j v\rangle = \langle A, u, v\rangle$ and $|G| = |A|\,|\langle u, v\rangle|$. Therefore, $G = A \times \langle u, v\rangle$ and then obviously, $U = (A \cap U) \times \langle u, v\rangle \cong (A \cap U) \times H$. But A (being abelian) is s-self-dual and so there is a subgroup N of A such that $A/N \cong A \cap U$. Then $G/N \cong (A/N) \times H \cong (A \cap U) \times \langle u, v\rangle = U$. □

Lemma 192.6. *Let $G = H \times B$ be a p-group, where $H \cong \mathrm{S}(p^3)$, $p > 2$, (defined in Theorem 192.1) and B is elementary abelian. Then G is s-self-dual.*

Proof. Obviously, H is s-self-dual. In fact, each maximal subgroup V of H is isomorphic to H/H'. Now let U be a given maximal subgroup of G.

If $B \leq U$, then $U = B \times (U \cap H) \cong G/H'$.

If $B \nleq U$, then there is an element $b \in B - U$ so that $G = \langle b\rangle \times U$. Hence, U is isomorphic to $G/\langle b\rangle$ and by Lemma 192.4 we are done. □

Lemma 192.7. *Let G be an s-self-dual nonabelian p-group with $\mathrm{d}(G) = 2$. Then either $G \cong \mathrm{M}(n, n)$, $n \geq 2$, or $G \cong \mathrm{S}(p^3)$, $p > 2$, where these groups are defined in Theorem 192.1.*

Proof. (i) Suppose $\exp(G) = p^n$, $n \geq 2$. Choose an element $g \in G$ of order p^n. Then there is a normal subgroup N of G such that $G/N \cong \langle g \rangle$. Let $d \in G - N$ such that $\langle d \rangle$ covers G/N. It follows $G = N\langle d \rangle$ with $N \cap \langle d \rangle = \{1\}$ since $o(d) = p^n = \exp(G)$.

If N is not cyclic, then there is a G-invariant subgroup M of N so that $N/M \cong \mathrm{E}_{p^2}$. Acting with $\langle d \rangle$ on N/M, we see that $\langle d^p \rangle \cong \mathrm{C}_{p^{n-1}}$ centralizes N/M and so $(N\langle d^p \rangle)/M \cong \mathrm{E}_{p^2} \times \mathrm{C}_{p^{n-1}}$ and $\mathrm{d}(N\langle d^p \rangle) \geq 3$, contrary to $\mathrm{d}(G) = 2$ and Lemma 192.3. Thus N is cyclic of order $\leq p^n$ because $\exp(G) = p^n$.

Set $N = \langle a \rangle$ and let T be the subgroup of of index p in $G' \neq \{1\}$, where $T \trianglelefteq G$. Since $(G/T)' = G'/T \cong \mathrm{C}_p$, it follows that G/T is minimal nonabelian (Lemma 65.2 (a)). Then $(N\langle d^p \rangle)/T$ is abelian and so $(N\langle d^p \rangle)/T \cong \mathrm{C}_{p^i} \times \mathrm{C}_{p^{n-1}}$, where $p^i = |N/T|$. Since G is s-self-dual, there is $L \trianglelefteq G$ such that $G/L \cong N\langle d^p \rangle$ and so there is a normal subgroup K in G such that $G/K \cong (N\langle d^p \rangle)/T \cong \mathrm{C}_{p^i} \times \mathrm{C}_{p^{n-1}}$. Hence, $G' \leq K$ and $|G/K| = p^{i+n-1}$. On the other hand, $|G/G'| = p^{i+n-1}$ and so $G' = K$ and $\exp(G/G') = p^n$ gives $i = n$. Hence, $T = \{1\}$, $G' \cong \mathrm{C}_p$ and $o(a) = p^n$. Thus G is metacyclic minimal nonabelian of order p^{2n} and for a suitable generator b of $\langle d \rangle$, we have $a^b = a^{1+p^{n-1}}$ so that $G \cong \mathrm{M}(n, n)$.

(ii) Suppose $\exp(G) = p$. Since G is nonabelian, we have $p > 2$. Let $S \cong \mathrm{S}(p^3)$ be a minimal nonabelian subgroup of G. If $\mathrm{C}_G(S) \nleq S$, then there is $b \in \mathrm{C}_G(S) - S$ of order p so that $\mathrm{d}(S \times \langle b \rangle) = 3$, contrary to $\mathrm{d}(G) = 2$ and Lemma 192.3. Hence, $\mathrm{C}_G(S) \leq S$ and then Proposition 10.17 implies that G is of maximal class.

Suppose that $\mathrm{K}_3(G) \neq \{1\}$. Then $\mathrm{K}_2(G)/\mathrm{K}_4(G) \cong \mathrm{E}_{p^2}$ and so (since $G/\mathrm{K}_2(G) \cong \mathrm{E}_{p^2}$) there is $a \in G - \mathrm{K}_2(G)$ such that a centralizes $\mathrm{K}_2(G)/\mathrm{K}_4(G)$. But then one has $\langle \mathrm{K}_2(G), a \rangle/\mathrm{K}_4(G) \cong \mathrm{E}_{p^3}$, and so $\mathrm{d}(\langle \mathrm{K}_2(G), a =\rangle) \geq 3$, contrary to $\mathrm{d}(G) = 2$ and Lemma 192.3. Hence, $\mathrm{K}_3(G) = \{1\}$ and so $|G| = p^3$ and $G \cong \mathrm{S}(p^3)$. □

The following lemma is the heart of the matter.

Lemma 192.8. *Let G be an s-self-dual p-group with* $\mathrm{d}(G) > 2$. *Then there exists an element $g \in G$ and a subgroup H of G such that $G = \langle g \rangle \times H$ and* $\mathrm{d}(H) < \mathrm{d}(G)$.

Proof. Let $k = \mathrm{d}(G) \geq 3$, $p^\alpha = \max\{|K| \mid K \leq G, \text{ and } \mathrm{d}(K) \leq k-1\}$. Choose a subgroup M of G of order p^α and $\mathrm{d}(M) = k-1$ and then find a normal subgroup N of G such that $G/N = \langle Nx_1, \dots, Nx_{k-1} \rangle \cong M$. Set $K = \langle x_1, \dots, x_{k-1} \rangle$ so that $G = NK$ and $M \cong K/(K \cap N)$. By the choice of M, $|K| \leq p^\alpha = |M|$ and so $K \cap N = \{1\}$ and $K \cong M$. Hence, G is a semidirect product of $N \trianglelefteq G$ and K, where $\mathrm{d}(K) = k-1$ and $|K| = p^\alpha$.

(i) Assume N is cyclic. In this case, we set $N = \langle g \rangle$, $H = K$ and $B = [N, H] \trianglelefteq G$. Then $G' = BH'$ and consider $\mathrm{C}_G(g) \trianglelefteq G$ so that $G' \leq \mathrm{C}_G(g)$. Notice that $\mathrm{C}_G(g) = \langle g \rangle \times T$, where $T = H \cap \mathrm{C}_G(g) \geq H'$. Since $[g, x] = [g, y]$, $(x, y \in G)$ if and only if $xy^{-1} \in \mathrm{C}_G(g)$, we have $|G/\mathrm{C}_G(g)| \leq |B|$. As G is s-self-dual, $\mathrm{C}_G(g)$ is isomorphic to a quotient group G/W of G. Hence, $|G' \cap W| \leq |W| = |G|/|\mathrm{C}_G(g)| \leq |B|$. Furthermore, since $(G/W)' \cong (\mathrm{C}_G(g))' = T'$, we get

$$|T'| = |G'|/|G' \cap W| \geq (|B|\,|H'|)/|B| = |H'|\,.$$

But T is a subgroup of H and so $T' = H'$.

Now, we show that B is trivial. Suppose $B \neq \{1\}$. Since $H = \langle x_1, \dots, x_{k-1}\rangle$ and $k \geq 3$, we may assume without loss of generality that $[x_1, g] \neq 1$ and then we define a new subgroup $L = \langle x_1, gx_2, \dots, gx_{k-1}\rangle$. Obviously, $G = NL$ and so for all $u, v \in T$, there are $y, z \in L$, and integers i, j such that $u = g^i y$ and $v = g^j z$. Since $T \leq C_G(g)$, we have $[y, z] = [g^{-i}u, g^{-j}v] = [u, v]$. Consequently, $T' \leq L'$. But $T' = H'$ and so $H' \leq L'$ and

$$1 \neq [x_1, g]^{x_2} = [x_1, x_2]^{-1}[x_1, gx_2] \in L \cap N\,.$$

Now, $G = NL$ and $H \cong G/N$. Therefore, $|H| = |L|/|L \cap N| < |L|$. This is a contradiction because L is generated by $k-1$ elements and hence its order cannot exceed $p^\alpha = |H|$. Thus B is trivial and so $G = \langle g\rangle \times H$ and $d(H) < d(G)$.

(ii) Assume N is not cyclic. Let V be a G-invariant subgroup of N such that $N/V \cong E_{p^2}$. So $C_G(N/V)$ contains a maximal subgroup of G. Set $U = C_G(N/V) \cap K$ so that $|K/U| \leq p$ and $N \cap K = \{1\}$ implies that $(NU)/V$ is isomorphic to $E_{p^2} \times U$. Together with Lemma 192.3, this gives $d(U) = k-2$ and $|K/U| = p$.

Choose an element $1 \neq h \in Z(G) \cap N$. Then $\langle h, U\rangle = \langle h\rangle \times U$ is generated by exactly $k-1$ elements and so has order p^α. By the same argument as in the beginning of this proof (before (i)), we can find a normal subgroup A and a subgroup H of G such that $G = AH$, $A \cap H = \{1\}$ and $H \cong \langle h\rangle \times U$. We may set $H = \langle y_1, \dots, y_{k-1}\rangle$, where one of these generators is in $Z(H)$ and $d(H) = k-1$. We claim that $[A, H] = \{1\}$.

Suppose $[A, H] \neq \{1\}$. Then there is $a \in A$ and a generator, say y_1, of H such that $[y_1, a] \neq 1$. The subgroup $H^* = \langle y_1, ay_2, \dots, ay_{k-1}\rangle$ satisfies the following conditions:

$$d(H^*) \leq k-1\,, \quad H^*/(H^* \cap A) \cong G/A \cong H\,.$$

Since one y's is in $Z(H)$, $[y_1, y_j] = 1$ for some $j \geq 2$ and hence

$$1 \neq [y_1, a]^{y_j} = [y_1, ay_j] \in H^* \cap A\,.$$

This implies $|H^*| > |H| = p^\alpha$, a contradiction. Thus $G = A \times H$. As $d(H) = k-1$, we conclude that A is cyclic. This completes the proof.

□

Lemma 192.9. *Let G be an s-self-dual p-group. If $G = H \times \langle a\rangle$, then H is also s-self-dual.*

Proof. Let V be a maximal subgroup of H. Then $V \times \langle a\rangle$ is a maximal subgroup of G and there is a normal subgroup N of G of order p such that $G/N \cong V \times \langle a\rangle$.

(i) Assume $N \leq H$. Then $G/N \cong H/N \times \langle a\rangle \cong V \times \langle a\rangle$. By Krull–Remak–Schmidt theorem ([Hup1], p.66), $H/N \cong V$.

(ii) Assume $N \cap H = \{1\}$. Since $G/N \cong V \times \langle a\rangle$, there are normal subgroups K/N and M/N of G/N such that

$$G/N = (K/N) \times (M/N)\,, \quad K/N \cong \langle a\rangle \text{ and } M/N \cong V\,.$$

Let $b \in K$ such that $K = N\langle b\rangle$. Since $[M, b] \leq [M, K] \leq M \cap K = N$, we have $[M, b] = \{1\}$, for otherwise, $N = [M, b] \leq G' \leq H$. Therefore, $b \in Z(G)$. If $\langle b\rangle \geq N$, then $o(b) = p \cdot o(a)$ and so $N = \langle b^{o(a)}\rangle$. On the other hand, for each $x \in G$, $x^{o(a)} \in H$, a contradiction. Therefore, $\langle b\rangle \cap N = \{1\}$, $o(b) = o(a)$ and $\langle b\rangle \cap M = \{1\}$. Thus $G = M \times \langle b\rangle = H \times \langle a\rangle$ and again, by the Krull–Remak–Schmidt theorem, $H \cong M$. But $M/N \cong V$ and so V is isomorphic to a homomorphic image of H. By Lemma 192.4, H is s-self-dual and we are done.

□

Corollary 192.10. *Let G be an s-self-dual p-group. If $G = H \times A$, where A is an abelian subgroup of G, then H is also s-self-dual.*

Proof. This is a consequence of Lemma 192.9 and the fact that A is a direct product of cyclic subgroups. □

Lemma 192.11. *If $G = \mathrm{M}(n, n) \times \langle g\rangle$, where $o(g) = p^\alpha$ with $\alpha \geq n \geq 2$, or $G = \mathrm{S}(p^3) \times \langle h\rangle$, where $o(h) = p^\beta > p > 2$, then G is not s-self-dual. Here $\mathrm{M}(n, n)$ and $\mathrm{S}(p^3)$ are defined in Theorem 192.1.*

Proof. First, we consider the case $G = \mathrm{M}(n, n) \times \langle g\rangle$, where $o(g) = p^\alpha$ with $\alpha \geq n \geq 2$ and

$$\mathrm{M}(n, n) = \langle a, b \mid a^{p^n} = b^{p^n} = 1,\ a^b = a^{1+p^{n-1}}\rangle\,.$$

By Lemma 65.2(a), any two noncommuting elements in G generate a minimal nonabelian subgroup with the commutator group $\langle a^{p^{n-1}}\rangle \cong \mathrm{C}_p$. Consider the minimal nonabelian subgroup $K = \langle ag, b\rangle$. Note that the socle $\Omega_1(\langle ag\rangle)$ of $\langle ag\rangle$ is not contained in $\mathrm{M}(n, n)$ and the socle of $\langle b\rangle$ is $\langle b^{p^{n-1}}\rangle$ and so $\Omega_1(K) \cong \mathrm{E}_{p^3}$ so that K is nonmetacyclic. According to Lemma 65.1, $K' = \langle a^{p^{n-1}}\rangle$ is a maximal cyclic subgroup in K. If G is s-self-dual, then there is a normal subgroup N of G such that $G/N \cong K$. But $N \cap \langle a\rangle = \{1\}$ since $G' = \langle a^{p^{n-1}}\rangle \nleq N$. The commutator group $(\langle a^{p^{n-1}}\rangle N)/N$ is obviously not a maximal cyclic subgroup in G/N since

$$(\langle a^{p^{n-1}}\rangle N)/N < (\langle a\rangle N)/N \cong \mathrm{C}_{p^n}\,, \quad n \geq 2\,,$$

a contradiction.

Now consider the case $G = \mathrm{S}(p^3) \times \langle h\rangle$, $o(h) = p^\beta > p > 2$ and we may set

$$\mathrm{S}(p^3) = \langle a, b \mid a^p = b^p = c^p = 1,\ c = [a, b],\ [c, a] = [c, b] = 1\rangle\,.$$

Here $\Omega_1(\mathrm{Z}(G)) = \langle c, h^{p^{\beta-1}}\rangle \cong \mathrm{E}_{p^2}$ and $|G| = p^{\beta+3}$. The minimal nonabelian subgroup $K = \langle ah, b\rangle$ with $K' = \langle c\rangle$ is of order $p^{\beta+2}$ and so K is a maximal subgroup of G. If G is s-self-dual, then there is a normal subgroup $N \leq \Omega_1(\mathrm{Z}(G))$ of order p such that $G/N \cong K$. Since $N \neq G' = \langle c\rangle$, it follows that $N \cap \langle a, b\rangle = \{1\}$. But G/N is not minimal nonabelian since it contains the minimal nonabelian proper subgroup $(\langle a, b\rangle N)/N \cong \langle a, b\rangle \cong \mathrm{S}(p^3)$, a contradiction. □

Proof of Theorem 192.1. Lemmas 192.5 and 192.6 show that the groups $\mathrm{M}(n,n) \times A$, $\mathrm{S}(p^3) \times B$ from Theorem 192.1 are indeed s-self-dual.

Now let G be a given s-self-dual nonabelian p-group. Applying Lemmas 192.8, 192.9, and Corollary 192.10, we can find an abelian subgroup K and a nonabelian subgroup H such that $G = H \times K$, $\mathrm{d}(H) = 2$ and H is self-dual. By Lemma 192.7, H is either isomorphic to $\mathrm{M}(n,n)$, $n \geq 2$, or to $\mathrm{S}(p^3)$, $p > 2$.

Since K is an abelian group, we have $K = L \times \langle g \rangle$, where $o(g) = \exp(K)$. Then $G = (H \times \langle g \rangle) \times L$, and Corollary 192.10 shows that $H \times \langle g \rangle$ is s-self-dual. Finally, according to Lemma 192.11, $\exp(K) = o(g) < p^n$ if $H \cong \mathrm{M}(n,n)$ and $\exp(K) = o(g) = p$ if $H \cong \mathrm{S}(p^3)$. Theorem 192.1 is proved. □

Lemma 192.12. *Let $G = \mathrm{M}(n,n) \times A$ be a p-group, where $n \geq 2$ and A is abelian. Then G is not self-dual.*

Proof. Set $\mathrm{M}(n,n) = \langle a, b \mid a^{p^n} = b^{p^n} = 1,\ a^b = a^{1+p^{n-1}} \rangle$, $n \geq 2$. Then $\Omega_1(G) = \langle a^{p^{n-1}} \rangle \times \langle b^{p^{n-1}} \rangle \times \Omega_1(A) \leq \mathrm{Z}(G)$. On the other hand, in the quotient group $\bar{G} = G/\langle b^p \rangle$, the element $\bar{b}$ is of order p but $\bar{b} \notin \mathrm{Z}(\bar{G})$ (since $[\bar{a}, \bar{b}] \neq 1$). Thus, $\bar{G}$ cannot be isomorphic to a subgroup of G. □

Lemma 192.13. *Let $G = \mathrm{S}(p^3) \times B$ be a p-group, $p > 2$, where B is elementary abelian. Then G is self-dual.*

Proof. Set $G = H \times B$, where $H \cong \mathrm{S}(p^3)$. By Lemma 192.6, G is s-self-dual. It remains to be proved that G is also q-self-dual. Let N be a given normal subgroup of G. Then we have either $H' \leq N$ or $H \cap N = \{1\}$ since $H' = \mathrm{Z}(H) \cong \mathrm{C}_p$ is a unique minimal normal subgroup in H.

If $H' \leq N$, then G/N is isomorphic to a homomorphic image of G/H'. But G/H' is elementary abelian and so is isomorphic to an elementary abelian maximal subgroup $T \times B$, where T is a maximal subgroup of H. This implies that the elementary abelian group G/N is isomorphic to a subgroup of $T \times B$.

If $H \cap N = \{1\}$, then $N \leq \Omega_1(\mathrm{Z}(G)) = H' \times B$ and so $G/N \cong \mathrm{S}(p^3) \times B_0$, where B_0 is elementary abelian and $|B_0| \leq |B|$. But then there is an elementary abelian subgroup B_1 of B such that $G/N \cong H \times B_1$, and we are done. □

Proof of Corollary 192.2. This corollary is a direct consequence of Theorem 192.1 and Lemmas 192.12 and 192.13. □

Exercise 1. Let a p-group $G = A \times B$, where $A \cong B$. Prove that G possesses a proper subgroup H such that $|H'| = p^2$, provided A is one of the following groups: (a) A is nonabelian group of order p^3. (b) $A \cong \mathrm{M}_{p^n}$.

Solution (for $A \cong \mathrm{S}(p^3)$). Let $A = \langle a_1, a_2 \rangle$ and $B = \langle b_1, b_2 \rangle$. Write $H = \langle a_1, a_2 b_1, b_2 \rangle$. As $\mathrm{d}(H) = 3 < 4 = \mathrm{d}(G)$, it follows that $H < G$. Next, $[a_1, a_2 b_1] = [a_1, a_2]$ is a generator of A' and $[a_2 b_1, b_2] = [b_1, b_2]$ is a generator of B'. It follows that $H' =$

$A' \times B' = G'$ is of order p^2. (As any proper epimorphic image of G has derived subgroup of order $\leq p$, the group G is not s-self-dual.)

Exercise 2. Let G be an s-self-dual minimal nonabelian p-group. Prove, using only Lemmas 65.1 and 192.4 that $G \in \{S(p^3), M(n, n)\}$.

Exercise 3. Is it true that all minimal nonabelian subgroups of a nonabelian s-self-dual p-group are isomorphic?

Exercise 4. Is it true that all minimal nonabelian subgroups of a nonabelian s-self-dual p-group are its direct factors?

Problem 1. Study the p-groups all of whose noncyclic normal subgroups are complemented.

Problem 2. Classify the p-groups G such that, for any $H < G$ of class ≤ 2 there is $N \triangleleft G$ such that $G/N \cong H$. Consider in detail the case when $\exp(G) = p$.

It follows from Theorem 192.1 that s-self-duality is inherited by subgroups.

Problem 3. Classify the p-groups all of whose proper subgroups are s-self-dual.

Problem 4. Suppose that a p-group G possesses a subgroup of order p^k and exponent p (k is fixed). Study the structure of G provided all such subgroups are complemented.

Problem 5. Study the nonmetacyclic p-groups G such that for every metacyclic (maximal metacyclic) $M < G$ there is $N \triangleleft G$ such that $G/N \cong M$.

Problem 6. Study the p-groups all whose subgroups of class ≤ 2 are normally complemented.

Problem 7. Study the nonabelian p-groups G such that there is $N \triangleleft G$ such that (i) $G/N \cong \Phi(G)$, (ii) $G/N \cong G'$, (iii) $G/N \cong Z(G)$, (iv) $G/N \cong Z_2(G)$, (v) $G/N \cong \Omega_1(G)$, (vi) $G/N \cong \mho_1(G)$.

Problem 8. Study the p-groups all of whose maximal abelian subgroups are complemented.

Problem 9. Study the p-groups all of whose noncyclic normal subgroups are complemented.

Problem 10. Study the irregular p-groups G such that for every regular (maximal regular) $R < G$ there is $N \triangleleft G$ such that $G/N \cong R$.

Problem 11. Study the irregular p-groups G such that every regular quotient group of G is isomorphic to a subgroup of G.

Problem 12. Study the nonabelian p-groups G such that every abelian subgroup of G is isomorphic to a quotient group of G.

Problem 13. A p-group G is said to be a qn-self-dual group if for any $N \lhd G$ there is $H \lhd G$ such that $H \cong G/N$. Classify the p-groups that are qn-self-dual.

Problem 14. Classify the p-groups G such that, for any $H \lhd G$ there is $N \lhd G$ satisfying $N \lhd G/H$, i.e., for any epimorphic image of G there is a G-invariant subgroup which is isomorphic to this epimorphic image.

Problem 15. Classify the nonabelian metacyclic p-groups G such that for any (i) minimal nonabelian subgroup $A \leq G$ there is $N \lhd G$ satisfying $G/N \cong A$, (ii) minimal nonabelian epimorphic image $\bar{G}$ there is a subgroup $H \leq G$ satisfying $H \cong \bar{G}$.

Problem 16. Study the irregular p-groups G such that for any (i) subgroup $H < G$ of exponent p there is $N \lhd G$ satisfying $G/N \cong H$, (ii) epimorphic image $\bar{G}$ of exponent p there is a subgroup $H \leq G$ satisfying $H \cong \bar{G}$. (iii) Classify the p-groups of maximal class satisfying one of conditions (i) and (ii).

Problem 17. Study the p-groups G such that, whenever $H < G$ is of index p^2, there is $N \lhd G$ satisfying $G/N \cong H$. Consider in detail the case where G is (i) metacyclic, (ii) of exponent p.

§ 193 Classification of p-groups all of whose proper subgroups are s-self-dual

We recall that a group G is *s-self-dual* if every subgroup of G is isomorphic to a quotient group of G. According to Theorem 192.1, a nonabelian p-group G is s-self-dual if and only if either

$$G \cong \mathrm{M}(n,n) \times A\,, \quad n \geq 2,\ \text{where } \mathrm{M}(n,n) = \langle a, b \mid a^{p^n} = b^{p^n} = 1,\ [a,b] = a^{p^{n-1}} \rangle$$

and A is abelian of exponent $< p^n$, or

$$G \cong \mathrm{S}(p^3) \times B,\ \text{where } \mathrm{S}(p^3)\,, \quad p > 2\,,$$

is the nonabelian group of order p^3 and exponent p and B is elementary abelian.

Note that all abelian p-groups are s-self-dual and so the above result shows that s-self-duality is inherited by subgroups (but this property is inherited by epimorphic images if and only if $p > 2$). Therefore, it is of interest to classify *minimal non-s-self-dual* p-groups G, that is, p-groups G which are not s-self-dual but all proper subgroups of G are s-self-dual (Problem 3 in § 192). We see that each minimal nonabelian p-group which is not isomorphic to $\mathrm{M}(n,n)$, $n \geq 2$ or to $\mathrm{S}(p^3)$, $p > 2$, is minimal non-s-self-dual.

We shall determine here all minimal non-s-self-dual p-groups up to isomorphism. First, we prove an auxiliary result.

Lemma 193.1. *Let G be a p-group of class ≤ 3. Then, for any $a, b \in G$,*

$$(ab)^3 = a^3b^3[b,a]^3[b,a,a][b,a,b]^5\,.$$

Proof. We use the obvious identity $xy = yx[x,y]$, inclusion $\mathrm{K}_3(G) \leq \mathrm{Z}(G)$ and the identity

$$[xy,u] = [x,u]^y[y,u]$$

many times.

First, we get

$$(ab)^2 = a(ba)b = a(ab[b,a])b = a^2b(b[b,a])[b,a,b] = a^2b^2[b,a][b,a,b]\,.$$

This then gives

$$\begin{aligned}(ab)^3 &= a^2b^2[b,a][b,a,b]ab = a^2b^2a[b,a][b,a,a][b,a,b]b\\ &= a^2b^2ab[b,a][b,a,b][b,a,a][b,a,b] = a^2b^2ab[b,a][b,a,a][b,a,b]^2\,.\end{aligned}$$

Since

$$a^2b^2ab = a^2(ab^2[b^2,a])b = a^3b^3[b^2,a][b^2,a,b]\,,$$
$$[b^2,a] = [b,a]^b[b,a] = [b,a][b,a,b][b,a] = [b,a]^2[b,a,b]$$

and

$$[b^2, a, b] = [[b, a]^2[b, a, b], b] = [[b, a]^2, b]^{[b,a,b]}$$
$$= [[b, a]^2, b] = [b, a, b]^{[b,a]}[b, a, b] = [b, a, b]^2 ,$$

we obtain from the above,

$$(ab)^3 = a^3b^3 \cdot [b, a]^2[b, a, b] \cdot [b, a, b]^2 \cdot [b, a][b, a, a][b, a, b]^2$$
$$= a^3b^3[b, a]^3[b, a, a][b, a, b]^5 ,$$

as was to be shown. □

In particular, if $\mathrm{cl}(G) = 2$, then $(ab)^3 = a^3b^3[b, a]^3$, a partial case for the known formula for $(ab)^n$.

It follows from Lemma 193.1 that if G is a p-group of class > 3 and $a, b \in G$, then

$$(ab)^3 = a^3b^3[b, a]^3[b, a, a][b, a, b]^5c ,$$

where c is a product of commutators of weight > 3.

Exercise 1. Obtain a similar formula for $(ab)^4$, where $\mathrm{cl}(G) \le 3$.

Now we are ready to prove the main result of this section.

Theorem 193.2 (Janko). *Let G be a nonabelian p-group which is not minimal nonabelian and assume that G is not s-self-dual but each proper subgroups of G is s-self-dual. Then $\mathrm{d}(G) \le 4$, $|G'| \le p^3$, $\mathrm{cl}(G) \le 3$ and one of the following holds:*

(a) *$\mathrm{d}(G) = 4$, $|G'| = p$, and G is extraspecial of order p^5 and exponent $p > 2$.*

(b) *$\mathrm{d}(G) = 2$, $|G'| = p^2$, $\mathrm{cl}(G) = 3$, and*

(b1)
$$p > 2 ,$$
$$G = \langle a, b \mid a^p = b^p = 1, [a, b] = c, [a, c] = d, [b, c] = d^{-1},$$
$$c^p = d^p = [d, a] = [d, b] = 1\rangle ,$$

where

$$|G| = p^4 , \quad G' = \langle c, d\rangle \cong \mathrm{E}_{p^2} , \quad \mathrm{K}_3(G) = \langle d\rangle \cong \mathrm{C}_p ,$$

all nonabelian maximal subgroups are isomorphic to $S(p^3)$ and the unique maximal abelian subgroup is elementary abelian if $p \ge 5$ and is of type $(9, 3)$ if $p = 3$;

(b2)
$$p = 2 , \quad G = \langle a, b \mid a^8 = b^4 = 1, \ a^b = a^{-1+4\eta}\rangle , \quad \eta = 0, 1 ,$$

is metacyclic, where $|G| = 2^5$ and $G' = \langle a^2\rangle \cong \mathrm{C}_4$;

(b3)
$$p = 2 , \quad G = \langle a, v \mid a^8 = v^4 = 1, \ a^v = a^{-1}e, \ e^2 = [e, a] = [e, v] = 1\rangle ,$$

where

$$|G| = 2^6 , \quad \Omega_1(G) = \mathrm{Z}(G) = \langle a^4, v^2, e\rangle \cong \mathrm{E}_8 \text{ and } G' = \langle a^2e\rangle \cong \mathrm{C}_4 .$$

(c) $$p > 3\,,\quad \mathrm{d}(G) = 2\,,\quad G' \cong \mathrm{E}_{p^3}\,,\quad \mathrm{cl}(G) = 3\,,$$

and

$$G = \langle a, b \mid [b, a] = c\,,\quad [c, a] = d\,,\quad [c, b] = e\,,$$
$$a^p = b^p = c^p = d^p = e^p = [d, a] = [d, b] = [e, a] = [e, b] = 1\rangle\,,$$

where

$$|G| = p^5\,,\quad \exp(G) = p\,,\quad G' = \langle c, d, e\rangle\,,\quad \mathrm{E}_{p^2} \cong \mathrm{K}_3(G) = \langle d, e\rangle = \mathrm{Z}(G)\,.$$

(d) $$\mathrm{d}(G) = 3\,,\quad G' \cong \mathrm{E}_{p^2} \text{ or } \mathrm{E}_{p^3}\,,\quad \mathrm{cl}(G) = 2\,,\quad \Phi(G) = \mathrm{Z}(G)$$

and

(d1) $p > 2$, $$G = \langle a, b, c \mid [a, c] = d\,,\quad [b, c] = e\,,\quad a^p = b^p = c^p = d^p = e^p = 1\,,$$
$$[a, b] = [d, a] = [d, b] = [d, c] = [e, a] = [e, b] = [e, c] = 1\rangle\,,$$

where

$$|G| = p^5\,,\quad \exp(G) = p\,,\quad \text{and } G' = \Phi(G) = \mathrm{Z}(G) = \langle d, e\rangle \cong \mathrm{E}_{p^2}\,.$$

(d2) $p > 2$, $G = \langle a, b, c\rangle$ *with*

$$[a, b] = d\,,\quad [a, c] = e\,,\quad [b, c] = f\,,\quad a^p = b^p = c^p = d^p = e^p = f^p = 1\,,$$
$$[d, a] = [d, b] = [d, c] = [e, a] = [e, b] = [e, c] = [f, a] = [f, b] = [f, c] = 1\rangle\,,$$

where

$$|G| = p^6\,,\quad \exp(G) = p\,,\quad G' = \Phi(G) = \mathrm{Z}(G) = \langle d, e, f\rangle \cong \mathrm{E}_{p^3}\,.$$

(d3) $p = 2$, *G is the minimal nonmetacyclic group of order 2^5 (Theorem 66.1).*

(d4) $$G = \langle a, b, c \mid a^{p^n} = b^{p^n} = c^{p^n} = [a, b] = 1,\ a^c = a^{1+p^{n-1}},\ b^c = b^{1+p^{n-1}}\rangle\,,\quad n \geq 2\,,$$

where

$$|G| = p^{3n}\,,\quad G' = \langle a^{p^{n-1}}, b^{p^{n-1}}\rangle \cong \mathrm{E}_{p^2}\,,$$
$$\mathrm{Z}(G) = \mho_1(G) = \Phi(G) = \langle a^p\rangle \times \langle b^p\rangle \times \langle c^p\rangle \cong \mathrm{C}_{p^{n-1}} \times \mathrm{C}_{p^{n-1}} \times \mathrm{C}_{p^{n-1}}\,.$$

In addition, if $p > 2$ or $p = 2$ and $n \geq 3$, then G is a modular p-group (see Theorem 73.15).

Conversely, all the above groups satisfy the assumptions of our theorem.

Proof. Let G be a p-group satisfying the assumptions of our theorem. Let H be a nonabelian maximal subgroup of the maximal possible exponent p^m, $m \geq 1$, in G. If $m = 1$, then $H \cong \mathrm{S}(p^3) \times B$, where $p > 2$ and B is elementary abelian and in this case each nonabelian maximal subgroup is isomorphic to H. Suppose $m \geq 2$ and then

$H \cong \mathrm{M}(m, m) \times A$, where A is abelian with $\exp(A) < p^m$. Here $\mathrm{Z}(H) = \Omega_{m-1}(H)$ and $H/\mathrm{Z}(H) \cong \mathrm{E}_{p^2}$. Hence, if K is any other nonabelian maximal subgroup of G, then $H \cap K$ is a maximal subgroup of H and so $H \cap K \nleq \Omega_{m-1}(H)$ so that $\exp(K) = p^m$. In both cases, $\exp(G) \leq p^{m+1}$ and if $\exp(G) = p^{m+1}$, then G must possess an abelian maximal subgroup L with $\exp(L) = p^{m+1}$.

(∗) We have proved the following facts which will be used often in what follows. All nonabelian maximal subgroups H are of the same exponent p^m, $m \geq 1$ (m fixed), $|H : \mathrm{Z}(H)| = p^2$ and $|H'| = p$. If $m = 1$, then each minimal nonabelian subgroup of G is isomorphic to $\mathrm{S}(p^3)$, $p > 2$, and if $m \geq 2$, then each minimal nonabelian subgroup of G is isomorphic to $\mathrm{M}(m, m)$. In addition, $\exp(G) \leq p^{m+1}$ and if $\exp(G) = p^{m+1}$, then G must possess an abelian maximal subgroup of exponent p^{m+1}.

We are now in a position to use Theorem 151.1 about nonabelian p-groups which are not minimal nonabelian and all of whose nonabelian maximal subgroups have the largest possible center. By that theorem, we must have in our case one of the following possibilities:

(i) $G = M_1\mathrm{Z}(G)$, where M_1 is minimal nonabelian and $M_1 < G$.

(ii) $G = M_1 * M_2$, where M_1 and M_2 are minimal nonabelian with $M_1' = M_2'$.

(iii) $\mathrm{d}(G) = 2$ and $|G'| = p^2$.

(iv) $p > 2$, $\mathrm{d}(G) = 2$, $\mathrm{cl}(G) = 3$, $G' \cong \mathrm{E}_{p^3}$ and $\mho_1(G) \leq \mathrm{Z}(G)$.

(v) $\mathrm{d}(G) = 3$, $\mathrm{cl}(G) = 2$, $G' \cong \mathrm{E}_{p^2}$ or E_{p^3}, $\Phi(G) = \mathrm{Z}(G)$.

To finish the proof, we have only to investigate systematically the above possibilities (i)–(v).

(i) Suppose that $G = M_1\mathrm{Z}(G)$, where M_1 is minimal nonabelian and $M_1 < G$.

(i1) First, we consider the case $M_1 \cong S(p^3)$, $p > 2$, where we may set $M_1 = \langle a, b\rangle$ and then

$$\langle [a, b]\rangle = \mathrm{Z}(M_1) = G' = M_1 \cap \mathrm{Z}(G) \cong \mathrm{C}_p\,.$$

If $\exp(\mathrm{Z}(G)) = p$, then Theorem 192.1 implies that G is s-self-dual, a contradiction. Hence there is $x \in \mathrm{Z}(G)$ with $o(x) = p^2$. But then $\langle ax, b\rangle$ is minimal nonabelian of exponent p^2 (see Lemma 65.2 (a)), contrary to the facts in (∗).

(i2) Now assume $M_1 \cong \mathrm{M}(n, n)$, $n \geq 2$. Then

$$M_1 = \langle a, b \mid a^{p^n} = b^{p^n} = 1,\ [a, b] = a^{p^{n-1}}\rangle \text{ and}$$
$$G' = \langle a^{p^{n-1}}\rangle \cong \mathrm{C}_p\,, \quad M_1 \cap \mathrm{Z}(G) = \mathrm{Z}(M_1) = \langle a^p\rangle \times \langle b^p\rangle < \mathrm{Z}(G)\,.$$

If $\exp(\mathrm{Z}(G)) = p^{n-1}$, then $\langle a^p, b^p\rangle$ has a complement in $\mathrm{Z}(G)$ and so, by Theorem 192.1, G is s-self-dual, a contradiction. It follows that there is $g \in \mathrm{Z}(G)$ with $o(g) = p^n$. By Theorem 192.1, $M_1\langle g\rangle$ is not s-self-dual and so we get $G = M_1\langle g\rangle$ and $\mathrm{Z}(G) = \langle a^p, b^p, g\rangle$.

First assume that the socle $\langle g^{p^{n-1}}\rangle$ of $\langle g\rangle$ is not contained in M_1 and then we have

$$\mathrm{Z}(G) = \langle a^p\rangle \times \langle b^p\rangle \times \langle g\rangle\,.$$

But then $\langle ag, b\rangle$ is minimal nonabelian with

$$\langle ag, b\rangle' = \langle a^{p^{n-1}}\rangle \text{ and } \Omega_1(\langle ag, b\rangle) = \langle a^{p^{n-1}}g^{p^{n-1}}, b^{p^{n-1}}, a^{p^{n-1}}\rangle \cong \mathrm{E}_{p^3},$$

and so $\langle ag, b\rangle$ is nonmetacyclic of order p^{2n+1}, contrary to our results in $(*)$.

Now assume that the socle $\langle g^{p^{n-1}}\rangle$ of $\langle g\rangle$ is contained in M_1. If $p > 2$, then $\mho_1(M_1) \geq M_1' = \langle a^{p^{n-1}}\rangle$ and if $p = 2$ and $n \geq 3$, then $\mho_2(M_1) \geq M_1'$. Hence in both cases, M_1 is a powerful p-group and so by Proposition 26.10 (and a remark following that proposition) and Proposition 26.23, we get

$$\mho_{n-1}(M_1) = \langle a^{p^{n-1}}, b^{p^{n-1}}\rangle = \{x^{p^{n-1}} \mid x \in M_1\}.$$

Suppose that there is $x \in M_1 - \mho_1(M_1)$ such that $x^{p^{n-1}} = g^{-p^{n-1}}$. Then (noting that $g \in \mathrm{Z}(G)$),

$$(xg)^{p^{n-1}} = x^{p^{n-1}}g^{p^{n-1}} = 1 \text{ and so } o(xg) \leq p^{n-1} \text{ with } 1 \neq xg \notin \mathrm{Z}(G).$$

Hence, there is $y \in G$ such that $[xg, y] \neq 1$. Thus, $\langle xg, y\rangle$ is minimal nonabelian with a generating element xg of order $\leq p^{n-1}$. But then $\langle xg, y\rangle \ncong \mathrm{M}(n, n)$, a contradiction.

We have proved that $p = 2$ and $n = 2$ so that $M_1 \cong \mathrm{M}(2, 2) = \mathcal{H}_2$ and g^2 is not a square in M_1. Then $g^2 = a^2b^2$ because a^2b^2 is the only involution in $\mathcal{H}_2$ which is not a square in $\mathcal{H}_2$. Since

$$(bg)^2 = b^2g^2 = b^2(a^2b^2) = a^2 \text{ and } [a, bg] = [a, b] = a^2,$$

one has $\langle a, bg\rangle \cong \mathrm{Q}_8$. But Q_8 is minimal nonabelian and $\mathrm{Q}_8 \ncong \mathrm{M}(2, 2)$, contrary to the results in $(*)$.

(ii) Assume that $G = M_1 * M_2$, where M_1 and M_2 are minimal nonabelian and $M_1' = M_2'$.

(ii1) First suppose $M_1 \cong M_2 \cong \mathrm{M}(n, n)$, $n \geq 2$. We have

$$M_1' \leq M_1 \cap M_2 \leq \mho_1(M_1) = \mathrm{Z}(M_1) = \Omega_{n-1}(M_1).$$

Let x be an element of order p^n in M_2. But then $M_1 * \langle x\rangle < G$ and $x \notin M_1$ and so $M_1 * \langle x\rangle$ should be s-self-dual which contradicts Theorem 192.1.

(ii2) Suppose that $M_1 \cong M_2 \cong \mathrm{S}(p^3)$, $p > 2$, so that $M_1 \cap M_2 = G' = \mathrm{Z}(G) \cong \mathrm{C}_p$ and G is extraspecial of order p^5 and exponent p as given in part (a) of our theorem.

Conversely, each maximal subgroup of the group from part (a) of our theorem is isomorphic to $\mathrm{S}(p^3) \times \mathrm{C}_p$ and so, by Theorem 192.1, is s-self-dual.

(iii) Suppose that $\mathrm{d}(G) = 2$ and $|G'| = p^2$.

(iii1) First assume that all nonabelian maximal subgroups of G are of exponent p. Then $G' \cong \mathrm{E}_{p^2}$ and $p > 2$ so that Proposition 145.5 implies that G has exactly one abelian maximal subgroup A and $\mathrm{cl}(G) \leq 3$. Since nonabelian maximal subgroups of G generate G, we have $\Omega_1(G) = G$. For any $x, y \in G$ of order p, we obtain by the Hall–Petrescu formula (Theorem A.1.3),

$$(xy)^p = x^p y^p c_2^{\binom{p}{2}} c_3^{\binom{p}{3}} = c_3^{\binom{p}{3}}, \text{ where } c_i \in \mathrm{K}_i(G), \quad i = 2, 3 .$$

Hence, if $p \geq 5$, then $(xy)^p = 1$ and so $\exp(G) = p$. If $p = 3$, then $(xy)^3 = c_3 \in \mathrm{K}_3(G)$ and so, if $\exp(G) = 9$, then for some x, y of order 3, we have $(xy)^3 \neq 1$ and then $(xy)^3 = c_3$, where $\langle c_3 \rangle = \mathrm{K}_3(G) \cong \mathrm{C}_3$. On the other hand, if our 3-group G has an element y of order 9, then

$$y \in A, \ G' < A, \quad |\mho_1(A)| = 3 \text{ and so } \mho_1(G) = \mho_1(A) = \mathrm{K}_3(G) < G' .$$

Hence, in any case we have $\Phi(G) = G' \cong \mathrm{E}_{p^2}$ and since $\mathrm{d}(G) = 2$, we obtain $|G| = p^4$. If $G' \leq \mathrm{Z}(G)$, then G is minimal abelian and so $|G'| = p$, a contradiction. Hence $\mathrm{cl}(G) = 3$.

Note that all elements in $G - A$ are of order p. Choose generators a', b of G so that $a' \in A - G'$ and $b \in G - A$. Set $a = a'b^{-1}$ and then

$$G = \langle a, b \rangle, \quad o(a) = o(b) = p, \quad \text{ and } ab \in A .$$

Set $[a, b] = c$ so that

$$1 = [ab, c] = [a, c]^b [b, c] = [a, c][b, c] .$$

Hence, setting $[a, c] = d$, we get $[b, c] = d^{-1}$ and $\langle d \rangle = \mathrm{K}_3(G) = \mathrm{Z}(G) \cong \mathrm{C}_p$. We have obtained the groups stated in part (b1) of the theorem.

Conversely, let G be a group from part (b1) of the theorem. If $p \geq 5$, then, by the Hall–Petrescu formula, $\exp(G) = p$, and so all nonabelian maximal subgroups are isomorphic to $S(p^3)$ and the unique abelian maximal subgroup is elementary abelian. However, if $p = 3$, then by Lemma 193.1,

$$(ba)^3 = b^3 a^3 [a, b]^3 [a, b, b][a, b, a]^5 = dd^{-5} = d^{-4} \neq 1$$

and so $\langle ba, c \rangle$ is abelian of type $(9, 2)$. Since $c^a = cd^{-1}$, we obtain

$$[a^2, b] = [a, b]^a [a, b] = c^a c = cd^{-1}c = c^2 d^{-1}$$

and so Lemma 193.1 implies

$$\begin{aligned}(ba^2)^3 &= b^3 a^6 [a^2, b]^3 [a^2, b, b][a^2, b, a^2]^5 = [c^2 d^{-1}, b][c^2 d^{-1}, a^2]^5 \\ &= [c^2, b][c^2, a^2]^5 = [c, b]^2 [c, a]^{20} = d^2 d^{-20} = d^{-18} = 1 .\end{aligned}$$

Hence, in case $p = 3$, all three nonabelian maximal subgroups

$$\langle a, G' \rangle, \quad \langle b, G' \rangle \text{ and } \langle ba^2, G' \rangle$$

are isomorphic to $S(3^3)$.

(iii2) Suppose that all nonabelian maximal subgroups of G are of exponent p^n, where $n \geq 2$, and let M be one of them. Then $M = K \times A$, where $K \cong \mathrm{M}(n, n)$ and A is abelian of exponent $< p^n$. Set $H = K' = M' \cong C_p$. Since $d(G/H) = 2$ and $(G/H)' = G'/H \cong C_p$, it follows (by Lemma 65.2 (a)) that G/H is minimal nonabelian. By the structure of G/H (see Lemma 65.1), the abelian group M/H is of rank ≤ 3. Since $d(K/H) = 2$, we get $d(A) \leq 1$ and so A is cyclic of order $\leq p^{n-1}$. Suppose $|A| = p^{\alpha}$, where $2 \leq \alpha \leq n-1$. Then M/H is abelian of type $(p^{n-1}, p^n, p^{\alpha})$. In particular, G/H is nonmetacyclic minimal nonabelian and so (by Lemma 65.1) $(G/H)' = G'/H$ (noting that $G' < M$) must be a maximal cyclic subgroup of G/H (Lemma 65.1). It follows that at least one of the invariants p^{n-1}, p^n, p^{α} of the abelian group M/H must be equal p. This gives $n = 2$ and $\exp(A) < p^2$, a contradiction. We have proved that $|A| \leq p$ and so M/H is abelian of type $(p^{n-1}, p^n, |A|)$.

(iii2a) Suppose that $|A| = \{1\}$, that is, G is an A_2-group with $|G'| = p^2$, $d(G) = 2$, $|G| = p^{2n+1}$, $n \geq 2$, G has at most one abelian maximal subgroup and all other maximal subgroups of G are minimal nonabelian and they are isomorphic to $\mathrm{M}(n, n)$ with a fixed $n \geq 2$.

First, suppose that G is nonmetacyclic. If G is minimal nonmetacyclic, then by Theorems 66.1 and 69,1, we obtain a contradiction because $|G| \geq p^5$ and $d(G) = 2$. Hence, G is not minimal nonmetacyclic and so G must possess exactly one abelian maximal subgroup and this one is nonmetacyclic. It follows that G must be a group of Proposition 71.3 and so $p > 2$, $G' \cong E_{p^2}$ and $G' \nleq Z(G)$. Let $k \in G' - Z(G)$. Then there is $g \in G$ such that $[g, k] \neq 1$. But $\langle g, k\rangle < G$ (since $k \in G' \leq \Phi(G)$) and so $\langle g, k\rangle' \cong C_p$ and Lemma 65.2 (a) implies that $\langle g, k\rangle$ is minimal nonabelian. But $\langle g, k\rangle$ has a generating element k of order p, contrary to $\langle g, k\rangle \cong \mathrm{M}(n, n)$, a contradiction (by our results in $(*)$).

Hence, G must be a metacyclic A_2-group from Proposition 71.2. First suppose that G is a group of Proposition 71.2 (a). Then all maximal subgroups of G are minimal nonabelian and so they should be pairwise isomorphic. By Proposition A.58.3, there is no such a group. Now assume that G is a group of Proposition 71.2 (b). Then $p = 2$ and exactly one maximal subgroup is abelian and the other two maximal subgroups are isomorphic to $\mathrm{M}(n, n)$. By Proposition A.58.4, we get $\epsilon = 0$ and $n = 2$ and this is a group of order 2^5 from part (b2) of our theorem. Conversely, if G is such a group, then G has exactly one abelian maximal subgroup and the other two maximal subgroups are isomorphic to $\mathrm{M}(2, 2) = \mathcal{H}_2$.

(iii2b) Suppose that $|A| = p$, that is, each nonabelian maximal subgroup of G is isomorphic to $\mathrm{M}(n, n) \times C_p$, where $n \geq 2$ is fixed. Then $|G| = p^{2n+2}$, $\exp(G) \leq p^{n+1}$ and $\mathrm{cl}(G) \leq 3$.

(1) Assume that $\mathrm{cl}(G) = 3$. Let $k \in G' - Z(G)$ so that there is $g \in G$ such that $[g, k] \neq 1$. Since $\langle g, k\rangle < G$, we have $\langle g, k\rangle' = \langle [g, k]\rangle = K_3(G) \cong C_p$ and so $\langle g, k\rangle$ is minimal nonabelian and $\langle g, k\rangle \cong \mathrm{M}(n, n)$. It follows that $n = 2$, $o(k) = p^2$, $o(g) = p^2$ and $\langle k\rangle = G' \cong C_{p^2}$. We may choose a generator g in $\langle g, k\rangle$ so that for a maximal subgroup

M of G containing $\langle g, k\rangle$, we may set $M = K \times A$, where

$$K = \langle g, k \mid g^{p^2} = k^{p^2} = 1,\ [k, g] = k^p\rangle\,, \quad \langle k\rangle = G' \cong C_{p^2}\,, \quad A = \langle c\rangle \cong C_p$$

and $|G| = p^6$ with $K \trianglelefteq G$.

First, we consider the case $p = 2$. Then each minimal nonabelian subgroup of G is isomorphic to $M(2, 2) = \mathcal{H}_2$. We are in a position to use Theorem 57.3. Since $d(G) = 2$, $G' \cong C_4$ and $cl(G) = 3$, G must be a group in case (a) of that theorem. Hence, $\exp(G) = 8$, G has a unique abelian maximal subgroup A^*, all elements in $G - A^*$ are of order 4 and

$$E = \Omega_1(A^*) = \Omega_1(G) = Z(G)$$

is of order ≥ 4. By Lemma 1.1, $|Z(G)| = 8$ and so $E \cong E_8$. Since

$$\Omega_1(M) = \langle k^2, g^2, c\rangle \cong E_8, \text{ we have } Z(G) = Z(M) = \Omega_1(M)\,.$$

But $M/Z(M) \cong E_4$ and so $G/Z(G) \cong D_8$ and $A^*/Z(G) \cong C_4$ is the unique cyclic subgroup of index 2 in $G/Z(G)$. Let $a \in A^*$ be an element of order 8 in A^* so that $\langle a\rangle$ covers $A^*/Z(G)$. Since $d(G) = 2$, we get

$$\Phi(G) = \mho_1(G) = Z(G)\langle a^2\rangle \cong C_4 \times C_2\,.$$

Let v be an element of order 4 in $G - A^*$ so that $C_{A^*}(v) = E = Z(G)$ and (by Theorem 57.3 (a)) v inverts $\Phi(A^*) = \langle a^2\rangle$ and A^*/E. Since $1 \neq v^2 \in E$, v inverts $\langle a^2\rangle$ and Q_8 is not a subgroup in G, we have $v^2 \in E - \langle a^4\rangle$ and the nonabelian maximal subgroup $\Phi(G)\langle v\rangle$ is isomorphic to $\mathcal{H}_2 \times C_2$. Write

$$a^v = a^{-1}e \text{ for some } e \in E\,.$$

We have

$$\mho_1(A^*) = \langle a^2\rangle\,, \quad \mho_1(\Phi(G)\langle v\rangle) = \langle v^2, a^4\rangle \text{ and } \mho_1(\Phi(G)\langle va\rangle) = \langle (va)^2, a^4\rangle\,,$$

where $\mho_1(\Phi(G)\langle va\rangle) \cong \mathcal{H}_2 \times C_2$ is the other nonabelian maximal subgroup of G. We compute

$$(va)^2 = vava = v^2a^va = v^2(a^{-1}e)a = v^2e\,,$$

and since $\Phi(G) = \mho_1(G)$, we must have

$$e \in E - \langle v^2, a^4\rangle \text{ so that } E = \langle v^2, a^4, e\rangle\,.$$

Because

$$[a, v] = a^{-1}a^v = a^{-1}(a^{-1}e) = a^{-2}e\,,$$

we get $G' = \langle a^2e\rangle \cong C_4$. We have obtained the group of order 2^6 stated in part (b3) of our theorem. Obviously, this group satisfies the assumptions of our theorem.

Now we consider the case $p > 2$. By Proposition 145.4, each of the $p + 1$ maximal subgroups of G is nonabelian and so is isomorphic to $M(2, 2) \times C_p$. In particular, G

is of exponent p^2. For any $x, y \in G$, $\langle x, y^p\rangle < G$ and so if $[x, y^p] \neq 1$, then $\langle x, y^p\rangle$ is minimal nonabelian. But $o(y^p) = p$ and so $\langle x, y^p\rangle \ncong \mathrm{M}(2, 2)$, contrary to our results in (∗). Hence, $[x, y^p] = 1$ and then Lemma 188.2 implies

$$1 = [x, y^p] = [x, y]^{\binom{p}{1}}[x, y, y]^{\binom{p}{2}} = [x, y]^p .$$

This is a contradiction since $G' \cong \mathrm{C}_{p^2}$.

(2) Assume that $\mathrm{cl}(G) = 2$. Since $\mathrm{d}(G) = 2$, $\mathrm{cl}(G) = 2$ and $|G'| = p^2$, we have $G' \cong \mathrm{C}_{p^2}$.

If $p > 2$, then Proposition 145.4 implies that each maximal subgroup of G is nonabelian and so $\exp(G) = p^n$, $n \geq 2$ (fixed) and each minimal nonabelian subgroup is isomorphic to $\mathrm{M}(n, n)$. For any $x, y \in G$, $\langle x, y^p\rangle < G$ and so if $[x, y^p] \neq 1$, then $\langle x, y^p\rangle$ is minimal nonabelian, a contradiction (since $o(y^p) \leq p^{n-1}$). Hence, $[x, y^p] = [x, y]^p = 1$, contrary to $G' \cong \mathrm{C}_{p^2}$.

We have proved that in this case we must have $p = 2$. Note that $\Omega_n(G) = G$ since G has at least two distinct nonabelian maximal subgroups. Suppose that $\exp(G) = 2^n$. Then, for any $x, y \in G$, one has $\langle x, y^2\rangle < G$ and so if $[x, y^2] \neq 1$, then $\langle x, y^2\rangle$ is minimal nonabelian, contrary to $\langle x, y^2\rangle \cong \mathrm{M}(n, n)$. Hence, we have for any $x, y \in G$,

$$1 = [x, y^2] = [x, y]^2, \text{ contrary to } G' \cong \mathrm{C}_4 .$$

Hence, we must have $\exp(G) = 2^{n+1}$ and so G has exactly one abelian maximal subgroup (of exponent 2^{n+1}). For any $x, y \in G$ such that $o(x) \leq 2^n$, $o(y) \leq 2^n$, we have

$$(xy)^{2^n} = x^{2^n} y^{2^n} [y, x]^{\binom{2^n}{2}} = [y, x]^{2^{n-1}(2^n-1)} .$$

If $n \geq 3$, then $(xy)^{2^n} = 1$ and so $\exp(G) = 2^n$, a contradiction. It follows that we must have $n = 2$ and so each minimal nonabelian subgroup of G is isomorphic to $\mathrm{M}(2, 2) = \mathcal{H}_2$ and G has exactly one abelian maximal subgroup of exponent 8. By Theorem 57.3 (a), $\mathrm{cl}(G) \geq 3$, contrary to our assumption.

(iv) Suppose that $p > 2$, $\mathrm{d}(G) = 2$, $\mathrm{cl}(G) = 3$, $G' \cong \mathrm{E}_{p^3}$ and $\mho_1(G) \leq \mathrm{Z}(G)$.

Since $\mathrm{d}(G) = 2$, the quotient group $G'/\mathrm{K}_3(G)$ is cyclic and so $\mathrm{E}_{p^2} \cong \mathrm{K}_3(G) \leq \mathrm{Z}(G)$. All $p + 1$ maximal subgroups M_i of G are nonabelian and $\{M_i' \mid i = 1, \dots, p + 1\}$ is the set of $p + 1$ pairwise distinct subgroups of order p in $\mathrm{K}_3(G)$. Here we have used Exercise 1.69 (a) implying $|G' : (M_i'M_j')| \leq p$ for any $i \neq j$, $i, j = 1, \dots, p + 1$.

Let $k \in G' - \mathrm{K}_3(G)$ so that there is $g \in G$ such that $[k, g] \neq 1$. But $\langle k, g\rangle < G$ and so $\langle k, g\rangle' \cong \mathrm{C}_p$, where $[k, g] \in \mathrm{K}_3(G) \leq \mathrm{Z}(G)$. Thus, $\langle k, g\rangle$ is minimal nonabelian and so by our results in (∗), $\langle k, g\rangle \cong \mathrm{S}(p^3)$ and each maximal subgroup of G (being nonabelian) is of exponent p. Hence, $\exp(G) = p$, $\Phi(G) = \mho_1(G)G' = G' \cong \mathrm{E}_{p^3}$ and so $|G| = p^5$.

Setting

$$G = \langle a, b\rangle , \quad [b, a] = c , \quad [c, a] = d , \quad [c, b] = e ,$$
$$\text{with} \quad a^p = b^p = c^p = d^p = e^p = 1 ,$$

we have

$$K_3(G) = \langle d, e\rangle \cong E_{p^2} \quad \text{and} \quad G' = \langle c, d, e\rangle \cong E_{p^3} .$$

However, if $p = 3$, then Lemma 193.1 gives

$$(ab)^3 = [b, a, a][b, a, b]^{-1} = de^{-1} \neq 1 ,$$

a contradiction. Hence, we must have $p \geq 5$ and in that case for any $x, y \in G$ with $o(x) \leq p$, $o(y) \leq p$, the Hall–Petrescu formula (Theorem A.1.3) gives

$$(xy)^p = x^p y^p k_2^{\binom{p}{2}} k_3^{\binom{p}{3}} = 1 , \quad \text{where } k_i \in K_i(G),\ i = 2, 3 ,$$

and so G is indeed of exponent p. It follows that we have obtained the groups of order p^5 stated in part (c) of our theorem.

Conversely, let G be a group of part (c) of our theorem. Then each maximal subgroup of G is nonabelian of order p^4 and exponent p containing $E_{p^2} \cong K_3(G) \leq Z(G)$ and so is isomorphic to $S(p^3) \times C_p$.

(v) Finally, assume $d(G) = 3$, $cl(G) = 2$, $G' \cong E_{p^2}$ or E_{p^3} and $\Phi(G) = Z(G)$.

Let M be any nonabelian maximal subgroup of G. Then $M/Z(G) \cong E_{p^2}$, where $\Phi(G) = Z(G)$, and so $Z(M) = Z(G)$. If $G' \cong E_{p^2}$, then Lemma 146.7 implies that G has exactly one nonabelian maximal subgroup. However, if $G' \cong E_{p^3}$, then, by Exercise 1.69 (a), all maximal subgroups M_i of G are nonabelian and $\{M_i \mid i = 1, \dots, p^2 + p + 1\}$ is the set of pairwise distinct subgroups of order p in G'.

If G is an A_2-group, then Propositions A.58.6, A.58.7, and A.58.8 show that G is the minimal nonmetacyclic group of order 2^5 as given in part (d3) of our theorem. Conversely, such a group has exactly one abelian maximal subgroup of type $(4, 4)$ and all other maximal subgroups are isomorphic to $M(2, 2) = \mathcal{H}_2$.

From now on we may assume that G is not an A_2-group. This implies that each nonabelian maximal subgroup of G is not minimal nonabelian.

(v1) Suppose that all minimal nonabelian subgroups of G are isomorphic to $S(p^3)$. $p > 2$.

By $(*)$, each nonabelian maximal subgroup of G is of exponent p and note that G is generated by nonabelian maximal subgroups. If $x, y \in G$ are of orders $\leq p$, then

$$(xp)^p = x^p y^p [y, x]^{\binom{p}{2}} = 1 ,$$

and so G is of exponent p. In this case,

$$\Phi(G) = \mho_1(G)G' = G' \text{ and so } |G| = p^5 \text{ or } p^6 .$$

In case $G' \cong E_{p^2}$, the group G has a unique abelian maximal subgroup A. We may choose generators a, b, c of G so that $a, b \in A$. Then we set $[a, c] = d$ and $[b, c] = e$, where

$$E_{p^2} \cong G' = \langle d, e\rangle \leq Z(G) .$$

We have obtained a group of order p^5 stated in part (d1) of our theorem. Conversely, if G is such a group, then G has exactly one abelian maximal subgroup (isomorphic to E_{p^4}) and all other maximal subgroups are isomorphic to $S(p^3) \times C_p$.

If $G' \cong E_{p^3}$, then we may set

$$G = \langle a, b, c\rangle , \quad [a, b] = d , \quad [a, c] = e , \quad [b, c] = f ,$$
$$\text{where} \quad E_{p^3} \cong \langle d, e, f\rangle = G' = Z(G) = \Phi(G)$$

and we have obtained a group of order p^6 stated in part (d2) of our theorem. Conversely, if G is such a group, then each maximal subgroup of G is isomorphic to $S(p^3)\times E_{p^2}$.

(v2) Assume that all minimal nonabelian subgroups of G are isomorphic to $M(n, n)$, $n \geq 2$ (n fixed).

By our results in (∗), all nonabelian maximal subgroups of G are of exponent p^n and since they generate G, we have $\exp(G) = p^n$. Indeed, if $x, y \in G$ are of orders $\leq p^n$, then (since $n \geq 2$)

$$(xy)^{p^n} = x^{p^n} y^{p^n} [y, x]^{\binom{p^n}{2}} = 1 .$$

Any two noncommuting elements in G generate a minimal nonabelian subgroup isomorphic to $M(n, n)$ and so all elements in $G - Z(G)$ are of order p^n. On the other hand, if M is a nonabelian maximal subgroup of G, then $\exp(Z(M)) = p^{n-1}$ and since $Z(M) = Z(G)$, we have $\exp(Z(G)) = p^{n-1}$, giving $\Omega_{n-1}(G) = Z(G)$.

First assume $p = 2$ and $n = 2$ so that each minimal nonabelian subgroup of G is isomorphic to $M(2, 2) = \mathcal{H}_2$ and $\exp(G) = 4$. By Theorem 57.3 (and noting that we have assumed that G is not an A_2-group), we see that G is isomorphic to a group of order 2^6 from part (d4) of our theorem with $p = 2$ and $n = 2$. Conversely, any nonabelian maximal subgroup of this group of order 2^6 is isomorphic to $\mathcal{H}_2 \times C_2$.

In what follows we shall always assume that either $p > 2$ or $p = 2$ and $n \geq 3$. If $p > 2$, then $G' \leq \mho_1(G)$ and if $p = 2$ and $n \geq 3$, then $G' \leq \mho_2(G)$. In any case, G is a powerful p-group (see § 26). Also, each minimal nonabelian subgroup of G (being isomorphic to $M(n, n)$) is powerful.

Let M be any nonabelian maximal subgroup in G so that $M = K \times A$, where

$$K = \langle a, b \mid a^{p^n} = b^{p^n} = 1,\ [a, b] = a^{p^{n-1}}\rangle \cong M(n, n) ,$$
$$n \geq 2, \text{ and if } p = 2, \text{ then } n \geq 3 ,$$

and $A \neq \{1\}$ (noting that G is not an A_2-group) is abelian of exponent $< p^n$. By Propositions 26.13 and 26.26, $d(M) \leq d(G) = 3$. But $d(K) = 2$ and so $\{1\} \neq A = \langle d\rangle$ is cyclic of order $\leq p^{n-1}$. On the other hand,

$$Z(M) = Z(G) = \langle a^p\rangle \times \langle b^p\rangle \times \langle d\rangle = \Phi(G) = \mho_1(G)G' = \mho_1(G) ,$$

since $G' \leq \mho_1(G)$. By Propositions 26.10 and 26.23, there exists $c \in G - M$ such that $c^p = d$. By the above, all elements in $G - Z(G)$ are of order p^n (since $\Omega_{n-1}(G) = Z(G)$)

and so $o(d) = p^{n-1}$. We have proved that each nonabelian maximal subgroup of G is isomorphic to $\mathrm{M}(n, n) \times \mathrm{C}_{p^{n-1}}$ and so $|G| = p^{3n}$.

Suppose, by way of contradiction, that $G' \cong \mathrm{E}_{p^3}$. In this case each maximal subgroup M of G is nonabelian. We may set

$$G = \langle a, b, c \rangle\,, \quad M = \langle a, b \mid a^{p^n} = b^{p^n} = 1,\ [a, b] = a^{p^{n-1}} \rangle \times \langle c^p \rangle\,,$$
$$\text{where} \quad G' = \langle a^{p^{n-1}} \rangle \times \langle b^{p^{n-1}} \rangle \times \langle c^{p^{n-1}} \rangle = \Omega_1(\mathrm{Z}(G)) = \Omega_1(G)\,.$$

Any two distinct maximal subgroups of G have distinct commutator subgroups of order p. Moreover, for any $x, y \in G - \mathrm{Z}(G)$ such that $|(\langle x, y \rangle \mathrm{Z}(G)) : \mathrm{Z}(G)| = p^2$, we have

$$1 \neq [x, y] \in \langle x^{p^{n-1}}, y^{p^{n-1}} \rangle \text{ since } \langle x, y \rangle \cong \mathrm{M}(n, n)\,.$$

It follows tat

$$[a, c] \in \langle a^{p^{n-1}}, c^{p^{n-1}} \rangle - \langle a^{p^{n-1}} \rangle\,,$$

and so choosing a suitable generator $c \in \langle c \rangle$, we may set

$$[a, c] = a^{rp^{n-1}} c^{p^{n-1}} \text{ with some fixed integer } r \pmod p\,.$$

We consider all p maximal subgroups $\langle a, b^i c \rangle \mathrm{Z}(G)$ of G containing $\langle a \rangle \mathrm{Z}(G)$ which are distinct from $M = \langle a, b \rangle \mathrm{Z}(G)$. For all integers $i \pmod p$,

$$[a, b^i c] = a^{ip^{n-1}} [a, c] = a^{ip^{n-1}} a^{rp^{n-1}} c^{p^{n-1}} = a^{(i+r)p^{n-1}} c^{p^{n-1}}$$

must be an element contained in $\langle a^{p^{n-1}}, b^{ip^{n-1}} c^{p^{n-1}} \rangle$. Hence, for $i \equiv -r \pmod p$, we obtain

$$c^{p^{n-1}} \in \langle a^{p^{n-1}},\ b^{-rp^{n-1}} c^{p^{n-1}} \rangle\,,$$

which implies $r \equiv 0 \pmod p$ and so $[a, c] = c^{p^{n-1}}$. For all integers $i \pmod p$, we must have

$$a^{ip^{n-1}} c^{p^{n-1}} \in \langle a^{p^{n-1}},\ b^{ip^{n-1}} c^{p^{n-1}} \rangle\,.$$

In particular, for $i \equiv 1 \pmod p$, one has

$$a^{p^{n-1}} c^{p^{n-1}} \in X = \langle a^{p^{n-1}}, b^{p^{n-1}} c^{p^{n-1}} \rangle \cong \mathrm{E}_{p^2}\,.$$

But $c^{p^{n-1}} \in X$, which implies that also $b^{p^{n-1}} \in X$ and this forces

$$X = \langle a^{p^{n-1}}, b^{p^{n-1}}, c^{p^{n-1}} \rangle \cong \mathrm{E}_{p^3}\,,$$

a contradiction.

We have proved that $G' \cong \mathrm{E}_{p^2}$. Since

$$\Phi(G) = \mathrm{Z}(G) = \mho_1(G) = \Omega_{n-1}(G) \cong \mathrm{C}_{p^{n-1}} \times \mathrm{C}_{p^{n-1}} \times \mathrm{C}_{p^{n-1}}\,,$$

we may choose generators a', b', c' of $\Phi(G)$ so that

$$\Phi(G) = \langle a' \rangle \times \langle b' \rangle \times \langle c' \rangle \text{ and } G' = \langle (a')^{p^{n-2}}, (b')^{p^{n-2}} \rangle \cong \mathrm{E}_{p^2}\,.$$

By Propositions 26.10 and 26.23, there are $a, b, c \in G - \Phi(G)$ so that $a^p = a'$, $b^p = b'$, $c^p = c'$, where

$$G' = \langle a^{p^{n-1}}, b^{p^{n-1}} \rangle \quad \text{and} \quad \Omega_1(G) = \langle a^{p^{n-1}}, b^{p^{n-1}}, c^{p^{n-1}} \rangle \cong \mathrm{E}_{p^3} .$$

Set $H = \langle a, b \rangle \trianglelefteq G$. If $[a, b] = 1$, then $H \cong \mathrm{C}_{p^n} \times \mathrm{C}_{p^n}$ and if $1 \neq [a, b] \in \langle a^{p^{n-1}}, b^{p^{n-1}} \rangle$, then $H \cong \mathrm{M}(n, n)$. Since the socle $\langle c^{p^{n-1}} \rangle$ of the subgroup $\langle c \rangle$ is not contained in H, we get that $H\langle c \rangle$ is a semidirect product of $H \trianglelefteq G$ by $\langle c \rangle \cong \mathrm{C}_{p^n}$ and so $|G| = |H|\,|\langle c \rangle| = p^{3n}$ giving $G = \langle a, b, c \rangle$. If $H \cong \mathrm{M}(n, n)$, then we may choose the generators a, b of H so that $a^b = a^{1+p^{n-1}}$. In any case, $|H'| \leq p$ and so (since $G' \cong \mathrm{E}_{p^2}$) $[H, c] \neq \{1\}$.

Let $h \in H - \Phi(H)$ (noting that $\Phi(H) \leq Z(G)$) be such that $[h, c] \neq 1$. Since $\langle h, c \rangle$ is minimal nonabelian, we have $\langle h, c \rangle \cong \mathrm{M}(n, n)$ (of order p^{2n}) and so

$$[h, c] \in H \cap \langle h, c \rangle = \langle h \rangle , \quad \text{which implies } \langle [h, c] \rangle = \langle h^{p^{n-1}} \rangle .$$

We may choose a generator c of $\langle c \rangle$ so that $h^c = h^{1+p^{n-1}}$. Let $k \in H - \Phi(H)$ be such that

$$H = \langle h, k \rangle \text{ so that } k^c = k^{1+ip^{n-1}}$$

for some integer $i \pmod p$. Then we obtain

$$(hk)^c = h^c k^c = h^{1+p^{n-1}} k^{1+ip^{n-1}} .$$

On the other hand, by the above argument:

$$(hk)^c = (hk)(hk)^{jp^{n-1}} \text{ for some integer } j \pmod p .$$

This gives

$$h^{jp^{n-1}} k^{jp^{n-1}} = h^{p^{n-1}} k^{ip^{n-1}} ,$$

and so $j \equiv 1 \pmod p$ and $i \equiv 1 \pmod p$. We have proved that $k^c = k^{1+p^{n-1}}$ and so for all $x \in H$, $x^c = x^{1+p^{n-1}}$.

Assume that $H = \langle a, b \rangle$ is nonabelian. Then from $a^b = a^{1+p^{n-1}}$, we obtain

$$a^{b^{-1}} = aa^{-p^{n-1}} \text{ and so } a^{b^{-1}c} = (aa^{-p^{n-1}})^c = aa^{p^{n-1}} a^{-p^{n-1}} = a .$$

On the other hand,

$$(b^{-1}c)^{p^{n-1}} = b^{-p^{n-1}} c^{p^{n-1}} [c, b^{-1}]^{\binom{p^{n-1}}{2}} = b^{-p^{n-1}} c^{p^{n-1}} ,$$

noting that $n \geq 2$ and in case $p = 2$, we have $n \geq 3$. It follows that the socle of $\langle b^{-1}c \rangle$ is not contained in H and so G is a semidirect product of $H \trianglelefteq G$ with $\langle b^{-1}c \rangle \cong \mathrm{C}_{p^n}$. Now we act on H with $c^* = b^{-1}c$ (instead of c) and note that $[a, c^*] = 1$ so that we must have $[b, c^*] \neq 1$. But (by the above argument) we may choose a generator c^* of $\langle c^* \rangle$ so that $x^{c^*} = x^{1+p^{n-1}}$ for all $x \in H$, a contradiction. Hence, $H = \langle a, b \rangle$ is abelian and $a^c = a^{1+p^{n-1}}$ and $b^c = b^{1+p^{n-1}}$ so that we have obtained the groups stated in part (d4) of our theorem.

Conversely, let G be a p-group given in part (d4) of our theorem, where $n \geq 2$ and if $p = 2$, then $n \geq 3$. Clearly, $A = \langle a \rangle \times \langle b \rangle \times \langle c^p \rangle$ is a unique abelian maximal subgroup of G and

$$\langle a^p \rangle \times \langle b^p \rangle \times \langle c^p \rangle = \mathrm{Z}(G) = \Phi(G) = \Omega_{n-1}(G) .$$

Set $H = \langle a \rangle \times \langle b \rangle$ so that for each $h \in H$, $h^c = h^{1+p^{n-1}}$. Let $x = hc^{pi} \in A$, where $h \in H$ and i is an integer. Then

$$x^c = h^c c^{pi} = hh^{p^{n-1}} c^{pi} = (hc^{pi})h^{p^{n-1}} = hc^{pi}(hc^{pi})^{p^{n-1}} = x^{1+p^{n-1}} , \quad \text{for all } a \in A .$$

Hence for any integer $r \pmod p$ we have $x^{c^r} = x^{1+rp^{n-1}}$, for all $x \in A$. Let $M \in \Gamma_1 - \{A\}$. Then $M = \langle xc^r, y \rangle \mathrm{Z}(G)$ with $r \not\equiv 0 \pmod p$, $x \in A$ and $y \in A - \mathrm{Z}(G)$ so that $o(y) = p^n$. The socle of $\langle y \rangle$ is contained in $\Omega_1(\langle a^{p^{n-1}}, b^{p^{n-1}} \rangle)$ and the socle of $\langle xc^r \rangle$ is $\langle x^{p^{n-1}} c^{rp^{n-1}} \rangle$ since

$$(xc^r)^{p^{n-1}} = x^{p^{n-1}} c^{rp^{n-1}} [c^r, x]^{\binom{p^{n-1}}{2}} = x^{p^{n-1}} c^{rp^{n-1}} .$$

In particular, $\langle xc^r \rangle \cap \langle y \rangle = \{1\}$. Also,

$$y^{xc^r} = y^{c^r} = y^{1+rp^{n-1}} \text{ and so } [y, xc^r] = y^{rp^{n-1}} , \text{ and } K = \langle y, xc^r \rangle \cong \mathrm{M}(n, n) .$$

On the other hand,

$$\mathrm{Z}(K) = \langle y^p, (xc^r)^p \rangle \cong \mathrm{C}_{p^{n-1}} \times \mathrm{C}_{p^{n-1}}$$

has a complement $L \cong \mathrm{C}_{p^{n-1}}$ in $\mathrm{Z}(G)$ and $K \times L$ is of order p^{3n-1} so that $M = K \times L$ is s-self-dual and we are done. The proof is complete. □

Problem 1. Classify the p-groups all of whose minimal nonabelian subgroups (i) are isomorphic to either $S(p^3)$ or $\mathrm{M}(n, n)$ where n is a positive integer, (ii) have order p^3 or $\mathrm{M}(n, n)$, where n is as in (i).

Theorem 193.2 deals with a partial case of Problem 1. Note that p-groups all of whose minimal nonabelian subgroups are metacyclic, are not classified yet. The same holds for p-groups, $p > 2$, all of whose minimal nonabelian subgroups have order p^3.

Problem 2. Classify the p-groups all of whose $\mathcal{A}_2$-subgroups are s-self-dual. (See §§ 65, 71.)

Problem 3. Classify the p-groups all of whose subgroups of index p^2 are s-self-dual.

Problem 4. Classify the non-Dedekindian p-groups all of whose nonnormal subgroups are s-self-dual.

Problem 5. Study the non-Dedekindian p-groups in which the normalizers of all nonnormal subgroups are s-self-dual.

Problem 6. Study the non-Dedekindian p-groups G such that $G/\mathrm{Z}(G)$ is s-self-dual.

Problem 7. Classify the p-groups all of whose maximal abelian subgroups have index $\leq p^2$. (All groups of Theorem 193.2 satisfy this condition.)

Problem 8. Classify the p-groups G such that, for all $H \leq G$, (i) $|H'| \neq p^2$ (any $\mathcal{A}_2$-group all of whose maximal subgroups are nonabelian and also any group of Theorem 193.2 satisfy this condition), (ii) $|\Phi(H)| \neq p^2$.

Problem 9. Study the p-groups G all of whose subgroups are q-self-dual. (A p-group is q-self-dual if any quotient group of G is isomorphic to its subgroup; see § 192.)

§ 194 *p*-groups all of whose maximal subgroups, except one, are *s*-self-dual

We recall that a group G is s-self-dual if every subgroup of G is isomorphic to a quotient group of G. Classification of s-self-dual groups is reduced to one of p-groups. Such p-groups were classified up to isomorphism in § 192. In particular, any minimal nonabelian subgroup of G is either $\cong S(p^3)$ or $\cong \mathrm{M}(n,n) = \langle x, y \mid o(x) = o(y) = p^n,\ n > 1,\ x^y = x^{1+p^{n-1}}\rangle$. It should be noted that any abelian group is s-self-dual. As the main result of § 192 shows, the subgroups of s-self-dual p-groups are s-self-dual. Next, $|G/\mathrm{Z}(G)| = p^2$ for any nonabelian s-self-dual p-group G.

The following theorem reduces the classification of the title groups to Theorem 137.7, where the p-groups all of whose proper subgroups have derived subgroup of order $\leq p$, described up to isomorphism.

Theorem 194.1. *Let G be a title p-group and let H be a unique maximal subgroup of G which is not s-self-dual. Then $|H'| = p$ and $|H/\mathrm{Z}(H)| \in \{p^2, p^4\}$.*

Proof. Let G be a p-group all of whose maximal subgroups, except one, are s-self-dual. Let H be a unique maximal subgroup of G which is not s-self-dual.

Suppose $|H'| > p$. By Theorem 192.1, all other maximal subgroups of G (distinct from H) have its commutator subgroup of order $\leq p$. We are in a position to use Theorem 145.8. By that theorem, $\mathrm{d}(G) = 2$, $p = 2$, G' is abelian of type $(4, 2)$,

$$\mathrm{K}_3(G) = \Omega_1(G') \leq \mathrm{Z}(G), \quad \Phi(G) = \mathrm{C}_G(G') \text{ is abelian and } \mho_2(G) \leq \mathrm{Z}(G)\,.$$

Let $\{H_1, H_2, H\}$ be the set of maximal subgroups of G. Then Theorem 145.8 also gives

$$H_1' = \langle z_1\rangle \text{ and } H_2' = \langle z_2\rangle \text{ are both of order } 2\,,$$
$$\langle z_1, z_2\rangle = \Omega_1(G') = H' \cong \mathrm{E}_4\,, \quad \mathrm{d}(H) = 3 \text{ and } \mho_1(G') = \langle z_1 z_2\rangle\,.$$

By Schreier inequality (Theorem A.25.1),

$$\mathrm{d}(H_i) \leq 1 + |G : H_i|(\mathrm{d}(G) - 1) = 3\,, \quad i = 1, 2\,.$$

Let $k \in G' - H'$ so that $o(k) = 4$ and $k \notin \mathrm{Z}(H_i)$, $i = 1, 2$, (since $\mathrm{C}_G(G') = \Phi(G)$). Hence, there is $h_i \in H_i$ such that $[k, h_i] \neq 1$ and so $[k, h_i] = z_i$ so that, by Lemma 65.2 (a), $\langle k, h_i\rangle$ is minimal nonabelian. By Theorem 192.1,

$$\langle k, h_i\rangle \cong \mathrm{M}(2,2) = \mathcal{H}_2\,, \quad H_i = \langle k, h_i\rangle \times A_i, \text{ where } A_i \text{ is elementary abelian}\,.$$

But $\mathrm{d}(H_i) \leq 3$, and so A_i is cyclic of order 1 or 2. Obviously, $|A_1| = |A_2|$ because $|G| = 2|H_i|$ and $|H_i| = 2^4|A_i|$, $i = 1, 2$.

Suppose that $|A_i| = 1$, $i = 1, 2$. Then $|H_i| = 2^4$ and so $|G| = 2^5$. But then $|H| = 2^4$ and since $H' \cong \mathrm{E}_4$, it follows that H is of maximal class (Proposition 1.6), a contradiction.

We have proved that $|A_i| = 2$, $i = 1, 2$, and so $H_i \cong \mathcal{H}_2 \times C_2$. In particular, $|G| = 2^6$ and so $|H| = 2^5$. But we know that $d(H) = 3$ and so $E_4 \cong H' = \Phi(H)$ and H/H' is elementary abelian, giving $\exp(H) = 4$ (noting that $G' \cong C_4 \times C_2$ is contained in H). Since also $\exp(H_1) = \exp(H_2) = 4$, we obtain $\exp(G) = 4$. On the other hand, we know from the above that $\Phi(G) = \mho_1(G)$ is abelian of order 2^4 and $\exp(\Phi(G)) = 4$ since $G' \leq \Phi(G)$.

For each $g \in G$, g^2 is of order ≤ 2 and so $g^2 \in \Omega_1(\Phi(G))$, where $\Omega_1(\Phi(G))$ is elementary abelian, a contradiction.

We have proved that $|H'| \leq p$. But abelian p-groups are s-self-dual and so we have $|H'| = p$. Let M be a maximal subgroup of G which is distinct from H. Set $H_0 = M \cap H$ so that H_0 is a maximal subgroup of H. Since M is s-self-dual, Theorem 192.1 implies that H_0 is also s-self-dual. If H_0 is abelian, then Lemma 1.1 implies that $|H/Z(H)| = p^2$ and we are done.

Suppose that H_0 is nonabelian. Then Theorem 192.1 implies $H_0 = K \times A$, where K is a certain minimal nonabelian subgroup and A is abelian. Since $|H'| = p$, we obtain

$$H = K * H_0^*, \quad \text{where } H_0^* = C_H(K) \text{ and } H_0^* \cap H_0 = Z(K) \times A$$

is an abelian subgroup of index 2 in H_0^*. If H_0^* is abelian, then $H_0^* = Z(H)$ and so again $|H/Z(H)| = p^2$. If H_0^* is nonabelian, then Lemma 1.1 implies that $|H_0^* : Z(H_0^*)| = p^2$ and since $Z(H_0^*) = Z(H)$, we obtain in this case $|H/Z(H)| = p^4$. The proof is complete. □

We suggest to the interested reader to finish the classification of groups of Theorem 194.1.

It follows from Theorem 194.1 that if $A < G$ is maximal abelian, then $|G : A| \leq p^4$.

Exercise 1. Study the p-groups G of Theorem 194.1 provided $G' \cong E_{p^2}$.

Exercise 2. Study the p-groups G of Theorem 194.1 provided $G' \cong E_{p^3}$ and $d(G) = 2$, in detail.

Exercise 3. Suppose that $L = \langle H' \mid H \in \Gamma_1\} < G'$. Then $d(G) = 2$.

Solution. As G' is generated by commutators, it follows that there is $a = [x, y] \in G' - L$. It follows that for any $H \in \Gamma_1$, $[x, y] = a \in H - H'$, and we conclude that $x, y \notin H$. In this case, $\langle x, y\rangle = G$, as was to be shown. (We have used this argument above in this section and in § 65).

Exercise 4. If all minimal nonabelian subgroups of a nonabelian p-group G, $p > 2$, are metacyclic, then $\Omega_1(G)$ is elementary abelian.

For a solution, see Exercise A.45.196. This is not true for $p = 2$ (in this case, there is in G a subgroup $\cong D_8$, unless G is a generalized quaternion group).

Problem 1. Study the p-groups containing two distinct s-self-dual subgroups of index p.

Problem 2. Classify the p-groups all of whose subgroups of index p^2 are s-self-dual.

Problem 3. Let L be a subgroup of order p of a p-group G. Study the structure of G provided all maximal subgroups of G containing L are s-self-dual.

Problem 4. Study the p-groups all of whose subgroups nonincident with G' (with $\Phi(G)$), are s-self-dual.

Problem 5. Let H be a subgroup of a p-group G. Study the structure of G provided all subgroups of G nonincident with H are s-self-dual. (This is a generalization of #3.)

§ 195 Nonabelian p-groups all of whose subgroups are q-self-dual

We recall that a p-group G is q-self-dual if every quotient of G is isomorphic to a subgroup of G. In contrast to s-self-duality (see §§ 192, 193, 194), there are many p-groups which are q-self-dual (see also [Yin2]). Therefore it seems impossible to classify all q-self-dual p-groups. Here we classify 2-groups G all of whose subgroups are q-self-dual (Theorem 195.3) and in case $p > 2$ we obtain a classification of such groups only under the additional assumptions that $\Omega_1(G)$ is abelian (Theorem 195.4). This solves partly Problem 9 stated in § 193. It is known that abelian p-groups are q-self-dual. We first prove two auxiliary lemmas.

Lemma 195.1. *Let G be a title p-group and let H be a minimal nonabelian subgroup of G. Then H is isomorphic to one of the following groups:* D_8, $\mathrm{S}(p^3)$ *(the nonabelian group of order p^3 and exponent $p > 2$),* M_{p^n}, *$n \geq 3$, where in case $p = 2$, $n \geq 4$.*

Proof. First of all, Q_8 is not q-self-dual since $\mathrm{Q}_8/\mathrm{Q}_8' \cong \mathrm{E}_4$ is not a subgroup of Q_8. Suppose that $\mathrm{Z}(H) = \Phi(H)$ is noncyclic. Then there is $z \in \mathrm{Z}(H) - H'$ and so $H/\langle z \rangle$ is nonabelian. This is a contradiction since H has no nonabelian subgroup of order $< |H|$. Hence, $\mathrm{Z}(H) = \Phi(H)$ is cyclic. We have $\mathrm{C}_p \cong H' \leq \mathrm{Z}(H)$. If $H' = \mathrm{Z}(H)$, then $|H| = p^3$ and so either $p > 2$ and H is isomorphic to one of $\mathrm{S}(p^3)$, M_{p^3} or $p = 2$ and $H \cong \mathrm{D}_8$. Suppose $H' < \mathrm{Z}(H) = \Phi(H)$ and then $\Phi(H) = H'\mho_1(H) = \mho_1(H)$ implies that H has a cyclic maximal subgroup and $|H| \geq p^4$. Hence, $H \cong \mathrm{M}_{p^n}$, $n \geq 4$ and we are done. □

Lemma 195.2. *Let G be a nonabelian 2-group all of whose subgroups are q-self-dual. If $a, b \in G$ do not commute, then either $\langle a, b \rangle \cong \mathrm{D}_{2^m}$, $m \geq 3$, or $\langle a, b \rangle \cong \mathrm{M}_{2^n}$, $n \geq 4$.*

Proof. Set $H = \langle a, b \rangle$ and let K be an H-invariant subgroup of index 2 in H'. By Lemma 65.2 (a), H/K is minimal nonabelian. Since H is q-self-dual, H possesses a subgroup H_0 isomorphic to H/K. By Lemma 195.1, $H_0 \cong \mathrm{D}_8$ or M_{2^n}, $n \geq 4$. In particular, H/K is metacyclic and so a result of N. Blackburn (Theorem 36.1) implies that H is metacyclic and therefore H' is cyclic. Let H^*/K be a cyclic subgroup of index 2 in H/K. It follows that H^* is a cyclic subgroup of index 2 in H and we may use Theorem 1.2. If $H/K \cong \mathrm{D}_8$, then G is of maximal class. Since H is Q_8-free (Lemma 195.1), we obtain $H \cong \mathrm{D}_{2^m}$, $m \geq 3$. If $H/K \cong \mathrm{M}_{2^n}$, $n \geq 4$, then $K = \{1\}$ and $H \cong \mathrm{M}_{2^n}$. Our lemma is proved. □

Theorem 195.3. *Let G be a nonabelian 2-group all of whose subgroups are q-self-dual. Then one of the following hold.*

(a) *G is quasidihedral (generalized dihedral), i.e., G has an abelian maximal subgroup A of exponent > 2 and an involution $t \in G - A$ such that t inverts each element in A.*

(b) *$G = M \times V$, where $M \cong \mathrm{M}_{2^{e+1}}$, $e \geq 3$, and* $\exp(V) \leq 2$.

Conversely, all the above groups satisfy the assumption of the theorem.

Proof. Let G be a nonabelian 2-group all of whose subgroups are q-self-dual. By Lemma 195.1, G is Q_8-free.

(i) First suppose that G is nonmodular, i.e., G is not D_8-free. We are in a position to use Theorem 79.7. According to that theorem, G must be isomorphic to a Wilkens 2-group of type (a), (b) or (c).

(i1) Suppose that G is a Wilkens group of type (a) with respect to N. Then $G = \langle x \rangle N$, where N is a maximal normal abelian subgroup of G with $\exp(N) > 2$, $\langle x \rangle$ is cyclic of order ≥ 2, $\langle x \rangle \cap N = \{1\}$ and, if t is the involution in $\langle x \rangle$, then every element in N is inverted by t.

Assume $\langle x \rangle > \langle t \rangle$. By Lemma 57.1, there is $n \in N$ such that $\langle x, n \rangle$ is minimal nonabelian. Then Lemma 195.1 implies that $\langle x, n \rangle \cong D_8$ or M_{2^n}, $n \geq 4$. In any case,

$$\Omega_1(\Phi(\langle x, n \rangle)) = \langle x, n \rangle' = \langle t \rangle ,$$

a contradiction since $G' \leq N$. We have proved that $\langle x \rangle = \langle t \rangle$ and so G is quasidihedral as stated in part (a) of our theorem.

Conversely, let G be a quasidihedral 2-group with an abelian maximal subgroup A of exponent > 2 and let t be any element in $G - A$ so that t is an involution which inverts each element of A. Since each subgroup of G is either quasidihedral or abelian, it is enough to show that G is q-self-dual. Let N be any normal subgroup in G. First suppose $N \nleq A$ so that

$$G = AN \text{ and } G/N \cong A/(A \cap N) \text{ is abelian.}$$

But $A/(A \cap N)$ is isomorphic to a subgroup of A and in this case we are done. Now suppose $N \leq A$. Then A possesses a subgroup A_0 such that $A/N \cong A_0$. An involution $t \in G - A$ inverts isomorphic abelian groups A/N and A_0. Hence,

$$A_0\langle t \rangle \cong (A\langle t \rangle)/N = G/N$$

and so G is q-self-dual.

(i2) Suppose that G is a Wilkens group of type (b) with respect to N. Then we have $G = \langle x \rangle N$, where N is a maximal elementary abelian normal subgroup of G and $\langle x \rangle$ is not normal in G. Since G is not D_8-free, G has a subgroup $D \cong D_8$. But $G/N \neq \{1\}$ is cyclic and so $D \cap N \neq \{1\}$. Suppose that $D \cap N \cong C_2$. Then $D/(D \cap N) \cong C_4$ and so D would be abelian, a contradiction. Hence, we have $D \cap N \cong E_4$. It follows that there are elements of order 2 and 4 in $D - N$ which do not centralize $D \cap N$ and so they do not centralize N. This implies $C_G(N) = N$ so that N is also a maximal normal abelian subgroup of G.

We note that each two-generator nonabelian subgroup of G is isomorphic to D_8 or M_{2^n}, $n \geq 4$. Indeed, by Lemma 195.2, we only have to show that $R \cong D_{16}$ is not a subgroup of G. If $R \leq G$, then $R' \cong C_4$ and $R' \leq G' \leq N$, a contradiction.

Suppose, by way of contradiction, that $|G/N| \geq 4$. Set $\Omega_1(\langle x \rangle) = \langle z \rangle$. Assume in addition that $z \notin N$, i.e., G splits over N. By Lemma 57.1, there is $n \in N$ such that $\langle x, n \rangle$

is minimal nonabelian. By Lemma 195.1,

$$\langle x, n\rangle \cong D_8 \text{ or } M_{2^n}, \quad n \geq 4, \text{ and so } \Omega_1(\Phi(\langle x, n\rangle)) = \langle z\rangle = \langle x, n\rangle',$$

contrary to the fact that $G' \leq N$. Hence, $\langle x\rangle \cap N = \langle z\rangle$ and $o(x) \geq 8$. By the previous paragraph, for each $n \in N$, we have either $[x, n] = 1$ or $\langle x, n\rangle \cong M_{2^n}$, $n \geq 4$, and in this case $[x^2, n] = 1$. It follows that x^2 centralizes N. This is a contradiction since N is a maximal normal abelian subgroup of G and $x^2 \notin N$.

We have shown that $|G : N| = 2$. Let y be an element of order 4 in $G - N$ and set $y^2 = z$ so that

$$G = \langle y\rangle N \quad \text{with } \langle y\rangle \cap N = \langle z\rangle \cong C_2 .$$

For any $n \in N$, we have either $[y, n] = 1$ or $\langle y, n\rangle \cong D_8$ and $[y, n] = z$. This gives that $N_0 = C_N(y)$ is of index 2 in N and if $n \in N - N_0$, then $\langle y, n\rangle = D \cong D_8$. Let V be a complement of $\langle z\rangle$ in N_0. Then $G = D \times V$.

We have proved that if G is a Wilkens group of type (b), then $G = D \times V$, where $D \cong D_8$ and $\exp(V) \leq 2$ and we note that this group is quasidihedral and so is included in part (a) of our theorem.

(i3) Suppose that G is a Wilkens group of type (c) with respect to N, x, t. We have $G = \langle N, x, t\rangle$, where N is an elementary abelian normal subgroup of G and t is an involution in $G - N$ with $[N, t] = \{1\}$. If $o(xN) = 2^k$, then

$$G/N \cong M_{2^{k+1}}, \quad k \geq 3, \quad x^{2^k} \neq 1 .$$

Furthermore, $[x^{2^{k-1}}, N] = \{1\}$ and $\langle t, x^{2^{k-1}}\rangle \cong D_8$.

By Proposition 79.10, the involution $z = x^{2^k}$ lies in $G' \cap Z(G)$ and $G/\langle z\rangle$ is a Wilkens group of type (b). Let G_0 be a subgroup of G such that $G_0 \cong G/\langle z\rangle$. Then $|G : G_0| = 2$ and G_0 is a Wilkens group of type (b). By (i2), $G_0 = D \times V$, where $D \cong D_8$ and $\exp(V) \leq 2$. In particular, $\exp(G_0) = 4$. But then $\exp(G) \leq 8$, contrary to $x^{2^k} \neq 1$, $k \geq 3$.

(ii) Assume that G is a nonabelian modular 2-group and so G is D_8-free. Since G is Q_8-free (Lemma 195.1), G is not a Hamiltonian group. We are in a position to use Theorem 73.15. By that theorem, G contains an abelian normal subgroup N with cyclic factor-group G/N and there is an element $t \in G - N$ such that $G = \langle N, t\rangle$ and a positive integer $s \geq 2$ so that $a^t = a^{1+2^s}$ for all $a \in N$.

We use many times the fact that each subgroup of G is q-self-dual. By Lemma 195.2, each two-generator nonabelian subgroup of G is minimal nonabelian and is isomorphic to M_{2^n}, $n \geq 4$. Hence, for any $x, y \in G$ with $[x, y] \neq 1$, $\langle x, y\rangle$ is minimal nonabelian and so

$$1 = [x^2, y] = [x, y]^2 .$$

On the other hand, $G' \leq N$ and N is abelian and so G' is elementary abelian. Since t centralizes G', we obtain $G' \leq Z(G)$ and $\mho_1(G) \leq Z(G)$.

Because G is nonabelian and $s \geq 2$, we have $\exp(N) = 2^e$ with $e \geq 3$. Let $a \in N$ with $o(a) = 2^e$ and $[a, t] \neq 1$. Then

$$a^t = a^{1+2^s} \text{ so that } 1 \neq [a, t] = a^{2^s} \text{ is an involution giving } s = e - 1 .$$

Thus, we have for all $a \in N$,

$$a^t = a^{1+2^{e-1}}, \text{ which implies } C_N(t) = \Omega_{e-1}(N) \le Z(G) .$$

Also, $t^2 \in Z(G)$ and so $Z(G) = \Omega_{e-1}(N)\langle t^2 \rangle$ and $A = \langle t^2 \rangle N$ is an abelian maximal subgroup of G.

Let $a \in N$ with $o(a) = 2^e$ so that $a^t = a^{1+2^{e-1}}$ and $[a, t] = a^{2^{e-1}}$. In particular, t normalizes $\langle a \rangle \cong C_{2^e}$ and $H = \langle a, t \rangle \cong M_{2^n}$, $n \ge 4$. Note that

$$Z(H) = \Phi(H) = \langle a^2, t^2 \rangle \text{ is cyclic and } \Omega_1(\Phi(H)) = \langle a^{2^{e-1}} \rangle = H' .$$

Hence the socle of any cyclic subgroup of composite order in H is equal H'.

(ii1) First assume $o(t) \le 2^e$. Then $\exp(H) = 2^e$ and therefore $\langle a \rangle$ is a cyclic maximal subgroup of H. Then there is an involution t' in $H - A$ which acts the same way on N as t does. Since G/N is cyclic, we obtain $|G : N| = 2$, $A = N$ and we replace t with t' and write t again (instead of t') so that $H = \langle a, t \rangle \cong M_{2^{e+1}}$. Suppose that there is $b \in N$ such that $o(b) = 2^e$ and $\langle a \rangle \cap \langle b \rangle = \{1\}$. Let $h \in H - \langle a \rangle$ be an element of order 4 so that $h^2 = a^{2^{e-1}}$. Then $b^h = b^{1+2^{e-1}}$ and so $\langle h, b \rangle \cong M_{2^m}$, $m \ge 4$. This is a contradiction since the socle of $\langle b \rangle$ is $\langle b^{2^{e-1}} \rangle$ and the socle of $\langle h \rangle$ is $\langle a^{2^{e-1}} \rangle$ and $\langle b^{2^{e-1}} \rangle \ne \langle a^{2^{e-1}} \rangle$. It follows that $N = \langle a \rangle \times N_0$, where $\exp(N_0) < 2^e$ so that $G = H \times N_0$, where $H \cong M_{2^{e+1}}$, $e \ge 3$, and N_0 is abelian of exponent $< 2^e$.

(ii2) Now suppose $o(t) \ge 2^{e+1}$, where $\exp(N) = 2^e$, $e \ge 3$. Since t centralizes $N \cap \langle t \rangle$, it follows $|N \cap \langle t \rangle| \le 2^{e-1}$. On the other hand, $\exp(H) = o(t)$ and so $\langle t \rangle$ is a maximal cyclic subgroup of H giving $a^2 \in \langle t \rangle$ and therefore $\langle a^2 \rangle = N \cap \langle t \rangle \cong C_{2^{e-1}}$. If there is $b \in N$ such that $o(b) = 2^e$ and $\langle a \rangle \cap \langle b \rangle = \{1\}$, then considering $\langle b, t \rangle \cong M_{2o(t)}$, we obtain

$$\langle t \rangle \cap N = \langle b^2 \rangle = \langle a^2 \rangle ,$$

contrary to $\langle a \rangle \cap \langle b \rangle = \{1\}$. It follows that $N = \langle a \rangle \times N_0$, where $\exp(N_0) < 2^e$ so that $G = H \times N_0$, where $H \cong M_{2o(t)}$, $o(t) \ge 2^{e+1}$ and N_0 is abelian of exponent $< 2^e$.

In both cases (ii1) and (ii2), we have obtained $G = M \times V$, where $M \cong M_{2^{e+1}}$, $e \ge 3$, and V is abelian of exponent $< 2^e$. We set

$$M = \langle a, t \mid a^{2^e} = t^2 = 1,\ [a, t] = a^{2^{e-1}} \rangle .$$

Now assume $\exp(V) > 2$. Let $x \in M - \langle a \rangle$ with $o(x) = 4$ and $y \in V$ with $o(y) = 4$. Then the socle of $\langle x \rangle$ is $M' = \langle a^{2^{e-1}} \rangle$ and the socle of $\langle xy \rangle$ is not contained in M. But $\langle a, xy \rangle$ is minimal nonabelian, where the socle of $\langle a \rangle$ is $\langle a, xy \rangle' = \langle a^{2^{e-1}} \rangle$ and the socle of $\langle xy \rangle$ is distinct from $\langle a^{2^{e-1}} \rangle$ so that $\langle a, xy \rangle$ cannot be isomorphic to some M_{2^n}, $n \ge 4$, a contradiction. It follows that $\exp(V) \le 2$ and so we have obtained the groups stated in part (b) of our theorem.

Conversely, let $G = M \times V$ be a group from part (b) of our theorem, where

$$M = \langle a, t \mid a^{2^e} = t^2 = 1,\ [a, t] = a^{2^{e-1}} \rangle \cong M_{2^{e+1}} , \quad e \ge 3 ,$$

and V is abelian of exponent ≤ 2. Here

$$Z(G) = \langle a^2 \rangle \times V\,, \quad A = \langle a \rangle \times V \text{ and for each } x \in A,\ x^t = x^{1+2^{e-1}}\,,$$

where A is an abelian maximal subgroup of G. Each subgroup of G is either abelian or of the form $\mathrm{M}_{2^{e+1}} \times \mathrm{E}_{2^s}$, $s \geq 0$, and so it is enough to show that G is q-self-dual. Let $N \neq \{1\}$ be any normal subgroup of G. If $N \nleq A$, then $G/N \cong A/(A \cap N)$ and there is a subgroup A_0 of A such that $A/(A \cap N) \cong A_0$ and we are done in this case. Now suppose $N \leq A$ and note that each subgroup of A is normal in G. First of all,

$$G/\langle a^{2^{e-1}} \rangle \cong \langle a^2 \rangle \times \langle t \rangle \times V = G_0\,,$$

and so each homomorphic image of $G/\langle a^{2^{e-1}} \rangle$ is isomorphic to a homomorphic image of G_0 which is then isomorphic to a subgroup of G_0. Let $N \leq A$ with $N \ngeq G' = \langle a^{2^{e-1}} \rangle$. Since $N \cap \langle a \rangle = \{1\}$, N is elementary abelian and so $N \leq \Omega_1(A) = \langle a^{2^{e-1}} \rangle \times V$. But then $G/N \cong \mathrm{M}_{2^{e+1}} \times \mathrm{E}_{2^s}$ with $|\mathrm{E}_{2^s}| \leq |V|$ and we are done. □

Theorem 195.4. *Let G be a nonabelian p-group, $p > 2$, all of whose subgroups are q-self-dual and assume that $\Omega_1(G)$ is abelian. Then*

$$G = M \times V\,, \quad \text{where } M \cong \mathrm{M}_{p^n},\ n \geq 3, \text{ and } \exp(V) \leq p\,.$$

Conversely, these groups satisfy the assumption of the theorem.

Proof. Let G be a nonabelian p-group, $p > 2$, all of whose subgroups are q-self-dual and $\Omega_1(G)$ is abelian. This implies that G is $\mathrm{S}(p^3)$-free and so G is modular (see § 73). Indeed, each subgroup of G is q-self-dual and G has no subgroup isomorphic to $\mathrm{S}(p^3)$.

Suppose $a, b \in G$ with $[a, b] \neq 1$. Set $H = \langle a, b \rangle$ and let K be a G-invariant subgroup of index p in H'. By Lemma 65.2 (a), H/K is minimal nonabelian. Since H/K is isomorphic to a subgroup of H, Lemma 195.1 implies $H/K \cong \mathrm{M}_{p^n}$, $n \geq 3$. Thus, H/K is metacyclic and so Theorem 36.1 gives that H is also metacyclic and consequently, H' is cyclic. Let H_0/K be a cyclic maximal subgroup of H/K. But then H_0 is a cyclic subgroup of index p in H and we may use Theorem 1.2. It follows $K = \{1\}$ and $H \cong \mathrm{M}_{p^n}$, $n \geq 3$. We have proved that each two-generator nonabelian subgroup of G is isomorphic to M_{p^n}, $n \geq 3$.

By Theorem 73.15, G has an abelian normal subgroup N of exponent p^e, $e \geq 2$, with cyclic G/N and there is an element $t \in G$ with $G = \langle N, t \rangle$ and a positive integer $s < e$ so that $a^t = a^{1+p^s}$ for all $a \in N$.

Let $x, y \in G$ with $[x, y] \neq 1$. Since $\langle x, y \rangle$ is minimal nonabelian, we obtain

$$1 = [x^p, y] = [x, y]^p\,,$$

and so $\mho_1(G) \leq Z(G)$ and $G' \leq Z(G)$ is elementary abelian. Let $a \in N$ with $o(a) = p^e$, where $p^e = \exp(N)$. Then

$$a^t = a^{1+p^s} \text{ and so } [a, t] = a^{p^s} \text{ is of order } p \text{ implying } s = e - 1\,.$$

Hence, $H = \langle a, t\rangle \cong \mathrm{M}_{p^n}$, $n \geq 3$, and for all $x \in N$, we have $x^t = x^{1+p^{e-1}}$.

(i) First suppose $o(t) \leq p^e$. Then $\exp(H) = p^e$ and therefore $\langle a\rangle$ is a cyclic maximal subgroup of H giving $t^p \in \langle a\rangle$ and so $|G : N| = p$. Suppose that there is $b \in N$ such that $o(b) = p^e$ and $\langle a\rangle \cap \langle b\rangle = \{1\}$. Let $h \in H - \langle a\rangle$ be an element of order p^2 so that $h^p = a^{p^{e-1}}$. Then $\langle h, b\rangle \cong \mathrm{M}_{p^m}$, $m \geq 3$. This is a contradiction since the socle of $\langle b\rangle$ is $\langle b^{p^{e-1}}\rangle$ and the socle of $\langle h\rangle$ is $\langle a^{p^{e-1}}\rangle$ and $\langle b^{p^{e-1}}\rangle \neq \langle a^{p^{e-1}}\rangle$. It follows that $N = \langle a\rangle \times N_0$, where $\exp(N_0) < p^e$ so that $G = H \times N_0$. Suppose that N_0 is not elementary abelian. Let $x \in H - \langle a\rangle$ with $o(x) = p^2$ and $y \in N_0$ with $o(y) = p^2$. Then the socle of $\langle x\rangle$ is $H' = \langle a^{p^{e-1}}\rangle$ and the socle of $\langle xy\rangle$ is not contained in H. But $\langle a, xy\rangle$ is minimal nonabelian, where the socle of $\langle a\rangle$ is $\langle a, xy\rangle' = \langle a^{p^{e-1}}\rangle$ and the socle of $\langle xy\rangle$ is distinct from $\langle a^{p^{e-1}}\rangle$ so that $\langle a, xy\rangle$ cannot be isomorphic to some M_{p^n}, $n \geq 3$, a contradiction. It follows that $\exp(N_0) \leq p$ and we have obtained the groups stated in our theorem.

(ii) Now suppose $o(t) \geq p^{e+1}$. Since t centralizes $N \cap \langle t\rangle$, it follows $|N \cap \langle t\rangle| \leq p^{e-1}$. On the other hand, $\exp(H) = o(t)$ and so $\langle t\rangle$ is a maximal cyclic subgroup of H, giving $a^p \in \langle t\rangle$ and therefore $\langle a^p\rangle = N \cap \langle t\rangle \cong \mathrm{C}_{p^{e-1}}$.

If there is $b \in N$ such that $o(b) = p^e$ and $\langle a\rangle \cap \langle b\rangle = \{1\}$, then considering $\langle b, t\rangle \cong \mathrm{M}_{po(t)}$, we obtain

$$\langle t\rangle \cap N = \langle b^p\rangle = \langle a^p\rangle ,$$

contrary to $\langle a\rangle \cap \langle b\rangle = \{1\}$. It follows that $N = \langle a\rangle \times N_0$, where $\exp(N_0) < p^e$ so that $G = H \times N_0$.

Suppose that N_0 is not elementary abelian. Let $x \in H - \langle t\rangle$ with $o(x) = p^2$ and $y \in N_0$ with $o(y) = p^2$. Then the socle of $\langle x\rangle$ is $H' = \langle a^{p^{e-1}}\rangle$ and the socle of $\langle xy\rangle$ is not contained in H. But $\langle t, xy\rangle$ is minimal nonabelian, where the socle of $\langle t\rangle$ is $\langle t, xy\rangle' = \langle a^{p^{e-1}}\rangle$ and the socle of $\langle xy\rangle$ is distinct from $\langle a^{p^{e-1}}\rangle$ so that $\langle t, xy\rangle$ cannot be isomorphic to some M_{p^n}, $n \geq 3$, a contradiction. It follows that $\exp(N_0) \leq p$ and we have obtained again the groups stated in our theorem.

Conversely, the groups stated in our theorem satisfy the assumptions of that theorem. The proof is the same as in case of groups in part (b) of Theorem 195.3. □

Problem 1. Classify the q-self-dual p-groups with cyclic derived subgroup.

Problem 2. Study the p-groups G all of whose normal subgroups are q-self-dual. In particular, study the p-groups all of whose (i) maximal subgroups are q-self-dual, (ii) members of the sets Γ_i are q-self-dual for all $i \leq \mathrm{d}(G)$.

Problem 3. Study the p-groups all of whose minimal nonabelian subgroups are q-self-dual (see Lemma 195.1).

Problem 4. Study the p-groups all of whose metacyclic subgroups are q-self-dual.

Problem 5. Study the p-groups G such that, whenever $N \trianglelefteq G$ is of exponent p, there exist $H < G$ with $H \cong G/N$. (Dihedral and semidihedral groups satisfy this condition.)

Problem 6. Study the p-groups G all of whose two-generator subgroups are q-self-dual. (See Lemma 195.2.)

Problem 7. Study the non-Dedekindian p-groups G all of whose nonnormal subgroups are q-self-dual.

Problem 8. Study the p-groups all of whose subgroups of index $> p$ are q-self-dual.

Problem 9. Classify the special p-groups such that, whenever $L < Z(G)$ is of order p, there is $H \in \Gamma_1$ with $G/L \cong H$.

§ 196 A p-group with absolutely regular normalizer of some subgroup

Let G be a p-group and $H < G$. Sometimes it is possible to deduce some useful information on the structure of G from knowledge of the normalizer $\mathrm{N}_G(H)$. There are some results of this type in our book. For example, if $\mathrm{N}_G(H)$ is of maximal class, then G is also of maximal class (Remark 10.5).

Recall that a p-group G is absolutely regular if $|G/\mho_1(G)| \leq p^{p-1}$. By Theorem 9.8 (a), any absolutely regular p-group G is regular so that $|\Omega_1(G)| < p^p$, by Theorem 7.2 (d). By Theorem 12.1 (a), if $|\Omega_1(G)| < p^p$, then a p-group G is either absolutely regular or of maximal class.

We suggest to the reader to look on Theorems 9.5, 9.6, 13.19 (dealing with p-groups of maximal class), and 9.11 (characterizing the metacyclic p-groups, $p > 2$). Recall that if a p-group G of maximal class has order $> p^p$, it is irregular; if, in addition, $|G| > p^{p+1}$, then exactly one member of the set Γ_1, which is denoted by G_1 and called the fundamental subgroup of G, is regular and satisfies $|G_1/\mho_1(G_1)| = p^{p-1}$, hence absolutely regular. Moreover, in the last case, all members of the set $\Gamma_1 - \{G_1\}$ are of maximal class.

If G is a p-group, $p > 2$, then it is metacyclic $\iff$ $|G/\mho_1(G)| \leq p^2$ (Theorem 9.11). We also use the following fact (see Exercise 10.10): If A is a subgroup of a p-group G and all subgroups of G containing A as a subgroup of index p are of maximal class, then G is also of maximal class.

Remark 1. Let G contain a proper subgroup H and let $M \trianglelefteq G$ be such that $M \nleq H$ and any proper G-invariant subgroup of M is contained in H. Then $|HM : H| = p$. Indeed, let N be a G-invariant subgroup of index p in M. Then $N \leq H$, by hypothesis, so that $M \cap H = N$, hence, by the product formula, $|HM| = \frac{|H|\,|M|}{|H \cap M|} = |H|\,|M : N| = p|H|$.

Remark 2. Let G contain a proper subgroup H and let $\exp(\Omega_1(H)) = p$. Suppose that $\Omega_1(\mathrm{N}_G(H)) = \Omega_1(H)$. Then G has no normal subgroup of order $p|\Omega_1(H)|$ and exponent p. Indeed, let $R \trianglelefteq G$ be of order $p|\Omega_1(H)|$ and exponent p and let $R_1 \leq R$ be G-invariant and such that $R_1 \nleq H$ is as small as possible satisfying the last relation. Then $|R_1H : H| = p$ (Remark 1) so that $R_1 \leq \mathrm{N}_G(H)$ and hence $\Omega_1(R_1H) > \Omega_1(H)$, contrary to the hypothesis.

Proposition 196.1. *Let H be an absolutely regular subgroup of a p-group G. If $\Omega_1(\mathrm{N}_G(H)) = \Omega_1(H)$, then $\Omega_1(H) \trianglelefteq G$ and G is either absolutely regular or of maximal class.*

Proof. Assume that G is neither absolutely regular nor of maximal class. Then there is $R \trianglelefteq G$ of order p^p and exponent p (Theorem 12.1 (a)). Let $R_0 \leq R$ be a G-invariant subgroups with is as small as possible such that $R_0 \nleq H$. Then $|R_0H : H| = p$ (Remark 1) so that $R_0 < N_G(H)$, a contradiction since $R_0 \nleq \Omega_1(H) = \Omega_1(N_G(H))$. Normality of $\Omega_1(H)$ follows from $\Omega_1(H) = \Omega_1(G_1)$. □

Corollary 196.2. *Let H be a noncyclic metacyclic subgroup of a p-group G, $p > 2$. If $N_G(H)$ is metacyclic, then G is either metacyclic or a 3-group of maximal class.*

Proof. This follows from Proposition 196.1, Theorem 7.2 (b), and Theorem 9.11. □

Exercise 1. Suppose that H is a proper subgroup of a p-group G, $p > 2$, such that there is in H a characteristic subgroup M with $H/M \cong M_{p^3}$. Then $N_G(H/M)$ is nonmetacyclic.

Solution. By hypothesis. $M \lhd N_G(H)$. Assume that $N_G(H/M)$ is metacyclic. Then the metacyclic group $N_G(H/M)$ contains a proper subgroup $H/M \cong M_{p^3}$, contrary to Proposition 10.19.

Exercise 2. Let $D \cong D_{2^n}$ be a nonnormal subgroup in a 2-group G. If $C_G(D)$ of order > 2 is cyclic and $N_G(D) = D * C_G(D)$ (central product), then there is in $N_G(D)$ a subgroup of maximal class and order 2^n whose normalizer in G is $> N_G(D)$. (**Hint.** Use Appendix 16.)

Let $e_n(X)$ be the number of subgroups of order p^n and exponent p in a p-group X.

Exercise 3. Let G be a p-group with $e_n(G) > 1$. If $R < G$ is of order p^n and exponent p, then there is $x \in G - R$ of order p such that $x \in N_G(R)$.

Solution. By hypothesis, $\Omega_1(G) > R$. If $\Omega_1(N_G(R)) = R$, then, being characteristic in $N_G(R)$, the subgroup R is normal in G. Therefore, there is $x \in G - R$ of order p, and x normalizes R. If R is not characteristic in $N_G(R)$, there is $y \in \Omega_1(N_G(R)) - R$ of order p which normalizes R.

Exercise 4. Suppose that H is a proper subgroup of a p-group G, $\exp(H) = p^e$ and $\Omega_e(G) > H$. Then there is in $G - H$ an element of order p^e normalizing H.

Hint. One may assume that H is not G-invariant. Then $\Omega_e(N_G(G)) > H$.

Exercise 5. If G is a group of exponent p^e, $e > 1$, then for each $n < e$ there is in G a normal subgroup of exponent p^n.

Exercise 6. Let a 2-group G contains a proper subgroup H of maximal class. If $N_G(H)$ is metacyclic, then G is of maximal class. (**Hint.** Use Proposition 10.19 and Remark 10.5.)

Exercise 7. Study a nonmetacyclic p-group G, $p > 2$, containing an element x of order p whose centralizer $C_G(x)$ is metacyclic of order $> p^2$.

Proposition 196.3. *Let H be a proper irregular subgroup of maximal class of a p-group G and $e_p(\mathrm{N}_G(H)) = e_p(H)$.*
(a) If $|H| = p^{p+1}$, then one of the following holds:
(a1) $e_p(H) = 1$ and $\Omega_1(H)$ is the unique normal subgroup of G of order p^p and exponent p.
(a2) G is of maximal class.
(b) If $|H| > p^{p+1}$, then G is of maximal class.

Proof. By Theorem 9.5, $|H| \geq p^{p+1}$. Write $N = \mathrm{N}_G(H)$; then $N > H$.

(i) Suppose that $|H| = p^{p+1}$. Then $e_p(H) \leq p$ since $\mathrm{d}(H) = 2$ and $\exp(H) = p^2$ (Theorem 7.1 (b)).

(i1) Let $e_p(H) = 0$. Then $e_p(N) = 0$, i.e., N has no normal subgroup of order p^p and exponent p. By Theorem 12.1 (a), N is of maximal class (being overgroup of H, N is irregular). In this case, G is of maximal class, by Remark 10.5.

(i2) Let $e_p(H) > 1$. Then $1 < e_p(N) \leq p$ so that $e_p(N) \not\equiv 1 \pmod{p}$. In this case, N is of maximal class, by Theorem 13.5. Then G is of maximal class again.

(i3) Let $e_p(H) = 1$. Then $e_p(N) = 1$, so that N (of order $\geq p^{p+2}$) is not of maximal class, by Theorem 9.6 (c). It follows from Theorem 9.6 (c) that all subgroups of G containing N are not of maximal class; then G is not of maximal class. By Theorem 13.5, there is in G a normal subgroup U of order p^p and exponent p. We claim that $U = \Omega_1(H) = \Omega_1(G)$. Assume that this is false. Let a G-invariant $U_0 < G$ of exponent p be as small as possible such that $U_0 \nleq H$. Then $|HU_0 : H| = p$ (Remark 1) so that $U_0 < HU_0 \leq N$. In that case, there is in the subgroup $\Omega_1(H)U_0$ of order p^{p+1} a subgroup $V > U_0$ of order p^p and exponent p, a contradiction since $V < N$. Thus, $U = \Omega_1(H)$ is the unique normal subgroup of order p^p and exponent p in G.

(ii) Let $|H| > p^{p+1}$. Assume that G is not of maximal class. Then G contains a normal subgroup, say U, of order p^p and exponent p. By Theorem 9.6 (c), $U \nleq H$. Let $U_0 \leq U$ be the least G-invariant subgroup such that $U_0 \nleq H$. By Remark 1, $|HU_0 : H| = p$. In this case, $U_0 < HU_0 \leq N$, and we get a contradiction as in (i3). Thus, U does not exist; then G is of maximal class, by Theorem 12.1 (a). □

§ 197 Minimal non-q-self-dual 2-groups

We recall that a p-group G is q-self-dual if every quotient of G is isomorphic to a subgroup of G. For example, all abelian p-groups are q-self-dual.

Nonabelian 2-groups G all of whose subgroups are q-self-dual are completely determined in Theorem 195.3. It turns out that such 2-groups G are either quasidihedral (nonmodular case) or G is of the form $G = M \times V$, where $M \cong M_{2^{e+1}}$, $e \geq 3$, and $\exp(V) \leq 2$ (modular case). However, the corresponding problem for p-groups G with $p > 2$ is open (see Theorem 195.4) and is very difficult. Therefore, it is of interest to classify the title groups (problem, posed by he first author, for $p = 2$), i.e., we shall determine up to isomorphism 2-groups G which are not q-self-dual but all proper subgroups of G are q-self-dual (Theorem 197.2). We get exactly three infinite classes of such 2-groups and an exceptional group of order 2^6. If we assume in addition that G has at least one nonmodular maximal subgroup, then G is a unique group of order 2^5 (Corollary 197.3).

First, we prove an auxiliary result which is very useful in the proof of our main result.

Lemma 197.1. *Let G be a nonabelian 2-group which is not minimal nonabelian. Assume that G is not q-self-dual but all proper subgroups of G are q-self-dual. Suppose in addition that G has a nonmodular maximal subgroup. Then G also has a nonabelian modular maximal subgroup. Each nonmodular maximal subgroup is isomorphic to $D_8 \times E_{2^s}$ $s \geq 1$, (fixed) and each nonabelian modular maximal subgroup is isomorphic to $M_{2^4} \times E_{2^{s-1}}$.*

All nonabelian maximal subgroups of G have the same derived subgroup of order 2 and we have either $G' \cong C_2$ or $|G'| = 4$ and $d(G) = 2$.

Proof. If each nonabelian maximal subgroup of G is quasidihedral, then each minimal nonabelian subgroup of G is isomorphic to D_8. But then Theorem 10.33 implies that G itself is quasidihedral. This is a contradiction because Theorem 195.3 asserts that quasidihedral groups are q-self-dual. We have shown that G has in any case some nonabelian modular maximal subgroup.

Let D be a fixed quasidihedral maximal subgroup of G and let F be any nonabelian modular maximal subgroup of G. Hence $F = M \times V$, where $M \cong M_{2^{e+1}}$, $e \geq 3$, and $\exp(V) \leq 2$. Note that F (being modular) is D_8-free and since each subgroup of D is either abelian or quasidihedral (in which case it is not D_8-free), it follows that $D_0 = D \cap F$ is an abelian maximal subgroup of D. On the other hand, $\exp(F) \geq 8$ and so $\exp(D_0) \geq 4$. But D has exactly one abelian maximal subgroup of exponent > 2 (which then must be equal D_0) and so all elements in $D - D_0$ are involutions which invert each element of D_0.

Suppose that D_0 contains an element a of order $2^e = \exp(F)$. Then there is an involution $u_0 \in F - D_0$ such that $a^{u_0} = aa^{2^{e-1}}$ and if t is an involution in $D - D_0$, then $a^t = a^{-1}$ and

$$a^{tu_0} = (a^{-1})^{u_0} = (a^{u_0})^{-1} = a^{-1}a^{2^{e-1}}, \quad e \geq 3 .$$

It follows $\langle a\rangle \trianglelefteq G$, where $o(tu_0) \geq 2$ and tu_0 induces on $\langle a\rangle$ a "semidihedral" automorphism of order 2 with $o(a) \geq 8$. Such a metacyclic subgroup $\langle a, tu_0\rangle$ cannot be contained in any maximal subgroup of G (Theorem 195.3) and so $G = \langle a, tu_0\rangle$ is metacyclic. But G contains a subgroup isomorphic to D_8 and then Proposition 10.19 implies that G would be of maximal class. Since Q_8 is not q-self-dual, Q_8 is not a subgroup in G and we get that G is dihedral, a contradiction.

We have proved that D_0 does not contain elements of order 2^e and so we obtain $D_0 = \Omega_{e-1}(F)$. Hence if $a \in F - D_0$, then $o(a) = 2^e$ and there is an involution $u \in D_0$ such that $a^u = aa^{2^{e-1}}$. Obviously, G is nonmetacyclic. Indeed, if G were metacyclic, then the fact that D_8 is a subgroup in G together with Proposition 10.19 implies that G would be of maximal class. But Q_8 is not a subgroup in G and so G is dihedral, a contradiction.

Suppose that $e \geq 4$ so that $o(a^2) \geq 8$. Then an element $ta \neq 1$ (with $t \in D - D_0$) inverts the cyclic subgroup $\langle a^2\rangle$. Since $\langle ta, a^2\rangle$ is metacyclic, we have $\langle ta, a^2\rangle < G$. It follows that $\langle ta, a^2\rangle$ must be contained in a quasidihedral maximal subgroup of G and so ta must be an involution. From $1 = (ta)^2 = tata$ we get $a^t = a^{-1}$. But t centralizes u and so tu is an involution and

$$a^{tu} = (a^{-1})^u = (a^u)^{-1} = (aa^{2^{e-1}})^{-1} = a^{-1}a^{2^{e-1}} .$$

It follows that $\langle a, tu\rangle$ is semidihedral and so $\langle a, tu\rangle$ contains a subgroup isomorphic to Q_8, a contradiction.

We have proved that $e = 3$, $\exp(D_0) = 4$, $M \cong M_{2^4}$ and so

$$D_0 = \langle a^2\rangle \times \langle u\rangle \times V , \quad D = \langle t, a^2\rangle \times \langle u\rangle \times V \text{ with } \langle t, a^2\rangle \cong D_8 ,$$
$$\exp(V) \leq 2 \text{ and } D' = F' = \langle a^4\rangle .$$

Hence we have $D \cong D_8 \times E_{2^s}$, $s \geq 1$ (fixed), and each nonabelian modular maximal subgroup F_i of G is isomorphic to $M_{2^4} \times E_{2^{s-1}}$ with $F_i' = D' \cong C_2$.

Now take a fixed nonabelian modular maximal subgroup $F = F_1$ of G and let D_i be any nonmodular maximal subgroup of G. By the above argument, $D_i \cong D_8 \times E_{2^s}$ and $D_i' = F' \cong C_2$ so that $D_i' = D' = F'$.

We have proved that all nonabelian maximal subgroups of G have the same derived subgroup D' of order 2. If $G' = D'$, then G/D' is abelian.

Assume that $G' > D'$. Then G/D' is nonabelian and each maximal subgroup of G/D' is abelian so that G/D' is minimal nonabelian. It follows in this case that $|G'/D'| = 2$ and $d(G/D') = 2$. But $D' \leq \Phi(G)$ and so $d(G) = 2$ and $|G'| = 4$ and we are done. □

Theorem 197.2. *Let G be a nonabelian 2-group which is not minimal nonabelian and assume that G is not q-self-dual but all proper subgroups of G are q-self-dual. Then* $d(G) = 2$, $|G'| = 4$, $cl(G) = 3$, *and G is isomorphic to one of the following groups:*

(a) $G = \langle a, e_1 \mid a^{2^{e+1}} = e_1^2 = 1,\ a^{2^e} = z,\ [a, e_1] = e_2,\ [a, e_2] = z,\ e_2^2 = [e_1, e_2] = 1\rangle$, *where* $e \geq 2$. *Here we have*

$$|G| = 2^{e+3}, \quad Z(G) = \langle a^4\rangle \cong C_{2^{e-1}}, \quad G' = \langle z, e_2\rangle \cong E_4, \quad E_8 \cong \langle z, e_1, e_2\rangle \trianglelefteq G.$$

(b) $G = \langle a, b \mid a^8 = 1,\ b^{2^n} = a^4,\ a^b = a^{-1+4\eta}\rangle$, *where* $n \geq 2$ *and* $\eta = 0, 1$. *Here*

$$|G| = 2^{n+3}, \quad G' = \langle a^2\rangle \cong C_4,$$

and the groups G *are metacyclic* A_2*-groups.*

(c) $$G = \langle a, b \mid a^8 = b^8 = 1,\ a^4 = b^4 = z,\ b^2 = a^2u,\ [a, b] = a^{-2}e,\ [u, b] = z,$$
$$u^2 = e^2 = [e, u] = [a, e] = [b, e] = [a, u] = 1\rangle.$$

Here we have

$$|G| = 2^6,\ G' = \langle a^2e\rangle \cong C_4,\ Z(G) = \langle e, b^2\rangle \cong C_2 \times C_4,$$

$E = \langle z, e, u\rangle$ *is a normal elementary abelian subgroup of order 8 with* $G/E \cong Q_8$ *and* $\Phi(G) = E\langle a^2\rangle \cong C_4 \times C_2 \times C_2$.

(d) $$G = \langle a, b \mid a^{2^e} = b^{2^e} = 1,\ b^2 = a^{-2+2^{e-2}}u,\ u^2 = [a, u] = [b, u] = 1,\ [a, b] = d,$$
$$d^4 = 1,\ d^2 = a^{2^{e-1}} = b^{2^{e-1}} = z,\ [a, d] = [b, d] = z\rangle, \text{ where } e \geq 4.$$

Here we have

$$|G| = 2^{e+3},\ G' = \langle d\rangle \cong C_4,\ Z(G) = \langle a^2, u\rangle \cong C_{2^{e-1}} \times C_2,\ G/Z(G) \cong D_8,$$
$$\Phi(G) = \langle a^2, u, (ab)^2\rangle \cong C_{2^{e-1}} \times C_2 \times C_2, \text{ and } \Omega_1(G) = \Omega_1(\Phi(G)) \cong E_8.$$

Conversely, all the above groups satisfy the assumptions of our theorem.

Proof. By Theorem 195.3 and Lemma 197.1, each nonabelian maximal subgroup H of G has the maximal possible center, i.e., $|H : Z(H)| = 4$. Therefore, we may use Theorem 151.1 and so it remains only to investigate the groups (i), (ii), (iii), and (v) stated in that theorem.

(i) Suppose $G = M_1Z(G)$, where M_1 is minimal nonabelian and $M_1 < G$. By Theorem 195.3 and Lemma 197.1, we may assume that

$$M_1 = \langle a, t \mid a^{2^e} = t^2 = 1,\ a^t = a^{1+2^{e-1}}\rangle \cong M_{2^{e+1}} \text{ with } e \geq 3.$$

We set $\langle a^{2^{e-1}}\rangle = \langle z\rangle = G'$ and we have

$$Z(G) \cap M_1 = \langle a^2\rangle \text{ with } \langle a^2\rangle < Z(G) \text{ and } \langle a^2\rangle \cong C_{2^{e-1}}.$$

First, we show that $Z(G)/\langle a^2\rangle \neq \{1\}$ is cyclic. Indeed, suppose that $Z(G)/\langle a^2\rangle$ is noncyclic and let $X_1 \neq X_2$ be two distinct maximal subgroups of $Z(G)$ containing $\langle a^2\rangle$. Then (by Theorem 195.3)

$$X_1 = \langle a^2\rangle \times V_1 \text{ and } X_2 = \langle a^2\rangle \times V_2, \text{ where } \exp(V_1) = \exp(V_2) = 2.$$

But $V_2 \nleq X_1$ and so there is an involution $t_2 \in V_2 - X_1$ so that

$$Z(G) = \langle a^2 \rangle \times V_1 \times \langle t_2 \rangle \text{ and } \langle V_1, t_2 \rangle \text{ is elementary abelian }.$$

In that case, $G = M_1 \times \langle V_1, t_2 \rangle$ is q-self-dual (Theorem 195.3), contrary to our assumptions.

Let X be a maximal subgroup of $Z(G)$ containing $\langle a^2 \rangle$. If $\langle a^2 \rangle = X$ is maximal in $Z(G)$, then $Z(G)$ is cyclic of order 2^e. Suppose that $\langle a^2 \rangle < X$ so that $X = \langle a^2 \rangle \times V$ with $\exp(V) = 2$. Since $X/\langle a^2 \rangle$ is cyclic, we have $|V| = 2$ and so in this case $Z(G)/\langle a^2 \rangle \cong C_4$.

(i1) Assume that $\langle a^2 \rangle$ is a maximal subgroup in $Z(G)$ so that $Z(G) \cong C_{2^e}$. Let $f \in Z(G) - \langle a^2 \rangle$ be such that $f^2 = a^{-2}$. Then $t' = af \in G - (M_1 \cup Z(G))$ is an involution with $[t, t'] = [t, af] = [t, a] = z$ so that $D = \langle t, t' \rangle \cong D_8$ and D covers $G/Z(G)$. It follows that $D \cap Z(G) = \langle z \rangle$ and Lemma 197.1 implies $e = 3$ and so $|G| = 2^5$ and $\langle a^2 \rangle \cong C_4$. But then the subgroup $D * \langle a^2 \rangle$ with $D \cap \langle a^2 \rangle = \langle z \rangle$ contains a subgroup $Q \cong Q_8$ which is not q-self-dual, a contradiction.

(i2) Now suppose that $Z(G)/\langle a^2 \rangle \cong C_4$. If $\langle a^2 \rangle < X < Z(G)$, then we know that $X = \langle a^2 \rangle \times \langle v \rangle$ with $\langle v \rangle \cong C_2$.

First, suppose that $\exp(Z(G)) = 2^{e-1}$. Then $Z(G)$ splits over $\langle a^2 \rangle$ and so $Z(G) = \langle a^2 \rangle \times \langle b \rangle$ with $\langle b \rangle \cong C_4$. But then $(tb)^2 = b^2$, $b^2 \notin \langle a \rangle$ and $[a, tb] = z$, so that $\langle a, tb \rangle$ is minimal nonabelian of order 2^{e+2}, where $\Phi(\langle a, tb \rangle) = \langle a^2 \rangle \times \langle b^2 \rangle$ is noncyclic, contrary to Theorem 195.3.

We have proved that $\exp(Z(G)) = 2^e$. Hence there is $h \in Z(G) - X$ such that $o(h) = 2^e$. Since $\langle h^4 \rangle = \langle a^4 \rangle$, we may choose $h \in Z(G) - X$ so that $h^4 = a^{-4}$. Thus,

$$ah \in G - (M_1 \cup Z(G)), \quad (ah)^4 = 1 \text{ with } (ah)^2 = a^2h^2 \in X - \langle a^2 \rangle$$

so that $(ah)^2 \neq 1$ is an involution in $X - \langle a^2 \rangle$. Because $[ah, t] = [a, t] = z$, $\langle ah, t \rangle$ is nonmetacyclic minimal nonabelian of order 2^4, contrary to Theorem 195.3.

(ii) Assume that $G = M_1 * M_2$, where M_1 and M_2 are minimal nonabelian subgroups of G with $M_1' = M_2' = G'$. Note that

$$Z(G) = Z(M_1) * Z(M_2) = \Phi(M_1) * \Phi(M_2) = \Phi(G) \text{ and } G/\Phi(G) = G/Z(G) \cong E_{2^4}.$$

Also, $M_1 \cap M_2 \leq Z(M_1) \cap Z(M_2)$, and so $G' \leq M_1 \cap M_2$ and $M_1 \cap M_2$ is cyclic.

Suppose that at least one of M_1, M_2 is isomorphic to D_8, say, $M_2 \cong D_8$. Then $M_1 \cap M_2 = M_2' = G' \cong C_2$ and in any case M_1 contains a subgroup $Z \cong C_4$ such that $Z > G' = M_2'$. But then $Z * M_2$ contains a subgroup $Q \cong Q_8$ which is not q-self-dual, a contradiction.

It follows that we may assume that both M_1 and M_2 are isomorphic to some M_{2^n}, $n \geq 4$. Let $\langle a \rangle$ be a cyclic subgroup of index 2 in M_2 so that $o(a) \geq 8$ and $\langle a \rangle > M_1 \cap M_2$ because $\langle a^2 \rangle = Z(M_2)$. Set $H = M_1 * \langle a \rangle$, where $|G : H| = 2$. Since H/M_1 is elementary abelian (Theorem 195.3), we must have $|H : M_1| = 2$ and so $\langle a^2 \rangle = M_1 \cap M_2 = Z(M_2)$. Similarly, considering a cyclic subgroup $\langle b \rangle$ of index 2 in M_1 and the maximal subgroup $K = M_2 * \langle b \rangle$ of G, we get $\langle b^2 \rangle = M_1 \cap M_2 = Z(M_1)$.

We have proved that

$$M_1 \cong M_2 \cong \mathrm{M}_{2^{e+1}}\,, \quad e \geq 3\,, \quad G = M_1 * M_2 \text{ with } M_1 \cap M_2 = \mathrm{Z}(M_1) = \mathrm{Z}(M_2) \cong \mathrm{C}_{2^{e-1}}\,,$$

and we set

$$M_i = \langle a_i, t_i \mid a_i^{2^e} = t_i^2 = 1,\ [t_i, a_i] = a_i^{2^{e-1}} = z\rangle\,, \quad i = 1, 2\,.$$

We may choose $a_2 \in \langle a_2\rangle$ so that $a_1^2 = a_2^{-2}$. Then set $t = a_1a_2$ so that $t \in G - (M_1 \cup M_2)$ is an involution and since $[t_1, t] = [t_1, a_1a_2] = [t_1, a_1] = z$, we get $\langle t_1, t\rangle = D \cong \mathrm{D}_8$. Set $M = \mathrm{C}_G(D)$ so that $|G'| = 2$ and $G/\mathrm{Z}(G) \cong \mathrm{E}_{2^4}$ implies $G = D * M$, where M is metacyclic minimal nonabelian with $D \cap M = D' = M' = G'$. But then M contains an element m of order 4 such that $\langle m\rangle > D'$ and so $D * \langle m\rangle$ contains a subgroup $Q \cong \mathrm{Q}_8$, which is not q-self-dual, a contradiction.

(iii) We suppose that $\mathrm{d}(G) = 2$ and $|G'| = 4$. By Theorem A.25.1, for each maximal subgroup H of G we get $\mathrm{d}(H) \leq 1 + (\mathrm{d}(G) - 1)|G : H| = 3$.

(iii1) First assume that G has a nonmodular maximal subgroup X. By Lemma 197.1, $X \cong \mathrm{D}_8 \times \mathrm{E}_{2^s}$, $s \geq 1$, and since $\mathrm{d}(X) \leq 3$, we have $s = 1$ and $|G| = 2^5$. Again by Lemma 197.1, G has a nonabelian modular maximal subgroup Y which is isomorphic to M_{2^4} and so $\exp(G) = 8$.

The subgroup $X \cap Y$ is the unique abelian maximal subgroup of X of exponent > 2, $X \cap Y \cong \mathrm{C}_4 \times \mathrm{C}_2$ and $X \cap Y$ is the unique noncyclic maximal subgroup of Y so that all elements in $Y - X$ are of order 8. We may set

$$Y = \langle a, t \mid a^8 = t^2 = 1,\ a^t = a^{1+4},\ a^4 = z\rangle\,, \quad \text{so that } X \cap Y = \langle a^2\rangle \times \langle t\rangle \cong \mathrm{C}_4 \times \mathrm{C}_2\,,$$

and there is an involution $u \in X - Y$ which inverts each element in $X \cap Y$ so that $\langle a^2, u\rangle \cong \mathrm{D}_8$ and $[u, t] = 1$. Since $X \cap Y \geq \Phi(G)$ and $\mathrm{d}(G) = 2$, it follows $X \cap Y = \Phi(G) = \mho_1(G)$.

Let W be the third maximal subgroup of G. If W is abelian, then Lemma 1.1 would imply $|\mathrm{Z}(G)| = 4$ and clearly $\mathrm{Z}(G) \leq \Phi(G)$. But then $\mathrm{Z}(G) \leq \mathrm{Z}(Y) = \langle a^2\rangle \cong \mathrm{C}_4$, and so $\mathrm{Z}(G) = \langle a^2\rangle$. On the other hand, the involution u inverts $\langle a^2\rangle$, a contradiction.

We have proved that W is nonabelian and since $\mathrm{Z}(G) \leq \Phi(G)$, $\mathrm{Z}(G) \leq \mathrm{Z}(Y) = \langle a^2\rangle$ and $\langle a^2\rangle \nleq \mathrm{Z}(G)$, we have $\mathrm{Z}(G) = \langle z\rangle \cong \mathrm{C}_2$ and so $\mathrm{cl}(G) = 3$. Because

$$\mho_1(G) = X \cap Y \quad \text{and} \quad \mho_1(X) = \langle z\rangle = X'\,, \quad \mho_1(Y) = \langle a^2\rangle\,,$$

we must have

$$W = \langle b, t \mid b^8 = t^2 = 1,\ b^t = bz\rangle \cong \mathrm{M}_{2^4}, \text{ and } b^2 = a^2t\,.$$

Indeed, if $W \cong \mathrm{D}_8 \times \mathrm{C}_2$, then $\mho_1(W) = W' = X' = \langle z\rangle$, and then $\mho_1(G) = \langle a^2\rangle$, a contradiction. In particular, all elements in $G - X$ are of order 8 and so $\Omega_2(G) = X \cong \mathrm{D}_8 \times \mathrm{C}_2$.

Using the results of § 52 (Theorems 52.1 and 52.2), we see that G is isomorphic to the group in Theorem 52.2 (a) for $n = 2$. We have obtained the group of order 2^5 stated in part (a) of our theorem for $e = 2$.

Conversely, let G be a group of order 2^5 stated in part (a) of our theorem for $e = 2$. Then the maximal subgroups of G are

$$\langle a^2, e_1, e_2\rangle \cong D_8 \times C_2, \quad \langle a, e_2\rangle \cong M_{2^4} \text{ and } \langle ae_1, e_2\rangle \cong M_{2^4}.$$

(iii2) Now assume that all nonabelian maximal subgroups H_i of G are modular. By Theorem 195.3 and the fact that $d(H_i) \le 3$, we have $H_i \cong M_{2^n} \times E_{2^s}$, where $n \ge 4$ and $s = 0$ or 1.

(iii2a) Assume that all nonabelian maximal subgroups of G are isomorphic to M_{2^n}, $n \ge 4$, i.e., G is an A_2-group of order $\ge 2^5$. By the results of § 71, G is metacyclic and so G is isomorphic to a group of Proposition 71.2.

If G is isomorphic to a group of Proposition 71.2 (a), then Proposition 205.3 gives a contradiction. If G is isomorphic to a group of Proposition 71.2 (b), then Proposition 205.4 implies $\epsilon = 1$ and we have obtained the metacyclic A_2-groups stated in part (b) of our theorem.

Conversely, if G is such a metacyclic A_2-group, then $\langle a, b^2\rangle$ is an abelian maximal subgroup of G and the other two maximal subgroups of G are isomorphic to $M_{2^{n+2}}$, $n \ge 2$.

(iii2b) Suppose that at least one nonabelian maximal subgroup H of G is isomorphic to $M_{2^n} \times C_2$, $n \ge 4$, i.e., G is not an A_2-group. We set

$$H = \langle a, t \mid a^{2^e} = t^2 = 1,\ [a, t] = a^{2^{e-1}} = z\rangle \times \langle u\rangle,$$

where $e \ge 3$, $\langle u\rangle \cong C_2$, and $E = \Omega_1(H) = \langle z, t, u\rangle \cong E_8$ with $E \trianglelefteq G$.

(1) First, assume that G/E is cyclic (of order 2^e). Since $|\Omega_2(H)| = 2^4$, $\Omega_2(H)$ is abelian of type (4, 2, 2) and $\mho_1(\Omega_2(H)) = \langle z\rangle$, we have $G = E\langle b\rangle$ with $o(b) = 2^{e+1}$ and $\langle b\rangle \cap E = \langle z\rangle$. Because $\Omega_2(G) = \Omega_2(H)$ and $G' \cong E_4$, we have obtained the groups stated in Theorem 52.2 (a) for $n \ge 3$, where a is written instead of b. These groups are stated in part (a) of our theorem for $e \ge 3$.

Conversely, let G be a group stated in part (a) of our theorem for $e \ge 3$. Then the maximal subgroups of G are

$$\langle a^2, e_1\rangle \times \langle e_2\rangle \cong M_{2^{e+1}} \times C_2, \quad \langle a, e_2\rangle \cong M_{2^{e+2}}, \quad \langle ae_1, e_2\rangle \cong M_{2^{e+2}}.$$

(2) Now assume that G/E is noncyclic. This implies that all elements in $G - H$ are of order $\le 2^e$ and so $\exp(G) = 2^e$. There is a normal subgroup N of G such that $E \le N < G$ and $G/N \cong E_4$. Since $d(G) = 2$, we get $N = \Phi(G)$ and so $E \le \Phi(G)$. We get $\Phi(G) = \langle a^2, t, u\rangle$ is abelian of type $(2^{e-1}, 2, 2)$ and each nonabelian maximal subgroup of G is isomorphic to $M_{2^{e+1}} \times C_2$, where apart from H there is at least another one $K \neq H$ with

$$K \cong M_{2^{e+1}} \times C_2 \text{ and } H \cap K = \Phi(G).$$

Since D_8 is not a subgroup of G, it follows that $E \le \Omega_1(G)$ is elementary abelian. Assume that $\Omega_1(G) > E$. We have in this case

$$G = H\Omega_1(G), \quad \text{where } H \cap \Omega_1(G) = \Omega_1(H) = E \text{ so that } \Omega_1(G) \cong E_{16}.$$

Let A be a maximal subgroup of G containing $\Omega_1(G)$ so that A must be abelian and

$$A = \Phi(G)\Omega_1(G) \text{ with } \mho_1(A) = \mho_1(\Phi(G)).$$

But we have

$$\mho_1(H) \cong C_{2^{e-1}}, \quad \mho_1(K) \cong C_{2^{e-1}} \quad \text{and} \quad \mho_1(A) = \mho_1(\Phi(G)) \le \mho_1(H)$$
$$\text{so that } \Phi(G) = \mho_1(G) = \langle \mho_1(H), \mho_1(K)\rangle,$$

a contradiction since $\Phi(G)$ is abelian of rank 3. We have proved that

$$\Omega_1(G) = E \cong E_8.$$

First, assume $\Omega_2(G) \le \Phi(G)$. Then $\Omega_2(G)$ is abelian of type $(4, 2, 2)$. Since $\Omega_2(G)/E$ is the only subgroup of order 2 in G/E and G/E is noncyclic, we have $G/E \cong Q_{2^n}$, $n \ge 3$. On the other hand, $E \le \Phi(G)$ and so we may use Theorem 52.5. If $n > 3$, then G/E contains a proper subgroup $G_0/E \cong Q_8$. Since G_0 is q-self-dual, G_0 possesses a subgroup $X_0 \cong Q_8$, a contradiction (because Q_8 is not q-self-dual). Hence we have $n = 3$, $G/E \cong Q_8$, $|G| = 2^6$ and since $|G'| = 4$, G is isomorphic to the group of Theorem 52.5 (b) for $n = 3$, which is stated in part (c) of our theorem.

Conversely, let G be the group of order 2^6 stated in part (c) of our theorem. Then $A = C_G(E) = E\langle a\rangle$ is a unique abelian maximal subgroup of G (of type $(8, 2, 2)$), where $E = \langle z, e, u\rangle$ and the other two maximal subgroups of G are

$$\langle b, u\rangle \times \langle e\rangle \cong M_{2^4} \times C_2 \quad \text{and} \quad \langle ab, u\rangle \times \langle e\rangle \cong M_{2^4} \times C_2.$$

It remains to consider the possibility $\Omega_2(G) \not\le \Phi(G)$. Let $c \in G - \Phi(G)$ be an element of order 4 so that $c^2 \in E$ and $c \in G - (H \cup K)$. Then $A = \Phi(G)\langle c\rangle$ is a maximal subgroup of G with $|\Omega_2(A)| > 2^4$ and so A is abelian. Hence G has exactly one abelian maximal subgroup A, which is of exponent 2^{e-1} and rank 3. For two nonabelian maximal subgroups of G we may set $H = E\langle a\rangle$ and $K = E\langle b\rangle$, where $o(a) = o(b) = 2^e$. We have

$$\Phi(G) = H \cap K = E\langle a^2\rangle = E\langle b^2\rangle, \quad \mho_1(A) = \langle c^2, a^4\rangle, \quad \mho_1(H) = \langle a^2\rangle,$$
$$\mho_1(K) = \langle b^2\rangle, \quad \text{and} \quad \Phi(G) = \langle \mho_1(A), \mho_1(H), \mho_1(K)\rangle = \langle a^2, b^2, c^2\rangle.$$

It follows that $b^2 \notin \langle a^2\rangle$ and so $b^2 = a^{2i'}u$, where i' is an odd integer and u is an involution in $E - \langle z\rangle$ with $z = a^{2^{e-1}} = b^{2^{e-1}}$ (noting that $\langle a^4\rangle = \langle b^4\rangle \ge \langle z\rangle$). Also, $c^2 \in E - \langle z, u\rangle$ implying that A is abelian of type $(2^{e-1}, 4, 2)$. All elements in $(H - \Phi(G)) \cup (K - \Phi(G))$ are of order $2^e = \exp(G)$ and all elements in $G - (H \cup K)$ are of order $\le 2^{e-1}$.

We have $G = \langle a, c\rangle = \langle b, c\rangle = \langle a, b\rangle$ and since $|G| = 2^{e+3}$, it follows that G is nonmodular and so G is minimal nonmodular. Indeed, if G were modular, then $\langle a\rangle\langle c\rangle = \langle c\rangle\langle a\rangle$ is a subgroup of order 2^{e+2} because $o(a) = 2^e$, $o(c) = 2^2$ and $\langle a\rangle \cap \langle c\rangle\{1\}$, contrary to $G = \langle a, c\rangle$ and $|G| = 2^{e+3}$.

Since G has an abelian maximal subgroup A and $|G'| = 4$, Lemma 1.1 implies that $|Z(G)| = 2^e$ and so $|G : Z(G)| = 2^3$. On the other hand, $Z(G) \le \Phi(G)$ and so $Z(G) = Z(H) = Z(K)$. But

$$\langle a^2\rangle \le Z(H)\,, \quad \langle b^2\rangle \le Z(K)\,, \quad \langle a^2\rangle \ne \langle b^2\rangle$$

and so $Z(G) = \langle a^2, b^2\rangle = \langle a^2\rangle \times \langle u\rangle \cong C_{2^{e-1}} \times C_2$.

By the proof of Lemma 1.1, $A/Z(G) \cong G'$ and since $A = Z(G) \times \langle c\rangle$ with $o(c) = 4$, we get $G' \cong C_4$.

Suppose that G is of class 2. Then $1 = [a^2, c] = [a, c]^2$ so that $G = \langle a, c\rangle$ implies that $[a, c]$ is the involution in $C_4 \cong G' > \langle z\rangle$. But then $[a, c] = z$ gives $G' = \langle z\rangle$, a contradiction. We have proved that $\mathrm{cl}(G) = 3$ and so $G/Z(G)$ is a nonabelian group of order 2^3. However, $\Phi(G)/Z(G)$ and $(\langle a\rangle Z(G))/Z(G)$ are two distinct subgroups of order 2 in $G/Z(G)$ and consequently $G/Z(G) \cong D_8$.

We may choose elements $a \in H - \Phi(G)$ and $b \in K - \Phi(G)$ of order 2^e so that $ab = c \in A - \Phi(G)$, where again

$$a^{2^{e-1}} = b^{2^{e-1}} = z\,, \quad b^2 = a^{2i}u\,,$$
$$i \text{ is an odd integer and } u \in E - \langle z\rangle \text{ and } c^2 \in E - \langle z, u\rangle\,.$$

We have

$$Z(G) = \langle a^2, b^2\rangle = \langle a^2\rangle \times \langle u\rangle \quad \text{and} \quad E = \Omega_1(\Phi(G)) = \Omega_1(G) = \langle z, u, c^2\rangle \cong E_8\,.$$

Set $[a, b] = d$ so that $\langle d\rangle = G' \cong C_4$ with $d^2 = z$ and $d^a = d^b = d^{-1}$. We compute

$$\begin{aligned} c^2 &= (ab)^2 = a(ba)b = a(ab[b, a])b = a^2bd^{-1}b = a^2b^2(d^{-1})^b \\ &= a^2b^2d = a^2a^{2i}ud = a^{2(1+i)}ud \end{aligned}$$

and so $1 = c^4 = a^{4(1+i)}u^2d^2 = a^{4(1+i)}z$. This implies $e \ge 4$ and then $a^{4(1+i)} = z = a^{2^{e-1}}$ which gives $4(1+i) \equiv 2^{e-1} \pmod{2^e}$. Hence we get $i = -1 + 2^{e-3} + k2^{e-2}$ for an integer k and so

$$b^2 = a^{-2+2^{e-2}+k2^{e-1}}u = a^{-2+2^{e-2}}(z^k u)\,,$$

where we may set $z^k u = u' \in \langle z, u\rangle - \langle z\rangle$ and we write again u instead of u' so that $b^2 = a^{-2+2^{e-2}}u$. We have obtained the groups stated in part (d) of our theorem.

Conversely, if G is a group stated in part (d) of our theorem, then the maximal subgroups of G are

$$\langle a, d\rangle \times \langle u\rangle \cong M_{2^{e+1}} \times C_2\,, \quad \langle b, d\rangle \times \langle u\rangle \cong M_{2^{e+1}} \times C_2\,, \quad e \ge 4\,,$$
$$\text{and } \langle a^2, u, ab\rangle \text{ (abelian of type } (2^{e-1}, 4, 2))\,.$$

(iv) There remains to consider the case $d(G) = 3$, $cl(G) = 2$, $G' \cong E_4$ or E_8 and $\Phi(G) = Z(G)$.

By Lemma 197.1, each maximal subgroup of G is modular. Since $d(G) = 3$, the results of § 78 imply that G cannot be minimal nonmodular. Hence G is modular and because $G' \cong E_4$ or E_8, G is not Hamiltonian.

We may use Theorem 73.15 of Iwasawa. By that theorem, G possesses an abelian normal subgroup N, an element $t \in G$ such that $G = \langle N, t\rangle$ and a positive integer $s \geq 2$ such that $a^t = a^{1+2^s}$ for all $a \in N$.

Since $t^2 \in \Phi(G) = Z(G)$, t^2 centralizes N so that $A = \langle t^2\rangle N$ is a unique abelian maximal subgroup of G and t induces on A an involutory automorphism. By a well known result of A.Mann, we have $G' \cong E_4$ and $G' = [N, t] \leq N$.

Any two noncommuting elements in G generate a minimal nonabelian subgroup which is isomorphic to some M_{2^n}, $n \geq 4$. Set $\exp(N) = 2^e$ so that $e \geq 3$ because $\Omega_2(N) \leq Z(G)$. Let $a \in N$ with $o(a) = 2^e$. If $[a, t] = 1$, then $s \geq e$ and then t centralizes N, a contradiction. Hence $s < e$ and $[a, t] \neq 1$ with $[a, t] = a^{2^s}$. Since a^{2^s} must be an involution, we obtain $s = e - 1$. We set $H = \langle a, t\rangle$ so that $H \cong M_{2^n}$ for some $n \geq 4$.

Suppose, by way of contradiction, that $o(t) > 2^e$. Note that $N \cap \langle t\rangle$ is centralized by t and so $|N \cap \langle t\rangle| \leq 2^{e-1}$. On the other hand, $\exp(H) = o(t) > 2^e$. It follows that $\langle t\rangle$ is a maximal subgroup of H and so $a^2 \in \langle t\rangle$ giving $\langle a^2\rangle = N \cap \langle t\rangle \cong C_{2^{e-1}}$ and $H \cong M_{2o(t)}$. Let N_0 be a complement of $\langle a\rangle$ in N. If $\exp(N_0) \leq 2^{e-1}$, then t centralizes N_0 and so $Z(G) = C_A(t) = \langle t^2\rangle \times \langle N_0\rangle$ with $|G : Z(G)| = 4$, contrary to $G' \cong E_4$. Hence $\exp(N_0) = 2^e$ and let $b \in N_0$ with $o(b) = 2^e$, where $\langle a\rangle \cap \langle b\rangle = \{1\}$. But then $K = \langle b, t\rangle \cong M_{2o(t)}$ is of exponent $o(t)$ and so $\langle t\rangle$ is a maximal subgroup of K giving $b^2 \in \langle t\rangle$ and $\langle b^2\rangle = N \cap \langle t\rangle \cong C_{2^{e-1}}$, contrary to $\langle a\rangle \cap \langle b\rangle = \{1\}$.

We have proved that $o(t) \leq 2^e$ and so $H \cong M_{2^{e+1}}$ because $\exp(H) = 2^e$. Then $t^2 \in \langle a\rangle \leq N$ and so $|G : N| = 2$ and $A = N$. We may replace t with an involution t' in $H - \langle a\rangle$ which acts the same way on $A = N$ as t does and we write again t (instead of t').

Since $G' \cong E_4$, we see that

$$N = \langle a\rangle \times \langle b\rangle \times N_0\,, \quad \text{where } \langle a\rangle \cong \langle b\rangle \cong C_{2^e} \text{ and } \exp(N_0) \leq 2^{e-1}\,.$$

Here we have $\langle a, t\rangle \cong \langle b, t\rangle \cong M_{2^{e+1}}$, $e \geq 3$. The subgroup $\langle a, t\rangle \times \langle b^2\rangle \cong M_{2^{e+1}} \times C_{2^{e-1}}$ is a proper subgroup of G because $d(G) = 3$, $d(\langle a, t\rangle) = 2$ and $b^2 \in \Phi(G)$. But this contradicts Theorem 195.3. Our theorem is finally proved. □

By inspecting the groups of Theorem 197.2, we see that in fact a group of Lemma 197.1 is unique. We obtain:

Corollary 197.3. *Let G be a nonabelian 2-group which is not minimal nonabelian. Assume that G is not q-self-dual but all proper subgroups of G are q-self-dual. Suppose in addition that G has a nonmodular maximal subgroup. Then G is a unique group of order 2^5 given in Theorem 197.2 (a) for $e = 2$.*

§ 198 Nonmetacyclic p-groups with metacyclic centralizer of an element of order p

Centralizers of noncentral elements impact on the structure of a finite group as the numerous results show.

In what follows G is a nonmetacyclic p-group containing an element x of order p such that its centralizer $C_G(x)$ is metacyclic. Our goal is to obtain, under that condition, some information on the structures of $C_G(x)$ and G and also on embedding of $C_G(x)$ in G. It appears that this condition is fairly restrictive. We show, among other things, the following: (i) $C_G(x)$ is either abelian of type (p^n, p) or $|G : C_G(x)| = p$, (ii) G has no normal subgroup of order p^{p+1} and exponent p (in case $p = 2$, such G are treated in § 50), (iii) if $p > 2$, then either $N_G(C_G(x))$ has no normal subgroup $\cong E_{p^3}$ or G is a 3-group of maximal class $\ncong \Sigma_9$. Note that if $|C_G(x)| = p^2$, then, by Proposition 1.8, G is of maximal class. Therefore, in Theorem 198.3 we confine on the case $|C_G(x)| > p^2$.

In what follows Blackburn's classification of minimal nonmetacyclic p-groups is important. This classification is presented in Theorems 66.1 and 69.1.

Exercise 1 (see also Theorem 10.4). Let G be a nonmetacyclic p-group, $p > 2$. If all minimal nonmetacyclic subgroups of G are nonabelian, then one of the following holds:

(a) $G = \Omega_1(G)C$, where C is cyclic and $\Omega_1(G) \cong S(p^3)$.

(b) G is a 3-group of maximal class and order $> 3^3$, $G \ncong \Sigma_{3^2}$.

Hint. The group G has no subgroup $\cong E_{p^3}$. Use Theorem 13.7.

Exercise 2. For a nonabelian p-group G the following assertions are equivalent:

(a) The centralizer of every noncentral element is cyclic.

(b) $G \cong Q_{2^n}$.

Solution. Clearly, (b) $\Rightarrow$ (a). Let us prove the reverse implication. If G is a 2-group of maximal class, then $G \cong Q_{2^n}$. Otherwise, there is in G a normal subgroup $R \cong E_{p^2}$. By hypothesis, $R \le Z(G)$. Let $L = \langle x \rangle < G$ be noncentral cyclic. Then $C_G(x) > R$ is noncyclic, a contradiction. Thus, R does not exist. By Lemma 1.4, G is a 2-group of maximal class, contrary to the assumption.

Exercise 3. Let $x \in G - Z(G)$ be an element of order p in a nonmetacyclic p-group G. Then $Z(C_G(x))$ is noncyclic.

Theorem 198.1. *Let G be a nonmetacyclic p-group, $p > 2$, containing an element x of order p such that $C = C_G(x)$ is metacyclic of order $> p^2$. Set*

$$N_1 = N_G(C), \quad N_2 = N_G(N_1), \quad R = \Omega_1(C).$$

(a) $R \cong E_{p^2}$, $C_G(R) = C$.

(b) *The subgroup N_1 is neither minimal nonabelian nor metacyclic and $|N_1 : C| = p$. If $A < N_1$ is minimal nonabelian and $A \nleq C$, then either $|A| = p^3$ or $A \cong M_{p^n}$.*

(c) *Either $N_1 = \Omega_1(N_1)Z$, where $\Omega_1(N_1) \cong S(p^3)$ and Z is cyclic of order $> p$, or G is a 3-group of maximal class $\not\cong \Sigma_9$. In the first case, C is abelian of type (p^n, p) for some $n > 1$.*

(d) *If R is characteristic in N_1, then $N_1 = G$, i.e., $|G : C| = p$.*

(e) *If R is not characteristic in N_1, then C is abelian of type (p^n, p) for some $n > 1$. In this case, C, being soft in G, is uniserial, i.e., there is only one maximal chain connecting C and G (see § 130).*

(f) *If M is a minimal nonmetacyclic subgroup of N_1, then M is either $\cong S(p^3)$ or a 3-group of maximal class and order 3^4 with $|\Omega_1(M)| = 9$.*

(g) *G has no normal subgroup of order p^{p+1} and exponent p.*

Proof. Suppose that G is a nonmetacyclic p-group with metacyclic centralizer $C = C_G(x)$ of some $x \in G$ of order p; then, by hypothesis, $|C| \geq p^3$. By Exercise 3, C is neither cyclic nor of maximal class; moreover, $E_{p^2} \cong R = \Omega_1(C) \leq Z(C)$. If C is nonabelian, it has no cyclic subgroup of index p. As C is metacyclic, one has

$$\langle x \rangle \not\leq C', \quad R \lhd N_1, \quad C_G(R) = C.$$

Then, by N/C-theorem,

$$|N_1 : C| = p \Rightarrow N_1 = N_G(C) = N_G(R).$$

Assume that there is in N_1 a subgroup $E \cong E_{p^3}$. Then $E \cap N_1 = R = C \cap E \Rightarrow C_G(x) \geq CE > C$, a contradiction. Thus,

(i) N_1 has no subgroup $\cong E_{p^3}$. Next, N_1 is nonmetacyclic. Indeed, this is true if $N_1 = G$, by hypothesis. Now assume that $N_1 < G$ and N_1 is metacyclic. Being characteristic in N_1, the subgroup $\Omega_1(N_1) = R \lhd N_G(N_1) = N_2$. Now, $|N_2 : C_{N_2}(R)| \leq p \Rightarrow |C_{N_2}(R)| > |C|$, a contradiction. Thus, N_1 is nonmetacyclic. By Theorem 13.7, Remark 10.5, Theorems 9.5 and 9.6 and Exercise 9.13, either $N_1 = \Omega_1(N_1)Z$, where $\Omega_1(N_1) \cong S(p^3)$ and Z is cyclic of order $> p$ or G is a 3-group of maximal class $\not\cong \Sigma_9$. In the first case, C is abelian of type $(|Z|, p)$.

Next we assume that $N_1 < G$.

The subgroup N_1 is not minimal nonabelian (otherwise, $\Omega_1(N_1) \cong E_{p^3}$, by Lemma 65.1, contrary to (i)).

Let $A < N_1$ be minimal nonabelian of order $> p^3$ and $A \not\leq C$. In this case, $|A \cap C| \geq p^3$, by the product formula. Assume, that $R < A$. As $A \not\leq C = C_G(R)$, it follows that A is nonmetacyclic (otherwise, $R \leq Z(A) \Rightarrow A \leq C$). In that case, $E_{p^3} \cong \Omega_1(A) \leq C_G(R)$, a contradiction since $C_G(R)$ is metacyclic. Thus, $R \not\leq A$ so that $C \cap A$ is cyclic of index p in A. It follows from Lemma 65.1 that $A \cong M_{p^n}$. If $|A| = p^3$, then $A \in \{M_{p^3}, S(p^3)\}$.

(ii) By (i), N_1 is a group of Theorem 13.7.

(ii1) Assume that N_1 is a 3-group of maximal class. Then, by Remark 10.5, G is also of maximal class. A 3-group of maximal class satisfies the hypothesis, since it is $\not\cong \Sigma_9$ (a group Σ_9 has no element x of order 3 such that $C_G(x)$ is metacyclic of order > 9).

(ii2) Suppose that N_1 is as in Exercise 1 (a). Then $C_G(x) = C$ is abelian of type (p^n, p) for some $n > 1$.

(iii) If $R = \Omega_1(C)$ is not characteristic in N_1, then C is abelian of type (p^n, p) for some $n > 1$. Indeed, there is $\phi \in \mathrm{Aut}(N_1)$ such that $R^\phi \neq R$. In this case, $C^\phi \neq C$. As C is maximal in N_1, it follows that $|N_1 : (C^\phi \cap C)| = p^2$. As $R \nleq C^\phi \cap C$, it follows that $|\Omega_1(C^\phi \cap C)| = p$, and we conclude that $C^\phi \cap C$ is cyclic (Proposition 1.3). As C has a cyclic subgroup of index p and $R \leq Z(C)$, C is abelian of type (p^n, p) for some $n > 1$.

(iv) Assume that R is characteristic in N_1 and $N_1 < G$. Then $R \triangleleft N_2$ and so $C = C_G(R)$ has index p in N_2, i.e., $|C| = |N_1|$, a contradiction. Thus, $G = N_1$ so that $|G : C| = p$, contrary to the assumption.

(v) Assume that $F \triangleleft G$ is of order p^{p+1} and exponent p. Set $H = \langle x, F\rangle$. As $C_H(x)$ is metacyclic, it follows that $C_H(x) \cong E_{p^2}$. By Proposition 1.8, H is of maximal class, contrary to Theorem 9.6.

(vi) Let $M < N_1$ be minimal nonmetacyclic. Then M, by (i) and Theorem 66.1, is either $S(p^3)$ or a group of maximal class and order 3^4 with $|\Omega_1(M)| = 3^2$. □

Exercise 4. If a nonmetacyclic p-group G contains a subgroup $R \cong E_{p^2}$ such that $C_G(R)$ is metacyclic, then $R^\#$ contains an element x such that $C_G(x)$ is metacyclic.

Solution. One has $R \nleq Z(G)$ since G is nonmetacyclic. Therefore, $R = \langle x\rangle \times \Omega_1(Z(G))$ for some $x \in R - Z(G)$. In the case under consideration, $C_G(x) = C_G(R)$ is metacyclic.

Exercise 5. If a noncyclic p-group G contains a cyclic subgroup L of order $> p$ such that $C_G(L) > L$ is cyclic, then G is a 2-group of maximal class.

Solution. Let $C_G(L) < M \leq G$, where $|M : C_G(L)| = p$; then M is a nonabelian group from Theorem 1.2. Assume that $M \cong M_{p|C_G(L)|}$. Then $L \leq \Phi(C_G(L)) \leq \Phi(M) \leq Z(M)$, a contradiction. Thus, by Theorem 1.2, M is a 2-group of maximal class. By Exercise 10.10, G is also of maximal class.

Exercise 6. Classify the nonabelian nonmetacyclic p-groups G, $p > 2$, in which the centralizer of any nonnormal cyclic subgroup is metacyclic.

Hint. If $E < G$ is isomorphic E_{p^3}, then $E \leq Z(G)$. Then the centralizer of any subgroup of G contains E so nonmetacyclic. Thus, G has no subgroups $\cong E_{p^3}$. Use Theorem 13.7.

Problem 1. Study the non-Dedekindian 2-groups in which the normalizer of any nonnormal cyclic subgroup is metacyclic.

Problem 2. Study the nonabelian p-groups containing a nonabelian metacyclic subgroup with metacyclic normalizer.

Problem 3. Study the p-groups G of exponent $> p$ such that $C_G(L)$ is abelian for every nonnormal cyclic $L < G$.

Problem 4. Study the p-groups G such that $\exp(C_G(x)) = o(x)$ for all $x \in G - Z(G)$.

Problem 5. Study the p-groups G such that $C_G(x) \cong M_{p^n}$ for some $x \in G$ of order p^2 and some $n > 3$.

Problem 6. Study the nonabelian p-groups containing an element x of order p such that $C_G(x)$ is minimal nonabelian.

Problem 7. Study the p-groups of exponent $> p$ in which the centralizers of all elements of order $> p$ are metacyclic.

Problem 8. Study the nonabelian p-groups in which the centralizers of all minimal nonabelian subgroups are cyclic.

Problem 9. Study the nonabelian p-groups in which the centralizers of all nonabelian subgroups are abelian. Consider in detail the groups of exponent p.

Problem 10. Study the nonmetacyclic p-groups in which the centralizers of all minimal nonmetacyclic subgroups are cyclic.

§ 199 *p*-groups with minimal nonabelian closures of all nonnormal abelian subgroups

The purpose of this section is to determine up to isomorphism the non-Dedekindian title p-groups G (Problem 805). It turns out that we must have

$$p = 2\,, \quad |G| \le 2^6\,, \quad \mathrm{cl}(G) = 3\,, \quad \exp(G) = 8$$

and there are exactly seven such groups (Theorem 199.1). This shows again that minimal nonabelian p-subgroups play a special role in p-group theory.

Theorem 199.1. *Let G be a non-Dedekindian p-group all of whose nonnormal abelian subgroups have minimal nonabelian closures. Then*

$$p = 2\,, \quad |G| \le 2^6\,, \quad \mathrm{cl}(G) = 3\,, \quad \exp(G) = 8\,, \quad G' \cong \mathrm{C}_4 \text{ or } \mathrm{C}_4 \times \mathrm{C}_2$$

and G is isomorphic to one of the following groups:

(a) $$\mathrm{D}_{16}\,, \quad \mathrm{Q}_{16} \text{ or } \mathrm{SD}_{16}\,.$$

(b) $$G = \langle a, b \mid a^8 = b^4 = 1,\ a^b = a^{-1}z^{\eta},\ z = a^4\rangle\,, \quad \eta = 0, 1\,.$$

These are two metacyclic A_2-groups of order 2^5 with $G' = \langle a^2\rangle \cong \mathrm{C}_4$.

(c) $$G = \langle a, b, c \mid a^8 = b^4 = c^4 = [a, c] = 1\,, \quad b^2 = c^2\,, \quad a^b = a^{-1}z^{\eta}\,,$$
$$z = a^4\,, \quad c^b = c^{-1}\rangle\,, \quad \eta = 0, 1\,.$$

Here

$$|G| = 2^6\,, \quad G' = \langle a^2, b^2\rangle \cong \mathrm{C}_4 \times \mathrm{C}_2\,,$$
$$\mathrm{Z}(G) = \langle z, b^2\rangle = \Omega_1(G) \cong \mathrm{E}_4\,, \quad \langle b, c\rangle \cong \mathrm{Q}_8\,.$$

Conversely, all the above groups satisfy the assumptions of our theorem.

Proof. Let G be a non-Dedekindian p-group all of whose nonnormal abelian subgroups have minimal nonabelian closures. Let A be a maximal normal abelian subgroup of G. Then $A < G$ and each subgroup of A is G-invariant, by hypothesis. Let B/A be a normal subgroup of order p in G/A so that B is nonabelian.

Suppose that each cyclic subgroup in B is normal in G. Then $p = 2$ and B is Hamiltonian. Since G is non-Dedekindian, we have $B < G$ and let C/B be a normal subgroup of order 2 in G/B. By Theorem 125.1, there is $g \in C - B$ so that $c \in \mathrm{Z}(C)$, contrary to $\mathrm{C}_G(A) = A$. We have proved that there is $b \in B - A$ such that $\langle b\rangle$ is not normal in G.

(i) The case $p > 2$. Let $g \in B - A$ be such that $\langle g\rangle$ is not normal in G. By Theorem 125.1, $\exp(A) = p^e$, $e \ge 2$, and

$$a^g = a^{1+p^{e-1}} \text{ for all } a \in A, \text{ where } \mathrm{Z}(B) = \mathrm{C}_A(g) = \Omega_{e-1}(A)\,.$$

There is $x \in G$ such that $[g^x, g] \ne 1$. Set $\langle g^*\rangle = \langle g^x\rangle$ so that

$$\langle g\rangle \cap \langle g^*\rangle = \langle g^p\rangle$$

because $\langle g^p \rangle \trianglelefteq G$ and $H = \langle g, g^* \rangle$ is minimal nonabelian. Since g does not centralize $H \cap A$, there is $h \in H \cap A$ of order p^e so that

$$\langle h \rangle \trianglelefteq G\,, \quad h^g = h^{1+p^{e-1}}\,.$$

Because $[g, h] \neq 1$, the subgroup $H = \langle h, g \rangle$ is metacyclic. We may choose a generator $g^* \in \langle g^* \rangle$ so that $g^p = (g^*)^{-p}$, and set $u = g^* g$. Then

$$u^p = (g^* g)^p = (g^*)^p g^p [g, g^*]^{\binom{p}{2}} = 1$$

because B is of class 2 with an elementary abelian commutator group. If $u \in A$, then $u \in Z(G)$ and therefore $[g^*, g] = [ug^{-1}, g] = 1$, a contradiction. Hence, $u \in H - A$ and so

$$H = \langle h, g \rangle = \langle h, u \rangle \cong M_{p^{e+1}}\,.$$

But $\langle h, u \rangle \trianglelefteq G$ and so

$$\Omega_1(\langle h, u \rangle) = \langle h^{p^{e-1}}, u \rangle \cong E_{p^2} \triangleleft G\,.$$

It follows that $\langle u \rangle \trianglelefteq G$ and so $u \in Z(G)$, a final contradiction.

(ii) The case $p = 2$. We have proved that there is $b \in B - A$ such that $\langle b \rangle$ is not normal in G. According to Theorem 125.1, $\exp(A) \geq 4$.

(ii1) Suppose $\exp(A) = 4$. Then Theorem 125.1 implies $o(b) \leq 4$, b inverts each element in A and $Z(B) = C_A(b) = \Omega_1(A)$. There is $x \in G$ such that $[b, b^x] \neq 1$. Then

$\langle b \rangle \cap \langle b^x \rangle = \langle b^2 \rangle$ with $o(b^2) \leq 2$ and $M = \langle b, b^x \rangle \trianglelefteq G$ is minimal nonabelian.

Since b does not centralize $A \cap M$, there is $a \in A \cap M$ with $o(a) = 4$. But

$$a^b = a^{-1}\,, \quad M' = \langle a^2 \rangle\,, \quad \langle a \rangle \trianglelefteq G\,,$$

and so the subgroup M is metacyclic.

Suppose, by way of contradiction, that $G > B$. If B/A is the only subgroup of order 2 in G/A, then G/A is either cyclic of order ≥ 4 or G/A is generalized quaternion. In any case, there is $y \in G - B$ such that $y^2 \in B - A$, where y^2 inverts each element of A. This is a contradiction since each subgroup of A is normal in G. Thus, there is another subgroup $C/A \neq B/A$ of order 2 in G/A. By Theorem 125.1, $c \in C - A$ (noting that C is nonabelian) inverts each element in A. But then $bc \notin A$ and bc centralizes A, a contradiction. We have proved that $G = B$ and so $x \in B$.

Note that $G' = B' = [A, b]$ is elementary abelian and $G' \leq Z(G)$. We have $b^x = ba_0$ for some $a_0 \in M \cap A$. But

$$a_0 = [b, x] \in Z(G) \Rightarrow [b, b^x] = [b, ba_0] = 1\,,$$

a contradiction.

(ii2) We have proved that $\exp(A) = 2^e$ with $e \geq 3$.

According to Theorem 125.1, first we assume that $a^b = a^{1+2^{e-1}}$ for all $a \in A$, i.e., b induces a "modular" involutory automorphism on A. Let $x \in G$ be such that $[b, b^*] \neq 1$, where

$$\langle b^* \rangle = \langle b^x \rangle, \quad b, b^* \in B - A \quad \text{and} \quad \langle b^* \rangle \cap \langle b \rangle = \langle b^2 \rangle,$$

because $\langle b^2 \rangle \trianglelefteq G$ and $H = \langle b, b^* \rangle \trianglelefteq G$ is minimal nonabelian. Since b does not centralize $H \cap A$, there is $a \in H \cap A$ of order 2^e so that

$$H = \langle b, a \rangle, \quad H' = \langle z \rangle, \quad \text{where } z = a^{2^{e-1}}.$$

We have $B' = [A, b]$ is elementary abelian, $B' \leq Z(G)$ and H is metacyclic. We may choose a generator b^* of $\langle b^* \rangle$ so that $b^2 = (b^*)^{-2}$ and $b^2(b^*)^2 = 1$.

Set

$$u = bb^*, \quad \text{where } u \in A \text{ and } u^2 = (bb^*)^2 = b^2(b^*)^2[b^*, b] = z.$$

Hence, u is an element of order 4 in A and so b centralizes u since

$$u^b = u^{1+2^{e-1}} = u, \quad e \geq 3.$$

But then

$$[b^*, b] = [b^{-1}u, b] = [u, b] = 1,$$

a contradiction.

Now, it is easy to see that $G = B$. Let U/A be a normal elementary abelian subgroup of order 4 in G/A. Then (by Theorem 125.1) three subgroups U_1/A, U_2/A, U_3/A of order 2 in U/A induce three distinct involutory automorphisms on A (inverting, semidihedral-inverting and modular). In particular, $U_i \trianglelefteq G$, $i = 1, 2, 3$, contrary to the previous paragraph. By Theorem 1.4, G/A is either cyclic or of maximal class. Assume that $G > B$ and note that in that case B/A is a square in G/A. By the previous paragraph, we must have for each $b \in B - A$ and $a \in A$,

$$a^b = a^{-1+\eta 2^{e-1}}, \quad \eta = 0, 1.$$

This is a contradiction because each subgroup of A is normal in G.

We have proved that $G = B$ and B/A induces on A an inverting or semi-dihedral inverting involutory automorphism, i.e., for each

$$b \in B - A \quad \text{and} \quad a \in A, \quad a^b = a^{-1+\eta 2^{e-1}}, \quad \eta = 0, 1.$$

Let $b \in B - A$ so that $o(b) \leq 4$ and $\langle b \rangle$ is not normal in G. Let $a \in A$ be of order 2^e, where $2^e = \exp(A)$, $e \geq 3$. Set

$$z = a^{2^{e-1}} \text{ so that } a^b = a^{-1}z^\eta, \quad [a, b] = a^{-2}z^\eta.$$

Then $H = \langle b \rangle^G$ is minimal nonabelian with $a^2 \in H$. But

$$|H'| = 2\,, \quad (a^2)^b = a^{-2} \Rightarrow [a^2, b] = a^{-4} \in H' \Rightarrow o(a^2) = 4\,, \quad e = 3\,, \quad H' = \langle z \rangle\,.$$

It follows that $H = \langle a^2, b \rangle$ is metacyclic minimal nonabelian and therefore

$$H \cong \mathrm{D}_8\,, \quad \mathrm{Q}_8 \quad \text{or} \quad \mathcal{H}_2\,,$$

where $\mathcal{H}_2$ is the metacyclic nonabelian group of order 16 and exponent 4.

(ii2a) First, assume that

$$b^2 \in \langle z \rangle \Rightarrow H \cong \mathrm{D}_8\,, \quad \text{or} \quad \mathrm{Q}_8\,.$$

Let V be a complement of $\langle a \rangle$ in A. Then $[V, b] \le V \cap H = \{1\}$ so that V is elementary abelian and $V \le \mathrm{Z}(G)$. Note that $\langle b \rangle \times V$ is abelian and so $\langle b \rangle \times V$ is not normal in G. Indeed, if $\langle b \rangle \times V \trianglelefteq G$, then $\langle b \rangle^G$ would be abelian, a contradiction. It follows that

$$H = \langle b \rangle^G \le (\langle b \rangle \times V)^G\,, \quad \text{and then} \quad (\langle b \rangle \times V)^G = H$$

since $(\langle b \rangle \times V)^G$ is minimal nonabelian giving $V \le H$ so that $V = \{1\}$. We have proved that in this case $A = \langle a \rangle \cong \mathrm{C}_8$ and so we have obtained the groups stated in part (a) of our theorem. Conversely, these groups obviously satisfy the assumptions of our theorem.

(ii2b) Suppose that $b^2 = u \notin \langle z \rangle$ so that

$$\langle a \rangle \cap \langle b \rangle = \{1\}\,, \quad \langle b \rangle \cong \mathrm{C}_4\,, \quad \langle a^2, b \rangle \cong \mathcal{H}_2\,.$$

Set

$$E = \Omega_1(A) \ge \langle z, b^2 \rangle \cong \mathrm{E}_4\,, \quad \text{where } E \le \mathrm{Z}(G)\,.$$

Then $E\langle b \rangle$ is abelian and $E\langle b \rangle$ is not normal in G (otherwise, $\langle b \rangle^G = H$ would be abelian). Then

$$H = \langle b \rangle^G \le (E\langle b \rangle)^G\,, \quad \text{where } (E\langle b \rangle)^G \text{ is minimal nonabelian}$$

and so

$$H = (E\langle b \rangle)^G \text{ so that } E \le H \text{ and } E = \langle z, b^2 \rangle\,.$$

In particular, A is of rank 2. If $A = \langle a \rangle \times \langle d \rangle$ with $o(d) = 8$, then

$$[A, b] = \langle a^2 \rangle \times \langle d^2 \rangle \cong \mathrm{C}_4 \times \mathrm{C}_4\,.$$

On the other hand, $[A, b] \le H \cap A \cong \mathrm{C}_4 \times \mathrm{C}_2$, a contradiction. Hence, we have either

$$A = \langle a \rangle \times \langle b^2 \rangle \text{ or } A = \langle a \rangle \times \langle c \rangle \text{ with } c^2 = b^2\,.$$

If $A = \langle a \rangle \times \langle b^2 \rangle$, then we have obtained the groups of order 2^5 from part (b) of our theorem. If $A = \langle a \rangle \times \langle c \rangle$ with $c^2 = b^2$, then we have obtained the groups of order 2^6 from part (c) of our theorem.

Conversely, let G be a group from part (b) or (c) of our theorem. Then $A = \langle a\rangle\times\langle b^2\rangle$ (in part (b)) or $A = \langle a\rangle \times \langle c\rangle$ (in part (c)) is a unique abelian maximal subgroup of G. In both cases, we have for each $x \in A$,

$$x^b = x^{-1+4\eta}\,, \quad \eta = 0, 1\,, \quad [a, b] = a^{-2}z^{\eta}\,, \quad b^a = ba^2z^{\eta}\,,$$

and in particular, each subgroup of A is normal in G. For any $x \in A$,

$$(xb)^2 = xbxb = xb^2x^b = xb^2x^{-1+4\eta} = b^2x^{4\eta}\,, \quad \text{where } x^{4\eta} \in \langle z\rangle\,.$$

Since

$$(xb)^a = (xb)a^2z^{\eta}\,, \quad \text{where } \langle z^{\eta}\rangle \le \langle a^2\rangle \text{ and } (a^2)^{xb} = a^{-2}\,,$$

it follows that $H = \langle xb, a^2\rangle \cong \mathcal{H}_2$. But

$$G' \le \langle a^2, b^2\rangle \Rightarrow H \trianglelefteq G\,, \quad \langle xb\rangle^G = \langle xb, z\rangle^G = H\,.$$

On the other hand,

$$\langle xb\rangle \cong C_4\,, \quad \langle xb, z\rangle \cong C_4 \times C_2$$

are the only nonnormal abelian subgroups of G. Hence, G satisfies the assumptions of our theorem. □

Problem 1. Classify the nonabelian p-groups G such that A^G is either abelian or minimal nonabelian for any abelian $A < G$.

Problem 2. Classify the p-groups G, that are $\mathcal{A}_n$-groups, $n > 2$, such that A^G is a $\mathcal{A}_k$-subgroup, $k \in \{1, 2\}$, for any minimal nonabelian $A_1 < G$.

§ 200 The nonexistence of p-groups G all of whose minimal nonabelian subgroups intersect $Z(G)$ trivially

It is asked in Problem 1392 that if there exists a nonabelian p-group G all of whose minimal nonabelian subgroups have trivial intersection with $Z(G)$. The following theorem answers this question.

Theorem 200.1 (Janko). *There is no nonabelian p-group G such that $M' \cap Z(G) = \{1\}$ for all minimal nonabelian $M < G$.*

This result follows immediately from the following lemma allowing to produce minimal nonabelian subgroups (see also Lemma 57.1 yielding another possibility to produce minimal nonabelian subgroups).

Lemma 200.2. *Let G be a nonabelian p-group and let $B/Z(G) \triangleleft G/Z(G)$ be normal of order p and let $B = \langle b\rangle Z(G)$, where $b \in B - Z(G)$. If $g \in G - C_G(b)$, then $\langle b, g\rangle$ is a minimal nonabelian subgroup.*

Proof. Write $C = B\langle g\rangle$; then C is nonabelian since $\langle b, g\rangle$ is nonabelian. The group $C/Z(G)$, being an extension of a normal subgroup $B/Z(G)$ by the cyclic group C/B, is abelian. As $C' \leq Z(G) \leq Z(C)$, it follows that $\mathrm{cl}(C) = 2$. We have $b^p \in Z(G)$ and so $[b, g]^p = [b^p, g] = 1$. Hence, by the above, $\langle b, g\rangle' = \langle [b, g]\rangle \cong C_p$ so that it follows from Lemma 65.2 (a) that the subgroup $\langle b, g\rangle$ is minimal nonabelian. □

Proof of Theorem 200.1. We use the same notation as in the statement of Lemma 200.2. Write $M = \langle b, g\rangle$. It suffices to show that $M \cap Z(G) > \{1\}$. By the proof of the lemma,

$$\langle b, g\rangle' \leq C' \leq Z(G) \Rightarrow \langle b, g\rangle \cap Z(G) \geq \langle b, g\rangle' > \{1\}$$

and we are done as the subgroup $\langle b, g\rangle$ is minimal nonabelian. □

In particular, if a p-group G is nonabelian, it contains a minimal nonabelian subgroup M whose core in G is $> \{1\}$. Next, for the same G, if $x \in \Omega_1(G' \cap Z(G))^\#$, there is in $G/\langle x\rangle$ an abelian subgroup $M/\langle x\rangle$ such that M is nonabelian. This is equivalent to Theorem 200.1 (indeed, if $A < M$ is minimal nonabelian, then $x \in A_G$.

If G is a nonabelian p-group with the center of order p, then there is in G a minimal nonabelian subgroup containing $Z(G)$.

Exercise 1 (= Problem 2985). Let G be a nonabelian p-group such that, whenever $X/Z(G) < G/Z(G)$ is abelian, then X is abelian. Prove that $G = MZ(G)$, where M is minimal nonabelian.

Solution. Let A be a maximal normal abelian subgroup of G so that $Z(G) < A < G$. Let B be a G-invariant subgroup such that $Z(G) < B \leq A$ and $|B : Z(G)| = p$. Let

$b \in B - Z(G)$ and $g \in G - C_G(b)$. By Lemma 200.2, $M = \langle b, g \rangle$ is minimal nonabelian with $1 \neq [b, g] \in Z(G)$. Since $MZ(G)/Z(G)$ is abelian but $MZ(G)$ is nonabelian, it follows by our hypothesis that $MZ(G) = G$ and we are done.

Exercise 2 (= Problem 3178). Let G be a p-group such that $A \leq Z(A^G)$ for any abelian $A < G$. Show that each two-generator subgroup of G is of class ≤ 2 and so G is of class ≤ 3 and if $p \neq 3$, then G is of class ≤ 2.

Solution. Let A be an abelian subgroup in G so that $A \leq Z(A^G)$. But then $Z(A^G) \trianglelefteq G$ and so $Z(A^G) = A^G$ is abelian. We have proved that the normal closure of each abelian subgroup is abelian. It follows that each maximal abelian subgroup is normal in G. Let X be a maximal abelian subgroup of G containing $\langle x \rangle$ and let Y be a maximal abelian subgroup of G containing $\langle y \rangle$. Then $[x, y] \in X \cap Y = Z(\langle X, Y \rangle)$ and so $[x, y]$ commutes with x and y. Hence $\langle x, y \rangle$ is of class at most 2 and $[x, y, y] = 1$ shows that G satisfies the second Engel condition. By [Hup1, III., Satz 6.5], G is of class at most 3 and if G is not a 3-group, then G is of class at most 2.

Exercise 3 (Problem 3252). Show that a nonabelian p-group G such that $C_G(x)$ is minimal nonabelian for each $x \in G - Z(G)$ does not exist.

Solution. Let G be a nonabelian p-group such that $C_G(x)$ is minimal nonabelian for each $x \in G - Z(G)$. Let A be any maximal abelian subgroup in G so that $Z(G) < A < G$. Let $a \in A - Z(G)$; then $M = C_G(a)$ is minimal nonabelian. Since A is also a maximal abelian subgroup in M, we get $|M : A| = p$. Also, $a \in Z(M) = \Phi(M) \leq \Phi(G)$. As a is an arbitrary element of $A - Z(G)$, it follows that $A = \langle A - Z(G) \rangle \leq \Phi(G)$. Thus, all maximal abelian subgroups of G, covering G, are contained in $\Phi(G) < G$, a final contradiction.

The first author stated the following

Problem. Study the p-groups G such that, whenever $S < G$ is minimal nonabelian, there is a cyclic $C < G$ independent of S such that $\langle S, C \rangle = G$.

The following exercise contains a remark due to the second author concerning the above problem.

Exercise 4 (Janko). Show that a nonabelian p-group G in the above problem has an abelian subgroup of index p.

Solution. Let A be a maximal normal abelian subgroup in G. Let $C = \langle c \rangle$ be a cyclic subgroup the existence of which is asserted in the above problem. Obviously, there is no minimal nonabelian subgroup containing C. Assume that $c \notin A$; then Lemma 57.1 implies that there is $a \in A$ such that $\langle c, a \rangle$ is minimal nonabelian contrary to the statement in the previous sentence. We have proved that $C \leq A$. Let $A < B \leq G$ be such that $|B : A| = p$. If $b \in B - A$, then Lemma 57.1 shows that there exists $a' \in A$ such that

$M = \langle b, a' \rangle$ is minimal nonabelian. But then

$$G = \langle M, C \rangle \leq \langle B, A \rangle = B \Rightarrow B = G\,,$$

and we are done.

Exercise 5. Let G be a non-Dedekindian p-group such that for any two distinct conjugate subgroups $A \neq A^x$ $(x \in G)$, one has $|A : (A \cap A^x)| = p$. If $p = 2$, then show that G is metabelian.

Solution. One may assume that G is nonabelian. Let $x, y \in G$. Then $|\langle x \rangle : (\langle x \rangle \cap \langle x^y \rangle) \leq 2$ so that $x^2 \in \langle x^y \rangle$. It follows that $\langle x^2 \rangle \trianglelefteq G$ since $\langle x^y \rangle$ is cyclic. It follows from commutativity of $\mathrm{Aut}(\langle x^2 \rangle)$ that $G' \leq C_G(x^2)$. Thus, G' centralizes x^2 for all $x \in G$ and hence G' centralizes $\Phi(G) = \mho_1(G)$. But $G' \leq \Phi(G)$ and so $G' \leq Z(\Phi(G))$ and we conclude that G' is abelian.

Problem 1. Does there exist an irregular p-group such that $L \cap Z(G) = \{1\}$ for any minimal irregular $L < G$?

By Theorem 200.1, this is not true for $p = 2$ since regular 2-groups are abelian so minimal irregular subgroups are minimal nonabelian.

Problem 2. Study the p-groups G such that $|M \cap Z(G)| = p^2$ for all minimal nonabelian $M < G$.

§ 201 Subgroups of order p^p and exponent p in p-groups with an irregular subgroup of maximal class and index $> p$

If H is a proper subgroup of a p-group G and suppose that any subgroup of G of order $p|H|$ containing H, is of maximal class, then G is also of maximal class (Exercise 10.10). In this section, we consider a p-group G containing an irregular subgroup H of maximal class such that $\langle H, x\rangle$ is of maximal class for any $x \in \mathrm{N}_G(H) - H$ of order p. If $\mathrm{N}_G(H) - H$ has no elements of order p, then, as a rule, G is of maximal class.

Let H be a proper irregular subgroup of maximal class of a p-group G such that $T = \langle x, H\rangle$ is of maximal class for any $x \in G - H$ of order p. Let $H < M \le G$ be such that $|M : H| = p$. If the set $M - H$ has an element of order p, then the subgroup M is of maximal class. Now assume that the set $M - H$ has no element of order p and let M be not of maximal class. Then $e_p(M) \equiv 1 \pmod p$ (Theorem 13.5). By assumption, $e_p(H) = e_p(M)$. Since $\mathrm{d}(H) = 2$ and $\exp(H) > p$, it follows that H has a normal subgroup of order p^p and exponent p (Theorem 12.1 (a)), and this implies that $|H| = p^{p+1}$. If $|H| > p^{p+1}$, then $e_p(M) = e_p(H) \equiv 0 \pmod p$ (Theorem 9.6 (c)), contrary to Theorem 12.1 (a). Thus, in the case under consideration, $|H| = p^{p+1}$ and $e_p(H) = 1$. If the subgroup U containing H as a subgroup of index p, is not of maximal class and if $V < H$ is U-invariant of index p, then U/V is abelian of type (p, p). Assume, by way of contradiction, that U/V is cyclic (of order p^2). Then a p-subgroup H of maximal class is a the unique subgroup of U containing V as a subgroup of index p, and therefore U is of maximal class, a contradiction.

Theorem 201.1. *Let G be a p-group and let $H \triangleleft G$ be irregular of maximal class. Suppose, in addition, that G is not of maximal class and for any $x \in G$ of order p the subgroup $\langle x, H\rangle$ is of maximal class. Then $|H| = p^{p+1}$, G has exactly one normal subgroup, say R, of order p^p and exponent p and $R = \Omega_1(H) = \Omega_1(G)$. In that case, H/R is maximal cyclic subgroup of the group G/R.*

Proof. By Theorem 12.1 (a), there is $R \triangleleft G$ of order p^p and exponent p.

(i) Suppose that $R \not\le H$. Let $R_0 \le R$ be G-invariant of the least order not contained in H. Then $|HR_0 : H| = p$ and, as $\exp(R_0) = p$, it follows that R_0H is of maximal class since $R_0H = \langle H, x\rangle$ for $x \in R_0 - H$. However $R_0H/(R_0 \cap H) = (R_0/(R_0 \cap H)) \times (H/(R_0 \cap H))$ is not two-generator since $H/(R_0 \cap H)$ is noncyclic in view $|H/(H \cap R_0)| > p$, a contradiction: a p-group of maximal class is two-generator.

(ii) Thus, $R < H$. By Theorem 9.6 (c), one has $|H| = p^{p+1}$. Let us show that R is the unique normal subgroup of G of order p^p and exponent p. Let $H < T \le G$, where $|T : H| = p$; then T is not of maximal class since $R < T$ and $|T| > p^{p+1}$ (Theorem 9.6 (c)). The set $T - H$ has no elements of order p (indeed, if $x \in T - H$ is of order p, then $T = \langle x, H\rangle$ is of maximal class). By Theorem 13.5, $e_p(T) \equiv 1 \pmod p$. By the above, $e_p(T) = e_p(H)$.

As H is irregular, it follows that $\exp(H) > p$ (Theorem 7.1 (b)), and we conclude that $e_p(H) < p+1$ since $d(H) = 2$. It follows that $e_p(H) = 1$ so that $\Omega_1(H)$ is a unique subgroup of H and T having order p^p and exponent p, and therefore $R = \Omega_1(H)$. If $x \in G - H$ is of order p, then $T_1 = \langle x, H\rangle$ is not of maximal class (Theorem 9.6 (c)), a contradiction. Thus, $R = \Omega_1(H) = \Omega_1(G)$. As in the paragraph preceding the theorem, $H/R < G/R$ is maximal cyclic. □

Corollary 201.2. *Let H be a proper irregular subgroup of order $> p^{p+1}$ of a p-group G and $\Omega_1(G) \nleq H$. If for all $x \in G - H$ of order p, the subgroup $\langle x, H\rangle$ is of maximal class, then G is also of maximal class.*

Proof. Assume that G is not of maximal class. Then there is $R \lhd G$ of order p^p and exponent p (Theorem 12.1 (a)). By Theorem 9.6 (c), $R \nleq H$. Let $R_0 \leq R$ be G-invariant of minimal order such that $R_0 \nleq H$. Then $|R_0H : H| = p$ so, by hypothesis, R_0H is of maximal class and so $d(R_0H) = 2$. As $R_0H/(R_0 \cap H) = (H/(R_0 \cap H)) \times (R_0/(R_0 \cap H))$ is not generated by two elements since $d(H/(R_0 \cap H)) = 2$, we obtain a contradiction. □

Corollary 201.3. *Suppose that H is a proper subgroup of a p-group G, $|H| > p^{p+1}$. Write $N = N_G(H)$.*

(a) *Let H be absolutely regular such that $|\Omega_1(H)| = p^{p-1} < |\Omega_1(G)|$. If for each $x \in N-H$ of order p the subgroup $\langle x, H\rangle$ is of maximal class, then G is also of maximal class.*

(b) *Let H be of maximal class and $\Omega_1(H) < \Omega_1(G)$. If for each $x \in N - H$ of order p, the subgroup $D = \langle x, H\rangle$ is of maximal class, then G is also of maximal class.*

Proof. (a) By hypothesis, G is not absolutely regular. Suppose that G is not of maximal class. In that case, G contains a normal subgroup R of order p^p and exponent p (Theorem 12.1 (a)); then $R \nleq H$. Let $R_0 \leq R$ be as small as possible G-invariant subgroup not contained in H. The subgroup $R_0\Omega_1(H) \leq \Omega_1(R_0H)$ has order p^p and exponent p so R_0H is not absolutely regular. As H contains an R_0H-invariant subgroup, say K, of order $|R_0|$ and $K \neq R_0$. it follows that R_0H is not of maximal class (Exercise 9.1 (b)). But $R_0H = \langle x, H\rangle$, where $x \in R_0 - H$ is of order p, so that R_0H is of maximal class, by hypothesis, a contradiction. Thus, R does not exist hence G is of maximal class (Theorem 12.1 (a)).

(b) Now suppose that H is of maximal class but G is not. Let R, R_0 be as in (a). Then, as in (a), on one hand, R_0H is not of maximal class, and on the other hand, it is of maximal class since $R_0H = \langle x, H\rangle$ for any $x \in R_0 - H$, a contradiction. □

Let G be a p-group. Suppose that $H < G$ of order $> p^{p+1}$ is either absolutely regular or of maximal class. If $\Omega_1(N_G(H)) = \Omega_1(H)$, then G is either absolutely regular or of maximal class. In particular, if $|\Omega_1(H)| = p^{p-1}$ and $N_G(H)$ is absolutely regular, then G is either absolutely regular or of maximal class since, as above, G has no normal subgroup of order p^p and exponent p (see Theorem 12.1 (a)).

Let H be a subgroup of index $> p^k$, $k > 1$, of a p-group G. If all subgroups of G of order $p^k|H|$, containing H, are of maximal class, then G is also of maximal class.

Indeed, let $H < M < G$, where $|M : H| = p^{k-1}$. Then, by hypothesis, all subgroups of G containing M as a subgroup of index p, are of maximal class, and the result follows from Exercise 10.10.

Suppose that H is a maximal regular subgroup of an irregular p-group G. If $|H| = p^p$, then G is of maximal class. Indeed, as all subgroups of G of order p^{p+1} containing H, are irregular so of maximal class (Theorem 7.1 (b)), the result follows from Exercise 10.10.

Problem 1. Study the nonabelian two-generator p-groups all of whose factors of the (i) lower central series, except of the first one, are cyclic, (ii) upper central series, except of the last one, are cyclic.

Problem 2. Study the p-groups all of whose nonabelian epimorphic images are of maximal class.

Problem 3. Study the nonabelian p-groups in which the center of any minimal nonabelian subgroup is cyclic.

Problem 4. Study the p-groups in which all factors of the lower (upper) central series have order $\leq p^2$.

Problem 5. Let H be a proper nonabelian metacyclic subgroup of a p-group G. Describe the structure of G provided (i) $N_G(H)$ is metacyclic, (ii) all subgroups of G containing H as a subgroup of index p, are metacyclic.

Problem 6. Classify the p-groups G containing a proper extraspecial (special) subgroup H such that $N_G(H)$ is extraspecial (special).

§ 202 p-groups all of whose $\mathcal{A}_2$-subgroups are metacyclic

§ 1. Problem 361 offers to classify the p-groups all of whose minimal nonabelian subgroups (= $\mathcal{A}_1$-subgroups) are metacyclic. This problem is very difficult as shows Theorem 90.1, where its partial case was solved. It seems that a classification of p-groups, all of whose $\mathcal{A}_2$-subgroups are metacyclic, must be fairly difficult. Fortunately, this is not a case as Theorem 202.1 shows.

If $L < H \leq G$ and L is characteristic in H, it follows from $\mathrm{N}_G(L) = H$ that $\mathrm{N}_G(H) = H$. If, in addition, G is a p-group, then $H = G$ so that $L \triangleleft G$. We use this fact in what follows.

Theorem 202.1 (Berkovich [BerZhang]). *Let a p-group G be an $\mathcal{A}_n$-group, $n > 1$. All $\mathcal{A}_2$-subgroups of G are metacyclic $\iff$ G is metacyclic.*[1]

Proof. It suffices to prove the implication $\Rightarrow$. In what follows, we suppose that all $\mathcal{A}_2$-subgroups of G are metacyclic. We have to prove that G is also metacyclic.

One may assume that $n > 2$ (otherwise, it is nothing to prove further). Let A be a minimal nonabelian subgroup of G whose order is as small as possible and let $A < B \leq G$, where $|B : A| = p$. Let us prove that B is an $\mathcal{A}_2$-group. Let C be a nonabelian maximal subgroup of B. Then there is in C a minimal nonabelian subgroup D. Since $|A| = |C| \geq |D| \geq |A|$, it follows that $|D| = |A| = |C|$ so that $C = D$ is minimal nonabelian. Thus, all maximal subgroups of B are either abelian or minimal nonabelian, i.e., B is an $\mathcal{A}_2$-group. It is easily seen that $B < G$ is an $\mathcal{A}_2$-subgroup of minimal order. Therefore, by hypothesis, B is metacyclic so is A. It follows that any subgroup of G containing A as a subgroup of index p must be a metacyclic $\mathcal{A}_2$-subgroup.

Suppose that G is a counterexample of minimal order. Then G is nonmetacyclic but every its (proper) $\mathcal{A}_2$-subgroup is metacyclic. Therefore, there is in G a minimal nonmetacyclic subgroup H. The minimal nonmetacyclic p-groups are classified in Theorems 66.1, 69.1, their orders are either p^3 or 3^4 for $p > 2$ and 2^3, or 2^4, or 2^5 for $p = 2$. The nonmetacyclic $\mathcal{A}_2$-groups of orders 3^4, 2^4 and 2^5 are nonmetacyclic $\mathcal{A}_2$-groups so these groups are not isomorphic to subgroups of G, and therefore we must have $|H| = p^3$ and $\exp(H) = p$.

Let us prove that there is in G a normal elementary abelian subgroup of order p^3.

Suppose that H is nonabelian; then H is minimal nonabelian of order p^3 and exponent $p > 2$. Let $H < M \leq G$, where $|M : H| = p$. In that case, the subgroup M is a nonmetacyclic $\mathcal{A}_2$-subgroup of G of order p^4, contrary to the hypothesis.

Thus, $H \cong \mathrm{E}_{p^3}$. In view of Theorem 10.4, G has a normal subgroup, say R which is $\cong \mathrm{E}_{p^3}$.

1 In [BerZhang] was assume that all $\mathcal{A}_2$-subgroups of minimal order are metacyclic. Given there proof is not sufficient. We do not know if the statement in [BerZhang] is correct].

Let $A < G$ be minimal nonabelian of the least order (as we know, A is metacyclic since it is contained in a metacyclic $\mathcal{A}_2$-subgroup of G of order $p|A|$) and let $L \le R \cap Z(G)$ be of order p. If $L \not\le A$, then $A \times L$ is a nonmetacyclic $\mathcal{A}_2$-group, contrary to the hypothesis. Thus, $L < A$ so that $A \cap R > \{1\}$. Let $A < B \le AR$, where $|B : A| = p$; then B is an $\mathcal{A}_2$-group, by the second paragraph of the proof, and so metacyclic, by hypothesis. By the modular law, $B = A(B \cap R)$. We claim that B is nonmetacyclic. Indeed, if $p > 2$, then $|\Omega_1(B)| > p^2$, and our claim follows. Now let $p = 2$. In that case, $B \cap R$ is B-invariant abelian of type $(2, 2)$ so B is not of maximal class since $|B| \ge 2^4$. By Proposition 10.19, $|A| > 2^3$ and so $\Omega_1(A) \cong E_4$ (Lemma 65.1). Then $|\Omega_1(B)| > 2^2$ which is a contradiction since the metacyclic 2-group B of order $> 2^3$ is not of maximal class. Thus, G is not a counterexample. □

Remark 1 (Janko). The last paragraph of the proof of Theorem 202.1 can be replaced as follows. Let $A < B \le AR$ such that $|B : A| = p$; then B is metacyclic so that $|R \cap B| \le p^2$ and $|A \cap R| = p$ since $A(R \cap B) = B > A$; in that case, $R \cap A \le Z(A)$. As $R \cap A$, $R \cap B$ are normal in B and $A/(R \cap A)$ is noncyclic since A is nonabelian (recall that $R \cap A \le Z(A)$) and $B/(R \cap A) = (A/(R \cap A)) \times (R/(R \cap A))$ has rank 3 so nonmetacyclic, we get a contradiction.

Remark 2. Let $e_3(G)$ be the number of subgroups of order p^3 and exponent p in a p-group G. Let us prove that if G is a nonmetacyclic p-group, $p > 3$, then $e_3(G) \equiv 1 \pmod{p}$. We proceed by induction on $|G|$. One may assume that $|G| > p^3$. By Hall enumeration principle (see § 5), one has

$$e_3(G) \equiv \sum_{H \in \Gamma_1} e_3(H) \pmod{p}. \tag{1}$$

If $H \in \Gamma_1$ is nonmetacyclic, then, by induction, $e_3(H) \equiv 1 \pmod{p}$. Therefore, if all members of the set Γ_1 are nonmetacyclic, it follows from (1) that $e_3(G) \equiv |\Gamma_1| \equiv 1 \pmod{p}$. Now assume that $H \in \Gamma_1$ is metacyclic. Then $G = H\Omega_1(G)$, where $\Omega_1(G)$ is of order p^3 and exponent p so that $e_3(G) = 1$. We suggest to consider $e_3(G) \pmod{p}$ for $p = 3$.

Theorem 202.2. *Let $k \ge 2$. Suppose that a p-group G, $p > 2$, is an $\mathcal{A}_n$-group for $n > k$. If all $\mathcal{A}_k$-subgroups of G are metacyclic, then G is metacyclic.*

Proof. Assume that G is nonmetacyclic; then $|G| \ge p^{k+3}$. In view of Theorem 202.1, one may assume that $k > 2$.

Let $H < G$ be a minimal nonmetacyclic; then $|H| \in \{p^3, 3^4\}$. Suppose that $|H|$ is as small as possible.

Assume that $|H| = 3^4$; then H is an $\mathcal{A}_2$-subgroup. Let $H < M \le G$, where $|M : H| = 3^{k-2}$ so that $|M| = 3^{k+2} < |G|$ and M is a nonmetacyclic $\mathcal{A}_k$-subgroup, contrary to the assumption. Thus, $|H| \ne 3^4$.

Now let $|H| = p^3$; then $\exp(H) = p$. Assume that H is nonabelian. If $H < M < G$, where $|M| = p^{k+2}$, then M is a nonmetacyclic $\mathcal{A}_k$-subgroup, a contradiction, Thus, $H \cong E_{p^3}$. In view of Theorem 10.4, one may assume that $H \triangleleft G$.

Let $M < G$ be an $\mathcal{A}_k$-subgroup of minimal possible order and let $L < M$ be an $\mathcal{A}_{k-1}$-subgroup. As M is metacyclic so is L. Let $H_0 < H$ be G-invariant and small as possible such that $H_0 \not\leq L$. Then $|H_0L : L| = p$, and it is easy to see that H_0L is an $\mathcal{A}_k$-subgroup. Since $H_0\Omega_1(L)$ is a subgroup of H_0L of order p^3 and exponent p, we conclude that H_0L is a nonmetacyclic $\mathcal{A}_k$-subgroup of G, contrary to the hypothesis. Thus, H does not exist so that G is metacyclic, completing the proof. □

By Theorem 72.1, a metacyclic p-group G is an $\mathcal{A}_k$-group $\iff |G'| = p^k$.

Corollary 202.3. *If a p-group G possesses only one proper $\mathcal{A}_2$-subgroup, say B, then B is nonmetacyclic.*

Proof. Assume that B is metacyclic; then G is metacyclic (Theorem 202.1). One has $B' \cong C_{p^2}$ (Theorem 72.1 or Corollary 65.3) and hence $\Omega_1(B') = \Omega_1(G')$. We also have $|G'| \geq p^3$. Let $M/\Omega_1(G') < G/\Omega_1(G')$ be minimal nonabelian ($<$ since $|(G/\Omega_1(G'))'| > p$ so that $G/\Omega_1(G)'$ is not an $\mathcal{A}_1$-group). As $|M'| = p^2$, it follows that M is an $\mathcal{A}_2$-subgroup, and we conclude that $M = B$. Therefore, $G/\Omega_1(G')$ contains only one (proper) minimal nonabelian subgroup, contrary to Remark 76.1. Thus, B is nonmetacyclic, as required. □

According to [ZZLS], there are many p-groups G containing only one proper $\mathcal{A}_2$-subgroup.

Some information on p-groups all of whose $\mathcal{A}_1$-subgroups are metacyclic, $p > 2$, is contained in the following exercise.

Exercise 1. Let G be a nonabelian p-group, $p > 2$. If all $\mathcal{A}_1$-subgroups of the least order are metacyclic, then $\Omega_1(G)$ is elementary abelian.

Solution. Let $E \triangleleft G$ be elementary abelian of maximal order. Assume that $x \in G - E$ is of order p. Write $H = \langle x, E\rangle$. Then H is nonabelian, by Theorem 10.1. By Lemma 57.1, there is $a \in E$ such that $M = \langle a, x\rangle$ is minimal nonabelian. As $\Omega_1(M) = M$, it follows that $|M| = p^3$ and $\exp(M) = p$ (Lemma 65.1), i.e., a nonmetacyclic $\mathcal{A}_1$-subgroup M has the least possible order, a contradiction. Thus, x does not exist so that $\Omega_1(G) = E$.

Exercise 2. Let a p-group G be an $\mathcal{A}_n$-group, $n > 1$, $p > 2$. If $\Omega_1(F)$ is elementary abelian for any $\mathcal{A}_2$-subgroup $F \leq G$ of the least possible order, then $\Omega_1(G)$ is elementary abelian.

Solution. Let E, x, a, H, M be as in the solution of Exercise 1. Let $M < T \leq H$, where $|T : M| = p$; then T is an $\mathcal{A}_2$-subgroup of the least possible order since $|T| = p^4$. As $\Omega_1(T) \geq M$ is nonabelian, we get a contradiction. Thus, x does not exist so that $\Omega_1(G) = E$ is elementary abelian.

Exercise 3. Let G be a nonabelian 2-group. If $\Omega_1(M)$ is abelian for any $\mathcal{A}_1$-subgroup M of G, then $\Omega_1(G)$ is abelian.

Solution. Assume that $u, v \in G$ are noncommuting involutions. Then $D = \langle u, v \rangle$ is dihedral. If $\mathrm{D}_8 \cong D_1 \leq D$, then $\Omega_1(D_1) = D_1$ is nonabelian, a contradiction. Thus, any two involutions from G are permutable so that $\Omega_1(G)$ is abelian.

Exercise 4. If all minimal nonabelian subgroups of a nonabelian 2-group G are nonmetacyclic, then $\Omega_1(G)$ is elementary abelian. (**Hint.** Any two involutions of G are permutable.)

§ 203 Nonabelian p-groups G in which the center of each nonabelian subgroup is contained in $Z(G)$

The purpose of this section is to solve Problem 1953: classify the nonabelian p-groups G in which the center of each nonabelian subgroup of G is contained in $Z(G)$. In fact, we shall characterize such p-groups.

In the proof we use the following obvious fact: If A, B are maximal abelian subgroups of a nonabelian p-group G and $\langle A, B\rangle = G$, then $A \cap B = Z(G)$.

Theorem 203.1 (Janko). *For a nonabelian p-group G the following conditions are equivalent:*

(a) $Z(M) \le Z(G)$ *for any nonabelian subgroup* $M \le G$.

(b) $A \cap B = Z(G)$ *for any two distinct maximal abelian subgroups* A, B *of* G.

If $p = 2$ and G satisfies condition (b) (and so condition (a)), then one of the following hold:

(i) G *possesses an abelian subgroup of index* 2.

(ii) $cl(G) = 2$, G' *is elementary abelian and* $\Phi(G) \le Z(G)$.

In the proof of the theorem, we use the following

Lemma 203.2. *Let G be a nonabelian p-group.*

(a) *(Theorem 91.2) We have $A \cap B = Z(G)$ for any distinct maximal abelian subgroups A, B of G if and only if $C_G(x)$ is abelian for each $x \in G - Z(G)$.*

(b) *(Theorem 91.3) If G satisfies (a) and $p = 2$, then one of the following hold:*

(b1) G *possesses an abelian subgroup of index* 2.

(b2) $cl(G) = 2$, G' *is elementary abelian and* $\Phi(G) \le Z(G)$.

Proof of Theorem 203.1. Suppose that G is a p-group satisfying condition (a). Let A, B be any two distinct maximal abelian subgroups of G. Then the subgroup $\langle A, B\rangle$ is nonabelian and $Z(\langle A, B\rangle) = A \cap B$ (see the paragraph preceding the theorem). By hypothesis, $A \cap B \le Z(G)$. Since the reverse inclusion $Z(G) \le A \cap B$ is obvious, one obtains $A \cap B = Z(G)$.

Assume that G is a nonabelian p-group such that $A \cap B = Z(G)$ for any two distinct maximal abelian subgroups A, B of G. By Lemma 203.2 (a), $C_G(x)$ is abelian for each $x \in G - Z(G)$. Let M be any nonabelian subgroup of G and suppose that $Z(M) \not\le Z(G)$. Take $x \in Z(M) - Z(G)$; then $M \le C_G(x)$ so that $C_G(x)$ is nonabelian, contrary to what has just been said. We have proved that G satisfies condition (a). Thus, (b) $\Rightarrow$ (a).

Suppose that G satisfies condition (b) and $p = 2$; then $A \cap B = Z(G)$ for any two distinct maximal abelian subgroups A, B of G. By Lemma 203.2 (b), G satisfies either (i) or (ii). The proof is complete. □

Exercise 1. Let A be an abelian subgroup of index p of a nonabelian p-group G. Then $C_G(x)$ is abelian for all $x \in G - Z(G)$.

Solution. Let $x \in G - Z(G)$. If $x \in A$, then $C_G(x) = A$ is abelian. If $x \in G - A$, then $C_G(x) = \langle x, Z(G)\rangle$ is also abelian.

Exercise 1 shows that the groups of Lemma 203.2 (b1) satisfy the condition of Lemma 203.2 (a).

Exercise 2. Suppose that G is a nonabelian p-group. Then the following conditions are equivalent:
(a) $G = HZ(G)$ for some $\mathcal{A}_1$-subgroup $H \le G$.
(b) $G = HZ(G)$ for any $\mathcal{A}_1$-subgroup $H \le G$.
(c) $C_G(x)$ is abelian for all $x \in G - Z(G)$.
(d) $A \cap B = Z(G)$ for any two distinct maximal abelian $A, B < G$.

Solution. Let (a) hold. If $Z(G) < A \in \Gamma_1$, then $A = Z(G)(A \cap H)$. As $A \cap H$ is abelian it follows that A is abelian. Next, $|G : Z(G)| = p^2$. Indeed, if $Z(G) < B \in \Gamma_1 - \{A\}$, then B is abelian, by what has just been proved. As $A \cap B = Z(G)$, our claim follows. Let $F \le G$ be an $\mathcal{A}_1$-subgroup of G. Set $M = FZ(G)$. Then

$$|M : Z(G)| \ge p^2 = |G : Z(G)| \Rightarrow M = G .$$

so that (a) $\Rightarrow$ (b). Clearly, (b) $\Rightarrow$ (a). Thus, (a) $\iff$ (b). Now, (a) and (b) imply (c) and (d) since all maximal abelian subgroups of G contain $Z(G)$ so belong to Γ_1. By Lemma 203.2 (a), (d) $\Rightarrow$ (c). Now let (c) holds. Assume that $A \cap B > Z(G)$. If $x \in (A \cap B) - Z(G)$, then $C_G(x) \ge \langle A, B\rangle$ is nonabelian, a contradiction. Thus, (c) $\Rightarrow$ (d).

Exercise 3. Let A be an abelian subgroup of index p of a nonabelian p-group G. If $U, V < G$ are maximal abelian and $U \ne A \ne V$, then $|U| = |V|$.

Hint. $A \cap U = Z(G) = A \cap V$. Make use of the product formula.

Problem 1. Study the nonabelian p-groups G such that $|(A \cap B) : Z(G)| \le p$ for any distinct maximal abelian $A, B < G$.

Problem 2. Study the nonabelian p-groups G such that $Z(A) \le Z(G)$ for any minimal nonabelian $A \le G$.

Problem 3. Given $n > 2$, study the $\mathcal{A}_n$-groups G satisfying $Z_2(M) \le Z_2(G)$ for any $\mathcal{A}_2$-subgroup $M < G$.

Problem 4. Study the nonabelian p-groups G such that $Z(M) \le Z_2(G)$ for any nonabelian $M \le G$.

Problem 5. Study the p-groups G that are neither abelian nor minimal nonabelian and such that $Z(M) \le Z_2(G)$ for any nonabelian $M \in \Gamma_1$. (This is a generalization of #4.)

Problem 6. Study the p-groups such that $C_G(x)$ is abelian for all (i) $x \in G - G'$, (ii) $x \in G - \Phi(G)$.

§ 204 Theorem of R. van der Waal on p-groups with cyclic derived subgroup, $p > 2$

It appears that if $p > 2$ and G is a p-group, $p > 2$, with cyclic derived subgroup, then the quotient group G/G', as a rule, is large. More exactly, the following nice theorem was proved by van der Waal [Waa1]:

Theorem 204.1. *If G is a p-group of order p^n with cyclic derived subgroup G', $p > 2$, then*

(1) $$p^{[n/2]+1} \mid |G/G'|$$

or, what is the same, $p^{[n/2]+1} \leq |G/G'|$.[1]

Proof. One may assume that G is nonabelian. Let

$$n = 2k + e\,, \quad e \in \{0, 1\}\,, \quad |G'| = p^s\,, \quad |G/G'| = p^{n-s} = p^{2k+e-s}\,.$$

Then $[n/2] = k$. Let G be a counterexample of minimal order. Then p^{k+1} does not divide $|G/G'|$, or, what is the same,

(2) $$p^k = p^{[n/2]} \geq |G/G'| = p^{2k+e-s}.$$

It follows from (2) that

(3) $$2k + e - s \leq k \Rightarrow s \geq k \Rightarrow 2k + e - s \leq s \Rightarrow |G'| \geq |G/G'| \Rightarrow |G| \leq |G'|^2\,.$$

By (3), there is $T < G'$ such that $|G'/T| = |G/G'|$. As G' is cyclic, $T \triangleleft G$, and

(4) $$G/G' \cong (G/T)/(G'/T) = (G/T)/(G/T)'\,.$$

Write $G_0 = G/T$; then G_0 is nonabelian since $T < G'$. In that case,

$$|G_0'| = |(G/T)'| = |G'/T| = |G/G'|\,,$$

and (4) can be rewritten as follows:

(5) $$G_0/G_0' \cong G/G'\,.$$

Write $|G_0'| = p^t$. By the choice of T (see (4), (5) and the line preceding (4)), one obtains $|G_0'| = |G/G'| = |G_0/G_0'|$ so that

(6) $$|G_0| = |G_0'|^2 = p^{2t}\,.$$

---

1 Here $[x]$ is the integer part of a real number x. The presented proof follows closely to [Waa1].

Since G_0 is nonabelian, it follows from (6) that $t \geq 2$.

As $p > 2$, a noncyclic p-group G_0 contains a normal subgroup $M \cong \mathrm{E}_{p^2}$ (Lemma 1.4). It is possible to choose M so that $\Omega_1(G_0') < M$. Set $G_1 = G_0/M$. As G_0' is cyclic of order $p^t > p = \exp(M)$, it follows that $G_0' \nleq M$ so that G_1 is nonabelian of order p^{2t-2} (see (6)). One has

$$(7) \qquad G_1' = G_0'M/M \Rightarrow G_1/G_1' = (G_0/M)/(G_0'M/M) \cong G_0/G_0'M$$

and G_1' is cyclic since G_1 is an epimorphic image of G_0. Let $|G_1/G_1'| = p^u$. Then, by (6) and (7), since $|G_0'M| = p^{t+1}$, by the product formula, one obtains

$$(8) \qquad p^u = |G_1/G_1'| = |G_0/G_0'M| = \frac{p^{2t}}{p^{t+1}} = p^{t-1} \Rightarrow u = t - 1 \Rightarrow t > u\,.$$

By induction applied to a group G_1 of order p^{2t-2}, we obtain

$$p^t = p^{[(2t-2)/2]+1} \mid |G_1/G_1'| = p^u \Rightarrow t \leq u\,,$$

contrary to (8). □

A particular case of Theorem 204.1 for metacyclic p-groups, $p > 2$, was presented in Theorem A.48.4.

As any 2-group of maximal class and order $2^n > 2^3$ shows, inequality (1) is not true for 2-groups with cyclic derived subgroup. If $p > 2$ and G is a p-group of maximal class and order $p^n \geq p^4$, then inequality (1) does not hold. Therefore, the assumption on cyclicity of G' in Theorem 204.1 is essential.

Exercise 1. Is it true that (1) holds for the 2-groups G with cyclic G' and without nonabelian sections of order 8?

Problem 1. Describe the 2-groups with cyclic derived subgroup for which (1) holds.

Problem 2. Classify the p-groups G of order p^n, $p > 2$, with cyclic derived subgroup for which $p^{[n/2]+1} = |G/G'|$.

§ 205 Maximal subgroups of $\mathcal{A}_2$-groups

This section is written by Qinhai Zhang.

In this section, all $\mathcal{A}_1$-subgroups of $\mathcal{A}_2$-groups are given up to isomorphism by using the classification of the $\mathcal{A}_2$-groups presented in Propositions J.1–J.5. In the list of the $\mathcal{A}_2$-groups there are some superfluous items. Therefore, we note these places. This is done in Remarks 2 and 3 and Proposition J.5 (b′). Note, that it is not asserted "if and only if" in Propositions 71.1–71.5 so stated there results are correct.

In the following five propositions, for convenience of the readers, all the $\mathcal{A}_2$-groups are listed.

Proposition J.1 (= Proposition 71.1). *Suppose that G is an $\mathcal{A}_2$-group of order $> p^4$. Then $|G'| = p$ if and only if G has exactly $p+1$ distinct abelian maximal subgroups and in that case one of the following hold:*

(a) $G = H \times C_p$, *where H is minimal nonabelian.*

(b) $G = \langle a, b, c\rangle$, *where*

$$a^{p^m} = b^{p^n} = c^{p^2} = 1\,, \quad m \ge n \ge 1\,, \quad m \ge 2\,, \quad [a,b] = d\,, \quad c^p = d\,,$$
$$[a,d] = [b,d] = [a,c] = [b,c] = 1\,.$$

Here $H = \langle a, b\rangle$ is nonmetacyclic minimal nonabelian, $H' = G' = \langle d\rangle$, $G = H * C$ (central product), where $C = \langle c\rangle$ is cyclic of order p^2 and $H \cap C = \langle d\rangle = \langle c^p\rangle = H'$.

Proposition J.2 (cf. Proposition 71.2). *Let G be a metacyclic $\mathcal{A}_2$-group of order $> p^4$. Then $G' \cong C_{p^2}$ and one of the following hold:*

(a) $G = \langle a, b \mid a^{p^m} = 1, m \ge 3, b^{p^n} = a^{\epsilon p^{m-1}}, n \ge 2, \epsilon = 0, 1, m+n \ge 5, a^b = a^{1+p^{m-2}}\rangle$, *where, in case $p = 2$, one has $m \ge 4$.*

(b) $p = 2, G = \langle a, b \mid a^8 = 1, b^{2^n} = a^{4\epsilon}, \epsilon = 0, 1, n \ge 2, a^b = a^{-1+4\eta}, \eta = 0, 1\rangle$.

Remark 1. In Proposition 71.2 (a), the condition given on n is just $n \ge 1$. However, if $n = 1$, then $\langle a\rangle$ is a maximal subgroup of G. By Theorem J.2, G is a 2-group of maximal class. Then $|G'| = 2^2$, $|G| = 2^4$, a contradiction since $|G| > 2^4$.

Proposition J.3 (= Proposition 71.3). *Let G be a nonmetacyclic $\mathcal{A}_2$-group of order $> p^4$ possessing exactly one abelian maximal subgroup M. Then $G' \cong E_{p^2}$ and assume, in addition, that $G' \nleq Z(G)$. In that case,*

$$p > 2\,, \quad d(G) = 2\,, \quad |K_3(G)| = p\,, \quad G/K_3(G) \text{ is nonmetacyclic minimal nonabelian}\,,$$

and one of the following hold:

(a) $G = \langle a, b\rangle$, *where*

$$a^p = b^{p^n} = 1\,, \quad n \ge 3\,, \quad [a,b] = c\,, \quad c^p = [a,c] = 1\,, \quad [b,c] = b^{p^{n-1}}\,.$$

Here

$$|G| = p^{n+2}\,, \quad G' = \langle c, b^{p^{n-1}}\rangle \cong E_{p^2}\,, \quad K_3(G) = \langle b^{p^{n-1}}\rangle,\ M = \langle a\rangle \times \langle b^p\rangle \times \langle c\rangle \in \Gamma_1\,,$$

and all members of the set $\Gamma_1 - \{M\}$ *are metacyclic.*

(b) $G = \langle a, b\rangle$, *where*

$$a^p = b^{p^n} = 1\,, \quad n \geq 2\,, \quad [a, b] = c\,, \quad [b, c] = d\,,$$
$$c^p = d^p = [a, c] = [d, a] = [d, b] = 1\,.$$

Here

$$|G| = p^{n+3}\,, \quad G' = \langle c, d\rangle \cong \mathrm{E}_{p^2}\,, \quad \mathrm{K}_3(G) = \langle d\rangle\,, \quad M = \langle a\rangle\times\langle b^p\rangle\times\langle c\rangle\times\langle d\rangle \in \Gamma_1\,.$$

(c) $G = \langle a, b\rangle$, *where*

$$a^{p^2} = b^{p^n} = 1\,, \quad n \geq 2\,, \quad [a, b] = c\,, \quad [c, b] = a^{sp}\,, \quad c^p = [a, c] = 1\,,$$

and $s = 1$ *or* s *is a fixed quadratic nonresidue* $\pmod p$. *Here*

$$|G| = p^{n+3}\,, \quad G' = \langle c, a^p\rangle \cong \mathrm{E}_{p^2}\,, \quad \mathrm{K}_3(G) = \langle a^p\rangle,\ M = \langle a\rangle \times \langle b^p\rangle \times \langle c\rangle \in \Gamma_1\,,$$

and all members of the set Γ_1 *are nonmetacyclic.*

Proposition J.4 (cf. Proposition 71.4). *Let* G *be a nonmetacyclic* $\mathcal{A}_2$*-group of order* $> p^4$ *and class 2 possessing exactly one abelian maximal subgroup. Then* $G' \cong \mathrm{E}_{p^2}$. *In that case* $\mathrm{d}(G) = 3$, $\mathrm{Z}(G) = \Phi(G)$, *and one of the following is true:*

(a) *If* G *has no normal elementary abelian subgroups of order* p^3, *then* $p = 2$ *and*

$$G = \langle a, b, c \mid a^4 = b^4 = [a, b] = 1,\ c^2 = a^2,\ a^c = ab^2,\ b^c = ba^2\rangle$$

is the minimal nonmetacyclic group of order 2^5 *(see Lemma J.7 (e), below). Here* G *is special,*

$$\Omega_1(G) = G' = \mathrm{Z}(G) = \Phi(G) = \langle a^2, b^2\rangle \cong \mathrm{E}_4\,,$$
$$\mathrm{C}_4 \times \mathrm{C}_4 \cong M = \langle a\rangle \times \langle b\rangle \in \Gamma_1 \text{ is abelian}\,,$$

and all other six members of the set Γ_1 *are metacyclic.*

(b) *If* G *has a normal elementary abelian subgroup* E *of order* p^3, *then* $E = \Omega_1(G)$ *and one of the following hold:*

(b1) $G = \langle a, b, d\rangle$ *with*

$$a^{p^\alpha} = b^p = d^{p^2} = 1\,, \quad \alpha \geq 2 \quad \text{and} \quad \alpha \geq 3$$
$$\text{if } p = 2\,, \quad [a, b] = d^p\,, \quad [d, b] = 1\,, \quad [a, d] = a^{p^{\alpha-1}}\,.$$

Here

$$|G| = p^{\alpha+3},\ G' = \langle d^p, a^{p^{\alpha-1}}\rangle \cong \mathrm{E}_{p^2},\ E = \Omega_1(G) = \langle b, d^p, a^{p^{\alpha-1}}\rangle \nleq \mathrm{Z}(G)\,,$$

and $\langle b\rangle \times \langle a^p\rangle \times \langle d\rangle \in \Gamma_1$.

(b2) $G = \langle a, b, c\rangle$ *with*

$$a^{p^\alpha} = b^{p^2} = c^{p^2} = [b, c] = 1\,, \quad \alpha \geq 1\,,$$
$$[a, b] = x\,, \quad [a, c] = y\,, \quad b^p = x^{\alpha'} y^\beta\,, \quad c^p = x^\gamma y^\delta\,,$$
$$x^p = y^p = [a, x] = [a, y] = [b, x] = [b, y] = [c, x] = [c, y] = 1\,,$$

where in case $p = 2$, $\alpha' = 0$, $\beta = \gamma = \delta = 1$, *and in case* $p > 2$, $4\beta\gamma + (\delta - \alpha')^2$ *is a quadratic nonresidue* $\pmod p$, *we have*

$$|G| = p^{\alpha+4}, \; G' = \langle x, y\rangle \cong \mathrm{E}_{p^2}, \; E = \Omega_1(G) = G'\langle a^{p^{\alpha-1}}\rangle\,.$$

Let $A \in \Gamma_1$ *be abelian. Then in case* $\alpha = 1$, *the subgroup* $A = \langle b, c\rangle$ *is abelian of type* (p^2, p^2) *and* $E \nleq Z(G)$, *and in case* $\alpha \geq 2$, *the subgroup* $A = \langle a^p, b, c\rangle$ *is abelian of type* $(p^{\alpha-1}, p^2, p^2)$ *and* $E \leq Z(G)$.

Remark 2. In Proposition 71.4 (b1), the following equality is given $[a, b] = d^p a^{\epsilon p^{\alpha-1}}$, where $\epsilon = 0$ unless $p = 2$ and $\alpha = 2$ in which case $\epsilon = 1$. Here we prove that the group of Proposition 71.4 (b1) with $\epsilon = 1$ is not an $\mathcal{A}_2$-group. Indeed, if $\epsilon = 1$, then $p = 2$ and $\alpha = 2$. We compute:

$$(ab)^2 = (db)^2 = d^2\,, \quad [ab, db] = [a, d][a, b] = d^2\,.$$

Thus, $\langle ab, db\rangle \cong \mathrm{Q}_8$, which is an $\mathcal{A}_1$-subgroup of index 2^2 in G so that G is not an $\mathcal{A}_2$-group.

Proposition J.5 (cf. Proposition 71.5). *Let* G *be a nonmetacyclic* $\mathcal{A}_2$*-group of order* $> p^4$ *all of whose maximal subgroups are nonabelian. Then we have one of the following possibilities:*

(a) $\mathrm{d}(G) = 3$, $p = 2$, *and* $G = \langle a, b, c\rangle$ *with*

$$a^4 = b^4 = c^4 = 1\,, \quad [a, b] = c^2\,, \quad [a, c] = b^2c^2\,, \quad [b, c] = a^2b^2\,,$$
$$[a^2, b] = [a^2, c] = [b^2, a] = [b^2, c] = [c^2, a] = [c^2, b] = 1\,,$$

where

$$G' = \langle a^2, b^2, c^2\rangle = Z(G) = \Phi(G) = \Omega_1(G) \cong \mathrm{E}_{2^3}\,, \quad \exp(G) = 4$$

and all members of the set Γ_1 *are nonmetacyclic* $\mathcal{A}_1$*-groups. (From [BJ2, Theorem 70.4] follows that* G *is isomorphic to the Suzuki 2-group of order* 2^6*.)*

(b) $\mathrm{d}(G) = 2$, $p > 2$, G *is of order* p^5 *and*

$$G = \langle a, x \mid a^{p^2} = x^{p^2} = 1, \; [a, x] = b, \; [a, b] = y_1, \; [x, b] = y_2\,,$$
$$b^p = y_1^p = y_2^p = [a, y_1] = [x, y_1] = [a, y_2] = [x, y_2] = 1, \; a^p = y_1^\alpha y_2^\beta, \; x^p = y_1^\gamma y_2^\delta\rangle\,,$$

where in case $p > 3$, $4\beta\gamma + (\delta - \alpha)^2$ *is a quadratic nonresidue* $\pmod p$ *and in case* $p = 3$, *one has*

$$\beta \not\equiv 0 \pmod 3\,, \quad \gamma \not\equiv 0 \pmod 3\,, \quad (\alpha - \delta)^2 + (1 + \beta - \gamma)^2 \equiv 1 \pmod 3\,.$$

Here

$$G' = \langle b, y_1, y_2\rangle = \Omega_1(G) \cong E_{p^3}, \qquad Z(G) = K_3(G) = \mho_1(G) = \langle y_1, y_2\rangle \cong E_{p^2}.$$

Remark 3. If $p > 3$, then those groups listed in Proposition J.5 (b) are $\mathcal{A}_2$-groups. If $p = 3$, then the conclusion is not true. For example, in that case, if, in addition, $\beta = 0$, then $\langle a, b\rangle$ is nonabelian of index 3^2. Thus, G is not an $\mathcal{A}_2$-group. In Proposition J.5 (b′), below, we deduce a criterion for G to be an $\mathcal{A}_2$-group if $p = 3$.

Proposition J.5 (b′). *Assume G is as in Proposition J.5 (b) and $p = 3$. Then G is an $\mathcal{A}_2$-group if and only if*

$$\beta \not\equiv 0 \pmod 3, \quad \gamma \not\equiv 0 \pmod 3, \quad (\alpha - \delta)^2 + (1 + \beta - \gamma)^2 \equiv 1 \pmod 3.$$

Proof. We prove this in three steps.

Step 1. The group G is an $\mathcal{A}_2$-group if and only if

$$|\langle a^\xi x^\eta, b\rangle| = 3^4, \text{ where } \xi \not\equiv 0 \pmod 3 \text{ or } \eta \not\equiv 0 \pmod 3.$$

Let $M \in \Gamma_1$. Then

$$M = \langle a^\xi x^\eta, \Phi(G)\rangle = \langle a^\xi x^\eta, b, a^p, b^p\rangle,$$

where either $\xi \not\equiv 0 \pmod 3$ or $\eta \not\equiv 0 \pmod 3$. It follows that $G \in \mathcal{A}_2$ if and only if

$$M \in \mathcal{A}_1 \iff M = \langle a^\xi x^\eta, b\rangle \iff |\langle a^\xi x^\eta, b\rangle| = 3^4.$$

Let $S = \langle a^\xi x^\eta, b\rangle$. Then $G \in \mathcal{A}_2$ if and only if $|S| = 3^4$.

Step 2. G is an $\mathcal{A}_2$-group if and only if

$$\begin{vmatrix} \alpha\xi + \gamma\eta + \eta\xi^2 & \beta\xi + \delta\eta - \xi\eta^2 \\ \xi & \eta \end{vmatrix} = \gamma\eta^2 - \beta\xi^2 + (\alpha - \delta)\eta\xi - \xi^2\eta^2 \not\equiv 0 \pmod 3. \quad (*)$$

Since $\mathrm{cl}(G) = 3$, it follows that

$$\begin{aligned}[a^\xi, x^{-\eta}] &= [a, x^{-\eta}]^{\binom{\xi}{1}}[a, x^{-\eta}, a]^{\binom{\xi}{2}} = ([a, x]^{\binom{-\eta}{1}}[a, x, x]^{\binom{-\eta}{2}})^{\binom{\xi}{1}}[a, x, a]^{-\eta\binom{\xi}{2}} \\ &= b^{-\eta\xi} y_2^{\xi\binom{\eta}{2}} y_1^{\eta\binom{\xi}{2}}.\end{aligned}$$

Thus,

$$\begin{aligned}(a^\xi x^\eta)^3 &= a^{3\xi}x^{3\eta}[a^\xi, x^{-\eta}]^{\binom{3}{2}}[a^\xi, x^{-\eta}, a^\xi][a^\xi, x^{-\eta}, x^{-\eta}] = a^{3\xi}x^{3\eta}[b^{-\eta\xi}, a^\xi][b^{-\eta\xi}, x^{-\eta}] \\ &= y_1^{\alpha\xi} y_2^{\beta\xi} y_1^{\gamma\eta} y_2^{\delta\eta} y_1^{\eta\xi^2} y_2^{-\xi\eta^2} = y_1^{\alpha\xi+\gamma\eta+\eta\xi^2} y_2^{\beta\xi+\delta\eta-\xi\eta^2}\end{aligned}$$

and

$$[a^\xi x^\eta, b] = [a, b]^\xi [x, b]^\eta = y_1^\xi y_2^\eta.$$

It follows that

$$\mho_1(S) = \langle y_1^{\alpha\xi+\gamma\eta+\eta\xi^2} y_2^{\beta\xi+\delta\eta-\xi\eta^2} \rangle\,, \quad S' = \langle y_1^{\xi} y_2^{\eta} \rangle\,.$$

Thus,

$$|S| = 3^4 \iff |\Phi(S)| = p^2 \iff y_1^{\xi} y_2^{\eta} \notin \mho_1(S) \iff (*) \text{ holds}\,.$$

It follows that G is an $\mathcal{A}_2$-group if and only if $(*)$ holds.

Step 3. G is an $\mathcal{A}_2$-group if and only if

$$\beta \not\equiv 0 \pmod 3\,, \quad \gamma \not\equiv 0 \pmod 3\,, \quad \text{and } (\alpha-\delta)^2 + (1+\beta-\gamma)^2 \equiv 1 \pmod 3\,.$$

If $\eta = 0$, then $(*)$ holds $\iff \beta \not\equiv 0 \pmod 3$. If $\xi = 0$, then $(*)$ holds if and only if $\gamma \not\equiv 0 \pmod 3$. If $\xi\eta = 1$, then $(*)$ holds if and only if $\gamma - \beta - 1 + (\alpha - \delta) \not\equiv 0 \pmod 3$. If $\xi\eta = -1$, then $(*)$ holds if and only if $\gamma - \beta - 1 - (\alpha - \delta) \not\equiv 0 \pmod 3$. It follows that $(*)$ holds if and only if

$$\beta \not\equiv 0 \pmod 3\,, \quad \gamma \not\equiv 0 \pmod 3$$

and

$$\gamma - \beta - 1 - (\alpha - \delta) \not\equiv 0 \pmod 3, \gamma - \beta - 1 - (\alpha - \delta) \not\equiv 0 \pmod 3\,.$$

Since

$$\gamma - \beta - 1 + (\alpha - \delta) \not\equiv 0 \pmod 3, \gamma - \beta - 1 - (\alpha - \delta) \not\equiv 0 \pmod 3$$

if and only if

$$\alpha - \delta \not\equiv 0 \pmod 3 \text{ when } 1 + \beta - \gamma \equiv 0 \pmod 3\,,$$
$$\alpha - \delta \equiv 0 \pmod 3 \text{ when } 1 + \beta - \gamma \not\equiv 0 \pmod 3$$

if and only if

$$(\alpha - \delta)^2 + (1 + \beta - \gamma)^2 \equiv 1 \pmod 3\,,$$

we get the $(*)$ holds if and only if

$$\beta \not\equiv 0 \pmod 3, \gamma \not\equiv 0 \pmod 3, (\alpha - \delta)^2 + (1 + \beta - \gamma)^2 \equiv 1 \pmod 3\,.$$

Thus, $G \in \mathcal{A}_2$ if and only if the conditions stated in the beginning of this step are fulfilled. □

In the following lemma, we gather some known results which are used in what follows.

Lemma J.6. *Suppose that a p-group G is nonabelian.*

(i) *(Proposition 10.19) If a metacyclic group G contains a nonabelian subgroup of order p^3, then G is of maximal class (so that, if $p > 2$, then $|G| = p^3$).*

(ii) *(Exercise 10.10) Let $A < G$. If all subgroups of G containing A as a subgroup of index p are of maximal class, then G is also of maximal class.*

(iii) *A 2-group of maximal class has no proper subgroup $\cong SD_{16}$.*

(iv) *(Proposition 10.17) If $B \leq G$ is a nonabelian subgroup of order p^3 and $C_G(B) < B$, then G is of maximal class.*

(v) *(Theorem 10.28) Any nonabelian p-group is generated by $\mathcal{A}_1$-subgroups.*

(vi) *(Remark 76.1) (see also [BJ2, Lemma 76.6]) If a p-group is neither abelian nor minimal nonabelian, then it contains at least p minimal nonabelian subgroups.*

Proof of Lemma J.6 (iii). Assume that G is a 2-group of maximal class containing a proper subgroup $H \cong SD_{16}$. To obtain a contradiction, one can assume, without loss of generality, that $|G : H| = 2$, i.e., $H \in \Gamma_1$ (indeed, all nonabelian subgroups of G are of maximal class). By [[Ber1, Theorem 1.2], all nonabelian members of the set Γ_1 are isomorphic either to D_{16} or Q_{16} which are not isomorphic to H, and this is a contradiction. □

In the following lemma, due to Blackburn, the minimal nonmetacyclic p-groups are classified. For proof, see [BJ2, Theorems 66.1, 69.1].

Lemma J.7. *Let G be a minimal nonmetacyclic p-group. Then one of the following hold:*

(a) $|G| = p^3$, $\exp(G) = p$.

(b) *G is a group of maximal class and order 3^4, $|\Omega_1(G)| = 3^2$.*

(c) *$G = Q \times C$, where $Q \cong Q_8$ and $|C| = 2$.*

(d) *$G = Q * C$ (central product) is of order 2^4, where $Q \cong Q_8$ and $C \cong C_{2^2}$.*

(e) *The group $G = \langle a, b, c\rangle$ with*

$$a^4 = b^4 = [a, b] = 1\,, \quad c^2 = a^2\,, \quad a^c = ab^2\,, \quad b^c = ba^2\,,$$

where G is special of order 2^5 with

$$\exp(G) = 4\,, \quad \Omega_1(G) = G' = Z(G) = \Phi(G) = \langle a^2, b^2\rangle \cong E_4\,,$$

$M = \langle a\rangle \times \langle b\rangle \cong C_4 \times C_4$ is the unique abelian maximal subgroup of G, and all other six maximal subgroups of G are isomorphic to $X = \langle x, y \mid x^4 = y^4 = 1, x^y = x^{-1}\rangle$ (which is the metacyclic minimal nonabelian group of order 2^4 and exponent 4).

It follows from Lemma J.7 (e), that the presented there group is an $\mathcal{A}_2$-group.

In Propositions 205.1–205.10, the maximal subgroups of all $\mathcal{A}_2$-groups are described.

Proposition 205.1. *Suppose that G is the group of Proposition J.1 (a), i.e., $G = H \times C_p$, where H is minimal nonabelian. Then $\alpha_1(G) = p^2$ and all $\mathcal{A}_1$-subgroups of G are isomorphic to H.*

Proof. By the modular law, all $p + 1$ members of the set Γ_1 containing C_p, are abelian. All p^2 remaining members of the set Γ_1 are isomorphic $G/C_p \cong H$ so are $\mathcal{A}_1$-groups. □

Proposition 205.2. *Suppose that G is a group as in Proposition J.1 (b). Then every $\mathcal{A}_1$-subgroup of G is one of the members of the set*

$$\Gamma = \{M_{ij} \mid M_{ij} = \langle ac^i, bc^j \rangle, \quad 0 \le i, j \le p-1\}.$$

Moreover, if $n \ge 2$, or $n = 1$ and $j = 0$, then $M_{ij} \cong M_p(m, n, 1)$; if $n = 1$ and $j > 0$, then $M_{ij} \cong M_p(2, m)$. In particular, all p^2 members of the set Γ are isomorphic if $n \ge 2$.

Proof. It follows from $d(G) = 3$ that $|\Gamma_1| = 1 + p + p^2$ and all members of the set Γ_1 are described as follows:

$$\begin{aligned} M &= \langle b, c, \Phi(G) \rangle, \\ M_i &= \langle ab^i, c, \Phi(G) \rangle, \text{ where } 0 \le i \le p-1, \\ M_{ij} &= \langle ac^i, bc^j, \Phi(G) \rangle, \text{ where } 0 \le i, j \le p-1. \end{aligned}$$

It is easy to see that $\Phi(G) \le Z(G)$ and $c \in Z(G)$. Thus, M and M_i are abelian for $0 \le i \le p-1$ and $M_{ij} = \langle ac^i, bc^j \rangle$ are $\mathcal{A}_1$-subgroups for $0 \le i, j \le p-1$ so that $\alpha_1(G) = p^2$.

Let $x = ac^i$ and $y = bc^j$ for a fixed i and j. Then $M_{ij} = \langle x, y \rangle$.

If $n \ge 2$, or $n = 1$ and $j = 0$, then

$$x^{p^m} = y^{p^n} = d^p = 1, \quad [x, y] = d, \quad [d, x] = [d, y] = 1.$$

By [Rob, Theorem 2.2.1], the group M_{ij} is isomorphic to a quotient group of $M_p(m, n, 1)$. Since $|M_{ij}| = p^{m+n+1} = |M_p(m, n, 1)|$, it follows that $M_{ij} \cong M_p(m, n, 1)$.

If $n = 1$ and $j > 0$, then

$$x^{p^m} = d^p = 1, \quad y^p = (bc^j)^p = c^{jp} = d^j, [x, y] = d, [d, x] = [d, y] = 1.$$

Let $x_1 = x^{-j}$. Then

$$[y, x_1] = [y, x^{-j}] = [y, x]^{-j} = d^j = y^p.$$

It follows that

$$M_{ij} = \langle y, x_1 \rangle, \text{ where } y^{p^2} = x_1^{p^m} = 1, \quad [y, x_1] = y^p$$

so M_{ij} is metacyclic and $M_{ij}/\langle y \rangle \cong C_{p^m}$, and we conclude that $M_{ij} \cong M_p(2, m)$. □

Proposition 205.3. *Suppose that G is a group as in Proposition J.2 (a). Then all maximal subgroups are $\mathcal{A}_1$-subgroups, and every $\mathcal{A}_1$-subgroup of G is exactly one of members of*

$$\Gamma = \{M, M_i \mid M = \langle a^p, b \rangle, M_i = \langle a^p, b^p, ab^i \rangle, \quad 0 \le i \le p-1\}.$$

Moreover,
(1) *If $\epsilon = 0$ and $m > n$, then $M \cong M_p(m-1, n)$ and $M_i \cong M_p(m, n-1)$.*
(2) *If $\epsilon = 0$ and $m \le n$, then $M_0 \cong M_p(m, n-1)$ and $M \cong M_i \cong M_p(m-1, n)$ for $i \ne 0$.*
(3) *If $\epsilon = 1$ and $m \ge n+2$, then $M \cong M_p(m-1, n)$ and $M_i \cong M_p(m, n-1)$.*

(4) *If* $\epsilon = 1$ *and* $m = n + 1$, *then*

$$M \cong M_i \cong M_p(m, m-2) \text{ for } i \neq p-1$$

and

$$M_{p-1} \cong M_p(m-1, m-1)\,.$$

Thus, $M_{p-1} \cong M_p(m-1, m-1)$, *all other* p *members of the set* Γ_1 *are isomorphic to* $M_p(m, m-2)$.

(5) *If* $\epsilon = 1$ *and* $m \leq n$, *then*

$$M_0 \cong M_p(n, m-1) \quad \text{and} \quad M \cong M_i \cong M_p(n+1, m-2) \text{ for } i \neq 0\,.$$

In particular, $M_0 \cong M_p(n, m-1)$, *all other* p *members of the set* Γ_1 *are isomorphic to* $M_p(n+1, m-2)$.

Proof. It follows from $\mathrm{d}(G) = 2$ that $|\Gamma_1| = 1 + p$. The members of the set Γ_1 are

$$M = \langle a^p, b\rangle \text{ and } M_i = \langle ab^i, a^p, b^p\rangle\,, \quad 0 \leq i \leq p-1\,.$$

Since

$$[a^p, b] = [a, b]^p = a^{p^{m-1}} \neq 1$$

and

$$[ab^i, b^p] = [a, b^p]^{b^i} = [a, b^p] = [a, b]^p [a, b, b]^{\binom{p}{2}} = [a, b]^p = a^{p^{m-1}} \neq 1\,,$$

all members of the set Γ_1 are $\mathcal{A}_1$-subgroups, i.e., $\Gamma = \Gamma_1$.

Next the isomorphism types of maximal subgroups will be determined.

Case 1. Let $M = \langle a^p, b\rangle$.

Write $x_1 = a^p$. Then $M = \langle x_1, b\rangle$, where

$$x_1^{p^{m-1}} = 1\,, \quad b^{p^n} = x_1^{\epsilon p^{m-2}}\,, \quad [x_1, b] = [a^p, b] = x_1^{p^{m-2}}\,.$$

If $\epsilon = 0$, then, regardless of $m > n$ or $m \leq n$, we have

$$x_1^{p^{m-1}} = b^{p^n} = 1, [x_1, b] = a^{p^{m-1}} = x_1^{p^{m-2}}\,.$$

By the same argument as in Proposition 205.2, one obtains $M \cong M_p(m-1, n)$.

If $\epsilon = 1$ and $m \geq n + 2$, then

$$x_1^{p^{m-1}} = 1\,, \quad b^{p^n} = x_1^{p^{m-2}}\,, \quad [x_1, b] = x_1^{p^{m-2}}\,.$$

Write $b_1 = bx_1^{-p^{m-n-2}}$. Then $M = \langle x_1, b_1\rangle$ with relations

$$[x_1, b_1] = [x_1, bx_1^{-p^{m-n-2}}] = [x_1, b] = x_1^{p^{m-2}}\,.$$

Since $n \geq 2$,

$$b_1^{p^n} = (bx_1^{-p^{m-n-2}})^{p^n} = b^{p^n}(x_1^{-p^{m-n-2}})^{p^n} = b^{p^n}x_1^{-p^{m-2}} = 1\,.$$

Thus,

$$x_1^{p^{m-1}} = b_1^{p^n} = 1\,, \quad [x_1, b_1] = x_1^{p^{m-2}}\,.$$

By the same argument as in Proposition 205.2, one obtains $M \cong M_p(m-1, n)$.

If $\epsilon = 1$ and $m < n+2$, then, regardless of $m = n+1$ or $m \leq n$, we have

$$x_1^{p^{m-1}} = 1\,, \quad b^{p^n} = x_1^{p^{m-2}}\,, \quad [x_1, b] = x_1^{p^{m-2}}\,.$$

Write $x_2 = x_1^{-1}b^{p^{n-m+2}}$. Then $M = \langle x_2, b\rangle$. We obtain

$$x_2^{p^{m-2}} = (x_1^{-1}b^{p^{n-m+2}})^{p^{m-2}} = x_1^{-p^{m-2}}b^{p^n} = 1\,,$$
$$[b, x_2] = [b, x_1^{-1}b^{p^{n-m+2}}] = [b, x_1^{-1}] = [x_1, b] = x_1^{p^{m-2}} = b^{p^n}\,.$$

It follows that

$$b^{p^{n+1}} = x_2^{p^{m-2}} = 1, [b, x_2] = b^{p^n}\,.$$

By the same argument as in Proposition 205.2, one obtains $M \cong M_p(n+1, m-2)$.

Case 2. $M_0 = \langle a, b^p\rangle$.

Let $y_1 = b^p$. Then $M_0 = \langle a, y_1\rangle$, where

$$a^{p^m} = 1\,, \quad y_1^{p^{n-1}} = a^{\epsilon p^{m-1}}\,, \quad [a, y_1] = [a, b^p] = a^{p^{m-1}}\,.$$

If $\epsilon = 0$, then, regardless of $m > n$ or $m \leq n$, we have $a^{p^m} = y_1^{p^{n-1}} = 1$ and $[a, y_1] = a^{p^{m-1}}$. Thus, $M_0 \cong M_p(m, n-1)$.

If $\epsilon = 1$ and $m \leq n$, then, letting $x_1 = a^{-1}y_1^{p^{n-m}}$, one obtains

$$x_1^{p^{m-1}} = (a^{-1}y_1^{p^{n-m}})^{p^{m-1}} = (a^{-1})^{p^{m-1}}(y_1^{p^{n-m}})^{p^{m-1}} = a^{-p^{m-1}}y_1^{p^{n-1}} = 1$$

and

$$[y_1, x_1] = [y_1, a^{-1}] = [a, y_1] = y_1^{p^{n-1}}\,.$$

Thus, $M_0 \cong M_p(n, m-1)$.

If $\epsilon = 1$ and $m > n$, then, regardless of $m \geq n+2$ or $m = n+1$, we get $y_2^{p^{n-1}} = 1$ and $[a, y_2] = a^{p^{m-1}}$ by letting $y_2 = y_1a^{-p^{m-n}}$. Thus, $M_0 \cong M_p(m, n-1)$.

Case 3. $M_i = \langle ab^i, a^p, b^p\rangle$, where $i \neq 0$.

If $\epsilon = 0$ and $m \leq n$, then $n \geq m \geq 3$. Let $x = a^p$ and $y = a^{i^{-1}}b$. Then $M_i = \langle x, y\rangle$ with relations

$$x^{p^{m-1}} = 1, y^{p^n} = (a^{i^{-1}}b)^{p^n} = (a^{i^{-1}})^{p^n}b^{p^n} = 1\,,$$
$$[x, y] = [a^p, a^{i^{-1}}b] = [a^p, b] = x^{p^{m-2}}\,.$$

Thus, $M_i \cong M_p(m-1, n)$.

If $\epsilon = 0$ and $m > n$, then, letting $y = b^p$ and $x = ab^i$ for a fixed i, we get $M_i = \langle x, y\rangle$ with relations

$$y^{p^{n-1}} = 1,\ [x,y] = [ab^i, b^p] = [a, b^p] = a^{p^{m-1}},$$
$$x^{p^{m-1}} = (ab^i)^{p^{m-1}} = a^{p^{m-1}} b^{ip^{m-1}} = a^{p^{m-1}}.$$

Thus, $M_i \cong M_p(m, n-1)$.

If $\epsilon = 1$ and $m \geq n+2$, then, letting $x = ab^i$ and $y = b^p a^{-p^{m-n}}$ for a fixed i, we get $M_i = \langle x, y\rangle$ with relations

$$y^{p^{n-1}} = b^{p^n} a^{-p^{m-1}} = 1, \quad [x,y] = [ab^i, b^p a^{-p^{m-n}}] = [a, b^p] = a^{p^{m-1}} = x^{p^{m-1}}.$$

Thus, $M_i \cong M_p(m, n-1)$.

If $\epsilon = 1$ and $m = n+1$, then, letting $x = a^p$ and $y = a^{i^{-1}} b$ for a fixed i, we get $M_i = \langle y, x\rangle$ with relations

$$x^{p^{m-1}} = 1, \quad [x,y] = [a^p, b] = x^{p^{m-2}}, \quad y^{p^n} = a^{i^{-1}p^{m-1}} b^{p^n} = a^{(1+i^{-1})p^{m-1}}.$$

If $1 + i^{-1} \equiv 0 \pmod p$, then $M_i \cong M_p(m-1, n)$. If $1 + i^{-1} \not\equiv 0 \pmod p$, then, letting $k = 1 + i^{-1}$ and $x_1 = x^{-k} y^p$, we get

$$y^{p^n} = a^{kp^{m-1}}, \quad x_1^{p^{n-1}} = x^{-kp^{n-1}} y^{p^n} = a^{-kp^n} a^{kp^n} = 1$$

and

$$[y, x_1] = [y, x^{-k}] = [a^{kp}, b] = a^{kp^{m-1}} = y^{p^n}.$$

Thus, $M_i \cong M_p(m, n-1)$.

If $\epsilon = 1$ and $m \leq n$, then, letting $x = a^p b^{-p^{n-m+2}}$ and $b_1 = a^{i^{-1}} b$ for a fixed i, we get $M_i = \langle b_1, x\rangle$ with relations

$$x^{p^{m-2}} = a^{p^{m-1}} b^{-p^n} = 1,\ [x, b_1] = [a^p, b] = a^{p^{m-1}},$$
$$b_1^{p^n} = a^{i^{-1}p^n} b^{p^n} = b^{p^n} = a^{p^{m-1}}.$$

Thus, $M_i \cong M_p(n+1, m-2)$. □

Proposition 205.4. *Suppose that G is a group as in group of Proposition J.2 (b). Then every $\mathcal{A}_1$-subgroup of G is a member of the set*

$$\Gamma = \{M, M_1 \mid M = \langle a^2, b\rangle, \quad M_1 = \langle a^2, b^2, ab\rangle\};$$
$$M \cong M_1 \cong M_2(2, n) \text{ if } \epsilon = 0;\ M \cong M_1 \cong M_2(n+1, 1) \text{ if } \epsilon = 1.$$

Proof. Since $d(G) = 2$, there are in G exactly three maximal subgroups, namely

$$M = \langle a^2, b\rangle, \quad M_i = \langle ab^i, a^2, b^2\rangle\ (i = 0, 1).$$

It follows from

$$o(a) = 8, \quad a^{-4} = a^4 \text{ and } G' = \langle a^2 \rangle \leq C_G(\langle a \rangle)$$

that

$$[a^2, ab] = [a^2, b] = [a, b]^2 = a^{-4} = a^4 \neq 1,$$
$$[a, b^2] = [a, b]^2[a, b, b] = a^4[a^{-2}, \quad b] = a^4[a, b]^{-2} = a^4a^{-4} = 1.$$

Thus, M_0 is abelian, M_1 and M are two $\mathcal{A}_1$-subgroups of G.

Notice that $n \geq 2$ and $(ab)^2 = b^2a^4$ or b^2, we get $(ab)^{2^n} = b^{2^n}$.

If $\epsilon = 1$, then $o(b) = o(ab) = 2^{n+1}$. It follows that $\langle b \rangle$ and $\langle ab \rangle$ are, respectively, cyclic subgroup of index 2 of M and M_1. Thus, $M, M_1 \cong M_2(n + 1, 1)$.

If $\epsilon = 0$, then $o(b) = o(ab) = 2^n$. It follows that $M, M_1 \cong M_2(2, n)$. □

Proposition 205.5. *Suppose that G is a group as in Proposition J.3. Then $\alpha_1(G) = p$. Moreover, if G is as in (a), then $\mathcal{A}_1$-subgroups of G are isomorphic to $M_p(n, 1)$; if G is as in (b) or (c), then $\mathcal{A}_1$-subgroups of G are isomorphic to $M_p(n, 1, 1)$.*

Proof. By Proposition J.3, $d(G) = 2$ and G possesses exactly one abelian maximal subgroup. It follows that the number of $\mathcal{A}_1$-subgroups of G is p. It follows from $c \in \Phi(G)$ and $[ba^i, c] \neq 1$ that the $\mathcal{A}_1$-subgroups of G are

$$M_i = \langle ba^i, \Phi(G) \rangle = \langle ba^i, c \rangle, \text{ where } 0 \leq i \leq p - 1.$$

It follows from $p > 2$ and $\exp(G') = p$ that $(ba^i)^{p^2} = b^{p^2}$, and therefore

$$o(ba^i) = o(b) = p^n = \exp(G).$$

If G is as in (a), then, letting $x = ba^i$ for a fixed i, we get

$$x^{p^2} = (ba^i)^{p^2} = b^{p^2} \text{ and } [x, c] = [ba^i, c] = [b, c] = b^{p^{n-1}} = x^{p^{n-1}}.$$

Thus, $M_i = \langle ba^i, c, b^p \rangle = \langle x, c \rangle \cong M_p(n, 1)$.

If G is as in (b) or (c), then $\Phi(G) \cong C_{p^{n-1}} \times C_p \times C_p$, which is not metacyclic so that M_i is nonmetacyclic. It follows from $\exp(M_i) = \exp(G) = p^n$ and $|M_i| = p^{n+2}$ that $M_i \cong M_p(n, 1, 1)$. □

Proposition 205.6. *Suppose that G is a group as in Proposition J.4 (a). Then G has exactly six $\mathcal{A}_1$-subgroups, which are isomorphic to $M_2(2, 2)$.*

Proof. By Proposition J.4, $d(G) = 3$ and G possesses exactly one abelian maximal subgroup, hence $\alpha_1(G) = 6$. Since G is minimal nonmetacyclic, the $\mathcal{A}_1$-subgroups of G are metacyclic. Since $\exp(G) = 2$ and $|G| = 2^5$, the $\mathcal{A}_1$-subgroups of G are isomorphic to $M_2(2, 2)$. □

Proposition 205.7. *Suppose that G is a group as in Proposition J.4 (b1). Then every $\mathcal{A}_1$-subgroup of G belongs to the set*

$$\Gamma = \{M_i, M_{ij} \mid M_i = \langle ab^i, d\rangle, \quad M_{ij} = \langle ad^i, bd^i\rangle, 0 \le i, j \le p-1\}.$$

Moreover,

$$M_i \cong M_p(\alpha, 2), \quad M_{i0} \cong M_p(\alpha, 1, 1) \text{ and } M_{ij} \cong M_p(2, \alpha) \text{ for } j \neq 0.$$

Proof. Since $\mathrm{d}(G) = 3$, G has $1 + p + p^2$ maximal subgroups, namely,

$$\begin{aligned} M &= \langle b, d, \Phi(G)\rangle, \\ M_i &= \langle ab^i, d, \Phi(G)\rangle, \text{ where } 0 \le i \le p-1, \\ M_{ij} &= \langle ad^i, bd^j, \Phi(G)\rangle, \text{ where } 0 \le i, j \le p-1. \end{aligned}$$

By Proposition J.4, G possesses exactly one abelian maximal subgroup $\langle b, d, a^p\rangle = M$. It follows that $\alpha_1(G) = p + p^2$, and the $\mathcal{A}_1$-subgroups are M_i and M_{ij}, $0 \le i, j \le p-1$.

Since $G' \le Z(G)$ and $\exp(G') = p$, the group G is p^2-abelian and p-abelian if $p > 2$.[1] It follows that $(ab^i)^{p^2} = a^{p^2}$ and $(ad^i)^{p^2} = a^{p^2}$.

Now we determine the isomorphism types of these nonabelian maximal subgroups.

Case 1. $M_i = \langle ab^i, d\rangle$, where $0 \le i \le p-1$.

Write $x = ab^i$ for a fixed i. Then

$$x^{p^\alpha} = 1, \quad [x, d] = [ab^i, d] = [a, d] = x^{p^{\alpha-1}}.$$

Thus, $M_i = \langle x, d\rangle \cong M_p(\alpha, 2)$.

Case 2. $M_{i0} = \langle ad^i, b\rangle$, where $0 \le i \le p-1$.

Write $z = d^p$ and $x = ad^i$ for a fixed i. Then

$$x^{p^\alpha} = a^{p^\alpha} d^{ip^\alpha} = 1 = z^p, [x, b] = [a, b] = z, [z, x] = [z, b] = 1.$$

Thus $M_{i0} = \langle x, b, z\rangle \cong M_p(\alpha, 1, 1)$.

Case 3. $M_{ij} = \langle ad^i, bd^j\rangle$, where $0 \le i \le p-1$ and $1 \le j \le p-1$.

Write $x = ad^i$ and $y = b^{j^{-1}}d$ for a fixed i and j. Then

$$\begin{aligned} x^{p^\alpha} = a^{p^\alpha} d^{ip^\alpha} = 1, \quad y^p = b^{j^{-1}p} d^p = d^p, \\ [x, y] = [ad^i, b^{j^{-1}} d] = [a, b]^{j^{-1}} [a, d] = d^{j^{-1}p} a^{p^{\alpha-1}}. \end{aligned}$$

1 A group G is said to be n-abelian if $(ab)^n = a^n b^n$ for all $a, b \in G$.

Let $x_1 = x^j$. Then

$$[x_1, y] = [x^j, y] = [x, y]^j = d^p a^{jp^{\alpha-1}} .$$

If $\alpha \geq 3$, then, letting $y_1 = ya^{jp^{\alpha-2}}$, we obtain

$$[x_1, y_1] = [x_1, ya^{jp^{\alpha-2}}] = [x_1, y] = d^p a^{jp^{\alpha-1}} = y_1^p ,$$
$$x_1^{p^\alpha} = x^{jp^\alpha} = 1 , \quad y_1^{p^2} = d^{p^2} a^{jp^\alpha} = 1 .$$

Thus, $M_{ij} = \langle x_1, y_1 \rangle \cong M_p(2, \alpha)$.

If $\alpha = 2$, then $p > 2$. Since $\Omega_1(M_{ij}) = \langle a^p, d^p \rangle$, the subgroup M_{ij} is metacyclic. It follows from $|M_{ij}| = p^4$ and $\exp(M_{ij}) = p^2$ that $M_{ij} \cong M_p(2, 2) = M_p(\alpha, 2)$. □

Proposition 205.8. *Suppose that G is as in Proposition J.4 (b2). Then every $\mathcal{A}_1$-subgroup of G belongs to the set*

$$\Gamma = \{M_i, M_{ij} \mid M_i = \langle ab^i, c \rangle, \ M_{ij} = \langle ac^i, bc^i \rangle, \ 0 \leq i, j \leq p - 1\} .$$

Moreover,

(1) *If $\alpha \geq 2$, then $M_i \cong M_{ij} \cong M_p(\alpha, 2, 1)$.*
(2) *If $\alpha = 1$ and $p = 2$, then $M_0 \cong M_{00} \cong M_{01} \cong M_2(2, 1, 1)$ and $M_1 \cong M_{10} \cong M_{11} \cong M_2(2, 2)$.*
(3) *If $\alpha = 1$ and $p > 2$, then $M_0 \cong M_{0j} \cong M_p(2, 1, 1)$ and $M_i \cong M_{ij} \cong M_p(2, 2)$ for $i \neq 0$ and all j.*

Proof. It follows from $d(G) = 3$ that $|\Gamma_1| = 1 + p + p^2$. The members of Γ_1 are:

$$M = \langle b, c, \Phi(G) \rangle .$$
$$M_i = \langle ab^i, c, \Phi(G) \rangle , \quad \text{where } 0 \leq i \leq p - 1 .$$
$$M_{ij} = \langle ac^i, bc^j, \Phi(G) \rangle , \quad \text{where } 0 \leq i, j \leq p - 1 .$$

By Proposition J.4, G possesses exactly one abelian maximal subgroup $\langle b, c, a^p \rangle = M$ so that $|\Gamma| = \alpha_1(G) = p + p^2$, and $\Gamma = \{M_i, M_{ij}, 0 \leq i, j \leq p - 1\}$.

Now we determine the isomorphism types of the $\mathcal{A}_1$-subgroups of G.

Since $G' \leq Z(G)$ and $\exp(G') = p$, G is p^2-abelian and p-abelian if $p > 2$.

If $\alpha \geq 2$, then

$$(ab^i)^{p^2} = a^{p^2} b^{ip^2} = a^{p^2} , \quad (ac^i)^{p^2} = a^{p^2} c^{ip^2} = a^{p^2}$$

It follows that $o(ab^i) = o(ac^i) = p^\alpha$. Hence

$$\exp(M_i) = \exp(M_{ij}) = p^\alpha = \exp(G) .$$

It follows from $|G| = p^{\alpha+4}$ that $|M_i| = |M_{ij}| = p^{\alpha+3}$. Since $\Phi(G) = \langle a^p, b^p, c^p \rangle$ is not metacyclic, M_i and M_{ij}, which are of order $p^{\alpha+3}$ and exponent p^α, are also nonmetacyclic. Thus, $M_i \cong M_{ij} \cong M_p(\alpha, 2, 1)$.

If $\alpha = 1$ and $p > 2$, then

$$|M_i| = |M_{ij}| = p^4 , \quad \exp(M_i) = \exp(M_{ij}) = p^2 = \exp(G) .$$

Notice that G is p-abelian, we get

$$\mho_1(M_i) = \langle b^{ip}, c^p \rangle , \quad \mho_1(M_{ij}) = \langle c^{ip}, b^p c^{jp} \rangle .$$

If $i = 0$, then

$$\mho_1(M_0) = \langle c^p \rangle \cong C_p , \quad \mho_1(M_{0j}) = \langle b^p c^{jp} \rangle \cong C_p .$$

Hence,

$$|M_0/\mho_1(M_0)| = |M_{0j}/\mho_1(M_{0j})| = p^3 > p^2 .$$

It follows that M_0 and M_{0j}, which are of order p^4 and exponent p^2, are not metacyclic. Thus, $M_0 \cong M_{0j} \cong M_p(2, 1, 1)$ for all j.

If $i \neq 0$, then

$$\mho_1(M_i) = \langle b^{ip}, c^p \rangle \cong E_{p^2} , \quad \mho_1(M_{ij}) = \langle c^{ip}, b^p c^{jp} \rangle \cong E_{p^2} .$$

Hence, $|M_i/\mho_1(M_i)| = |M_{ij}/\mho_1(M_{ij})| = p^2$. It follows that M_i and M_{ij}, which are of order p^4 and exponent p^2, are metacyclic. Thus, $M_i \cong M_{ij} \cong M_p(2, 2)$.

If $\alpha = 1$ and $p = 2$, then

$$a^2 = c^4 = y^2 = 1, [a, c] = y, [y, a] = [y, b] = 1 .$$

Thus, $M_0 = \langle c, a \rangle \cong M_2(2, 1, 1)$.

By computation,

$$[abc, c] = [a, c] = y , \quad (abc)^2 = a^2 b^2 c^2 [a, b][a, c][b, c] = yxyxy = y ,$$

we have $M_1 = \langle ab, c \rangle = \langle abc, c \rangle \cong M_2(2, 2)$.

By computation,

$$[a, b] = x , \quad a^2 = 1 , \quad (b)^2 = y ,$$

we have $M_{00} = \langle a, b \rangle \cong M_2(2, 1, 1)$.

By computation,

$$[a, bc] = [a, b][a, c] = xy , \quad a^2 = 1 , \quad (bc)^2 = b^2 c^2 = yxy = x ,$$

we have $M_{01} = \langle a, bc \rangle \cong M_2(2, 1, 1)$.

By computation,

$$[ac, b] = [a, b][c, b] = x , \quad b^2 = y , \quad (ac)^2 = a^2 c^2 [a, c] = xyy = x ,$$

we have $M_{10} = \langle ac, b \rangle \cong M_2(2, 2)$.

By computation,

$$[ac, bc] = [a, b][a, c][c, b] = xy , \quad (ab)^2 = a^2 b^2 [a, b] = yx , \quad (bc)^2 = b^2 c^2 = yxy = x ,$$

we have $M_{11} = \langle ac, bc \rangle = \langle ab, bc \rangle \cong M_2(2, 2)$. □

Proposition 205.9. *Suppose that G is as in Proposition J.5 (a). Then* $\alpha_1(G) = 7$ *and all* $\mathcal{A}_1$*-subgroups are isomorphic to* $M_2(2, 2, 1)$.

Proof. By Proposition J.5, $d(G) = 3$ and all maximal subgroups of G are nonabelian. It follows that $\alpha_1(G) = 7$. Since

$$G' = \langle a^2, b^2, c^2 \rangle = Z(G) = \Phi(G) \cong E_{2^3},$$

all maximal subgroups of G are nonmetacyclic. Since $|G| = 2^6$ and $\exp(G) = 2^2$, all maximal subgroups are of order 2^5 and exponent 4. Thus, all maximal subgroups are isomorphic to $M_2(2, 2, 1)$. □

Proposition 205.10. *Suppose that G is a group as in Proposition J.5 (b). Then all* $1 + p$ *maximal subgroups of G are isomorphic to* $M_p(2, 1, 1)$.

Proof. By Proposition J.5, $d(G) = 2$ and all maximal subgroups are nonabelian. Thus, G has exactly $1 + p$ $\mathcal{A}_1$-subgroups. Since

$$G' = \langle b, y_1, y_2 \rangle = \Phi(G) \cong E_{p^3},$$

all maximal subgroups of G are not metacyclic. Since $|G| = p^5$ and $\exp(G) = p^2$, all maximal subgroups are of order p^4 and exponent p^2. Thus, all maximal subgroups are isomorphic to $M_p(2, 1, 1)$. □

Additional inspection of the obtained results yields the following

Corollary 205.11. *Let G be an* $\mathcal{A}_2$*-group of order* $> p^4$. *Then*
(a) *G has no three pairwise nonisomorphic* $\mathcal{A}_1$*-subgroups.*
(b) *G contains at least* $p - 1$ *pairwise isomorphic minimal nonabelian subgroups.*[2]

It is easy to present, using Propositions 205.1–205.10, the list of those nonmetacyclic $\mathcal{A}_2$-groups (of prime power order) all of whose $\mathcal{A}_1$-subgroups are metacyclic.

2 For $p = 2$ this is trivial.

§ 206 p-groups all of whose minimal nonabelian subgroups are pairwise nonisomorphic

A nonabelian p-group is said to be an $\mathcal{A}_n$-group provided all its subgroups of index p^n are abelian but it possesses a nonabelian subgroup of index p^{n-1} (see § 72).

The $\mathcal{A}_2$-groups are classified in § 71, their nonabelian maximal subgroups (which are $\mathcal{A}_1$-groups) are listed by Qinhai Zhang (see § 205).

Suppose that a p-group G is an $\mathcal{A}_n$-group with $n > 1$. Then there are in G at least p distinct minimal nonabelian subgroups of equal order. Indeed, let $A < G$ be an $\mathcal{A}_1$-subgroup such that $|A|$ is as small as possible and let $A < H \leq G$, where $|H| = p|A|$. Then H is an $\mathcal{A}_2$-subgroup (see § 202). By Remark 76.1, there are in H at least p distinct $\mathcal{A}_1$-subgroups which are maximal in H.

It is natural, therefore, to classify the p-groups all of whose $\mathcal{A}_1$-subgroups are pairwise non-isomorphic; we call them $(*)$-groups. All $\mathcal{A}_1$-subgroups of $(*)$-groups are characteristic; therefore, by Theorem 10.28, all nonabelian subgroups are characteristic there. Clearly, classification of $(*)$-groups must depend on classification of $\mathcal{A}_2$-groups. We also prove (see Theorem 206.2) that if a p-group G is an $\mathcal{A}_n$-group, $n > 1$, then it contains at least $p - 1$ pairwise isomorphic minimal nonabelian subgroups.

Let $\mathcal{A}_1(G)$ denote the set of all $\mathcal{A}_1$-subgroups in G and set $\alpha_1(G) = |\mathcal{A}_1(G)|$. If a p-group G is a metacyclic $\mathcal{A}_1$-group, then either $G \cong \mathrm{Q}_8$ or

$$G = \mathrm{M}_p(m, n) = \langle a, b \mid o(a) = p^m,\ o(b) = p^n,\ a^b = a^{1+p^{m-1}} \rangle \,.$$

If a p-group G is nonmetacyclic $\mathcal{A}_1$-group, then

$$\begin{aligned} G &= \mathrm{M}_p(m, n, 1) \\ &= \langle a, b \mid o(a) = p^m,\ o(b) = p^n,\ [a, b] = c,\ c^p = 1,\ [a, c] = [b, c] = 1 \rangle \,. \end{aligned}$$

See Exercise 1.8a or Lemma 65.1.

In this section, the following two theorems are proved.

Theorem 206.1. *Suppose that a p-group G is an $\mathcal{A}_n$-group, $n > 1$. If all $\mathcal{A}_1$-subgroups of G are pairwise nonisomorphic, then $p = 2$ and the group G is metacyclic. If G contains a subgroup isomorphic to SD_{16}, then $G \cong \mathrm{SD}_{16}$.*

Theorem 206.2. *Suppose that a p-group G is an $\mathcal{A}_n$-group, $n > 1$. Then G has $p - 1$ isomorphic $\mathcal{A}_1$-subgroups of minimal order. If G has an $\mathcal{A}_1$-subgroup of order p^3 then $|G| = p^4$.*

Remark 1. If $G = \langle a, b \mid a^{2^m} = 1,\ b^{2^n} = a^{2^{m-1}},\ n \geq 2,\ m \geq 4,\ a^b = a^{1+2^{m-2}} \rangle$, i.e., G is as in Theorem 205.3(5) for $p = 2$, then all three $\mathcal{A}_1$-subgroups of G are pairwise nonisomorphic.

In Propositions 71.1–71.5 only $\mathcal{A}_2$-groups of order $> p^4$ are considered. To prove Theorem 206.1, we have to consider also $\mathcal{A}_2$-groups of order p^4; this will be done in the following remark and Lemma 206.3.

Remark 2. We claim that if all $\mathcal{A}_1$-subgroups of an $\mathcal{A}_2$-group G of order p^4 are pairwise nonisomorphic, then $G \cong \mathrm{SD}_{16}$. Note that there are exactly two nonisomorphic $\mathcal{A}_1$-groups of order p^3. Therefore, if A, B, C are pairwise distinct nonabelian groups of order p^3, then they are not pairwise nonisomorphic. By Lemma 76.6, $\alpha_1(G) \geq p$. Therefore, if $p > 2$, there are in G two different isomorphic $\mathcal{A}_1$-subgroups. Thus, $p = 2$. If G is not of maximal class, it follows from Proposition 10.17 that $G = AZ(G)$ so G contains exactly four nonabelian subgroups of order 8, and at least two of them are isomorphic, a contradiction. Thus, G is of maximal class and order 16, and now the result follows from Theorem 1.2: $G \cong \mathrm{SD}_{16}$.

Now we are ready to prove Theorem 206.1.

Proof of Theorem 206.1. By hypothesis, $|G| \geq p^4$. In view of Remark 2, it remains to consider the case when $|G| > p^4$. Let $A < G$ be minimal nonabelian such that $|A|$ is as small as possible and let $A < H \leq G$ with $|H : A| = p$. Then H is an $\mathcal{A}_2$-subgroup (check or see § 202).

Let $|H| = p^4$; then $H \cong \mathrm{SD}_{16}$, by Remark 2. Let $K < H$ be nonabelian (of order 2^3). Assume that $C_G(K) \nleq K$. If $K < L = KC_G(K)$, where L is of order 16, then L contains two isomorphic $\mathcal{A}_1$-subgroups (Remark 2), contrary to the hypothesis. Therefore, $C_G(K) < K$ so G is of maximal class (Proposition 10.17). It follows from $|G| > 2^4$ that $\alpha_1(G) \geq 2^2$ and so there are in G two isomorphic $\mathcal{A}_1$-subgroups, a contradiction. In particular, if G contains a subgroup isomorphic to SD_{16}, then $G \cong \mathrm{SD}_{16}$.

Now let $|H| > p^4$. Then, by Theorems 205.1–205.10, the $\mathcal{A}_2$-subgroup H is as in Theorem 205.3(5) with $p = 2$ so that H is a metacyclic 2-group. Thus, all $\mathcal{A}_2$-subgroups of G are metacyclic. It follows from Theorem 202.1 that G is also a metacyclic 2-group, completing the proof. □

In the proof of Theorem 206.2, we use the following

Lemma 206.3. *Let G be an $\mathcal{A}_2$-group of order p^4. If G has no p isomorphic minimal nonabelian subgroups, then it is of maximal class and one of the following holds:*

(a) $G \cong \mathrm{SD}_{16}$.

(b) $p = 3$, $\mathcal{A}_1(G) = \{M_1, M_2, M_3\}$, *where*

  (b1) $G \cong \Sigma_9$, $M_1 \cong M_3(1,1,1)$ *and* $M_2 \cong M_3 \cong M_3(2,1)$,

  (b2) $M_1 \cong M_2 \cong M_3(1,1,1)$ *and* $M_3 \cong M_2(2,1)$,

(c) $p > 3$, $\Omega_1(G) \cong M_p(1,1,1)$, *exactly* $p-1$ *nonabelian maximal subgroups of G are isomorphic to* $M_p(2,1)$.

Proof. Note that there is in Γ_1 an abelian member. Let $A < G$ be minimal nonabelian.

Assume that $C_G(A) \nleq A$. Then $G = AZ(G)$. If $Z(G)$ is noncyclic, then $G = A \times C$, where $|C| = p$, and all $p^2 > p$ minimal nonabelian subgroups of G are isomorphic to A, contrary to the hypothesis. For $Z(G)$ is cyclic, we consider two cases separately. If $p = 2$, then G contains $3 > 2$ minimal nonabelian subgroups $\cong D_8$ (Appendix 16), a contradiction. Now let $p > 2$. In this case, $d(G) = 3$ and all $p^2 - 1 > p$ members of the set $\mathcal{A}_1(G) - \{\Omega_1(G)\}$ are isomorphic to $M_p(2, 1)$ so this group does not satisfy the hypothesis.

Thus, $C_G(A) < A$. Then, by Proposition 10.17, G is of maximal class so that $|\mathcal{A}_1(G)| = p$ since G has a unique abelian subgroup, say A, of index p. If $p = 2$, then $G \cong SD_{16}$ (Theorem 1.2) is as in (a). Now let $p > 2$. Then G is nonmetacyclic since $|G'| = p^2$.

Let G be regular (Theorem 7.1 (b)). If $\exp(G) = p$, then all $\alpha_1(G) = p$ minimal nonabelian subgroups are isomorphic to $M_p(1, 1, 1)$ (Lemma 65.1), a contradiction. If $\exp(G) > p$, then $|\Omega_1(G)| = p^3$ (Theorem 7.2 (b), Corollary 36.6). Assume that $\Omega_1(G)$ is abelian. Then all p members of the set $\mathcal{A}_1(G) = \Gamma_1 - \{\Omega_1(G)\}$ are isomorphic to $M_p(2, 1)$, contrary to the hypothesis. Thus, $\Omega_1(G) \cong M_p(1, 1, 1)$. In that case, all $p - 1$ members of the set $\mathcal{A}_1(G) - \{\Omega_1(G)\}$ are isomorphic to $M_p(2, 1)$, and G is as in (c).

It remains to consider the case when G is an irregular 3-group (see Theorem 7.1 (b)) so that $\exp(G) = 3^2$ (Theorems 7.1 (b) and 9.5). If $|\Omega_1(G)| = 3^2$, then all 3 nonabelian members of the set $\mathcal{A}_1(G)$ are isomorphic to $M_3(2, 1)$, contrary to the hypothesis. Thus, $|\Omega_1(G)| \geq 3^3$. If $A \cong E_{3^3}$, then, by Exercise 9.13, $G \cong \Sigma_{3^2} \in \mathrm{Syl}_3(S_{3^2})$. In this case, G has exactly one subgroup isomorphic to $M_3(1, 1, 1)$ and exactly two subgroups isomorphic to $M_3(2, 1)$ so G is as in (b1). If $\Omega_1(G) \cong M_3(1, 1, 1)$, then exactly 2 members of the set $\mathcal{A}_1(G)$ are $\cong M_3(2, 1)$ and one is isomorphic to $\Omega_1(G) \cong M_3(1, 1, 1)$, so that G is as in (b2). It is possible that there is also G such that $M_1 \cong M_2 \cong M_3(1, 1, 1)$ and $M_2 \cong M_3(2, 1)$ (here we have not used the classification of groups of order 3^4). □

Proof of Theorem 206.2. As the case $p = 2$ has considered in Theorem 206.1, one may assume that $p > 2$. In view of Lemma 206.3, one may assume that $|G| > p^4$.

Let $H \leq G$ be an $\mathcal{A}_2$-subgroup whose order is as small as possible. Then all $\mathcal{A}_1$-subgroups of H have minimal possible order (see the proof of Theorem 206.1). It follows from Lemma 206.3 and Theorems 205.1–205.10 that H contains $p - 1$ isomorphic minimal nonabelian subgroups of order $\frac{1}{p}|H|$.

It remains to prove the second assertion of the theorem without assumption that $|G| > p^4$. Let $A \in \mathcal{A}_1(G)$ be of order p^3. If A is not normal in G, there are in G at least $|G : N_G(A)| \geq p$ subgroups isomorphic to A, a contradiction. Thus, all nonabelian subgroups of order p^3 are G-invariant. If $G - A$ has an element x of order p centralizing A, then all p^2 minimal nonabelian subgroups of $A \times \langle x \rangle$ are isomorphic, a contradiction. Thus, $C_G(A)$ is cyclic. Assume that $|G| > p^4$. Then $C_G(A) \nleq A$ (note that $|\mathrm{Aut}(A)|_p = p^3$ and p^2 does not divide $|\mathrm{Aut}(A)|$) and therefore, by Lemma 206.3, there are in $A\Omega_2(C_G(A))$ exactly $p^2 - 1 > p - 1$ minimal nonabelian subgroups $\cong M_p(2, 1)$, a contradiction. Thus, $|G| = p^4$, as was to be shown. □

Of course, Theorem 206.1 follows from Theorem 206.2.

We use in Remark 3 the following fact. If all proper subgroups of a nonsolvable group G are solvable, then $G/\Phi(G)$ is nonabelian simple. Indeed, assume that $N/\Phi(G)$ is a nontrivial normal subgroup of $G/\Phi(G)$. Then N is solvable. If $M < G$ is maximal, then MN is solvable so that $N < M$. It follows that $N \leq \Phi(G)$, contrary to the choice of N.

Remark 3. Let G be a nonnilpotent group. If all minimal nonabelian subgroups of G are normal, then G is solvable. Assume that G is a counterexample of minimal order. Then, by induction, all maximal subgroups of G are solvable so that $G/\Phi(G)$ is nonabelian simple. It follows that all proper G-invariant subgroups are contained in $\Phi(G)$. Assume that $P \in \mathrm{Syl}_p(G)$ is nonabelian. Then P is generated by minimal nonabelian subgroups (Theorem 10.28) so that $P \lhd G$ and hence, by the above, $P \leq \Phi(G)$. By the Schur–Zassenhaus theorem, $G = PM$, where M is a p'-Hall subgroup of G. Then $G = PM \leq \Phi(G)M < G$, a contradiction. Thus, all Sylow subgroups of G are abelian. In this case all minimal nonabelian subgroups of G are minimal nonnilpotent. It follows that all minimal nonnilpotent subgroups of G lie in $\Phi(G)$ which is nilpotent, a final contradiction. The same argument shows that if all minimal nonabelian subgroups of G are subnormal, then G is solvable.

Problem. Classify the p-groups without $p + 1$ isomorphic $\mathcal{A}_1$-subgroups.

§ 207 Metacyclic groups of exponent p^e with a normal cyclic subgroup of order p^e

Let G be a finite metacyclic p-group. Then there is in G a normal cyclic subgroup C such that G/C is cyclic. It is interesting to know when there exists in G a subgroup B such that $G = CB$ and $C \cap B = \{1\}$. In many cases this is not true. Let us present a simplest example. Let G be an abelian group of type (p^3, p). Then $\Phi(G) \cong C_{p^2}$ and G has exactly p cyclic subgroups of order p^2. If $C_0 < G$ is cyclic of order p^2 and $C_0 \neq \Phi(G)$, then G/C_0 is cyclic and C_0 is not complemented in G (otherwise, G will be abelian of type (p^2, p^2)). Note that C_0 is a maximal cyclic subgroup of G (indeed, if $C_0 < Z < G$, where Z is cyclic, then $C_0 = \Phi(Z) = \Phi(G)$, a contradiction).

Below G is a metacyclic group of exponent p^e. In the case when G possesses a normal cyclic subgroup A of order p^e, we prove (see Theorem 207.2) that, as a rule, there exists in G a cyclic $B < G$ such that $G = AB$ and $A \cap B = \{1\}$ (this holds always provided $p > 2$).

A metacyclic group of exponent p^e has order $\leq p^{2e}$. Indeed, there is in G a normal cyclic subgroup C such that G/C is cyclic. As $|C| \leq p^e$, $|G/C| \leq p^e$, the assertion follows.

Lemma 207.1. *If a metacyclic group G of exponent p^e contains a normal cyclic subgroup A of order p^e, then the quotient group G/A is cyclic.*

Proof. Assume that G is a counterexample of minimal order; then G is nonabelian (Exercise 1.4), G/A is noncyclic, $e > 1$, $|G| > p^3$ and G has no cyclic subgroup of index p (Theorem 1.2). As for any metacyclic group of order p^4 and exponent p^2 the conclusion holds, we get $e > 2$. Next, $R = \Omega_1(G) \cong E_{p^2}$ (Lemma 1.4 since G is not a 2-group of maximal class, and Proposition 10.19). One has $\exp(G/R) = p^{e-1}$. Write $\bar{G} = G/R$. The subgroup $\bar{A} = AR/R \cong A/(A \cap R)$ is $\bar{G}$-invariant cyclic of order $p^{e-1} = \exp(\bar{G})$, i.e., the (metacyclic) quotient group $\bar{G}$ satisfies the hypothesis with $e-1$ instead of e. Therefore, by induction, $\bar{G}/\bar{A} \cong (G/R)/(AR/R) \cong G/AR$ is cyclic. But $G/AR \cong (G/A)/(AR/A)$, Therefore, being an extension of the group AR/A of order p by the cyclic group G/AR, the noncyclic group G/A (by assumption) is abelian, and so, as a cyclic subgroup, say $B/A(\cong G/AR)$, has index p in G/A, we get $G/A = (B/A) \times (C/A)$, where $|C/A| = p$ (Introduction, Exercise 4). Since

$$|B| > |A| = p^e = \exp(G) \text{ and } |C| > |A| = \exp(G)\,,$$

the subgroups B and C are noncyclic. Note that G has no nonabelian subgroup of order p^3 (Proposition 10.19) so that $E_{p^2} \cong R \leq B \cap C = A$, a contradiction since A is cyclic. Thus, G/A is cyclic. □

Now we are ready to prove the main result of this section.

Theorem 207.2. *Suppose that a metacyclic group G of exponent p^e possesses a normal cyclic subgroup A of order p^e. Suppose, in addition, that for $p = 2$, G has no nonabelian section of order 8. Then G contains a cyclic subgroup B such that $G = AB$ and $A \cap B = \{1\}$.*

Proof. We proceed by induction on $|G|$. One may assume that G is nonabelian (otherwise, by Exercise 4 in Introduction, A is a direct factor of G; in this case, by the basic theorem on abelian groups, G/A is cyclic) and G has no cyclic subgroup of index p (Theorem 1.2). Since the assertion (under above assumption) is true for all metacyclic groups of order p^4 and exponent p^2, one may also assume that $e > 2$. Then $R = \Omega_1(G) \cong \mathrm{E}_{p^2}$ (Lemma 1.4 and Proposition 10.19). Write $\bar{G} = G/R$. In that case, $\exp(\bar{G}) = p^{e-1}$ and $AR/R = \bar{A} \lhd \bar{G}$ is cyclic of order $p^{e-1} = \exp(\bar{G})$, $A \cap R = \Omega_1(A)$ is of order p. By induction, $\bar{G} = \bar{A}\bar{B}$, where $\bar{B}$ is cyclic (here we use Lemma 207.1) and

$$\bar{A} \cap \bar{B} = \{\bar{1}\} \Rightarrow AR \cap B = R, \quad G = (AR)B = A(RB) = AB\,.$$

One has $\Omega_1(B) = R$, and so $B = B_1R$, where B_1 is cyclic of index p in B. Also $|AR| = p|A|$ and $|B| = p|B_1|$. Therefore, by the product formula,

$$(1) \qquad |G| = |(AR)B| = |(AR)(B_1R)| = \frac{|AR|\,|B_1R|}{|AR \cap B_1R|} = \frac{p|A|p|B_1|}{p^2} = |A| \cdot |B_1|$$

so that $|B_1| = |G : A| > p$ ($>$, by assumption). Next,

$$AR \cap B_1R = R \Rightarrow |A \cap B_1| \le |AR \cap B_1| = |R \cap B_1| = p$$

since B_1 is cyclic. We claim that $(\Omega_1(G) =)R < AB_1$. If this is false, then, $|\Omega_1(AB_1)| = p$, and, being noncyclic, AB_1 is a generalized quaternion group so G has a nonabelian section of order 8, contrary to the hypothesis.

If $A \cap B_1 = \{1\}$, then $G = AB_1$, by (1) and the product formula, so that B_1 is the required subgroup. Now assume that $|A \cap B_1| = p$. Then, by the assumption and (1),

$$|G| = |AB| = |A(B_1R)| = |(AB_1)R| = |AB_1| = \frac{|A|\,|B_1|}{|A \cap B_1|} = \frac{|A|\,|B_1|}{p} = \frac{1}{p}|G|\,,$$

a contradiction. Thus, $A \cap B_1 = \{1\}$ so that $G = AB_1$ is the semidirect product with kernel A and complement B_1. □

Proposition 207.3. *Suppose that G is a metacyclic group of order p^{2e} and exponent p^e. Then*

(a) *$G = AB$, where $A \lhd G$ and $B < G$ are cyclic of order p^e so that $A \cap B = \{1\}$.*
(b) *$Z(G)$ is noncyclic.*
(c) *Every noncyclic subgroup of G contains $\Omega_1(G)$.*
(d) *If G is nonabelian, then every minimal nonabelian subgroup, say H, of G has no cyclic subgroup of index p and $\Omega_2(H) = \Omega_2(G)$.*

Proof. One may assume from that G is nonabelian (see Exercise 4 in Introduction).

(a) Let a cyclic $A \lhd G$ be such that G/A is cyclic. Assume that $|A| < p^e$; then $|G/A| > p^e$. If a cyclic $B < G$ cover G/A, then $|B| > p^e = \exp(G)$, a contradiction. Thus, $|A| = p^e$.

In that case, any cyclic subgroup of G, say B again, that cover G/A, has order p^e; then $G = AB$ and $A \cap B = \{1\}$, completing the proof of (a).

(b) Note that $|\mathrm{Aut}(A)| = (p-1)p^{e-1}$ (where A is as in (a)). Therefore, $C_G(A) > \Omega_1(B) \Rightarrow \Omega_1(A) \times \Omega_1(B) \le Z(G)$ so $Z(G)$ is noncyclic.

(c) Let $U < G$ be noncyclic. If $R = \Omega_1(G) \not\le U$, then $|\Omega_1(U)| = p$ so that U is a generalized quaternion group (Proposition 1.3). Then, by Proposition 10.19, G is a 2-group of maximal class, hence $|G| = 2^{e+1} < 2^{2e}$ since $e > 1$, a contradiction.

(d) Let $H \le G$ be minimal nonabelian. As $\Omega_1(H) = \Omega_1(G) \le Z(H)$, by (c) and (b), H has no cyclic subgroup of index p (otherwise, H will be abelian). Then $|\Omega_2(H)| = p^4$ (here we apply twice Lemma 1.4 to G and $G/\Omega_1(G)$, a metacyclic group of order $p^{2(e-1)}$ and exponent p^{e-1}, and part (c) of the theorem to noncyclic $H/\Omega_1(G)$) so that $\Omega_2(H) = \Omega_2(G)$ (compare orders). □

Exercise 1. Suppose that G is a metacyclic minimal nonabelian group of order p^{2e} and exponent p^e. If $A < G$ is cyclic of order p^e, then G contains a cyclic subgroup B of order p^e such that $G = AB$; then $A \cap B = \{1\}$.

Solution. In view of Proposition 207.3 (a), one may assume that A is not normal in G. By the same proposition, there is in G a normal cyclic subgroup C of order p^e such that G/C is cyclic. Note that $G' = \Omega_1(C)$. If $A \cap C = \{1\}$, then $G = AC$, by the product formula. If $A \cap C > \{1\}$, then $G' = \Omega_1(C) \le A \Rightarrow A \triangleleft G$, contrary to the choice of A.

As in § 124, if G is metacyclic, then

$$w(G) = \max\{i \mid |\Omega_i(G)| = p^{2i}\}, \quad R(G) = \Omega_{w(G)}(G).$$

By Lemma 1.4, $G/R(G)$ is either cyclic or a 2-group of maximal class.

Exercise 2. Let G be a metacyclic minimal nonabelian group of exponent $p^e > p^{w(G)} > 1$; then $G/R(G)$ is cyclic since $G' < R(G)$. Then for any cyclic $A < G$ of order p^e the group G contains a cyclic subgroup B satisfying $G = AB$ and $A \cap B = \{1\}$ (in that case, $|B| = p^{w(G)}$).

Solution. The subgroup $A \cap R(G) = A_1$ is cyclic of order $p^{w(G)}$, by the product formula, so that $G = AR(G)$ since $|G : A| = p^{w(G)}$. By Exercise 1, there is in $R(G)$ a cyclic subgroup B of order $p^{w(G)}$ such that $R(G) = A_1B$ and $\{1\} = A_1 \cap B = A \cap B$. Therefore, $G = AR(G) = A(A_1B) = AB$.

Exercise 3. Let A be a cyclic subgroup of order p^e of a metacyclic minimal nonabelian group G of exponent p^e. Suppose, in addition, that if $p = 2$, then G has no nonabelian sections of order 8. Then the group G contains a cyclic subgroup B satisfying $G = AB$ and $A \cap B = \{1\}$.

Solution. If $R(G) = \{1\}$, then G is cyclic, by hypothesis and Lemma 1.4, a contradiction. Now let $R(G) > \{1\}$. Then, again by Lemma 1.4, the quotient group $G/R(G)$ is cyclic. Now the result follows from Exercise 2.

Exercise 4. Let G be a metacyclic 2-group such that $w(G) = 1$ and $G/R(G)$ is of maximal class. Then G/G' is abelian of type $(4, 2)$.

Solution. One has $|G : R(G)G'| = 4$. As $|G' \cap R(G)| = 2$, we get $|G : G'| = 8$ (here we use the product formula), and the result follows since the noncyclic G/G' is abelian metacyclic.

Exercise 5. Let G be a metacyclic 2-group such that $G/R(G)$ is of maximal class. Is it true that $|G' \cap R(G)| = 2^{w(G)}$?

Hint. One has $|G : G'R(G)| = 4$ so the cyclic subgroup $G' \nleq R(G)$. Next, $|\Omega_w(G')| = 2^w$.

Exercise 6. If any minimal nonabelian subgroup of a nonabelian metacyclic p-group G has a cyclic subgroup of index p, then G has an abelian subgroup of index p.

Solution. If $\Omega_1(G) \ncong \mathrm{E}_{p^2}$, the result follows from Lemma 1.4. If $R = \Omega_1(G) \cong \mathrm{E}_{p^2}$, then $|G : \mathrm{C}_G(R)| \leq p$. If $H \leq \mathrm{C}_G(R)$ is minimal nonabelian, then H has no cyclic subgroup of index p, contrary to the hypothesis. It follows that $H \nleq \mathrm{C}_G(R)$ so that $\mathrm{C}_G(R) \in \Gamma_1$ is abelian since $\mathrm{C}_G(R)$ has no minimal nonabelian subgroups.

Exercise 7. (a) If a regular p-group G has exponent p^e and $|\Omega_1(G)| = p^k$, then $|G| \leq p^{ke}$.

(b) Let a p-group G of exponent p^e has a normal absolutely regular subgroup A such that G/A is absolutely regular. Then $|G| \leq p^{2e(p-1)}$.

As I. D. Macdonald [Macd14] has proved, there exists a p-group G of exponent $p^e > p$ such that $\Omega_e^*(G) \leq \Phi(G)$.

Remark 1. Let us prove that if a p-group G is metacyclic, then $\exp(\Phi(G)) < \exp(G)$. One may assume that G is nonabelian and $R(G) > \{1\}$ (if $R(G) = \{1\}$, the result follows from Lemma 1.4). If $G = R(G)$, then $\Phi(G) = \mho_1(G)$ has order $p^{2(e-1)}$ and exponent p^{e-1}, completing this case. Now let $R(G) < G$. Write $\Phi/R(G) = \Phi(G/R(G))$; then $\Phi \geq \Phi(G)$. Now $\exp(\Phi/R(G)) < \exp(G/R(G)) \Rightarrow \exp(\Phi(G)) < \exp(G)$.

Remark 2. Suppose that a nonabelian p-group G of exponent $p^e > p$ is such that $|G/\mho_i(G)| = p^{2i}$ for all $i \leq e$. By hypothesis, $\mathrm{d}(G) = 2$. If $p > 2$, then G is metacyclic (Theorem 9.11). In this case, $G = R(G)$, and every such G satisfies the hypothesis. Now let $p = 2$. (i) Suppose that $e = 2$; then $|G| = 2^4$. If G is minimal nonabelian, then it is either metacyclic or $\cong \mathrm{M}_2(2, 1, 1)$ (Lemma 65.1). Now suppose that G is not minimal nonabelian. Then G contains a proper minimal nonabelian subgroup A of order 2^3. By Theorem 12.12 (a), G is a 2-group of maximal class; then $|G/\mho_2(G)| = 2^3 < 2^4$ (see Proposition 1.6, Theorem 1.2), contrary to the hypothesis. (ii) Now let $e > 2$. Then, $G/\mho_2(G)$, being of order 2^4, is either metacyclic or $|\Omega_1(G/\mho_2(G))| = 8$, by (i). By Supplement to Corollary 36.6, G is metacyclic in the first case and then $G = R(G)$. In the second case, classification of such G is a difficult problem.

Remark 3. Let G be a p-group of exponent $p^e > p$ such that $|\Omega_i(G)| = p^{2i}$ for all $i \leq e$. It is easy to show that if $e = 2$, then G is either metacyclic, or $Q_8 \times C_2$, or the minimal nonmetacyclic 3-group of maximal class. Now let $e > 2$. Let us prove that then G is metacyclic. Note that if X is minimal nonmetacyclic p-group, then $\exp(X) \leq p^2$ (see Theorems 66.1, 69.1). If $\Omega_2(G)$ is of maximal class and order 3^4, then $G = \Omega_2(G)$ (Theorem 13.19 (a)) and so, in the case under consideration, $|G| = 3^4$, contrary to the assumption. Now let $p = 2$. If $\Omega_2(G)$ is metacyclic, then G has no minimal nonmetacyclic subgroups (Theorems 66.1 and 69.1) so it is metacyclic (in that case, $G = R(G)$). Now assume that $\Omega_2(G) \cong Q_8 \times C_2$. Then G is the group of order $2^5 \neq \exp(G)^2$, by Theorem 52.1 (e), which is a contradiction. Thus, G is metacyclic if $p = 2$ and $e > 2$.

Remark 4. Let G be a metacyclic group of exponent p^e and suppose that $\Omega_e^*(G) = \langle x \in G \mid o(x) = p^e \rangle < G$. Then $G/R(G)$ is a 2-group of maximal class nonisomorphic to Q_8. Indeed, if $G = R(G)$, then $\Omega_{e-1}(G) = \Phi(G)$ so $\Omega_e^*(G) = \langle G - \Phi(G) \rangle = G$. If $G/R(G) > \{1\}$ is cyclic, then $|G : \Omega_{e-1}(G)| = p$, and so $\Omega_e^*(G) = \langle G - \Omega_{e-1}(G) \rangle = G$. Now let $G/R(G)$ be a 2-group of maximal class and let $T/R(G)$ be cyclic of index 2 in $G/R(G)$. If $G/R(G) \ncong Q_8$, then it is easily seen that $\Omega_e^*(G) = T < G$. If $G/R(G) \cong Q_8$, then $\Omega_e^*(G) = G$ since, in the case under consideration, all elements of the set $G - \Phi(G)$ have order 2^e.

Problem 1. Classify the metacyclic p-groups containing a minimal nonabelian subgroup of index p. (This is solved in [QYXA, ALQZ, QXA, AHZ, QZGA].)

Problem 2. Let C be a normal cyclic subgroup of a noncyclic p-group G of exponent p^e such that G/C is cyclic. State a necessary and sufficient condition guaranteeing the existence of $B < G$ satisfying $G = CB$ and $C \cap B = \{1\}$.

Problem 3. Classify the 2-groups of exponent 2^e such that $|G/\mho_i(G)| = 2^{2i}$ for $i \leq 2$.

Problem 4. Classify the nonabelian metacyclic p-groups all of whose minimal nonabelian subgroups have no cyclic subgroups of index p.

§ 208 Non-Dedekindian p-groups all of whose nonnormal maximal cyclic subgroups are maximal abelian

The purpose of this section is to determine completely the structure of the title p-groups (= Problem 2222).

Theorem 208.1 (Janko). *Let G be a non-Dedekindian p-group all of whose nonnormal maximal cyclic subgroups are maximal abelian. Then $p = 2$ and $G \cong Q_{2^n}$ with $n \geq 4$.*

Conversely, each such group $G \cong Q_{2^n}$ with $n \geq 4$ satisfies the assumptions of the theorem.

Proof. Let G be a non-Dedekindian p-group all of whose nonnormal maximal cyclic subgroups are maximal abelian.

First, assume that G does not have a normal abelian subgroup of type (p, p). Since G is non-Dedekindian, Lemma 1.4 implies that $p = 2$ and G is of maximal class distinct from Q_8. If G has a noncentral involution i, then $\langle i \rangle$ is a nonnormal maximal cyclic subgroup in G but $\langle i \rangle$ is not maximal abelian, a contradiction. It follows that G has only one involution (which is central) and so $G \cong Q_{2^n}$, $n \geq 4$.

Now suppose that G has a normal abelian subgroup U of type (p, p). Let S be a nonnormal maximal cyclic subgroup in G. Since S is a maximal abelian subgroup in G, we have $Z(G) < S$ is cyclic and $Z(G) \cap U = \langle z \rangle$ is of order p. We set

$$U = \langle u, z \rangle \cong E_{p^2} \text{ and } G_0 = C_G(U) \text{ so that } |G : G_0| = p .$$

Let T be a maximal cyclic subgroup in G which contains $\langle u \rangle$. Because $\langle u \rangle \nleq Z(G)$, T is not normal in G. By hypothesis, T must be maximal abelian in G. This implies $T \nleq G_0$. But then

$$C_G(u) \geq \langle G_0, T \rangle = G \text{ and so } u \in Z(G) ,$$

a contradiction.

Conversely, it is obvious that the groups Q_{2^n}, $n \geq 4$, satisfy the assumptions of our theorem. □

The following proof of Theorem 208.1 which is due to the first author, is not so elementary as the previous one.

The second proof (Berkovich). The center $Z(G)$ is contained in any nonnormal maximal cyclic subgroup of G so it is cyclic.

Assume that $p = 2$ and all minimal nonabelian subgroups of G are isomorphic to Q_8. Then $G = Q_{2^m} \times E$, where $m > 3$ since G is non-Dedekindian and $\exp(E) \leq 2$ (Corollary A.17.3). Since $Z(G)$ is cyclic, it follows that $E = \{1\}$ so that $G \cong Q_{2^m}$, $m > 3$.

Now assume that G is not generalized quaternion. Then it contains a minimal nonabelian subgroup $A \not\cong Q_8$ so A is non-Dedekindian. In that case, A contains a nonnormal cyclic subgroup U. Let $U \leq S < G$, where S is a maximal cyclic subgroup of G; then S is not normal in G (otherwise, $U \lhd G$). By Lemma 65.1, A contains a cyclic subgroup V such that $U \cap V = \{1\}$. Let $V \leq T < G$, where T is a maximal cyclic subgroup of G. One has $Z(G) < S$ since S is maximal abelian, by hypothesis. The subgroup T is normal in G (otherwise, $Z(G) < T$, a contradiction since $Z(G) > \{1\}$ and $S \cap T = \{1\}$). Then $\Omega_1(T) \leq Z(G)$, a final contradiction since $Z(G) < S$ and $S \cap T = \{1\}$. □

The third proof (Berkovich). As above, $Z(G)$ is cyclic. Assume that $G \not\cong Q_{2^m}$. Then there are in G two distinct subgroups U, V of order p. At least one of the subgroups U, V, say U, is nonnormal in G. Let $U \leq S < G$. Then S is nonnormal in G. It follows that $Z(G) < S \Rightarrow U \lhd G$, a contradiction. □

Problem 1. Classify the non-Dedekindian p-groups G of exponent $> p^2$ all of whose nonnormal cyclic subgroups have order p^2. (All generalized quaternion groups of order > 8 satisfy the above condition.)

Problem 2. Classify the non-Dedekindian p-groups G all of whose nonnormal maximal cyclic subgroups of order $> p$ are maximal abelian.

§ 209 p-groups with many minimal nonabelian subgroups, 3

It is clear that minimal nonabelian subgroups in a nonabelian p-group G play an important role. In particular, if there is a relatively big number of minimal nonabelian subgroups in G, then this fact in many cases determines the structure of G.

Let G be a nonabelian p-group and let A be a maximal normal abelian subgroup in G. Then a very useful Lemma 57.1 asserts that for any $g \in G - A$, there is $a \in A - Z(G)$ such that $\langle g, a\rangle$ is minimal nonabelian. It is interesting to see what can we say about the structure of G if for any two noncommuting elements $g \in G - A$ and $a \in A - Z(G)$, the subgroup $\langle g, a\rangle$ is minimal nonabelian (= Problem 1980 (ii)). We can show that in this case G has two important properties (Theorem 209.1 (i)) and we prove that in case $p = 2$ these two properties in fact characterize such a group G (Theorem 209.1 (ii)).

Theorem 209.1 (Janko). *Let G be a nonabelian p-group and let A be a maximal normal abelian subgroup of G.*

(i) *If for any two noncommuting elements $g \in G - A$ and $a \in A - Z(G)$, the subgroup $\langle g, a\rangle$ is minimal nonabelian, then the factor-groups G/A and $A/Z(G)$ are of exponent p.*

(ii) *Conversely, if G/A and $A/Z(G)$ are of exponent p and $p = 2$, then any two noncommuting elements $g \in G - A$ and $a \in A - Z(G)$ generate a minimal nonabelian subgroup $\langle g, a\rangle$.*

Proof. Let G be a nonabelian p-group and let A be a maximal normal abelian subgroup of G.

(i) Suppose that for any two noncommuting elements $g \in G - A$ and $a \in A - Z(G)$, the subgroup $\langle g, a\rangle$ is minimal nonabelian.

(i1) Let a be a fixed element in $A - Z(G)$ and x any element in $G - A$. Then either $[a, x] = 1$ or $[a, x] \neq 1$ and then $\langle a, x\rangle$ is minimal nonabelian and this gives $[a^p, x] = 1$. Hence, in any case, $[a^p, x] = 1$ for all $x \in G - A$ and this implies $a^p \in Z(G)$. We have proved that $A/Z(G)$ is elementary abelian.

(i2) Let g be a fixed element in $G - A$ and y be any element in A. Then either $[g, y] = 1$ or $[g, y] \neq 1$ and then $\langle g, y\rangle$ is minimal nonabelian so that $[g^p, y] = 1$. Hence, in any case, g^p centralizes each element $y \in A$ which gives $g^p \in A$. We have proved that $\exp(G/A) = p$.

(ii) Assume that $p = 2$ and both G/A and $A/Z(G)$ are elementary abelian. Let $a \in A - Z(G)$ and $g \in G - A$ be such that $[a, g] \neq 1$. Since G is metabelian, we may use Lemma 188.2 and we get: (1) $1 = [a, g^2] = [a, g]^2[a, g, g]$, (2) $1 = [a^2, g] = [a, g]^2[a, g, a]$. Since $[a, g] \in A$, we have $[a, g, a] = 1$ and so (2) implies $[a, g]^2 = 1$. Then (1) gives $[a, g, g] = 1$. It follows that $1 \neq [a, g]$ commutes with a and g and so $\langle a, g\rangle' = \langle [a, g]\rangle$, where $o([a, g]) = 2$. By Lemma 65.2 (a), $\langle a, g\rangle$ is minimal nonabelian and our theorem is proved. □

Theorem 209.2 (= Problem 1841). *A nonabelian p-group G has the property that whenever $A \trianglelefteq G$ is maximal abelian and $x \in G - A$, then $C_A(x)$ is cyclic if and only if $Z(G)$ is cyclic and G has an abelian maximal subgroup.*

Proof. Suppose that whenever $A \trianglelefteq G$ is maximal abelian and $x \in G - A$, then $C_A(x)$ is cyclic. If G has no normal abelian subgroup of type (p, p), then Lemma 1.4 implies that G is a 2-group of maximal class and so $Z(G)$ is cyclic and G has an abelian maximal subgroup.

Now assume that G possesses a normal abelian subgroup U of type (p, p). Let A be a maximal normal abelian subgroup of G containing U. Set $G_0 = C_G(U)$ so that $A \le G_0$ and $|G : G_0| \le p$. By hypothesis, $A = G_0$ (indeed, if $A < G_0$ and $x \in G - A$, then $C_A(x) \ge U$ is noncyclic) and so $|G : A| = |G : G_0| = p$, i.e., A is an abelian maximal subgroup of G.

Conversely, let G be a nonabelian p-group with the cyclic center $Z(G)$ and an abelian maximal subgroup A. Then for each $x \in G - A$, the centralizer $C_A(x) = Z(G)$ is cyclic. Let A_0 be any other maximal abelian subgroup of G. Then $|A_0 : (A_0 \cap A)| = p$, by the product formula. Therefore, there is $g \in A_0 - (A_0 \cap A)$ such that $A_0 = Z(G)\langle g\rangle$; in that case, $g \in G - A$ and $g^p \in Z(G)$. Let $y \in G - A_0$. If $y \in A$, then $C_G(y) = A$ so that $C_{A_0}(y) = C_G(y) \cap A_0 = A \cap A_0 = Z(G)$. Now let $y \in G - A$. Then $C_G(y) = \langle y, Z(G)\rangle$ and $y^p \in Z(G)$. It follows that $|C_G(y)| = p|Z(G)|$. Therefore, $C_{A_0}(y) = Z(G)$ since $y \notin C_{A_0}(y)$. □

Problem 1. Classify the nonabelian p-groups G containing a maximal abelian subgroup A such that, whenever $a \in A$ and $b \in G - C_G(a)$, then the subgroup $\langle a, b\rangle$ is of class 2.

Problem 2. Classify the nonabelian p-groups G containing a minimal nonabelian subgroup A such that, whenever $a \in A$ and $b \in G - C_G(a)$, then the subgroup $\langle a, b\rangle$ is minimal nonabelian.

Problem 3. Classify the nonabelian p-groups G containing a maximal subgroup H such that, whenever $a \in H$ and $b \in G - C_G(a)$, then the subgroup $\langle a, b\rangle$ is (i) minimal nonabelian (ii) metacyclic, (iii) either minimal nonabelian or metacyclic.

Problem 4. Classify the nonabelian p-groups G containing a maximal abelian subgroup A such that, whenever $a \in A$, there is $b \in G - A$ such that $\langle a, b\rangle$ is (i) either $\mathcal{A}_1$- or $\mathcal{A}_2$-subgroup, (ii) of class 2, (iii) metacyclic, (iv) nonabelian minimal nonmetacyclic.

§ 210 A generalization of Dedekindian groups

A group G is said to be *Dedekindian* if all its subgroups are normal. In this case, if G is nonabelian, then $G = P \times A$, where A is abelian of odd order, $P = Q \times E \in \mathrm{Syl}_2(G)$, where $Q \cong \mathrm{Q}_8$ and $\exp(E) \leq 2$ (see Theorem 1.20). It is natural to study a group in which only some of its subgroups are normal. For example, Passman [Pas] has classified the non-Dedekindian p-groups all of whose nonnormal subgroups have order p (see Theorem 1.25). Continuing Passman's research, the p-groups in which all noncyclic subgroups are normal were classified in § 16 (note that Passman [Pas] himself has classified such groups of order $> 2^6$). In both these cases, all "large" subgroups of p-groups are normal. Below we study the groups G all of whose subgroups of "small" orders are normal.

In this section we describe the groups G satisfying the following condition:

$(*)$ Whenever $H < G$ and $|H|^2 < |G|$, then $H \lhd G$.

Such G we call a *$(*)$-group*. Identity group is also considered as a $(*)$-group. Subgroups of $(*)$-groups are $(*)$-groups. Indeed, let $M < G$ and let $H < M$ be such that $|H|^2 < |M|$. Then $|H|^2 < |G|$ so that $H \lhd G \Rightarrow H \lhd M$. As follows from Theorem 210.1, all proper subgroups of $(*)$-groups are Dedekindian. The nonnilpotent group $G = B \cdot A$, where $A \lhd G$ is of order 3 and B is cyclic of order 4, is a $(*)$-group. The metacyclic group of order p^4 and exponent p^2 is a $(*)$-group but $M_p(m, 1) \cong M_{p^{m+1}}$ (see below) is not. Moreover, all non-Dedekindian p-groups of Passman's Theorem 1.25 are not $(*)$-groups. Note that quotient groups of $(*)$-groups are not necessarily $(*)$-groups (indeed, $Q \cong \mathrm{Q}_{16}$ is a $(*)$-group but the quotient group $Q/Z(Q) \cong \mathrm{D}_8$ is not).

Example 1. Let $G = P \cdot A$ (semidirect product with kernel A and complement P), where $P = \langle d \mid d^{16} = 1 \rangle$, $A = \langle a \mid a^9 = 1 \rangle$, $\mathrm{C}_A(d) = \{1\}$ and $|G : \mathrm{C}_G(A)| = 2$. Then G is a nonnilpotent $(*)$-group.

Obviously, one can consider property $(*)$ for subgroups H satisfying some additional conditions (e.g., for H is abelian, noncyclic abelian, regular, metacyclic).

1°. In this subsection, we consider the $(*)$-groups of prime power order.

We begin with the following

Definition 1. Let X be a minimal nonabelian p-group. If a nonmetacyclic group X is such that X/X' is abelian of type (p^m, p^n), then it is said to be an $\mathrm{M}_p(m, n, 1)$-group. A metacyclic group X is said to be $\mathrm{M}_p(m, n)$-group if $X = B \cdot A$, where $A \cap B = \{1\}$, $A \lhd G$, $A \cong \mathrm{C}_{p^m}$ and $B \cong \mathrm{C}_{p^n}$.

If G is minimal nonabelian p-group, then $G \in \{M_p(m, n, 1), M_p(m, n), \mathrm{Q}_8\}$ (see Exercise 1.8a or Lemma 65.1). Note that $M_p(m, n, 1)$-group contains nonnormal cyclic subgroups of orders p^m and p^n.

In Theorem 210.1, the $(*)$-groups of prime power order are classified up to isomorphism.

Theorem 210.1. *If a p-group G is a $(*)$-group, then G is either Dedekindian or $G \in \{Q_{16}, M_p(m,m),\ m > 1\}$.*

Proof. It is easy to check that groups Q_{16} and $M_p(m,m)$ are $(*)$-groups. For example, if $G = M_p(m,m)$ and $|H|^2 < |G| = p^{2m}$, then $|H| < p^m$ so that $H < \Omega_{m-1}(G)) = Z(G)$ so that $H \triangleleft G$.

Assume that the $(*)$-group G is non-Dedekindian.

(i) Let G be minimal nonabelian. If a nonmetacyclic $G = M_p(m,n,1)$, where, for definiteness, $m \le n$, then there is in G a nonnormal cyclic subgroup C of order p^m. It follows from $|C|^2 = p^{2m} < p^{m+n+1} = |G|$ that G is not a $(*)$-group, a contradiction. Now let

$$G = \langle a, b \mid o(a) = p^m,\ o(b) = p^n,\ a^b = a^{1+p^{m-1}} \rangle \cong M_p(m,n)$$

be metacyclic. Since there is in G a nonnormal cyclic subgroup of order p^n, then $p^{2n} \ge |G| = p^{m+n} \Rightarrow n \ge m$. If $n > m$, then the cyclic subgroup $\langle ab^{p^{n-m}} \rangle$ of order p^m is not b-invariant so nonnormal in G. Since, in the case under consideration, $p^{2m} < p^{m+n} = |G|$, G is not a $(*)$-group. If $n = m$, then $G = M_p(m,m)$ is a $(*)$-group (see the first paragraph of the proof). Thus, if G is a non-Dedekindian minimal nonabelian $(*)$-group, then $G \cong M_p(m,m)$, $m > 1$, proving the theorem for minimal nonabelian p-groups.

It follows from (i) that if a p-group G is minimal nonabelian, $H < G$ and $|H|^2 \le |G| \Rightarrow H \triangleleft G$, then $G \cong Q_8$. Indeed, the group $M_p(m,m)$ does not satisfy that condition.

(ii) Assume that G is not minimal nonabelian and let $H < G$ be minimal nonabelian. Take $A < H$ such that $|A|^2 \le |H|$. Then $|A|^2 \le |H| < |G|$ so that A is G-invariant, by $(*)$, and hence $A \triangleleft H$. By the previous paragraph, $H \cong Q_8$. Thus, $p = 2$ and all minimal nonabelian subgroups of G are isomorphic to Q_8. In that case, $G = Q \times E$, where $Q \cong Q_{2^n}$ and $\exp(E) \le 2$ (Corollary A.17.3 or Theorem 90.1). Since, by assumption, G is non-Dedekindian, $n > 3$, and so there is in G a nonnormal subgroup of order 4 so that $|G| \le 4^2 \Rightarrow E = \{1\}$ and therefore $G \cong Q_{16}$. Thus, if a 2-group G is not minimal nonabelian, then either $G = Q_8 \times E$ ($\exp(E) = 2$) is Dedekindian or $G \cong Q_{16}$. □

Let G be a nonabelian p-group. It follows from Corollary A.17.3 that if all subgroups of all minimal nonabelian subgroups of G are G-invariant, then G is Dedekindian.

Exercise 1. Classify the minimal nonabelian p-groups G such that, whenever $H < G$ and $p|H|^2 < |G|$, then $H \triangleleft G$.

Solution. One may assume that $G \not\cong Q_8$. Let $G = M_p(m,n,1)$, where $m \le n$, As G has a nonnormal cyclic subgroup of order p^m, one obtains $p^{2m+1} \ge p^{m+n+1} \Rightarrow m \ge n$, and so $m = n$. The group $G = M_p(m,m,1)$ satisfies the hypothesis since $\Omega_{m-1}(G) = Z(G)$. Now let $G = M_p(m,n)$ is metacyclic and

$$G = \langle a, b \mid o(a) = p^m,\ o(b) = p^n,\ a^b = a^{1+p^{m-1}} \rangle .$$

The group $M_p(m, m)$ satisfies the hypothesis. If $m < n$, then the subgroup $\langle ab^{p^{n-m}}\rangle$ of order p^m is not G-invariant. It follows that $p^{2m+1} \geq p^{m+n} \Rightarrow m \geq n-1$ so that $m = n-1$. The group $M_p(m, m+1)$ satisfies the hypothesis since all its subgroups of order $< p^m$ are central. Now let $m \geq n+1$. Then $p^{2n+1} \geq p^{m+n} \Rightarrow n+1 \geq m \Rightarrow m = n+1$. It is easy to check that $M_p(m, m-1)$ satisfies the hypothesis since all its nonnormal subgroups have order p^{m-1} and $p^{2(m-1)+1} = p^{2m-1} = |G|$ and all its subgroups of orders $< p^{m-1}$ are central.

2°. In this subsection, we consider the nonnilpotent $(*)$-groups. We use the following

Lemma 210.2 (= Theorem A.22.1). *Let H be a minimal nonnilpotent group. Then $|H| = p^a q^\beta$, where $p \neq q$ are primes. Let $P_0 \in \mathrm{Syl}_p(H)$, $Q \in \mathrm{Syl}_q(H)$.*

(a) *One of subgroups P_0, Q (say P_0) coincides with H'. Let b be the order of p (mod q).*
(b) *If P_0 is abelian, then $P_0 \cap Z(H) = \{1\}$, P_0 is a minimal normal subgroup of H and $P \cong E_{p^b}$.*
(c) *Q is cyclic and $|Q : (Q \cap Z(H))| = q$.*
(d) *If P_0 is nonabelian, then P_0 is special and $|P_0/P_0'| = p^b$ (so that, if $p < q$, then $b > 1$), $P_0 \cap Z(H) = Z(P_0) = P_0'$, P_0/P_0' is minimal normal subgroup of H/P_0', P_0' is elementary abelian. If $p > 2$, then* $\exp(P_0) = p$.

In this subsection, we suppose that the order of a $(*)$-group G is not a prime power. If G is nilpotent, then G must be Dedekindian. Indeed, let a non-Dedekindian nilpotent $(*)$-group $G = P \times T$, where $P \in \mathrm{Syl}_p(G)$ is non-Dedekindian, and $T > \{1\}$. As we know, P is a $(*)$-group so that $P \in \{M_p(m, m), Q_{16}\}$, by Theorem 210.1. In the first case, P contains a nonnormal cyclic subgroup, say C, of order p^m. We have $|C|^2 = |P| < |G|$ since $T > \{1\}$, a contradiction. In the second case, P contains a nonnormal subgroup of order 4 and $4^2 < |G|$, a contradiction.

In what follows, we assume that the $(*)$-group G is nonnilpotent. Then G contains a minimal nonnilpotent subgroup $H = Q \cdot P_0$ as in Lemma 210.2. Assume that $|P_0| > p$. Then all subgroups of P_0 of order p are G-invariant since $|G| > p^2$ so that P_0 is nonabelian and $\exp(P_0) > p$ (Lemma 210.2 (b)). By Lemma 210.2 (d), $p = 2$; then $|G| \geq 2^3 \cdot 3 = 24$. In this case, G has a nonnormal cyclic subgroup of order 4 (Lemma 210.2 (d)), which implies $16 = 4^2 \geq |G|$, a contradiction. Thus, $|P_0| = p$. In this case, $p \equiv 1 \pmod q$ so that $p > q$. Next, $|Q|^2 > |G| = |Q|\,|P_0)| \Rightarrow |Q| > |P_0| = p$. Thus, if $H \leq G$ be minimal nonnilpotent, then, for some distinct primes p, q, $H = Q \cdot P_0$, where $P_0 = H' \in \mathrm{Syl}_p(H)$ is of order p, $Q \in \mathrm{Syl}_q(G)$ is cyclic, $p \equiv 1 \pmod q$ and $|Q| > p$.

Theorem 210.3. *Let G be a nonnilpotent $(*)$-group without a nontrivial nilpotent direct factor. Then $G = P \cdot A$, a semidirect product with kernel A and a cyclic complement $P = \langle d\rangle \in \mathrm{Syl}_p(G)$. Set $|C_P(A)| = p^s$. Then*

(a) $C_A(d) = \{1\}$.

(b) $p^{2(s+1)} > |G|$.

(c) *All subgroups of A are G-invariant and A is abelian.*

Conversely, any group $G = P \cdot A$, satisfying conditions (a), (b) and (c), is a $(*)$-group.

Proof. Let $|G| = \prod_{i=1}^{k} p_i^{a_i}$ be a standard decomposition, $P_i \in \mathrm{Syl}_{p_i}(G)$ and $|P_1| < |P_2| < \cdots < |P_k|$. For all $i < k$, one has $P_i \triangleleft G$ since $|P_i|^2 < |P_i|\,|P_{i+1}| \le |G|$. Write

$$A = P_1 \times \cdots \times P_{k-1}\,, \quad P = P_k\,, \quad p = p_k\,.$$

Then $G = P \cdot A$, semidirect product with nilpotent kernel $A > \{1\}$ and a non-G-invariant complement $P \in \mathrm{Syl}_p(G)$.

(i) We claim that the subgroup P is abelian. Let $R < P$ be such that $|R|^2 \le |P| < |P|\,|A| = |G|$ ($<$ since $|A| > 1$) so that, by $(*)$, $R \triangleleft G$, and therefore $R \triangleleft P$. By Theorem 210.1, P is Dedekindian. Clearly, $|G| = |P|\,|A| \ge 2^3 \cdot 3 = 24$. Assume that P is nonabelian. Then $P = P_0 \times E_0$, where $P_0 \cong Q_8$ and $\exp(E_0) \le 2$ (Theorem 1.20). As $4^2 < 24 \le |P|\,|A| = |G|$, it follows that all subgroups of P of order ≤ 4 are G-invariant. Then $P = \Omega_2(P) \triangleleft G$, a contradiction. Thus, P is abelian.

(ii) By $(*)$, $|P|^2 \ge |G| = |P|\,|A| \Rightarrow |P| > |A|$ since $|P| \ne |A|$. In particular, $|A|^2 < |A|\,|P| = |G|$. It follows that if $A_0 < A$, then $|A_0|^2 < |A|^2 < |G|$, and we conclude that $A_0 \triangleleft G$, by $(*)$. Thus, all subgroups of A are G-invariant, proving the first part of (c).

(iii) We claim that A is abelian. Assume that this is false. Then, by (iii), $A = (Q \times E) \times P_2 \times \cdots \times P_{k-1}$ is nonabelian Dedekindian, where $Q \cong Q_8$, $\exp(E) \le 2$ (here $Q \times E = P_1$). As $|P| \ge 3$ is odd (indeed, $|A|$ is even), P centralizes all (P-invariant) cyclic subgroups of A of order 4 (indeed, $\mathrm{Aut}(C_4) \cong C_2$) so that P centralizes $\Omega_2(P_1) = P_1$ since cyclic subgroups of order 4 generate P_1). Then $P_1 > \{1\}$ is a nilpotent direct factor of G, a contradiction. Thus, A is abelian, completing the proof of (c).

(iv) We claim that the subgroup P is cyclic. Assume that this is false. Then, by the basic theorem on abelian groups, $P = Z_1 \times \cdots \times Z_t$, where all Z_i are cyclic; by assumption, $t > 1$. Let $Z_1 \ge \cdots \ge Z_t$. By $(*)$, the subgroups $Z_2, \ldots, Z_t$ are G-invariant (moreover, these subgroups are central since they centralize A and P is abelian). In that case, $G = (Z_1 A) \times (Z_2 \times \cdots \times Z_t)$, where $Z_2 \times \cdots \times Z_t > \{1\}$ is an abelian so nilpotent direct factor of G, contrary to the hypothesis.

(v) By (iii) and the basic theorem on abelian groups, $A = L_1 \times \cdots \times L_n$, where all cyclic subgroups L_i of prime power orders are G-invariant, by (ii).

(vi) If $P = \langle d \rangle$ (see (iv)), then $C_A(d) = \{1\}$. Indeed, let $U = C_A(d)$. Then $U = A \cap Z(G)$, so, by Fitting's Lemma [BZ, Corollary 1.18] U is an abelian direct factor of G, and we conclude that $U = \{1\}$. Assertion (a) of the theorem is established.

(vii) As $|C_P(A)| = p^s$, there is in P a non-G-invariant subgroup K of order p^{s+1}; then $p^{2(s+1)} = |K|^2 > |G|$ ($>$ since $A > \{1\}$ is a p'-subgroup), proving (b).

Now let our nonnilpotent group $G = P \cdot A$ satisfy conditions (a–c). It is easy to see that G is a $(*)$-group. Indeed, let $L < G$ be non-G-invariant. Then $p^{s+1} \mid |L|$ so that

$|L|^2 > |G|$, by (b). If p^{s+1} does not divide $M < G$, then M is G-invariant abelian, and we conclude that G is a $(*)$-group. □

Corollary 210.4. *Any nonidentity $(*)$-group has nontrivial center.*

Proof. Assume that a nonidentity $(*)$-group G has trivial center. Then G is nonnilpotent and has no nontrivial nilpotent direct factor so, by Theorem 210.3, $G = P \cdot A$ is a group with a cyclic complement $P \in \mathrm{Syl}_p(G)$ and the abelian kernel $A = Z_1 \times \cdots \times Z_k$, where $Z_1, \dots, Z_k$ are prime power cyclic P-invariant primary p'-subgroups and, for all i, one has

$$|P|^2 > |G| = |P|\,|A| \Rightarrow |P| > |A| \geq |Z_i| > |\mathrm{Aut}(Z_i)| \text{ for all } i\,.$$

It follows that the action of P on Z_i is not faithful so that $\Omega_1(P)$ centralizes Z_i for all i, and this implies that $\Omega_1(P)$ centralizes A. We conclude that $\{1\} < \Omega_1(P) \leq Z(G)$, contrary to the assumption. □

3⁰ Problems.

Problem 1. Classify the p-groups G such that, whenever $H < G$ and $p|H|^2 < G$, then $H \lhd G$. State a similar problem for arbitrary finite groups.[1]

Problem 2. Produce a proof of Theorem 210.1 without using minimal nonabelian subgroups.

Problem 3. Study the groups G all of whose subgroups H satisfying $|H|^2 < |G|$ are quasinormal.

Problem 4. Classify the p-groups all of whose subgroups of order $> p$ are quasinormal.

Problem 5. Study the non-Dedekindian p-groups G all of whose subgroups $H < G$ satisfying $|H|^2 \geq |G|$, are normal.

Problem 6. Classify the minimal prime power non-$(*)$-groups G. (The minimal nonmetacyclic group of order 2^5 satisfies that condition.)

Problem 7. Study the nonabelian p-groups G such that, whenever $M \leq G$ is minimal nonabelian, then $M \in \{M_{p^n} = M_p(n-1, 1), Q_8, M_2(m, m)\}$.

Problem 8. Classify the nonabelian p-groups G such that, whenever $H < G$ is abelian and $|H|^2 < |G|$, then H is normal.

1 This problem is solved in [Mal7].

§ 211 Nonabelian p-groups generated by the centers of their maximal subgroups

The purpose of this section is to initiate a classification of the title groups (problem 2447 (i)). We prove here the following result.

Theorem 211.1 (Janko). *Let G be a nonabelian p-group which is generated by the centers of its maximal subgroups. Then G is of class 2 and $\Phi(G) \leq Z(G)$ so that G' is elementary abelian.*

If in addition G has an abelian maximal subgroup, then $G = MZ(G)$, where M is minimal nonabelian and all such groups satisfy the assumption of the theorem.

Proof. Suppose that G is a nonabelian p-group which is generated by the centers of its maximal subgroups.

Assume, by way of contradiction, that $\Phi(G) \nleq Z(G)$. Set $C = C_G(\Phi(G))$ so that $\Phi(G)C < G$. Let $H \in \Gamma_1$ be such that $H \geq \Phi(G)C$. If $Z(H_i) \leq H$ for all $H_i \in \Gamma_1$, then our hypothesis gives a contradiction. Hence, there is $H_j \in \Gamma_1$ so that $Z(H_j) \nleq H$. Let $x \in Z(H_j) - H$ so that

$$C_G(x) \geq H_j > \Phi(G) \Rightarrow x \in C_G(\Phi(G)) = C \leq H,$$

contrary to the choice of x.

We have proved that $\Phi(G) \leq Z(G)$ and so G is of class 2. For any $x, y \in G$,

$$[x, y]^p = [x^p, y] = 1,$$

and so G' is elementary abelian. By Proposition 65.2 (a), any two noncommuting elements in G generate a minimal nonabelian subgroup.

Suppose in addition that G possesses an abelian maximal subgroup A. If the center of each maximal subgroup of G is contained in A, then our main hypothesis gives a contradiction. Hence, there is $g \in G - A$ such that $g \in Z(B)$, where $B \in \Gamma_1$ and $B \neq A$. It follows

$$Z(G) = C_A(g) = A \cap B \text{ and so } |A : Z(G)| = p.$$

But then $G/Z(G) \cong E_{p^2}$ and $p + 1$ maximal subgroups of G containing $Z(G)$ are abelian and they generate G. It follows that such a group G satisfies the main assumption of our theorem. Since G is nonabelian, there are $a, b \in G - Z(G)$ such that $[a, b] \neq 1$. By the above, $M = \langle a, b \rangle$ is minimal nonabelian. Because

$$M \text{ covers } G/Z(G), \text{ it follows that } G = MZ(G),$$

completing the proof. □

To complete the classification of the title groups, we must assume that in our groups G with $\Phi(G) \leq Z(G)$ all maximal subgroups are nonabelian.

Exercise 1. If a p-group G be such that $|G : Z(G)| = p^2$, then $G = MZ(G)$ for any minimal nonabelian subgroups $M \leq G$.

Solution. Take $M \leq G$ be minimal nonabelian. Then $|M : (M \cap Z(G)| > p$ since M is nonabelian. It follows that $MZ(G) = G$.

Exercise 2. If a p-group G is such that $\bar{G} = G/Z(G)$ is metacyclic, then $G = AB$, where A and B are abelian.

Solution. One has $\bar{G} = \bar{A}\bar{B}$, where $\bar{A}$, $\bar{B}$ are cyclic. Then A, B are abelian and $G = AB$.

Exercise 3. Study the p-groups generated by centers of their nonabelian maximal subgroups.

Solution. If there is no abelian member in the set Γ_1, then G satisfies the hypothesis of Theorem 211.1. Now assume that there is an abelian $A \in \Gamma_1$. Clearly, G is not minimal nonabelian. By hypothesis, there is a nonabelian $H \in \Gamma_1$ such that $Z(H) \nleq A$. Let $g \in Z(H) - A$. Then

$$Z(G) = C_A(g) = C_G(g) \cap A = H \cap A ,$$

so that $|G : Z(G)| = p^2$. It follows that $|H : Z(G)| = p$ so that H is abelian, contrary to the choice of H. Thus, A does not exist. Thus, all maximal subgroups of G are nonabelian so that G satisfies the hypothesis of Theorem 211.1.

Problem 1. Study the p-groups $G = MZ(G)$, where $M \leq G$ is minimal nonabelian.

Problem 2. Study the p-groups $G = M\mathcal{N}(G)$, where $M \leq G$ is minimal nonabelian and $\mathcal{N}(G)$ the norm of G.

Problem 3. Study the p-groups $G = MZ_2(G)$, where $M \leq G$ is minimal nonabelian.

Problem 4. Classify the $\mathcal{A}_n$-groups, $n > 1$, of prime power order generated by centers (norms) of their nonabelian maximal subgroups.

Problem 5. Study the p-groups generated by norms of their non-Dedekindian maximal subgroups.

Problem 6. Classify the p-groups generated by centers (norms) (i) of their $\mathcal{A}_2$-subgroups, (ii) of their $\mathcal{A}_1$-subgroups.

§ 212 Nonabelian p-groups generated by any two nonconjugate maximal abelian subgroups

The purpose of this section is to initiate a study of the title groups (problem 2540 where it was asked to classify the nonabelian p-groups G in which any two nonconjugate maximal abelian subgroups generate G). We prove here the following result.

Theorem 212.1 (Janko). *Let G be a nonabelian p-group which is generated by any two nonconjugate maximal abelian subgroups. Then G is metabelian and $Z(G)$ is the intersection of any two distinct abelian maximal subgroups (and so G is a group appearing in Theorems 91.2 and 91.3). Moreover, if A is a maximal normal abelian subgroup of G, then G/A is elementary abelian and for each $g \in G - A$, $g^p \in Z(G)$.*

Proof. Let G be a nonabelian p-group which is generated by any two nonconjugate maximal abelian subgroups. Let A be a maximal normal abelian subgroup of G so that $Z(G) < A < G$. Let B be any maximal abelian subgroup of G which is distinct from A. Since A is not conjugate to B, we get $G = AB$ and $A \cap B = Z(G)$. Indeed,

$$A \cap B \geq Z(G) \text{ and } C_G(A \cap B) \geq \langle A, B \rangle = G \Rightarrow A \cap B \leq Z(G) \Rightarrow A \cap B = Z(G)\,.$$

Let $g \in G - A$ and assume that g commutes with $a \in A - Z(G)$. Let X be a maximal abelian subgroup of G containing the abelian subgroup $\langle g, a \rangle$. But then $G = AX$ with $A \cap X \geq \langle Z(G), a \rangle$, contrary to the previous paragraph. We have proved that $C_A(g) = Z(G)$ for each $g \in G - A$.

Let $U \neq B$ be another maximal abelian subgroup of G which is distinct from A. Suppose that $B \cap U > Z(G)$ so that there is $x \in (B \cap U) - Z(G)$ (by the first paragraph of the proof, $x \notin A$). But then

$$\begin{aligned} &C_G(x) \geq \langle B, U \rangle > B \text{ and } \langle B, U \rangle (A \cap \langle B, U \rangle) B \\ &\Rightarrow A \cap \langle B, U \rangle > Z(G) \text{ and so } C_A(x) > Z(G)\,, \end{aligned}$$

contrary to the result of the previous paragraph. We have proved that the intersection of any two distinct maximal abelian subgroups of G is equal to $Z(G)$. It follows that G is a group appearing in Theorems 91.2 and 91.3.

Let $g \in G - A$, where, as above, A is a maximal normal abelian subgroup of G. By Lemma 57.1, there is $a \in A - Z(G)$ such that $\langle g, a \rangle$ is minimal nonabelian. Then g^p commutes with a and so, by the above, $C_G(a) = A$ which implies $g^p \in A$. If

$$g^p \in A - Z(G), \text{ then } C_G(g^p) > \langle A, g \rangle > A\,,$$

a contradiction. Hence, $g^p \in Z(G)$ for any $g \in G - A$. It follows that G/A is elementary abelian. The proof is complete. □

Problem 1. Classify the irregular p-groups generated by any two distinct (nonconjugate) maximal regular subgroups. (See #2540 and § 212.)

Problem 2. Classify the irregular p-groups generated by any two distinct (nonconjugate) maximal metacyclic subgroups.

Problem 3. Classify the nonabelian p-groups generated by any two nonconjugate minimal nonabelian subgroups. (Any $\mathcal{A}_2$-group satisfies the above condition.)

Problem 4. Classify the nonabelian p-groups G such that $G = \langle A, B \rangle$, where $A < G$ is maximal abelian, $B \leq G$ is minimal nonabelian and $A \nleq B$.

Problem 5. Classify the nonabelian p-groups of exponent $> p$ generated by any two nonnormal (nonconjugate) maximal cyclic subgroups of order $> p$.

§ 213 p-groups with $A \cap B$ being maximal in A or B for any two nonincident subgroups A and B

The purpose of this section is to solve a partial case of Problem 2372. We prove here the following result.

Theorem 213.1 (Janko). *Let G be a nonabelian p-group such that $A \cap B$ is maximal in A or B for any two nonincident subgroups A and B in G. Then G is one of the following groups:*

(a) *minimal nonabelian groups: D_8, Q_8, $S(p^3)$ (nonabelian group of order p^3 and exponent p), $p > 2$, M_{p^n}, $n \geq 3$ with $n \geq 4$ in case $p = 2$;*
(b) *2-groups of maximal class and order $\geq 2^4$;*
(c) *$G \cong Q_8 \times C_2$;*
(d) *$G = Q * S$ with $Q \cong Q_8$, $S \cong C_4$ and $Q \cap S = Z(Q)$;*
(e) *a regular p-group G of order p^4 and exponent p^2, $p > 2$, with $\mho_1(G) \cong C_p$ and $\Omega_1(G) \cong S(p^3)$;*
(f) *an irregular 3-group G of maximal class, $|G| = 3^4$ and $\exp(G) = 9$, where G has no elementary abelian subgroup of order 27.*

Conversely, all the above p-groups satisfy the assumption of the theorem.

Proof. Let G be a p-group such that $A \cap B$ is maximal in A or B for any two nonincident subgroups A and B in G. We note that this hypothesis is hereditary for subgroups and factor-groups.

(i) Let A be an abelian subgroup of G. Then A is either cyclic or elementary abelian with $p^2 \leq |A| \leq p^3$ or $A \cong C_p^e \times C_p$, $e \geq 2$. From now on, we assume that G is nonabelian.

Indeed, first suppose that A is elementary abelian. If $|A| \geq p^4$, then A has a subgroup $A_1 \times A_2$ with $A_1 \cong A_2 \cong E_{p^2}$, contrary to our hypothesis. Hence, $|A| \leq p^3$. Now suppose that A is not elementary abelian so that $\exp(A) = p^e$, $e \geq 2$. Let A_1 be a cyclic subgroup of order p^e in A and let A_2 be a complement of A_1 in A. By the hypothesis, $|A_2| \leq p$ and we are done.

(ii) Let M be a minimal nonabelian subgroup of G. Then

$$M \cong D_8,\ Q_8,\ S(p^3),\ p > 2, \quad \text{or} \quad M \cong M_{p^n},\ n \geq 3 \text{ with } n \geq 4 \text{ in case } p = 2.$$

All these minimal nonabelian subgroups satisfy the hypothesis of our theorem. From now on we assume that G is not minimal nonabelian.

Indeed, D_8, Q_8, and $S(p^3)$ satisfy the hypothesis of our theorem. Let

$$M = \langle a, b \mid a^{p^u} = b^{p^v} = c^p = 1,\ [a,b] = c,\ [a,c] = [b,c] = 1\rangle, \quad \text{where } u \geq 2,\quad v \geq 1.$$

Then, we obtain

$$\langle a\rangle \cap \langle b, c\rangle = \{1\} \text{ with } |\langle a\rangle| \geq p^2 \text{ and } |\langle b, c\rangle| \geq p^2 ,$$

a contradiction. Let

$$M = \langle a, b \mid a^{p^u} = b^{p^v} = 1,\ [a, b] = a^{p^{u-1}}\rangle, \quad \text{where } u \geq 2,\ v \geq 2 .$$

But then

$$\langle a\rangle \cap \langle b\rangle = \{1\} \quad \text{with } |\langle a\rangle| \geq p^2 \text{ and } |\langle b\rangle| \geq p^2 ,$$

a contradiction. Hence, we have $u \geq 2$, $v = 1$ and so $M \cong \mathrm{M}_{p^n}$, $n \geq 3$ with $n \geq 4$ in case $p = 2$.

Conversely, let

$$\mathrm{M}_{p^{u+1}} \cong M = \langle a, b \mid a^{p^u} = b^p = 1,\ [a, b] = a^{p^{u-1}}\rangle, \quad \text{where } u \geq 2 \text{ and } u \geq 3 \text{ if } p = 2 .$$

Let X_1, X_2 be any two nonincident subgroups of M such that setting $X_1 \cap X_2 = X$, we have $|X_1 : X| \geq p^2$ and $|X_2 : X| \geq p^2$. If both X_1 and X_2 are noncyclic, then $X_1 \geq S$, $X_2 \geq S$, where $S = \Omega_1(M) \cong \mathrm{E}_{p^2}$. In this case, X_1 and X_2 are incidents since M/S is cyclic, a contradiction. We may assume that X_1 is cyclic so that $|X_1 : \mho_1(X_1)| = p$ and $\mho_1(X_1) > X$. Since $X_1 \geq M' \cong \mathrm{C}_p$, we have $X_1 \trianglelefteq M$. Suppose that $\mho_1(X_2) \nleq X$. Then $\mho_1(X_1X_2)$ is noncyclic, contrary to the fact that $\mho_1(M)$ is cyclic. Hence, $\mho_1(X_2) \leq X$. But $M' \leq \mho_1(X_1)$ and so $\Phi(X_1X_2) \leq \mho_1(X_1)$ giving $\mathrm{d}(X_1X_2 \geq 3$, contrary to the fact that each subgroup Y of M possesses a cyclic subgroup of index p and so $\mathrm{d}(Y) \leq 2$. We have proved that our group M satisfies the hypothesis of our theorem.

(iii) Suppose that G has no abelian normal subgroup of type (p, p). Then G is a 2-group of maximal class and order $\geq 2^4$. All these 2-groups of maximal class satisfy the hypothesis of our theorem. In the sequel we assume that G has a normal abelian subgroup U of type (p, p).

Let G be a 2-group of maximal class and order $\geq 2^4$. Then G possesses a unique cyclic subgroup $\langle a\rangle \cong \mathrm{C}_{2^n}$, $n \geq 3$, of index 2 which is a unique abelian maximal subgroup of G. Set $\langle z\rangle = \langle a^{2^{n-1}}\rangle$ so that for each $x \in G - \langle a\rangle$, one has

$$x^2 \in \langle z\rangle \text{ and } \mathrm{C}_{\langle a\rangle}(x) = \langle z\rangle .$$

Let X_1 and X_2 be two nonincident subgroups of G and set $X = X_1 \cap X_2$. Assume, by way of contradiction, $|X_1 : X| \geq 4$ and $|X_2 : X| \geq 4$. Since $X_i \cap \langle a\rangle \neq \{1\}$, $i = 1, 2$, we get $X_i \geq \langle z\rangle$ and so $X \geq \langle z\rangle$ implying $|X_i| \geq 8$. Assume that one of X_i, $i = 1, 2$, is contained in $\langle a\rangle$, say $X_1 \leq \langle a\rangle$. Since X_1 and X_2 are nonincident, we have $X_2 \nleq \langle a\rangle$ and $X_2 \cap \langle a\rangle \ngeq X_1$ so that $X_2 \cap \langle a\rangle = X_2 \cap X_1 = X$. But then $|X_2 : X| = 2$, a contradiction. We have proved that both X_1 and X_2 are not contained in $\langle a\rangle$ and since $|X_i| \geq 8$, $i = 1, 2$, we see that both X_1 and X_2 are of maximal class. On the other hand, $X_1 \cap \langle a\rangle$ and

$X_2 \cap \langle a \rangle$ are incidents of orders ≥ 4 and so we may assume without loss of generality that $X_1 \cap \langle a \rangle \leq X_2 \cap \langle a \rangle$. We may set

$$X_1 = (X_1 \cap \langle a \rangle)\langle x_1 \rangle\,, \quad \text{where } x_1 \in G - \langle a \rangle \text{ and } x_1^2 \in \langle z \rangle\,.$$

Since $X_1 \nleq X_2$, we have $x_1 \notin X_2$ and $X = X_1 \cap X_2 = X_1 \cap \langle a \rangle$ so that $|X_1 : X| = 2$, a contradiction.

(iv) (Berkovich) If G possesses an elementary abelian subgroup E of order p^3, then $\exp(G) = p$.

Suppose that this is false. Then there is a cyclic subgroup $C \cong C_{p^2}$. By hypothesis, $|C \cap E| = p$. Then $E = (C \cap E) \times R$, where $C \cap R = \{1\}$ and $|C| = |R| = p^2$, contrary to our hypothesis.

(v) Let A be a maximal normal abelian subgroup of G containing U. Then we have

$$A = \langle a \rangle \times \langle u \rangle\,,$$

$$\text{where } \langle a \rangle \cong C_{p^e}\,, \quad e \geq 2\,, \quad \langle u \rangle \cong C_p\,, \quad a^{p^{e-1}} = z, \langle z \rangle \leq Z(G)\,, \quad U = \langle z, u \rangle$$

and G has no elementary abelian subgroups of order p^3.

Let A be a maximal normal abelian subgroup of G containing U so that $A < G$. Assume that G has an elementary abelian subgroup of order p^3. By (iv), $\exp(G) = p$ and so $p > 2$. By (i), we get either $A = U \cong E_{p^2}$ or $A \cong E_{p^3}$. However, if $A = U$, then $|G| = p^3$ and $G \cong M_{p^3}$ or $G \cong S(p^3)$, contrary to our assumption that G is not minimal abelian. Hence, we have $A \cong E_{p^3}$. Let U_0 be a subgroup of order p in U such that $U_0 \leq Z(G)$ and let $g \in G - A$ so that $o(g) = p$ and $S = \langle U_0, g \rangle \cong E_{p^2}$ with $S \cap A = U_0$. Let T be a complement of U_0 in A so that $A = T \times U_0$. But then $T \cap S = \{1\}$ and $|T| = |S| = p^2$, contrary to our hypothesis.

We have proved that G has no elementary abelian subgroups of order p^3. If $A = U$, then G is minimal nonabelian of order p^3, contrary to our assumption in (ii). By (i), A is abelian of type (p^e, p) with $e \geq 2$ and we are done.

(vi) If $C_G(U) > A$, then $G \cong Q_8 \times C_2$ and so G is a group appearing in part (c) of our theorem. From now on, we may assume that $C_G(U) = A$ so that $|G : A| = p$ and $\exp(G) = p^e$.

Indeed, assume that $C_G(U) > A$. Since G has no elementary abelian subgroups of order p^3, we have

$$\Omega_1(C_G(U)) = U \leq Z(C_G(U))\,.$$

By Lemma 57.1, for each $x \in C_G(U) - A$, there is $y \in A - U$ such that $M = \langle x, y \rangle$ is minimal nonabelian, where $\Omega_1(M) \leq Z(M)$. By (ii), this forces $p = 2$ and $M \cong Q_8$. Hence, $o(x) = o(y) = 4$ and $x^2 = y^2 = z$ (see (v)). All elements $x \in C_G(U) - A$ are of order 4 with $x^2 = z$ and since for an element y of order 4 in $A - U$,

$$y^x = y^{-1} = yz \text{ and } (yu)^x = (yu)z = (yu)^{-1}\,,$$

it follows that x inverts $\Omega_2(A) \cong C_4 \times C_2$ which gives $|C_G(U) : A| = 2$. If $e > 2$, then

$$|\langle a\rangle| \geq 8\,, \quad \langle x, u\rangle \cong C_4 \times C_2 \text{ with } \langle a\rangle \cap \langle x, u\rangle = \langle z\rangle\,,$$

contrary to our hypothesis. Hence, $e = 2$ and so $C_G(U) = Q \times \langle u\rangle \cong Q_8 \times C_2$, where $Z(Q) = \langle z\rangle$ and $A \cong C_4 \times C_2$. If $G = C_G(U)$, then we are done. Therefore we may assume that $G > C_G(U)$ so that $|G : C_G(U)| = 2$ and $|G| = 2^5$. Suppose that there is $g \in G - C_G(U)$ such that $g^2 \in \langle z, u\rangle$. Since $D = \langle z, u\rangle\langle g\rangle \cong D_8$, we have $g^2 \in \langle z\rangle$. But then $Q \cap D = \langle z\rangle$, contrary to our hypothesis. We have proved that for each $g \in G - C_G(U)$, $g^2 \in C_G(U) - \langle z, u\rangle$ and so $\Omega_2(G) \cong Q_8 \times C_2$. By Theorem 52.1, G is isomorphic to a uniquely determined group of order 2^5 given in part (A2)(a) of Theorem 49.1. In particular, there is an element $y \in G - C_G(U)$ of order 8 such that $\langle y\rangle \cap Q = Z(Q)$, contrary to our hypothesis.

Henceforth, we may assume that $C_G(U) = A$ so that $|G : A| = p$ and $\exp(G) = p^e$. Indeed, if $\exp(G) = p^{e+1}$, then G has a cyclic subgroup of index p and so G is either a 2-group of maximal class or $G \cong M_{p^n}$ which is minimal nonabelian. But all these groups are excluded by our assumptions in (ii) and (iii).

(vii) If there are elements of order $\leq p^2$ in $G - A$, then $e = 2$, $A \cong C_{p^2} \times C_p$, $|G| = p^4$ and G is isomorphic to a group given in parts (d), (e) or (f) of our theorem. Conversely, all groups from parts (c) to (f) of our theorem satisfy the hypothesis of that theorem.

Assume that there is $g \in G - A$ such that $g^p \in \langle z, u\rangle = \Omega_1(A)$ and so $g^p \in \langle z\rangle$, where

$$\langle g, u\rangle \cong M_{p^3} \text{ or } S(p^3) \text{ with } p > 2 \text{ or } \langle g, u\rangle \cong D_8\,.$$

If $e > 2$, then

$$|\langle a\rangle : \langle z\rangle| \geq p^2 \text{ and } |\langle g, u\rangle : \langle z\rangle| \geq p^2, \text{ where } \langle a\rangle \cap \langle g, u\rangle = \langle z\rangle\,,$$

contrary to our hypothesis. Thus, $e = 2$, $A \cong C_{p^2} \times C_p$ and $|G| = p^4$.

First, consider the case $p = 2$. Since G is not of maximal class, a result of O. Taussky implies $G' = \langle z\rangle$. Hence, $D = \langle g, u\rangle \cong D_8$ with $D' = \langle z\rangle$ so that $G = D * S$, where $|S| = 4$ and $D \cap S = \langle z\rangle$. But G has no elementary abelian subgroups of order 8 and so $S \cong C_4$. Since $D_8 * C_4 \cong Q_8 * C_4$, we have obtained the group from part (d) of the theorem.

Now we consider the case $p > 2$, where

$$D = \langle g, u\rangle \cong M_{p^3} \text{ or } S(p^3)\,.$$

By (vi), $\exp(G) = p^2$ and so for each $x \in G - A$, $x^p \in \langle z\rangle$ and this implies $\mho_1(G) = \langle z\rangle$. We may use Theorem 74.1. If G is a group of part (a) of that theorem, then Proposition 149.1 implies that G is minimal nonabelian, a contradiction. Hence, G is a group of order p^4 from parts (c) or (d) of Theorem 74.1. We have obtained the groups stated in parts (e) and (f) of our theorem.

Conversely, let G be any group of order p^4 stated in parts (c), (d), (e), (f) of our theorem. Note that in all these cases, $\mho_1(G) \cong C_p$ and G has no elementary abelian subgroups of order p^3. We claim that G satisfies the hypothesis of the theorem. Indeed, let X_1, X_2 be two nonincident subgroups of G. If X_1 or X_2 is a maximal subgroup of G, say X_1, then $|X_2 : (X_1 \cap X_2)| = p$. We may assume, by way of contradiction, that $|X_1| = |X_2| = p^2$ and $X_1 \cap X_2 = \{1\}$. Set $\mho_1(G) = \langle z\rangle \cong C_p$, where $\langle z\rangle \leq Z(G)$. If X_1 or X_2, say X_1, is elementary abelian of order p^2, then $z \in X_1$ because G has no elementary abelian subgroups of order p^3. But then $\mho_1(X_2) = \{1\}$ and so $\langle z\rangle \times X_2 \cong E_{p^3}$, a contradiction. It follows that both X_1 and X_2 are cyclic of order p^2, which contradicts the fact that $\mho_1(G) \cong C_p$.

(viii) Finally, we consider the remaining case that $\Omega_2(G) = \Omega_2(A) \cong C_{p^2} \times C_p$ and we shall obtain in this case a contradiction.

Indeed, in this case we must have $e \geq 3$ and so $|G| = p^{e+2} \geq p^5$ and by (vi), G is of exponent p^e.

First, we consider the case $p = 2$. Since G is not minimal nonabelian, it follows that G is not an L_2-group and so Lemma 42.1 implies that

$$G = \langle a, b \mid a^{2^e} = b^8 = 1,\ a^b = a^{-1},\ a^{2^{e-1}} = b^4 = z\rangle,\ e \geq 3\,.$$

But then

$$o(a) \geq 8\,,\quad o(b) = 8 \quad \text{and} \quad \langle a\rangle \cap \langle b\rangle = \langle z\rangle \cong C_2\,,$$

contrary to the hypothesis.

Now suppose $p > 2$. Since $|G| = p^{e+2} \geq p^5$ and G is of exponent p^e, $e \geq 3$, we may use Theorem 74.1. Hence, G is one of the groups of parts (a) or (b) of that theorem. However, if G is metacyclic, then the fact that A is an abelian maximal subgroup of G implies together with Proposition 149.1 that G is minimal nonabelian, a contradiction. Hence, G is an L_3-group. But this contradicts our assumption that $\Omega_2(G) = \Omega_2(A) \cong C_{p^2} \times C_p$. The theorem is proved. □

Exercise 1. Classify the nonabelian p-groups G such that, whenever $A, B \leq G$ are distinct of equal order, then $A \cap B$ has index p in A.

Problem 1. Classify the p-groups G such that $|A : (A \cap B)| = p$ for any two distinct subgroups (abelian subgroups, maximal abelian subgroups, maximal metacyclic subgroups) of G of equal order. Consider in detail the case $\exp(G) = p$.

Problem 2. Classify the p-groups G such that $|A : (A \cap B)| = p$ for any two distinct nonnormal subgroups (nonnormal abelian subgroups) of G. Consider in detail the cases (i) $\exp(G) = p$, (ii) $|A| = |B|$.

§ 214 Nonabelian p-groups with a small number of normal subgroups

Let G be a group of maximal class and order p^m. Then G contains only one normal subgroup of order p^k for any $k \in \{1, \dots, m-2\}$ (Exercise 9.1 (b)). Conversely, a noncyclic group of order $p^m > p^2$ satisfying the above condition, is of maximal class (Lemma 214.1).

The two-generator groups of order $p^m > p^3$ containing only one normal subgroup of order p^k for any $k \in \{1, \dots, m-3\}$ are classified in Theorem 214.2. Next, if a noncyclic group G of order $p^m \geq p^{p+1}$ contains only one normal subgroup of index p^k for $k \in \{2, \dots, p+1\}$, then it is of maximal class (Theorem 214.3).

Remark 1. If a nonabelian p-group G contains only one normal subgroup of order p^k for $k = 1, 2$, then $|Z(G)| = p$. Indeed, assume that $|Z(G)| > p$. Since $Z(G)$ has only one subgroup of order p, it is cyclic. In that case, $Z(G)$ contains a cyclic subgroup of order p^2 which is G-invariant. Therefore, by hypothesis G has no normal abelian subgroup of type (p, p), so, by Lemma 1.4, G is a 2-group of maximal class hence $|Z(G)| = 2$, contrary to the assumption.

A starting point for this section is the following

Lemma 214.1 (= Lemma 9.1). *If a noncyclic group of order $p^m > p^2$ has only one normal subgroup of order p^k for any $k \in \{1, \dots, m-2\}$, then it is of maximal class.*

Theorem 214.2. *If a two-generator noncyclic group G of order $p^m > p^3$ contains only one normal subgroup of order p^k for each $k \in \{1, \dots, m-3\}$, then one of the following holds:*

(a) $G \cong M_{p^4}$.

(b) *G is of maximal class.*

Proof. By hypothesis, $Z(G)$ is cyclic. It follows that G is nonabelian. By Remark 1, $|Z(G)| = p$ if $|G| > p^4$. We proceed by induction on m.

(i) Let $m = 4$. If G is minimal nonabelian, then $G \cong M_{p^4}$ since $Z(G)$ is cyclic. Next we assume that G is not minimal nonabelian. If $G/Z(G)$ is abelian, it is of type (p^2, p); then G is minimal nonabelian (Lemma 65.2 (a)), contrary to the assumption. If $G/Z(G)$ is nonabelian, then G is of maximal class since $|Z(G)| = p$ in this case.

In what follows we assume that $m > 4$.

(ii) Let $m = 5$. Then the noncyclic $G/Z(G)$ is one of groups (a), (b) of order p^4. Assume that $G/Z(G) \cong M_{p^4}$. Let $A/Z(G)$ and $B/Z(G)$ be two distinct cyclic subgroups of index p in $G/Z(G)$. Then $A, B \in \Gamma_1$ are distinct abelian so that $A \cap B = Z(G)$, and we conclude that G is minimal nonabelian since $d(G) = 2$. As the center of our (minimal nonabelian p-group) G has order p, we obtain $m = 3$, a contradiction. Now let $G/Z(G)$ be of maximal class. Then G is of maximal class since $|Z(G)| = p$.

(iii) Now let $m > 5$. By induction, $G/Z(G)$ is of maximal class; then G is also of maximal class since $|Z(G)| = p$. □

Proposition 214.3 (= Corollary 12.11). *Suppose that, for any $k \in \{2, \dots, p+1\}$, a nonabelian group G of order $p^m \geq p^{p+1}$, $p > 2$, contains only one normal subgroup of index p^k. Then G is of maximal class.*

Proof. Obviously, $d(G) = 2$ and $|G : G'| = p^2$. Assume that G is not of maximal class. Then $m > p+1$, by Lemma 214.1. Let $T < G'$ be G-invariant of index p^{p+1} in G. By Lemma 214.1 again, the quotient group G/T, being noncyclic, is of maximal class. As, by hypothesis, T is the unique normal subgroup of index p^{p+1} in G, Theorem 12.9 implies that G is of maximal class. □

Corollary 214.4. *Let $p > 2$. Suppose that, for any $k \in \{3, \dots, p+2\}$, a two-generator nonabelian group G of order $p^m \geq p^{p+2}$, $p > 2$, contains only one normal subgroup of index p^k. Then G is of maximal class.*

Proof. It is easily seen that $|G/G'| \leq p^3$. In view of Theorem 214.2, one may assume that $|G| > p^{p+2}$. Let $T < G'$ be G-invariant of index p^{p+2} in G. By Theorem 214.1, the quotient group G/T, being nonabelian, is of maximal class. By hypothesis, T is the unique G-invariant subgroup of index p^{p+2} in G. Therefore, if $S \lhd G$ is of index p^{p+1}, then $T < S$. As G/T is of maximal class, it follows that it has only one normal subgroup of order p, and this subgroup coincides with S. We conclude that S if the unique normal subgroup of index p^{p+1} in G. As G/S and a nonabelian epimorphic image of the p-group G/T of maximal class is itself of maximal class, it follows from Theorem 12.9 that G is of maximal class. □

Theorem 214.5. *Suppose that G is a two-generator nonabelian p-group. If the center of any nonabelian epimorphic image of G is cyclic, then one of the following holds:*
(a) *G is of maximal class.*
(b) *$G \cong M_{p^n}$.*

Proof. We proceed by induction on $|G|$. By hypothesis, $Z(G)$ is cyclic.

Let $|G'| = p$. Then, by Lemma 65.2 (a), G is minimal nonabelian. In that case, since $Z(G)$ is cyclic, one has either $|G| = p^3$ or $G \cong M_{p^n}$. In the first case, G is of maximal class.

Next we assume that $|G'| > p$.

Let $L \lhd G$ be of order p. Then, by induction, either G/L is of maximal class or $G/L \cong M_{p^n}$. Assume that $G/L \cong M_{p^n}$. Let A/L and B/L be cyclic subgroups of index p in G/L. Then A, B are distinct abelian subgroups of index p in G so that $A \cap B = Z(G)$. In that case, G is minimal nonabelian since $d(G) = 2$. As in (i), $G \cong M_{p^{n+1}}$, a contradiction since, by assumption, $|G'| > p = |(M_{p^{n+1}})'|$. Now let G/L be of maximal class. If $|Z(G)| = p$, then G is of maximal class. Now assume that $|Z(G)| > p$; then $Z(G) \cong C_{p^2}$. By Lemma 1.4, there is in G a normal abelian subgroup of type (p, p). Clearly, $R \cap Z(G) = L$. Then $RZ(G)/L \cong E_{p^2}$ is a central subgroup of the nonabelian group G/L, contrary to the assumption. The proof is complete. □

Problem 1. Study the two-generator groups of order $p^m > p^4$ containing only one normal subgroup of order p^k for any $k \in \{1, \dots, m-4\}$.

Problem 2. Let $\nu_k(G)$ be the number of normal subgroups of index p^k in a p-group G. Study the p-groups G satisfying $\nu_k(G) \le p+1$ for all k.

Problem 3. Study the two-generator groups of order $p^m > p^{p+2}$, containing, for any $k \in \{3, \dots, p+1\}$, only one normal subgroup of index p^k.

Problem 4. Study the nonabelian two-generator p-groups G of exponent $> p > 2$, such that any nonabelian epimorphic image of G has no normal cyclic subgroup of order p^2.

Problem 5. Study the nonabelian two-generator p-groups all of whose factors of their upper (lower) central series, apart of the last one (first one), are cyclic.

§ 215 Every p-group of maximal class and order $\geq p^p$, $p > 3$, has exactly p two-generator nonabelian subgroups of index p

By Theorem 5.8 (b), any p-group G, $p > 3$, such that $|G/\mho_1(G)| \geq p^p$ (in particular, if G is irregular) has a maximal subgroup H that is not generated by two elements. If a 2-group G and all its maximal subgroups are two-generator, then G is metacyclic (Corollary 36.6).

Below we prove the result stated in the title.

Theorem 215.1. *The number $\delta_2(G)$ of two-generator nonabelian subgroups of index p in a p-group of maximal class and order $\geq p^p$, $p > 3$, is equal to p.*

Proof. By Theorem 12.1 (b), the set Γ_1 has no two-generator abelian member.

As $|G'| > p$, the group G has at most one abelian subgroup of index p. Assume that there are distinct $U, V \in \Gamma_1$ such that $d(U) > 2$ and $d(V) > 2$. Let $\Phi(U) \leq R \triangleleft G$ and $\Phi(V) \leq S \triangleleft G$, where $|U : R| = p^3 = |V : S|$ (R and S exist since $\Phi(U)$ and $\Phi(V)$ are normal in G). Then $R = S$, by Exercise 9.1 (b). The group G/R has two abelian subgroups (namely, U/R and V/R) of index p. It follows that $(U/R) \cap (V/R) = Z(G/R)$, i.e., $cl(G/R) = 2$. But G/R, as an epimorphic image of a p-group G/R of maximal class and order p^4, is of class $3 > 2$, which is a contradiction. Thus, G has at most one non-two-generator maximal subgroup of index p.

It remains to show that G has a subgroup of index p that does not generated by two elements. By Theorems 9.5 and 9.8 (a), $|G/\mho_1(G)| \geq p^{p-1} \geq p^4$. By Theorem 5.8 (b), the number of two-generator subgroups of index p in $G/\mho_1(G)$ is equal to p. Therefore, there is in $G/\Phi(G)$ a subgroup $A/\mho_1(G)$ that is not generated by two elements. Then $d(A) > 2$, as required. □

If G is a 3-group of maximal class, then all its maximal subgroups are two-generator, unless $G \cong \Sigma_9$ so that the theorem is not true for $p = 3$.

If, in Theorem 215.1, $|G| > p^{p+1}$, then Theorem 9.6 (e) shows that the fundamental subgroup of G is not two-generator.

Remark 1. The same argument shows that if G is a regular p-group of maximal class such that $|G| \geq p^5$ (then, by Theorem 9.5 and 9.8 (a), must be $p > 3$), then $\delta_2(G) = p$.

Exercise 1. Let G be a p-group of maximal class and order $> p^3$, $p > 3$. Then there is in the set Γ_1 only one member G such that $|H/H'| > p^2$.

Problem 1. Study the p-groups containing exactly one maximal subgroup that is not generated by two elements.

Problem 2. Study the two-generator p-groups all of whose maximal subgroups are not generated by two elements.

Problem 3. Study the p-groups all of whose nonabelian maximal subgroups are two-generator.

§ 216 On the theorem of Mann about p-groups all of whose nonnormal subgroups are elementary abelian

One of more important topics of p-group theory is investigation of groups with given structures of nonnormal subgroups (see Theorems 1.20, 1.25, §§ 16, 59, 63, 112 and so on). In this direction, Mann [Man18] has proved the following nice result:

Theorem 216.1. *If all nonnormal subgroups of a p-group G are elementary abelian, then one of the following holds:*

(a) $\exp(G) = p$.

(b) $\mathrm{cl}(G) \leq 2$.

(c) $G \cong D_{16}$.

In fact, as Theorem 216.6 shows, if G is nonabelian as in Theorem 216.1 (a), then either G is of maximal class and order p^4 or $|G'| = p$. If G is as in Theorem 216.1 (b), then $|G'| = p$.

Definition 1. A nonabelian p-group all of whose minimal nonabelian subgroups are normal is said to be an *$\mathcal{M}_p$-group*.

If G is an $\mathcal{M}_p$-group, then it follows from Lemma J(i), below, that any its nonnormal subgroup is abelian. The classification of $\mathcal{M}_p$-groups is very nontrivial (see [FA] where such groups are called metahamiltonian). The $\mathcal{M}_p$-groups of exponent p are classified in Theorem 216.2, and the proof of that result is easy and short. Main results of this section are taken from [Ber53].

Definition 2. A p-group G is said to be a *$\mathcal{K}_p$-group* (see § 63) if $\exp(G) > p$ and all its cyclic subgroups of orders $> p$ are normal.

Lemma J contains some known results which are used in what follows.

Lemma J. *Let G be a nonabelian p-group.*

(i) *Proposition* 10.28. *The group G is generated by minimal nonabelian subgroups.*

(ii) *Lemma* 65.1. *If G is minimal nonabelian, then $d(G) = 2$ and $|G'| = p$. If, in addition, $\Omega_1(G) = G$, then $|G| = p^3$. If G is metacyclic $\ncong Q_8$, then $G = A \cdot B$ (semidirect product with cyclic kernel B and cyclic complement A).*

(iii) *Lemma* 65.1. *If all nonnormal subgroups of a $\mathcal{A}_1$-group G are elementary abelian, then $G \in \{D_8, S(p^3), M_{p^n}\}$. Next, G has only one maximal elementary abelian subgroup, unless $G \cong D_8$.*

(iv) *Theorem* 63.3. *If $p = 2$ and G is a nonabelian $\mathcal{K}_2$-group, then either $|G'| = 2$ or $|G : H_2(G)| = 2$. (Here $H_2(G) = \langle x \in G \mid o(x) > 2\rangle$ is the H_2-subgroup of G.)*

(v) *Lemmas 4.3, 4.2. Let $|G'| = p$ and let $E < G$ be minimal nonabelian. Then $G = E * C_G(E)$. In particular, $G = (E_1 * \cdots * E_n)Z(G)$, where $E_1, \dots, E_n$ are minimal nonabelian.*

(vi) *Corollary 10.2. Let $p > 2$, $E < B < G$, where $E \triangleleft G$ and B are elementary abelian, $|B| = p|E|$. Then there is $B_1 \triangleleft G$, where $B_1 > E$ is elementary abelian of order $|B|$.*

(vii) *Lemma 1.1. If $A < G$ is abelian of index p, then $|G| = p|G'||Z(G)|$.*

(viii) *Theorem 1.17(b). If $n > 1$ and G has exactly one cyclic subgroup of order p^n, then G is a 2-group of maximal class.*

(ix) *Lemma 57.1; see also Appendix 87. If $A \triangleleft G$ is maximal abelian, then for any $x \in G - A$ there is $a \in A$ such that $\langle a, x\rangle$ is minimal nonabelian.*

(x) *Lemma 65.2 (a). If $d(G) = 2$ and $G' \leq \Omega_1(Z(G))$, then G is minimal nonabelian.*

Proof of Lemma J (iii). Assume that G is as in Lemma J (ii) and satisfies condition (iii). If G is metacyclic, then it has a cyclic subgroup of index p, and hence $G \in \{D_8, M_{p^n}\}$ (Lemma J (ii)). Now let G be nonmetacyclic and $\exp(G) = p^e > p$. Then G has a nonnormal cyclic subgroup of order $p^e > p$ so it does not satisfy the hypothesis. Thus, if G is nonmetacyclic, then $\exp(G) = p$ whence $G \cong S(p^3)$ (Lemma J (ii)). □

Remark 1. Suppose that G is a p-group, $p > 2$, all of whose minimal nonabelian subgroups are metacyclic. We claim that then $\Omega_1(G)$ is elementary abelian. Indeed, let E be a normal elementary abelian subgroup of G of maximal order. Assume that there is $x \in G - E$ of order p. Set $H = \langle x, E\rangle$. By Lemma J (vi)), the subgroup H is nonabelian. By Lemma J (ix), there is $a \in E$ such that $M = \langle a, x\rangle$ is minimal nonabelian. As $\Omega_1(M) = M$, we get $M \cong S(p^3)$ (Lemma J (ii)), a contradiction since the group $S(p^3)$ is nonmetacyclic. Thus, x does not exist so that $\Omega_1(G) = E$.

Remark 2. If $E \triangleleft G$ is of order p^2, G is a nonabelian p-group and the quotient group G/E is cyclic, then G is minimal nonabelian. Indeed, $G' < E$ is of order p. By Lemma J (x), G is minimal nonabelian since $d(G) = d(G/G') = 2$ and $G' \leq \Omega_1(Z(G))$.

Our first result is the following

Theorem 216.2. *A nonabelian group G of order $> p^4$ and exponent p is an $\mathcal{M}_p$-group if and only if $|G'| = p$.*

A minimal nonabelian p-group of exponent p has order p^3 (Lemma J (ii)).

Any nonabelian p-group of order $\leq p^4$ is an $\mathcal{M}_p$-group.

Next we use the following obvious fact: a group is not covered by two proper subgroups.

Theorem 216.2 is a consequence of the following stronger result.

Theorem 216.3. *Let G be a nonabelian $\mathcal{M}_p$-group of order $> p^4$, $p > 2$. If there is in G a minimal nonabelian subgroup of order p^3, then $|G'| = p$.*

Proof. Assume that $|G'| > p$; then G is not minimal nonabelian (Lemma J (ii)). Let $A, B < G$ be distinct minimal nonabelian subgroups such that $|A| = p^3$ (A and B exist, by hypothesis and Lemma J (i)). All containing A subgroups of G are generated by minimal nonabelian subgroups (Lemma J (i)) so G-invariant, and hence the quotient group G/A is abelian since $p > 2$ (Theorem 1.20). Similarly, G/B is also abelian. Set $D = A \cap B$; then G/D, as a subgroup of the direct product of the abelian groups G/A and G/B, is abelian, and we conclude that $G' \leq D$. Since $|D| < |A| \leq p^3$ and $|G'| \geq p^2$. it follows that $G' = D$ is of order p^2. By the above, all minimal nonabelian subgroups of G contain $G'(= D)$. Moreover, since any nonabelian p-group contains a minimal nonabelian subgroup, it follows that

(∗) All nonabelian subgroups of G contain $D = G'$.

Therefore, to obtain a contradiction, it suffices to find in G a nonabelian subgroup not containing D. Let $E \neq D$ be a maximal subgroup of A. Write $C = C_G(E)$. Then $A \nleq C \Rightarrow C < G$.

(i) Assume that $E \lhd G$; then all maximal subgroups of A are G-invariant (to see this, it suffices to consider the action of G by conjugation on the set of $p - 1$ maximal subgroups of A that are $\neq D, E$). In that case, $C \in \Gamma_1$ since $|\mathrm{Aut}(E)|_p = p$. If $x \in G - (C \cup A)$, then the subgroup $U = \langle x, E\rangle$ is nonabelian and $U \neq A$. Therefore, to obtain a contradiction, it suffices to show that $D \nleq U$. Assume, however, that $D < U$. Then $A = DE < U$ ($<$ since $x \notin U - A$) so that $|U| > |A| = p^3$. Next, U/E is cyclic as an epimorphic image of $\langle x\rangle$, $|U/E| > p$ and $|E| = p^2$. By Remark 2, U is minimal nonabelian so that $DE = A \nleq U$, contrary to what has been said above.[1]

(ii) Thus, all p maximal subgroups of A, except of D, are not G-invariant so that E is nonnormal in G. In that case, $N = N_G(E) \in \Gamma_1$ (recall that $A \lhd G$ has exactly p non-G-invariant subgroups of index p). It follows from $A \nleq C = C_G(E)$ and $A < N$ that $C < N$ so that $|N : C| = p$. As $A, C < N$, it follows that $C \cup A \neq N$. Therefore, there is $y \in N - (C \cup A)$. The subgroup $F = \langle y, E\rangle \neq A$ is nonabelian and F/E is cyclic. By (∗), $D < F$; then $A = DE < F$ ($<$ since $x \notin A$) so that $|F| > p^3$. By Remark 2, F is minimal nonabelian, a contradiction.

Thus, the assumption $|G'| > p$ is contradictory. Therefore, $|G'| = p$, as was to be shown. □

Proof of Theorem 216.2. If G is a p-group with $|G'| = p$ and $A \leq G$ is minimal nonabelian, then $|A'| = p = |G'|$ so that $G' = A' < A \Rightarrow A \lhd G$, i.e., G is an $\mathcal{M}_p$-group.

Let G be an $\mathcal{M}_p$-group of exponent p and order $> p^4$. A minimal nonabelian group of exponent p has order p^3 (Lemma J (ii)). Therefore, G satisfies the hypothesis of Theorem 216.3 which implies that $|G'| = p$, completing the proof. □

---

1 Here we offer another proof. By (∗), the subgroup $C = C_G(E) \in \Gamma_1$ is abelian since $D \nleq C$. Let $E_1 < A$ be maximal and $E_1 \notin \{D, E\}$. Then $C_1 = C_G(E_1) \in \Gamma_1$ is also abelian (recall that $E_1 \lhd G$). Now, $C_1 \neq C$ since $E \nleq C_1$. In that case, $C \cap C_1 = Z(G)$ has index p^2 in G, and so, by Lemma J (vii), $|G'| = \frac{1}{p}|G : Z(G)| = p < p^2$, contrary to the assumption.

Now we present a supplement of Proposition 63.1.

Proposition 216.4. *Let G be a nonabelian $\mathcal{K}_p$-group of exponent $> p$. Suppose, in addition, that, provided $p = 2$, then G is Q_8-free and has no subgroup $\cong D_8$. Then $|G'| = p$. In particular, if G is a nonabelian p-group of exponent $> p > 2$ all of whose nonnormal subgroups have exponent p, then $|G'| = p$.*

Proof. By hypothesis, all cyclic subgroups of order $> p$ are G-invariant. Assume that $|G'| > p$. Let $Z < G$ be cyclic of order p^2 and $C_0 = \Omega_1(Z)$; then $C_0 \lhd G$ and G/C_0 is nonabelian, since $|G'| > p$. Therefore, since G is Q_8-free, G/C_0 is non-Dedekindian. In that case, there exists in G/C_0 a nonnormal cyclic subgroup K/C_0. By hypothesis, the nonnormal abelian subgroup K of order $> p$ is noncyclic with cyclic subgroup of index p; then K is abelian of type $(|K/C_0|, p)$. All p (cyclic) complements to C_0 in K generate K so, since K is not G-invariant, these complements must have order p, and so $K \cong E_{p^2}$. Write $H = KZ$; then H is either abelian of type (p^2, p) or $p > 2$ and $H \cong M_{p^3}$ since, by hypothesis, the noncyclic $H \ncong D_8$. Then $H \lhd G$ since H is generated by its p normal in G cyclic subgroups of order p^2. In that case, $\Omega_1(H)$, being characteristic in H, is normal in G. But $\Omega_1(H) = K$, a contradiction since K, by the above, is not normal in G. Thus, $|G'| = p$. □

It remains to prove that if G is a non-Dedekindian 2-group of Theorem 216.1, then $|G'| = 2$, unless $G \cong D_{16}$. This is done in the following

Theorem 216.5. *If G is a non-Dedekindian 2-group all of whose nonnormal subgroups are elementary abelian, then either $|G'| = 2$ or $G = D_{16}$.*

Proof. Clearly, G is a $\mathcal{K}_2$-group (see Definition 2). Assuming that $|G'| > 2$, one has to prove that $G \cong D_{16}$.

In the case under consideration, by Lemma J (iv), the H_2-subgroup $H = H_2(G)$ has index 2 in G so that $G = \langle x \rangle \cdot H$, where x is an involution inverting H. Since H is abelian (Burnside), one has, by basic theorem on abelian p-groups,

$$H = Z_1 \times \cdots \times Z_k, \quad |Z_1| \geq \cdots \geq |Z_k|, \quad \text{where } Z_1, \ldots, Z_k \text{ are cyclic}.$$

Next, $|Z_1| > 2$ since G is nonabelian.

If $k = 1$, then $G = \langle x, Z_1 \rangle \cong D_{2|Z_1|}$ and so $|G| = 16$ (if $|G| > 16$, there is in G a nonnormal subgroup $\cong D_8$ of exponent > 2).

Next we assume that $k > 1$.

Assume that $|Z_2| > 2$. Then $D_i = \langle x \rangle \cdot Z_i \cong D_{2|Z_i|}$, $i = 1, 2$. In that case, by hypothesis, $D_i \lhd G$, $i = 1, 2$. It follows from $Z_1 \cap Z_2 = \{1\}$ that $\langle x \rangle = D_1 \cap D_2 \lhd G$ hence $G = \langle x \rangle \times H$ is abelian, contrary to the hypothesis. Thus, $|Z_2| = 2$ and $|Z_1| > 2$. It follows that $|Z_i| = 2$ for $i > 1$ and $G = D_1 \times E$, where $E = Z_2 \times \cdots \times Z_k \cong E_{2^{k-1}}$. Then $|D_1'| = |G'| > 2$ so that $|D_1| \geq 16$. If $|D_1| > 16$, then D_1 contains a nonnormal subgroup $\cong D_8$ which is not elementary abelian. Thus, $D_1 \cong D_{16}$ and $G = D_{16} \times E$, where E is of exponent ≤ 2.

Let us prove that the group $G = D \times E$, where $D \cong \mathrm{D}_{16}$ and $\exp(E) = 2$, does not satisfy the hypothesis. Let $T < D$ be cyclic of index 2 and $H = T \times E$. Then $H = \mathrm{H}_2(G)$ is the H_2-subgroup of G. Let $C < H$ be cyclic of order 4 such that $C \nleq T$ (C exists by Lemma J (viii)). Then $C \lhd G$ since all subgroups of $H = \mathrm{H}_2(G)$ are G-invariant. As $\exp(G/T) = 2$, it follows that $C \cap T = \Omega_1(T)$ is G-invariant of order 2 so that $C \cap T < T < D \Rightarrow C \cap T = \mathrm{Z}(D)$. In that case, $\bar{G} = G/C$ contains a subgroup

$$\bar{D} = DC/C \cong D/(D \cap C) = D/\mathrm{Z}(D) \cong \mathrm{D}_8 .$$

Let $\bar{U}$ be a nonnormal subgroup of order 2 in $\bar{D}$. Then $U(> C)$ is not normal in G and $\exp(U) \geq \exp(C) = 4 > 2$, i.e., the group G does not satisfy the hypothesis. Therefore, $E = \{1\}$ so that $G = D \cong \mathrm{D}_{16}$, completing the proof. □

Thus, the following result is obtained:

Theorem 216.6. *Suppose that all nonnormal subgroups of a non-Dedekindian p-group G, where $|G| > p^4$ provided $p > 2$, are elementary abelian. Then either $|G'| = p$ or $G = \mathrm{D}_{16}$.*[2]

A non-Dedekindian p-group G with cyclic $\mathrm{Z}(G)$ such that $\Omega_1(\mathrm{Z}(G)) = G'$ satisfies the hypothesis of Theorem 216.6. Indeed, let $L < G$ be nonnormal. Then $G' \nleq L$ so that $H = G' \times L \lhd G$, and hence $\Phi(H) = \Phi(L) \lhd G$. It follows that $\Phi(L) = \{1\}$ since all G-invariant subgroups $> \{1\}$ contain G' (G is a monolith), and we conclude that L is elementary abelian, as claimed.

Corollary 216.7. *If G is a p-group as in Theorem 216.6, then it is an $\mathcal{M}_p$-group.*

A partial case of Theorem 216.6 was considered by Passman (see Theorem 1.25; but Theorem 1.25 does not follows immediately from Theorem 216.6). Now we deduce Passman's theorem from Theorem 216.6. The presented proof is more technical than the original one.

Theorem 216.8 (= Theorem 1.25). *If all nonnormal subgroups of a non-Dedekindian p-group G have order p, then one and only one of the following holds:*
(a) *$G \cong \mathrm{M}_{p^n}$.*
(b) *$G = Z * G_0$, where Z is cyclic and $G_0 = \Omega_1(G)$ is nonabelian of order p^3 (i.e., $G_0 \in \{\mathrm{D}_8, S(p^3)\}$).*
(c) *$p = 2$ and $G = \mathrm{D}_8 * \mathrm{Q}_8$ is extraspecial of order 2^5.*

Proof. By Theorem 216.6, $|G'| = p$ since D_{16} has a nonnormal subgroup of order 4. Then, by Lemma J (v),

$$(**) \qquad G = (A_1 * \cdots * A_n)\mathrm{Z}(G), \quad A_1, \dots, A_n \text{ are minimal nonabelian.}$$

2 If $p > 3$ and G is a group of maximal class of order p^4 and exponent p, then all nonnormal subgroups of G are elementary abelian but $|G'| = p^2$.

Clearly, any non-Dedekindian subgroup of G satisfies the hypothesis.

(i) By Lemma J (iii), $A_i \in \{D_8, Q_8, S(p^3), M_{p^n}\}$, $i = 1, \dots, n$.

(ii) Assume that $E \le Z(G)$ is abelian of type (p, p). Let $L < G$ be nonnormal of order p. Then $L \not\le E$ so that $LE = L \times E \cong E_{p^3}$ and L, being the intersection of two G-invariant maximal subgroups of $L \times E$, is G-invariant, a contradiction, Thus, $Z(G)$ is cyclic. Also it follows from this that G has no subgroup $\cong E_{p^3}$ and the center of any non-Dedekindian subgroup of G is cyclic.

(iii) The extraspecial group $H \cong D_8 * D_8 \cong Q_8 * Q_8$ does not satisfy the hypothesis since it has a subgroup $\cong E_{2^3}$ (see (ii)). Next, if $p > 2$, then the decomposition $(**)$ contains at most one factor $\cong S(p^3)$ (indeed, having, by (ii), the cyclic center, the group $S(p^3) * S(p^3)$ is extraspecial so contains a subgroup $\cong E_{p^3}$).

(iv) There are among subgroups A_i at most one $\cong D_8$, at most one $\cong Q_8$ and at most one $S(p^3)$. This follows from (iii).

(v) Assume that $M_{p^m} = A_1 < G$; then $G = A_1 * C_G(A_1)$ (Lemma J (v)). Assume that $m > 3$. Let $L \le C_G(A_1)$ be a nonidentity cyclic subgroup not contained in A_1. Then the center $Z(A_1 * L) = Z(A_1)L$ of the non-Dedekindian group $A_1 * L$ is noncyclic so $A_1 * L$ does not satisfy the hypothesis (see the paragraph following (ii)). Thus, L does not exist so that $G = A_1$, i.e., G is as in (a). Now let $m = 3$; then $p > 2$. By (ii), $C_G(A_1)$ is cyclic. By Theorem 7.2 (d) and (ii), $\Omega_1(G) \cong S(p^3)$. Then $G = \Omega_1(G)Z(G)$ is as in (b).

Next we assume that $A_i \not\cong M_{p^n}$ so that $|A_i| = p^3$ for all i. It follows from (iv) that $n \le 2$.

(vi) Assume that $n = 2$. Then it follows from (iii) and (iv) that $p = 2$. In this case, by (iii), $G = HZ(G)$, where $H = A_1 * A_2 = D_8 * Q_8$. It is easy to see that the subgroup H satisfies the hypothesis. Assume that $H < G$. Then there is a cyclic subgroup $C \le Z(G)$ such that $C \not\le H$. Taking in mind our aim, one may assume, without loss of generality, that $G = H * C$ and $|C|$ is as small as possible. Then $C = Z(G)$ is cyclic of order 4 (here we use (ii)). Write $F = A_2 * C$ (here $A_2 \cong Q_8$); then F is of order 16, by (ii). The subgroup $F \cong Q_8 * C_4 \cong D_8 * C_4$ centralizes $A_1 \cong D_8$. In this case, G contains a subgroup $\cong D_8 * D_8$, contrary to (iv). Thus, if $H \le G$, then $G = H$ is as in (c).

(vii) Now let $n = 1$ and $A_1 < G$. Then $G = A_1Z(G)$ is not Dedekindian since $Z(G)$ is cyclic of order > 2 (Theorem 1.20). By (i) and (v), $A_1 \in \{S(p^3), D_8, Q_8\}$, and, by (ii), $Z(G)$ is cyclic so that $\exp(Z(G)) > p$. In that case, G is as in (b).

It is easily checked that groups G from (a–c) satisfy the hypothesis. □

Problem 1. Classify the p-groups all of whose non-quasinormal subgroups are cyclic (elementary abelian, abelian). (For a partial case, see Theorem 1.25 and § 16.)

Problem 2. Study the irregular p-groups all of whose maximal regular subgroups are normal.

Problem 3. Study the non-absolutely regular p-groups, $p > 2$, all of whose maximal absolutely regular subgroups are normal. (See § 9.)

Problem 4. Classify the nonabelian p-groups all of whose minimal nonabelian subgroups are isomorphic either to M_{p^n} or have order p^3.

Problem 5. Classify the non-Dedekindian p-groups G all of whose nonnormal abelian subgroups have orders $\leq p^3$. (See [ZLS].)

Problem 6. Study the p-groups all of whose nonnormal subgroups are absolutely regular. (For $p = 2$, see § 16.)

Problem 7. Classify the p-groups all of whose nonnormal abelian subgroups (nonabelian subgroups) have equal order.

Problem 8. Classify the non-Dedekindian p-groups all of whose nonnormal abelian subgroups are elementary abelian.

Problem 9. Is it true that an irregular p-group G contains a minimal irregular p-subgroup R such that $R_G \neq \{1\}$? (This is true for $p = 2$; see Theorem 200.1.)

Problem 10. Study the p-groups all of whose nonnormal nonabelian subgroups are minimal nonabelian. (Example: Any group of maximal class with abelian subgroup of index p and order p^5.)

§ 217 Nonabelian p-groups all of whose elements contained in any minimal nonabelian subgroup are of breadth < 2

The purpose of this section is to characterize the title groups (problem 2787). We prove here the following

Theorem 217.1 (Janko). *Let G be a nonabelian p-group such that, whenever $H \le G$ is minimal nonabelian and $x \in H$, then $|G : C_G(x)| \le p$. Then $|G'| = p$ and so each element in G is of breadth ≤ 1.*

Proof. Let G be a nonabelian p-group such that, whenever $H \le G$ is minimal nonabelian and $x \in H$, then $|G : C_G(x)| \le p$. Let A be a maximal normal abelian subgroup of G. If $g \in G - A$, then Lemma 57.1 implies that g is contained in a minimal nonabelian subgroup of G. It follows that $|G : C_G(g)| = p$.

Let $a \in A - Z(G)$ so that there is $h \in G - A$ such that $k = [a, h] \neq 1$; then $\langle a, h\rangle$ is nonabelian. Again, by Lemma 57.1, h is contained in a minimal nonabelian subgroup of G (Lemma 57.1) and therefore $|G : C_G(h)| = p$, by hypothesis, and this implies that $|A : C_A(h)| = p$. Write $F = \langle h, A\rangle$. By the above, $C_F(h) = Z(F)$ has index p^2 in F so that $k \in F' \le Z(F) \le C_A(h)$. Hence $k(\in Z(F))$ commutes with $a, h(\in F)$ and so

$$\langle k\rangle = \langle a, h\rangle' \le F' \cap \langle a, h\rangle \le Z(\langle a, h\rangle)\,.$$

It follows that $\mathrm{cl}(\langle a, h\rangle) = 2$, and so

$$k^p = [a, h]^p = [a^p, h] = 1 \Rightarrow |\langle a, h\rangle'| = p\,.$$

Therefore, by Lemma 65.2 (a), $\langle a, h\rangle$ is a minimal nonabelian subgroup, and so the hypothesis implies that $|G : C_G(a)| = p$. By this and the previous paragraph, the breadth of each noncentral element of G is equal to 1. Then, by Proposition 121.9, $|G'| = p$, completing the proof. □

This theorem one can consider as an improving of Theorem 121.9.

Problem. Classify the nonabelian p-groups G containing a minimal nonabelian subgroup H such that $|G : C_G(x)| \le p$ for all $x \in H - \{1\}$.

§ 218 A nonabelian two-generator p-group in which any nonabelian epimorphic image has the cyclic center

Let G be a p-group of maximal class. Then any proper nonabelian epimorphic image of G has the center of order p which is cyclic. The extraspecial p-groups, the groups $\mathcal{H}_{2,2}$ and M_{p^n} also satisfy the above property.

In this section we present two classifications of p-groups of maximal class: Theorems 218.1 and 218.3. Below we do not use deep properties of such groups proved in § 9.

Theorem 218.1. *Suppose that G is a nonabelian two-generator p-group. If the center of any nonabelian epimorphic image of G is cyclic, then one and only one of the following holds:*

(a) *G is of maximal class.*

(b) $G \cong M_{p^n}$, $n > 3$.

Next we use the following known result.

Lemma 218.2. *If G is a minimal nonabelian p-group with cyclic center, then* $G \in \{S(p^3), D_8, Q_8, M_{p^n}\}$.

Proof of Theorem 218.1. We proceed by induction on $|G|$. By hypothesis, $Z(G)$ is cyclic so that G is a monolith with the unique minimal normal subgroup $\Omega_1(Z(G))$.

Let $L < G'$ be G-invariant of index p. By Lemma 65.2 (a), the quotient group G/L is minimal nonabelian and, as its center is cyclic, it is one of the groups of Lemma 218.2. In the second and third cases, one has $|G : G'| = 4$ so that G is a 2-group of maximal class (Proposition 1.6), i.e., G is as in (a). Next we assume that $G/L \in \{S(p^2), M_{p^n}\}$.

(i) Suppose that $G/L \cong M_{p^n}$. Assume that $L > \{1\}$. Taking in mind our aim, one may assume that $|L| = p$. In that case, $|G'| = p^2$ and the quotient group $\bar{G} = G/L$ has two distinct cyclic subgroups A/L and B/L of index p. Then A and B are distinct abelian subgroups of index p in G so that $|G : Z(G)| = |G : (A \cap B| = p^2$ and, in view of $d(G) = 2$, G is minimal nonabelian hence $|G'| = p$ (Lemma 65.1), a contradiction. Thus, $L = \{1\}$, and G is as in (b).

(ii) Now let $G/L \cong S(p^3)$; then $p > 2$ and $|G : G'| = p^2$. We have to prove that in this case the group G is of maximal class. Assume that $|L| = p$; then $|G'| = p^2$, hence $|G| = p^4$. As G has an abelian subgroup of index p, it follows from Lemma 1.1 that $|Z(G)| = p$ so $Z(G) = L$, and we conclude that G is of maximal class.

Now let $|L| > p$. Let $R < L$ be G-invariant of order p. Then G/R is of maximal class, by induction. Assume that G is not of maximal class. Then $Z(G) \cong C_{p^2}$. By Lemma 1.4, there is in G a normal abelian subgroup $S \cong E_{p^2}$. Then $S \cap Z(G) = R$ since G is a monolith. However, $SZ(G)/R) \cong E_{p^2}$ is a central subgroup of G/R, contrary to the hypothesis. Thus, $R = Z(G)$ and G is of maximal class. □

The following theorem, in some sense, is dual to Theorem 218.1.

Theorem 218.3. *Let G be nonabelian p-group with $|G : G'| = p^2$. If all factors of the lower central series of G below G' are cyclic, then G is of maximal class.*

Proof. The theorem holds for $p = 2$, by Proposition 1.6.

In what follows we assume that $p > 2$. We proceed by induction on $|G|$.

If $\mathrm{cl}(G) = 2$, then $G' \leq Z(G)$ and $G/Z(G) \cong E_{p^2}$ which implies that G is minimal nonabelian hence $|G'| = p$ (Lemma 65.1), In that case, $|G| = |G : G'||G'| = p^3$ so that G is of maximal class.

Now let $\mathrm{cl}(G) > 2$; then $|G'| > p$. Let

$$G > K_2(G) = G' > \cdots > K_n(G) > K_{n+1}(G) = \{1\}$$

be the lower central series of G. Here $n = \mathrm{cl}(G) > 2$. In that case, $G/K_n(G)$ is of maximal class, by induction. By hypothesis, $K_n(G)$ is cyclic and $K_n(G) \leq Z(G)$.

Let $L \leq K_n(G)$ be of order p; then $L \triangleleft G$ and G/L is nonabelian since $\mathrm{cl}(G) > 2$.

If $L = K_n(G)$, then G is of maximal class (in that case, $Z(G) = L$).

Now assume that $L < K_n(G)$. By hypothesis, $K_n(G)$ is cyclic (of order $> p$). In that case, $G/L > K_2(G)/L > \cdots > K_n(G)/L > L/L$ is the lower central series of the quotient group G/L, and all factors of that series below $G'/L = K_2(G)/L = K_2(G/L)$ are cyclic. By induction, G/L is of maximal class. In that case, $K_n(G) = Z(G)$ is cyclic of order p^2. By Lemma 1.4, there is in G a normal abelian subgroup R of type (p, p); then $R > L$ since, in view of the cyclicicity of $Z(G)$, G is a monolith. It follows that $R \neq Z(G)$. In that case, $Z(G)/L$ and R/L are distinct normal subgroups of order p in G/L so that G/L is not of maximal class, a final contradiction. □

Problem 1. Study the nonabelian two-generator p-groups all of whose factors of the lower central series, except of the first one, are cyclic. (A partial case of this problem is solved in Theorem 218.3.)

Problem 2. Study the nonabelian two-generator p-groups all whose factors of the upper central series, except of the last one, are cyclic.

Problem 3. Study the nonabelian p-groups in which any minimal nonabelian subgroup has cyclic center.

Problem 4. Study the p-groups all of whose factors of the lower (upper) central series have orders $\leq p^2$.

Problem 5. Describe the representation groups of $\mathcal{A}_k$-groups. $k = 1, 2$.

Problem 6. Let G be a nonabelian two-generator p-group. Study the structure of G provided all factors of the lower central series of G below $\Phi(G)$ are cyclic.

§ 219 On “large” elementary abelian subgroups in p-groups of maximal class

By Exercise 9.13, if $p > 2$ and a p-group G of maximal class has a subgroup H with $d(H) > p-1$, then $G \cong \Sigma_{p^2} \in \mathrm{Syl}_p(\mathrm{S}_{p^2})$ and hence $H \cong \mathrm{E}_{p^p}$ (as any 2-group of maximal class and order $> 2^3$ shows, this is not true for $p = 2$). However, we know little on a p-group of maximal class, $p > 2$, if it has a subgroup H with $d(H) = p - 1$.

In this section, we consider, for $p > 3$, the p-groups G of maximal class containing an elementary abelian subgroup of order p^{p-1} or p^{p-2} (for $p \le 5$ all similar results are known) and show that then G contains a normal elementary abelian subgroup of order p^{p-1} and p^{p-2}, respectively. We present two different proofs of these results.[1]

Theorem 219.1. *Let $E \cong \mathrm{E}_{p^{p-1}}$ be a subgroup of a p-group $G \not\cong \Sigma_{p^2}$ of maximal class and order p^m, $p > 3$. Then, E is the unique elementary abelian subgroup of order p^{p-1} in G so that $E = \Omega_1(G_1) \lhd G$, where G_1 is the fundamental subgroup of G.*

We need the following

Lemma 219.2 (see also Theorem 13.19). *Let a p-group G of maximal class have order $p^m > p^{p+1}$. If $H < G$ is of order $> p^p$ and $H \not\le G_1$, then H is of maximal class.*

Proof. By Theorem 1.2, this is true for $p = 2$. Therefore, we assume in what follows that $p > 2$. We proceed by induction on m. If $m = p + 2$, the result follows from Theorem 9.6 (e) (indeed, then all members of the set $\Gamma_1 - \{G_1\}$ are of maximal class). Now let $m > p + 2$ and let $H \le M \in \Gamma_1$; then M is of maximal class (Theorem 9.6 (e)). Let $H \ne M$. The subgroup $M_1 = M \cap G_1$ is the fundamental subgroup of M since M_1 is the unique regular subgroup of index p in M (Theorem 9.6 (e) again). As $H \not\le M_1$, then H is of maximal class, by induction, applied to the pair $H < M$. □

Proof of Theorem 219.1. In view of Exercise 9.13, one may assume that E is a maximal elementary abelian subgroup of G. By hypothesis, $m \ge p$. If $m = p$, then $E \in \Gamma_1$ and, by Fitting's lemma (see Theorem 1.21 in Introduction), E is the unique elementary abelian subgroup of order p^{p-1} in G since $\mathrm{cl}(G) = p - 1 > 2$. Next we assume that $m \ge p + 1$.

(i) We first prove that there is in G a normal elementary abelian subgroup $\cong E$.

If $|G| = p^{p+1}$, there is in G a normal abelian subgroup, say E_1, of order p^{p-1} (Exercise 1.7). It follows from $|G : E_1| = p^2 = |G : \Phi(G)|$ that $E_1 = \Phi(G)$ (Exercise 9.1 (b)) so that $\Phi(G) = E_1$, and therefore, being of exponent p (Theorem 9.5), the subgroup $\Phi(G)$ is elementary abelian. Next we assume that $|G| > p^{p+1}$.

If $E \le G_1$, then $E = \Omega_1(G_1)$ is normal in G since it is characteristic in $G_1 \lhd G$ (see Theorem 9.6 (e)). In what follows we assume that $E \not\le G_1$.

---

1 As Mann has wrote, he is familiar with these results but do not know if they were published elsewhere.

If $E < H \le G$, where $|H : E| = p^2$, then $|H| = p^2|E| = p^{p+1} > p^p$ and $H \not\le G_1$ (see the previous sentence) so that H is of maximal class (Lemma 219.2). By Exercise 1.7, there is in H a normal abelian subgroup A of index $p^2 = |H : \Phi(H)|$ so that $A = \Phi(H)$ (Exercise 9.1 (b)). It follows from $\Phi(H) \le \Phi(G)$ that $A = \Phi(H) \le \Phi(G) < G_1$. However, $\exp(\Phi(H)) = p$ (Theorem 9.5), so that $A \cong \mathrm{E}_{p^{p-1}}$ is normal in G since $A = \Omega_1(G_1)$ (compare orders!).

(ii) It remains to prove that E is the unique elementary abelian subgroup of order p^{p-1} in G (if so, then $E = \Omega_1(G_1)$). By the result of the first paragraph and (i), one may assume that $m > p+1$. By (i) and Exercise 9.1 (b), one may assume that $E_1 = \Omega_1(G_1)$ is elementary abelian; then $E_1 \lhd G$. Assume, by way of contradiction, that $E \neq E_1$; then $E \not\le G_1$ so that $G = G_1E$ and hence $E \cap G_1 = E \cap \Omega_1(G_1) = E \cap E_1$ is of order p^{p-2} so that $H = EE_1$ is of order p^p, by the product formula; then E, E_1 are maximal subgroups of H. As H is regular (Theorem 7.1 (b)), we obtain $\exp(H) = p$ (Theorem 7.2 (b)), and we conclude that H is nonabelian (see the first paragraph of this section or Exercise 9.13). In that case, $E \cap E_1 = \mathrm{Z}(H)$. Let $H < T_1 < G$, where $|T_1 : H| = p$. Then, $T_1(\not\le G_1$ of order $p^{p+1})$ is of maximal class (Lemma 219.2) so $E \cap E_1 \lhd T_1$ since $E \cap E_1 = \mathrm{Z}(H)$ is a characteristic subgroup in $H \lhd T_1$, whence $E \cap E_1$ is characteristic in T_1 (Exercise 9.1 (b)). Let $T_1 < T_2 \le G$, where $|T_2 : T_1| = p$. Then, T_2 is of maximal class (Lemma 219.2) and $E \cap E_1$, being characteristic in $T_1 \lhd T_2$, is normal so characteristic in T_2 (Exercise 9.1 (b) again). Continuing so, we obtain $E \cap E_1 \lhd G$. It follows from $|E \cap E_1| = p^{p-2} \ge p^3 > |\mathrm{Z}_2(G)|$ that $\mathrm{Z}_2(G) < E \cap E_1$ (Exercise 9.1 (b)). In that case, $\mathrm{C}_G(\mathrm{Z}_2(G)) \ge G_1E = G$ so that $\mathrm{Z}_2(G) \le \mathrm{Z}(G)$, a final contradiction. □

Second proof of existence of a G-invariant subgroup $\cong E$ in Theorem 219.1. As above, one may assume that $|G| > p^{p+1}$, $E \not\le G_1$ and E is a maximal elementary abelian subgroup of G; then $\mathrm{Z}(G) < E$ since $\mathrm{Z}(G)E$ is elementary abelian. Note that $\mathrm{Z}_2(G) \cong \mathrm{E}_{p^2}$. As $\mathrm{C}_G(\mathrm{Z}_2(G)) = G_1$, we obtain $\mathrm{Z}_2(G) \not\le E$. Write $H = E\mathrm{Z}_2(G)$; then H is nonabelian (otherwise, $H > E$ is elementary abelian, contrary to the assumption). Then, $|H| = p^p$, by the product formula (recall that $E \cap \mathrm{Z}_2(G) = \mathrm{Z}(G)$), and, by the modular law, $\mathrm{C}_H(\mathrm{Z}_2(G)) = \mathrm{Z}_2(G)(E \cap \mathrm{C}_H(\mathrm{Z}_2(G))$ is elementary abelian of index p in H so the order of that centralizer is p^{p-1} since H is nonabelian. On the other hand, $\mathrm{C}_H(\mathrm{Z}_2(G)) \le H \cap \mathrm{C}_G(\mathrm{Z}_2(G)) = H \cap G_1$, being a subgroup of order $p^{p-1}(= |\Omega_1(G_1)|)$ and exponent p in (absolutely regular) G_1, coincides with $\Omega_1(G_1) \lhd G$, completing the proof. □

Corollary 219.3. *Suppose that a p-group $G \not\cong \Sigma_{p^2}$ of maximal class, $p > 3$, contains an abelian subgroup A of rank $p-1$. Then, $A \le G_1$.*

Proof. In view of Theorem 219.1, one may assume that $\exp(A) > p$. The result holds if $|G : A| = p$ (then $A = G_1$, by Theorems 9.5 and 9.6). Assume that $|G : A| > p$. Write $A_1 = \Omega_1(A)$; then $A_1 = \Omega_1(G_1)$ (Theorem 219.1). In that case, $\mathrm{Z}_2(G) < A_1 \Rightarrow A \le \mathrm{C}_G(A_1) \le \mathrm{C}_G(\mathrm{Z}_2(G)) = G_1$. □

Theorem 219.4. *If a p-group G of maximal class and order p^m, $p > 3$, possesses a subgroup $E \cong \mathrm{E}_{p^{p-2}}$, then there is in G a normal subgroup $\cong E$.*

Proof. One may assume that G has no subgroup $\cong E_{p^{p-1}}$ (see Theorem 219.1) and $m > p-1$. If $m = p$, there is in G a normal abelian subgroup A of index p^2 (Exercise 1.7). In that case, $A = \Phi(G)$ (Exercise 9.1 (b)) so $\exp(A) = p$ (Theorem 9.5) and therefore A is elementary abelian of order $|E|$. If $m = p+1$, then G has a normal abelian subgroup L of order p^{p-2} (Theorem 39.1) and L is elementary abelian (indeed, $L < \Phi(G)$ and $\exp(\Phi(G)) = p$, by Theorem 9.5). In what follows we assume that $m > p+1$.

Assume that $E < G_1$; then $E < \Omega_1(G_1)$. By Exercise 1.6, the number of subgroups $\cong E$ of order p^{p-2} in $\Omega_1(G_1)$ of order p^{p-1} is $\equiv 1 \pmod p$. As $\Omega_1(G_1) \lhd G$, there is in $\Omega_1(G_1)$ a G-invariant subgroup $\cong E$.

Next assume that $E \nleq G_1$. Let $E < H < G$, where $|H : E| = p^3$; then $|H| = p^{p+1}$. As $H \nleq G_1$, then H is of maximal class (Lemma 219.2). By Theorem 39.1, there is in H a normal so characteristic abelian subgroup A of index p^3 so of order p^{p-2}. By Exercise 9.1 (b), $A < \Phi(H) \Rightarrow A < \Omega_1(\Phi(H))$ so that $A \cong E_{p^{p-2}}$ and $A < \Phi(G)$ since $\Phi(H) < \Phi(G)$. In that case, $A < \Omega_1(\Phi(G))$. Arguing, word for word, as in the end of the previous paragraph with the pair $E < \Omega_1(\Phi(G))$, we see that there is in $\Omega_1(\Phi(G))$ a G-invariant subgroup $\cong A \cong E$. □

Second proof of Theorem 219.4. As in the first proof, one may assume that $|G| > p^{p+1}$ and that the fundamental subgroup G_1 has no subgroup $\cong E_{p^{p-2}}$. Also one may assume that E is a maximal elementary abelian subgroup of G (see Theorem 219.3); then $Z(G) < E$. Next, $C_G(Z_2(G)) = G_1 \Rightarrow Z_2(G) \nleq E$. Write $H = EZ_2(G)$; then $|H| = p^{p-1}$, by the product formula, $H \nleq G_1$ and H is nonabelian (otherwise $H \cong E_{p^{p-1}}$). Now the proof is finished by repeating with obvious changing, the second part of the second proof of Theorem 219.1. □

Remark 1. Suppose that G is a p-group of maximal class and order $> p^{p+1}$, $p > 3$. We claim that $d(G_1) > 2$. Write $\bar{G} = G/\mho_1(G)$; then $\bar{G}$ is of order $p^p \geq p^5$ and exponent p. By Theorem 5.8 (b), there is in $\bar{G}$ a subgroup $\bar{A}$ of index p with $d(\bar{A}) > 2$. Then, $d(A) > 2$ and $A \in \Gamma_1$. It follows from Theorem 9.6 (e) that $A = G_1$.

Proposition 219.5. *Suppose that a p-group G of maximal class, $p > 3$, and order $> p^{p+1}$ contains a subgroup H with $d(H) = p-1$. Then, $H \leq G_1$.*

Proof. As $d(G) = 2 < p-1 = d(H)$, it follows that $H < G$. If $H \in \Gamma_1$, the result follows from Theorem 9.6 (e). If $|H| = p^{p-1}$, then H is elementary abelian, and the result follows from Theorem 219.1. In what follows we assume that $\exp(H) > p$ and $|G : H| > p$. Then, $|H| \geq p^p$ and $|G| \geq p^2|H| = p^{p+2}$. We proceed by induction on $|G|$. By Lemma 219.2, if $|H| > p^p$, then $H < G_1$ since, in view of $d(H) = p-1 > 2$, the subgroup H is not of maximal class. Therefore, one may assume that $|H| = p^p$; then $|H'| \leq p$, by hypothesis, so that $cl(H) \leq 2$. By Corollary 219.3, if H is abelian, then $H \leq G_1$. Next we assume that H is nonabelian; then $cl(H) = 2$. Let $H < M < T \leq G$, where $|M : H| = |T : M| = p$. Then, M and T are of maximal class (Lemma 219.2). If $H \lhd T$, then $H = \Phi(T)$ (Exercise 9.1 (b)), which implies $H \leq \Phi(G) < G_1$. Now assume that H is not normal in T; then there is $x \in T$ such that $H^x \neq H$. In that case, $M =$

$HH^x = H^xH$. As $H^x \lhd M$ (indeed, if $y \in M$, then, since $xyx^{-1} \in M$, one has $(H^x)^{yx^{-1}} = H$ implies $((H^x)^y = H^x)$, we obtain, by Fitting's lemma (see Introduction, Theorem 21), $\mathrm{cl}(M) \le \mathrm{cl}(H) + \mathrm{cl}(H^x) = 2 + 2 = 4$. However, $\mathrm{cl}(M) = p > 4$ since M is of maximal class (Lemma 219.2) and order p^{p+1}, a contradiction. Thus, $H \le G_1$. □

Let us prove that if a 3-group G is of maximal class and order 3^m, then $\Phi(G)$ is abelian. One may assume that $m > 4$. The fundamental subgroup G_1 is absolutely regular (Theorem 9.6 (e)) so metacyclic (Theorem 9.11). As G_1 has a G-invariant abelian subgroup of type $(3, 3)$ (Lemma 1.4), one has $|G_1'| \le 3$ since $G_1' \lhd G$ is cyclic (see Exercise 9.1). If G_1 is nonabelian, it is minimal nonabelian (Lemma 65.2 (a)), and so $\Phi(G)$, being a proper subgroup of G_1, is abelian.

Let H be a subgroup of maximal class and index p in a p-group G. If G is not of maximal class, then all normal subgroups of H are G-invariant. Indeed, let $\{1\} < F \lhd H$. If $|H : F| > p$, then F is characteristic in H so normal in G. Now let $|H : F| = p$. If F is not normal in G, then $N_G(F) = H$ is of maximal class so G is of maximal class, by Remark 10.5, contrary to the hypothesis.

Suppose that a p-group G of order $> p^{p+1}$, contains an absolutely regular subgroup A of index p. Suppose, in addition, that G is neither absolutely regular nor of maximal class. Then the set Γ_1 has no member of maximal class. Indeed, assume that $H \in \Gamma_1$ is of maximal class. Then, $G/\mho_1(G)$ is of order p^{p+1} and exponent p (Theorem 12.12 (b)) so $A/\mho_1(G)$ of order p^p and exponent p is not absolutely regular, a contradiction.

Problem 1. Let $p \ge 11$. Assume that a p-group G of maximal class contains an elementary abelian subgroup E of order p^{p-3}. Is it true that G contains a normal subgroup $\cong E$? (Theorems 10.4 and 10.5 explain why $p \ge 11$.)

Problem 2. Is it true that if a p-group of maximal class contains an abelian subgroup of index p^4, then it contains a normal abelian subgroup of the same index?

Problem 3. Construct a p-group G of maximal class and order $> p^{p+1}$, $p > 3$, such that the subgroup $\Omega_1(\Phi(G))$ contains exactly $p + 1$ abelian subgroups of index p. (If such G exists, then $\Omega_1(\Phi(G)) \cong S(p^3) \times \mathrm{E}_{p^{p-4}}$.)

Problem 4. Study the p-groups of maximal class containing a subgroup H with $\mathrm{d}(H) = p - 1$.

Problem 5. Let m_G be the number of orders of maximal abelian subgroups in a p-group G of maximal class. Is it true that m_G is bounded on the set of all such G?

Problem 6. Given n, does there exist a p-group, containing exactly n maximal abelian subgroups of pairwise distinct orders? This problem is not solved even for metacyclic p-groups.

Problem 7. Study the p-groups G of maximal class such that $|G_1'| = p$.

§ 220 On metacyclic p-groups and close to them

In this section, which is a continuation of §§ 124, 148, metacyclic p-groups and close to them are considered.

There is only one nonabelian metacyclic group of order 16 and exponent 4; this group is denoted in our book by $H = \mathcal{H}_{2,2}$ (sometimes, we denote ir by $M_2(2,2)$); its defining relations are

$$\mathcal{H}_{2,2} = \langle a, b \mid a^4 = b^4 = 1, \ \ a^b = a^3 \rangle .$$

All its subgroups of order 2, namely,

$$H' = \langle a^2 \rangle , \quad U = \langle b^2 \rangle , \quad V = \langle a^2 b^2 \rangle ,$$

are characteristic. Indeed,

$$H/H' \cong C_4 \times C_2 , \quad H/U \cong D_8 , \quad H/V \cong Q_8$$

are pairwise nonisomorphic. Next, $R = \Omega_1(H) = \mho_1(H) = Z(H) \cong E_4 \cong H/R$. The group H has exactly two normal cyclic subgroups of order 4: $\langle a \rangle$ and $\langle ab^2 \rangle$, and therefore the maximal subgroup $\langle a, b^2 \rangle$ generated by these subgroups is characteristic in H. One has $c_2(G) = 6$ so that G has exactly four nonnormal cyclic subgroups of order 4: $\langle b \rangle$, $\langle a^2 b \rangle$, $\langle ab \rangle$, $\langle a^3 b \rangle$. Indeed, $b^a = a^2 b$, $(ab)^a = a^3 b$.

Also, for $p > 2$, there is only one nonabelian metacyclic group of order p^4 and exponent p^2. The two groups, from this and the previous paragraph, are minimal nonabelian, all their maximal subgroups are abelian of type (p^2, p).

It should be noted that if C is a normal cyclic subgroup of a nonabelian metacyclic p-group G such that G/C is cyclic, then the order of C, generally speaking, is not determined uniquely (example: $G = M_{p^n}$, $n > 3$). For the abelian group of type (p^m, p^n), $m \neq n$, the above assertion is obvious.

Remark 1. In what follows, we use freely the following fact. If $G > \{1\}$ is a metacyclic p-group such that $|\Omega_1(G)| \neq p^2$, then either G is cyclic or a 2-group of maximal class. Indeed, if $|\Omega_1(G)| < p^2$, then, by Proposition 1.3, G is either cyclic or a generalized quaternion group (which is of maximal class). Now suppose that $|\Omega_1(G)| > p^2$. If G has no normal abelian subgroup of type (p, p), it is a 2-group of maximal class (Lemma 1.4). Assume that G has a normal subgroup $R \cong E_{p^2}$. If $x \in G - R$ is of order p, then $H = \langle x, R \rangle$ is nonabelian of order p^3, and G is of maximal class, by Proposition 10.19. Obviously, in that case, $H \cong D_8$.

Let G be a metacyclic p-group. Set

$$p^{w(G)} = \max \{ i \mid |\Omega_i(G)| = p^{2i} \} , \quad R(G) = \Omega_{w(G)}(G)$$

(see §§ 124, 148). In that case, $G/R(G)$ is either cyclic or a 2-group of maximal class, by Remark 1. The characteristic subgroup $R(G)$ is important in treating of metacyclic p-groups (see §§ 124, 148).

Let G be a metacyclic p-group of exponent p^n. Then, $\Omega_n^*(G) = \langle x \in G \mid o(x) = 2^n\rangle = G$, unless $G/R(G) \in \{D_{2^k}, SD_{2^k}\}$.

Proposition 220.1. *Let $H = \mathcal{H}_{2,2}$ be a subgroup of a nonmetacyclic 2-group G, and suppose that $N_G(H)$ is metacyclic. Then, $R = \Omega_1(H) \lhd G$ and G/R is of maximal class.*

This is a partial case of Theorem A.79.5.

Exercise 1. Let H be a proper subgroup of maximal class of a p-group G. Study the structure of G provided the quotient group $N_G(H)/\Phi(H)$ is two-generator abelian.

Solution. Let $H < M \le G$ with $|M : H| = p$; then $M \le N_G(H)$. By hypothesis, $d(M) = d(M/\Phi(H)) = 2$ since $M/\Phi(H)$ is a noncyclic subgroup of the abelian two-generator group $N_G(H)/\Phi(H)$. By Theorem 12.12 (a), the subgroup M is of maximal class. Therefore, by Exercise 10.10, G is also of maximal class.

Exercise 2. Let $H = \langle a, b \mid o(a) = o(b) = 2^n,\ n > 2,\ a^b = a^{1+2^{n-1}}\rangle = R(H)$ be a proper metacyclic minimal nonabelian subgroup of a 2-group G. If $N_G(H)$ is metacyclic, then $\Omega_{n-1}(H) \lhd G$ and one of the following holds:

(a) G is metacyclic.

(b) $G/\mho_1(H)$ is of maximal class.

This is a partial case of Theorem A.79.5.

Exercise 3. Let $p > 2$ and let $H < G$ be a metacyclic subgroup of order p^{2n} and exponent p^n. Then, $N_G(H)$ is metacyclic $\iff$ G is. (**Hint.** Take into account that, by Theorem 7.1 (b), $N_G(H)$ is regular.)

Proposition 220.2. *Suppose that a nonabelian p-group G of exponent p^e, $p > 2$, satisfies $|\Omega_i(G)| = p^{2i}$ for $i \le e$. Then one of the following holds:*

(a) *G is metacyclic, $G = R(G)$ with $w(G) = e$.*

(b) *$e = 2$, G is a minimal nonmetacyclic group of order 3^4 (in that case, $cl(G) = 3$).*

Proof. If G is metacyclic, then $G = R(G)$ and all such G satisfy the hypothesis. In what follows, we assume that G is nonmetacyclic; then G is nonabelian. By hypothesis, G has no subgroup of order p^3 and exponent p.

Next we use Theorem 69.4. Let $H \le G$ be minimal nonmetacyclic; then $|\Omega_1(H)| = p^2$, $p = 3$. In that case, $H = \Omega_2(G)$, by hypothesis. One has $\Omega_1(G) = \Omega_1(H) \cong E_9$ and so G must be a 3-group of maximal class, by Theorem 12.1 (a), and then, by Theorem 13.19, $G = \Omega_2(G) = H$. □

The following proposition is a generalization of Theorem 138.2.

Proposition 220.3. *Let G be a nonabelian p-group of order $> p^3$, $p > 2$, such that $|\mathrm{N}_G(L) : L| = p$ for all nonnormal cyclic $L < G$. Then, G is metacyclic of order p^4 and exponent p^2.*

Proof. Assume that $E < G$ is elementary abelian of order p^3. Then, $E \le \mathrm{Z}(G)$ since all subgroups of E of order p are G-invariant so central, by hypothesis. If $L < G$ is nonnormal cyclic, then $EL \le \mathrm{N}_G(L)$ and $|EL/L| \ge p^2$, a contradiction. Thus, E does not exist. Then, G is one of the groups of Theorem 13.7 of order $> p^3$ (note that any nonabelian group of order p^3 satisfies the "normalizer" condition of the theorem).

By Theorem 1.2, G has no cyclic subgroup of index $p(> 2)$.

Assume that $L < G$ is nonnormal of order p. Then, by hypothesis, $\mathrm{N}_G(L)$ is of order p^2 so that G is of maximal class (Proposition 1.8). Let G_1 be the fundamental subgroup of G (see § 9). Then there is in G_1 a subgroup L_0 of order p, $L_0 \not\le \mathrm{Z}(G)$. In that case, $|\mathrm{N}_{G_1}(L_0)| > p^2$ since $|\mathrm{Z}(G_1)| \ge p^2$, a contradiction. Thus, all subgroups of G_1 of order p are central so $\Omega_1(G_1) \le \mathrm{Z}(G)$ has order p^2, a contradiction. Since, as have noted above, G is one of the groups of Theorem 13.7, it is metacyclic. By Proposition 10.19, all minimal nonabelian subgroups of G have orders $> p^3$. Let $A \le G$ be minimal nonabelian. It follows from Lemma 65.1 that

$$A = \langle a, b \mid o(a) = p^2,\ o(b) = p^n > p,\ a^b = a^{1+p} \rangle\,.$$

Assume that $n > 2$. Write $R = \Omega_2(A)$; then A/R is cyclic of order p^{n-2} and R is abelian of type (p^2, p^2). One has $R = U \times B$, where $U = \langle a \rangle$ and $B = \langle b^{p^{n-2}} \rangle$. Then, $L_1 = \langle ab^{p^{n-2}} \rangle$ is non-b-invariant cyclic of order p^2 in A and $L_1 \cap B = \{1\}$. As B of order p^2 centralizes L_1, we get a contradiction. Thus, $n = 2$. In that case, $A = \Omega_2(G) \lhd G$ since G is regular (Theorem 7.1 (c)), i.e., A is the unique minimal nonabelian subgroup in G. By Remark 76.1, $A = G$. □

Remark 2. If G is a minimal nonabelian 2-group of order $> 2^3$ and $|\mathrm{N}_G(L)/L| = 2$ for all nonnormal cyclic $L < G$, then $G \cong \mathcal{H}_{2,2}$.

Exercise 4. Let G be a nonabelian metacyclic 2-group of order $> 2^3$ such that $|\mathrm{N}_G(L) : L| = 2$ for all nonnormal cyclic $L < G$. Then either G is of maximal class or all minimal nonabelian subgroups of G are $\cong \mathcal{H}_{2,2}$.

Hint. All 2-groups of maximal class satisfy the hypothesis. Next, we assume that G is not of maximal class. Then, in view of Proposition 1.8, $\Omega_1(G) \le \mathrm{Z}(G)$. Therefore, if $A \le G$ is minimal nonabelian, then $A \cong \mathcal{H}_{2,2}$ (Remark 2).

Exercise 5. Suppose that a 2-group G contains two normal subgroups L, M of order 2 such that $G/L \cong \mathrm{D}_8$ and $G/M \cong \mathrm{Q}_8$. Prove that $G \cong \mathcal{H}_{2,2}$. State a similar result for p^2.

Solution. Our G as a subgroup of $(G/L) \times (G/M)$, has exponent 4. One has $|G| = 16$ and $Z(G) = L \times M \cong E_4$. If G is metacyclic, then $G \cong \mathcal{H}_{2,2}$. Now assume that G is nonmetacyclic. It follows from $G/M \cong Q_8$ that G has no subgroup $\cong E_8$. By Theorem 66.1, G has no minimal nonmetacyclic subgroup, a contradiction.

Exercise 6. Describe the structure of the automorphism group of the nonabelian metacyclic group of order p^4 and exponent p^2.

Exercise 7. Suppose that a 2-group G of order $> 2^4$ contains two normal subgroups L, M of order 2 such that $G/L \cong D_{2^n}$ and $G/M \cong Q_{2^n}$. Is it true that G is metacyclic?

Solution. Clearly, $\Omega_1(Z(G)) \cong E_4$. Assume that G is nonmetacyclic. Let $H \leq G$ be minimal nonmetacyclic. Since G/L and G/M are metacyclic, it follows that $L, M < H$. As $G/M \cong Q_{2^n}$, it follows that $H \notin \{E_8, D_8\}$ and $H \ncong D_8 * C_4$ so that $H/M \cong Q_8$ and $H \cong M \times Q_8$ (see Theorem 66.1). As H has no section $\cong D_8$, we get $H/L \cong E_8$ which is not isomorphic to a subgroup of $G/L \cong D_{2^n}$, a contradiction. Thus, H does not exist so G is metacyclic.

Exercise 8. Let G be a p-group, $p > 2$, containing two normal subgroups L, M of order p such that $G/L \cong S(p^3)$ and $G/M \cong M_{p^3}$. Prove that G is minimal nonabelian.

Hint. One has $|G| = p^4$, $Z(G) = L \times M \cong E_{p^2}$ and $\exp(G) = p^2$. As L, M are not direct factors of G, it follows that $L, M \leq \Phi(G)$ and so $d(G) = 2$. Therefore, $\Phi(G) = Z(G)$ implies that G is minimal nonabelian. One has $G = \langle a, b \mid a^{p^2} = b^p = 1,\ [a, b] = c,\ c^p = 1,\ [a, c] = [b, c] = 1\rangle$.

Exercise 9. Let a metacyclic 2-group G contain a nonnormal subgroup $H > R(H)$. Then, $R(H) \lhd G$.

Solution. One may assume that $R(H) > \{1\}$. Then, $\Omega_1(H) = \Omega_1(G) \lhd G$ (see Proposition 10.19). Therefore, one may assume that $\Omega_1(H) < R(H)$. By induction on $|R(H)|$, one obtains $R(H/\Omega_1(G)) \lhd G/\Omega_1(G)$, and so $R(H) \lhd G$.

Exercise 10. Let H be a proper absolutely regular subgroup of a p-group G, $p > 2$, such that $|\Omega_1(H)| = p^{p-1}$, $\exp(H) = p^e$ and $|H| = p^{e(p-1)}$. Prove that $H \lhd G$ provided $N_G(H)$ is absolutely regular.

Solution. By Theorems 7.2 (b), $H = \Omega_e(N_G(H))$ is characteristic in $N_G(H)$. It follows that $N_G(H) = G$, $H \lhd G$ so $G = N_G(H)$ is absolutely regular, by hypothesis.

Exercise 11. If a metacyclic p-group $G = AB$, where $A, B < G$ and $A \cap B = \{1\}$, then A, B are cyclic, unless $G \in \{D_{2^n}, SD_{2^n}\}$.

Solution. One may assume that G is not a 2-group of maximal class and $|G| > p^3$. Then, A, B are not of maximal class (Proposition 10.19). Assume that A is noncyclic. Then, $E_{p^2} \cong \Omega_1(A) = \Omega_1(G) \triangleleft G$ (here we use Proposition 10.19). In that case, $A \cap B \geq \Omega_1(G) \cap B > \{1\}$, a contradiction.

Problem 1. Classify the metacyclic 2-groups all of whose subgroups of order 2 (of index 2) are characteristic. In particular, classify the metacyclic 2-groups all of whose normal subgroups are characteristic.

Problem 2. Study the 2-groups G containing a proper nonabelian metacyclic subgroup $M \triangleleft G$ such that, whenever $M < H \leq G$ with $|H : M| = 2$, then $d(H) = 2$.

Problem 3. Study the 2-groups (metacyclic 2-groups) G containing a subgroup H isomorphic with $\mathcal{H}_{2,2}$ and such that $C_G(H) < H$.

Problem 4. Classify the 2-groups G such that $|G/\mho_k(G)| = 2^{2k}$ for all k. Consider in detail the case when $G/\mho_2(G)$ is nonmetacyclic minimal nonabelian of order 2^4.

Problem 5. Study the groups G of exponent p such that $|N_G(L)| = p^3$ for some $L < G$ of order p (of order p^2).

§ 221 Non-Dedekindian p-groups in which normal closures of nonnormal abelian subgroups have cyclic centers

If H is a subgroup of a group G, then its normal closure H^G in G is the intersection of all G-invariant subgroups containing H. The structure of G depends essentially on the normal closures of some its nonnormal subgroups (see, for example, § 62).

In this section we prove

Theorem 221.1. *The following conditions for a non-Dedekindian p-group G of order* > 8 *are equivalent:*

(a) *The normal closure of any nonnormal abelian subgroup of G has cyclic center.*

(b) *G is a 2-group of maximal class.*

Proof. (a) $\Rightarrow$ (b): Assume that G is not a 2-group of maximal class. Then, by Lemma 1.4, there is in G a normal abelian subgroup $R \cong \mathrm{E}_{p^2}$. If $C < R$ is non-G-invariant of order p, then the center of $C^G = R \cong \mathrm{E}_{p^2}$ is noncyclic, contrary to the hypothesis. Thus, all subgroups of R are G-invariant so that $R \leq \mathrm{Z}(G)$.

Let $L < G$ be nonnormal cyclic. Then $\mathrm{Z}(L^G)$ is cyclic, by hypothesis, so that R, being a noncyclic central subgroup, is not contained in L^G. As L is not normal in G, it follows that L^G is noncyclic. As $\mathrm{Z}(L^G)$ is cyclic, we conclude that L^G is nonabelian. Therefore, there is in R a subgroup K of order p which is not contained in L^G. Write $H = KL^G = K \times L^G$. The subgroup $\mathrm{Z}(H) = K \times \mathrm{Z}(L^G)$ is noncyclic. Since $(K \times L)^G = K \times L^G > K \times L$, it follows that $K \times L$ is a nonnormal abelian subgroup of G and its normal closure $K \times L^G$ has the noncyclic center, contrary to the hypothesis. Thus, R does not exist so that G is a 2-group of maximal class (Lemma 1.4).

The implication (b) $\Rightarrow$ (a) follows from the following fact. If a 2-group of maximal class and order > 8, then all normal subgroups of G have cyclic centers. □

For more general result see Theorem 236.1, where Problem 1, below, is solved (that proof is based on deep results from § 223).

Exercise 1. A non-Dedekindian 2-group in which any nonnormal abelian subgroup has the normal closure of maximal class, is a 2-group of maximal class of order > 8.

Problem 1. Classify the non-Dedekindian p-groups in which all nonnormal cyclic subgroups have normal closures with cyclic centers.

Problem 2. Classify the non-Dedekindian p-groups in which the normal closures of all nonnormal cyclic subgroups are minimal nonabelian.

Problem 3. Classify the primary non-Dedekindian $\mathcal{A}_n$-groups, $n > 2$, such that the normal closures of all their $\mathcal{A}_1$-subgroups are $\mathcal{A}_2$-subgroups.

Problem 4. Study the irregular p-groups, $p > 2$, in which the normal closures of all nonnormal $\mathcal{A}_1$-subgroups are (i) regular, (ii) irregular (iii) of maximal class.

Problem 5. Study the non-Dedekindian p-groups G such that, whenever $L < A < G$, where L is nonnormal cyclic and A is an $\mathcal{A}_1$-subgroup, then L^G is an $\mathcal{A}_1$-subgroup.

Problem 6. Study the p-groups in which normal closures of all cyclic subgroups are metacyclic.

Problem 7. Study the nonabelian p-groups covered by the normal closures of $\mathcal{A}_1$-subgroups. (This is a generalization of #860.)

§ 222 Characterization of Dedekindian p-groups, 2

A group is said to be Dedekindian if all its subgroups are normal. Such groups are classified by Dedekind (see Theorem 1.20; in particular, Dedekindian groups of odd order are abelian). It is interesting to study the groups all of whose "small" subgroups are normal. Some results of this type will be proved below. It is natural to replace in all such theorems normality by quasinormality. (Recall that $H \leq G$ is said to be quasinormal if it is permutable with all subgroups of G.)

It is proved in § 210 that if G is a p-group G and $|H|^2 < |G|$ implies $H \trianglelefteq G$, then G is either Dedekindian or $G \in \{M_p(m, m), Q_{16}\}$, where $M_p(m, m)$ is the metacyclic minimal nonabelian group of order p^{2m} and exponent p^m. In the following propositions, we establish other results of a similar type. For example, if $p^{\nu(G)}$ is a maximal order of an abelian subgroup of a nonabelian p-group G, then we describe, up to isomorphism, all such G in which all subgroups of orders $p^{\nu(G)-2}$ are normal. Minimal nonabelian subgroups play important role in our arguments.

If G is a minimal nonabelian p-group, then $G \in \{Q_8, M_p(m, n, 1), M_p(m, n)\}$, where

$$M_p(m, n, 1) = \langle a, b \mid o(a) = p^m,\ o(b) = p^n,\ [a, b] = c,\ o(c) = p,\ [a, c] = [b, c] = 1 \rangle ,$$
$$M_p(m, n) = \langle a, b \mid o(a) = p^m,\ o(b) = p^n,\ a^b = a^{1+p^{m-1}} \rangle .$$

Let $p^{\nu(G)} = p^{\nu} > p^2$ be the maximal order of an abelian subgroup of a nonabelian p-group G (in this case $|G| > p^3$).

Exercise 1. If all subgroups of G of order $\leq p^{\nu-1}$ are normal, then G is Dedekindian.

Solution. Clearly, $H < G \Rightarrow \nu(H) \leq \nu(G)$. Therefore, if $A < H$ is such that $|A| \leq p^{\nu(H)-1}$, then $|A| \leq p^{\nu(G)-1}$ so that $A \trianglelefteq G$. Let $H \leq G$ be minimal nonabelian. Then all subgroups of H of index $\leq p^2$ are G-invariant. In that case, A is Dedekindian so that $A \cong Q_8$. By Corollary A.17.3, $G = Q \times E$, where Q is a generalized quaternion group and $\exp(E) \leq 2$. In that case, $2^{\nu(G)} = \frac{1}{2}|G|$. It is easy to check that $Q \cong Q_8$ so that G is Dedekindian.

Let $p^{a(G)}$ be the maximal order of a minimal nonabelian subgroup of a nonabelian p-group G. If $H < G$ is nonabelian, then $p^{a(H)} \leq p^{a(G)}$.

Proposition 222.1. *Suppose that G is a non-Dedekindian p-group such that, whenever $H < G$ and $|H|^2 \leq p^{a(G)}$, then $H \trianglelefteq G$. Then $G = Q \times E$, there $Q \cong Q_{2^n}$, $n > 3$, and* $\exp(E) \leq 2$.

Proof. Let $M \leq G$ be minimal nonabelian of order $p^{a(G)}$ and let $H < M$ be such that $|H|^2 \leq |M| = p^{a(G)}$; then $H \trianglelefteq G$, by hypothesis, so $H \trianglelefteq M$. By Theorem 210.1, $M \cong Q_8$ (the group $M_p(m, m)$ does not satisfy the condition). It follows that $p = 2$ and all minimal nonabelian subgroups of our group G are $\cong Q_8$ (in that case, $a(G) = 3$). Now the result follows from Corollary A.17.3. □

Suppose that a p-group G all of whose abelian subgroups of order $p^{\nu(G)-2}$ are normal, contains a nonnormal abelian subgroup A of order $p^{\nu(G)}$. Then $\exp(G) \geq p^{\nu(G)-1}$, i.e., all nonnormal abelian subgroups of G have cyclic subgroups of index p. Indeed, by hypothesis, A is not generated by its G-invariant subgroups of index p^2 hence A has a cyclic subgroup of index p, and we conclude that $\exp(G) \geq p^{\nu(G)-1}$, as asserted.

Lemma 222.2. *Suppose that all subgroups of index $\geq p^3$ of a minimal nonabelian p-group G are normal. Then either $|G| = p^3$ or $G = \mathrm{M}_p(2, 2)$.*

Proof. Suppose that $|G| > p^3$. Let $G = \mathrm{M}_p(m, n, 1)$ be nonmetacyclic and $m \geq n$; then $m \geq 2$. In that case, G has a nonnormal cyclic subgroup C of order p^n. Then $|G : C| = p^{m+1} \geq p^3$, contrary to the hypothesis. Thus,

$$G = \mathrm{M}_p(m, n) = \langle a, b \mid o(a) = p^m,\ o(b) = p^n,\ a^b = a^{1+p^{m-1}} \rangle$$

is metacyclic and $m \geq 2$. There is in G a nonnormal cyclic subgroup D of order p^n. In that case, $|G : D| = p^m < p^3$ so that $m = 2$. Assume that $n > 2$. Then the cyclic subgroup $Z = \langle ab^{p^{n-2}} \rangle$ of order p^2 is non-b-invariant so non-G-invariant. In that case, $|G : Z| = p^n \geq p^3$ so $Z \lhd G$, which is a contradiction. Thus, $n = 2$ and hence $G = \mathrm{M}_p(2, 2)$. □

Proposition 222.3. *The following conditions for a nonabelian p-group G are equivalent:*
(a) *All abelian subgroups of G of order $\leq p^{a(G)-2}$ are normal.*
(b) *$G = Q \times E$, where Q is a generalized quaternion group and* $\exp(E) \leq 2$.

Proof. (a) $\Rightarrow$ (b): Let $M \leq G$ be minimal nonabelian of order $p^{a(G)}$. Then all subgroups of M of index $\geq p^2$ are G-invariant, by hypothesis, so that $M \cong \mathrm{Q}_8$ (Exercise 1), i.e., all minimal nonabelian subgroups of G are $\cong \mathrm{Q}_8$. By Corollary A.17.3, G is as in (b), establishing the required implication.

The reverse implication is obvious since, in (b), $a(G) = 3$ and $\Omega_1(G) = \mathrm{Z}(G)$ is elementary abelian. □

Theorem 222.4. *Let G be a nonabelian p-group, $p > 2$ and $\nu(G) > 2$ (so that $|G| > p^3$). If any subgroup of G of order $\leq p^{\nu(G)-2}$ is G-invariant, then $G \cong \mathrm{M}_p(2, 2)$.*

Proof. By hypothesis, $\Omega_1(G) \leq \mathrm{Z}(G)$. Let $M \leq G$ be minimal nonabelian. By Lemma 222.2, either $|M| = p^3$ or $M \cong \mathrm{M}_p(2, 2)$.

Let $|M| = p^3$. As $\nu(G) > 2$, all subgroups of M of order p are G-invariant so M-invariant, a contradiction since $p > 2$.

Now let $M \cong \mathrm{M}_p(2, 2)$. As M has a nonnormal subgroup of order p^2, it follows that $\nu(G) = 3$. Assume that $|\Omega_1(G)| > p^2$. Then there is in $\Omega_1(G) (\leq \mathrm{Z}(G))$ a subgroup L of order p such that $L \nleq M$. As $LM = L \times M$, one has $\nu(G) \geq 4$, contrary to what has just been said. Thus, $|\Omega_1(G)| = p^2$. As $\Omega_1(G) \leq \mathrm{Z}(G)$, it follows from Theorem 13.7 that G is metacyclic.

Assume that $M < G$; then $|G| \geq p^5$. There is in G a cyclic $Z \lhd G$ such that G/Z is cyclic. By assumption, $|G/Z| > p$. One has $p < |Z| \leq p^3$ since G is nonabelian and $\nu(G) = 3$. If $|Z| = p^2$, then $C_G(Z) \in \Gamma_1$ is abelian so $\nu(G) \geq 4$, a contradiction. Thus, $|Z| = p^3$. As G/Z is covered by a cyclic subgroup of order $\geq |G/Z|$, it follows that $|G/Z| \leq p^3$. As $\nu(G) = 3$, we get $C_G(Z) = Z$ so $|G/Z| = p^2$ since $|\mathrm{Aut}(Z)| = p^2(p-1)$. Hence $|G : M| = p$ (recall that $M \cong \mathrm{M}_p(2,2)$), $|G| = p^5$. Write $U = C_G(\mho_1(Z))$. If $|U : Z| = p$, then $|U| = p^4$ so U is nonabelian with the cyclic subgroup Z of index p, and hence $U \cong \mathrm{M}_{p^4}$ (Theorem 1.2) is minimal nonabelian, a contradiction since $\Omega_1(U) = \Omega_1(G) \leq Z(G)$ but $\Omega_1(U) \nleq Z(U)$. Thus, $U = G$, i.e. $\mho_1(Z) \leq Z(G)$. In that case, the metacyclic group $G/\mho_1(Z)$ of order p^3 has a cyclic subgroup $V/\mho_1(Z)$ of order p^2. Then the subgroup V of order $p^4 > p^3 = p^{\nu(G)}$ is abelian since $\mho_1(Z) \leq Z(G)$, a final contradiction. □

In the proof of Theorem 222.4, the oddness of p is important.

Remark 1. Let G be a non-Dedekindian 2-group with $\nu(G) > 2$. Suppose that all subgroups of G of order $\leq 2^{\nu(G)-2}$ are G-invariant. As above, $\Omega_1(G) \cong \mathrm{E}_4$ is contained in $Z(G)$, any $\mathcal{A}_1$-subgroup of G is isomorphic either Q_8 or $\mathrm{M}_2(2,2)$. If G has no subgroup $\cong \mathrm{M}_2(2,2)$, then $G = Q \times E$, where $Q \cong \mathrm{Q}_{2^n}$, $n > 3$, and $\exp(E) \leq 2$ (Corollary A.17.3). It is easy to see that we must have $E = \{1\}$ and $n = 4$. Next we assume that there is in G a subgroup $M \cong \mathrm{M}_2(2,2)$; then $\nu(G) = 3$. Assume that $M < G$. Then $R = \Omega_1(M) = \Omega_1(G)$. As $\nu(G) = 3$ and $R \leq Z(G)$, it follows that $\exp(G/R) = 2$. Let $A/R < G/R$ be of order 2. Then A is abelian of type $(4, 2)$ (indeed, A is not dihedral in view of $\Omega_1(G) \leq Z(G)$, and $C_G(A) = A$). As $\mathrm{Aut}(A) \cong \mathrm{D}_8$ and $\exp(G/A) = 2$, it follows that $G/A \cong \mathrm{E}_4$, and then $|G| = 2^5$. As $R = \Phi(M) \leq \Phi(G)$, it follows that $\Phi(G) = R$ so that $d(G) = 3$. It follows from $\nu(G) = 3$ that $Z(G) = R$. By Theorem 69.1, G is not minimal nonmetacyclic (indeed, the minimal nonmetacyclic group of order 2^5 has an abelian subgroup of order 2^4). Therefore, there is in the set Γ_1 a minimal nonmetacyclic subgroup $H \cong \mathrm{Q}_8 \times \mathrm{C}_2$ (see Theorem 66.1).

Problem 1. Classify the nonabelian p-groups G in which all abelian subgroups $A < G$ satisfying $|A| < p^{\nu(G)-2}$ are normal.

Problem 2. Classify the nonabelian 2-groups G in which $M \in \{\mathrm{Q}_8, \mathrm{M}_{p^n}, \mathrm{M}_2 2, 2\}$ for any minimal nonabelian $M \leq G$.

Problem 3. Classify the nonmodular p-groups G all of whose abelian subgroups of orders $\leq p^{a(G)-2}$ are quasinormal in G. In particular, classify the nonmodular p-groups G such that, whenever $A \leq G$ is minimal nonabelian, then all subgroups of A are quasinormal in G. (In that case all nonabelian subgroups of G are quasinormal, by Theorem 10.28.)

Problem 4. Study the nonmetacyclic p-groups such that, whenever $H \leq G$ is minimal nonmetacyclic, then all subgroups of H are normal in G. (In that case, any minimal nonmetacyclic subgroup of G is either $\cong \mathrm{E}_{p^3}$ or $\cong \mathrm{Q}_8 \times \mathrm{C}_2$, by § 66 and § 69.)

Problem 5. Study the nonmodular p-groups in which the intersection of any two quasinormal subgroups is quasinormal.

Problem 6. Classify the nonmodular p-groups all of whose nonquasinormal subgroups have equal order.

§ 223 Non-Dedekindian p-groups in which the normal closure of any nonnormal cyclic subgroup is nonabelian

The purpose of this section is to classify the title p-groups. It is surprising that we must have $p = 2$ and that the corresponding 2-groups can be determined up to isomorphism (Theorem 223.1). After that we determine such 2-groups in two special cases (Corollaries 223.4 and 223.5), where we use Propositions 223.2 and 223.3 about quasidihedral and quasi-generalized quaternion groups (see Definition 1).

Definition 1. Let M be a nonabelian 2-group possessing an abelian subgroup H of index 2 and exponent ≥ 4 such that there is an element $v \in M - H$ which inverts each element of H. Then $o(v) \leq 4$ since v inverts $\langle v^2 \rangle \leq H$.

If $o(v) = 2$, then all elements in $M - H$ are involutions and M is called "quasidihedral" (or generalized dihedral).

If $o(v) = 4$, then all elements in $M - H$ are of order 4 with the same square v^2 and then M is called "quasi-generalized quaternion."

In the proof of Theorem 223.1, we shall use very often Theorem 125.1 and therefore we state here explicitly that theorem for convenience.

Theorem 125.1. *Let G be a nonabelian p-group containing a maximal subgroup H such that all subgroups of H are G-invariant. Then there is an element $g \in G - H$ such that one of the following hold:*

(i) $p = 2$, *H is Hamiltonian, i.e., $H = Q \times V$, where $Q \cong \mathrm{Q}_8$, $\exp(V) \leq 2$, and $g \in \mathrm{Z}(G)$, $\mathrm{o}(g) \leq 4$.*

(ii) $p = 2$, *H is abelian of exponent 2^e, $e \geq 2$, and g either inverts each element in H, or $e \geq 3$ and $h^g = h^{-1+2^{e-1}}$ for all $h \in H$. In both cases $\mathrm{Z}(G) = \mathrm{C}_H(g) = \Omega_1(H)$ is elementary abelian and $\mathrm{o}(g) \leq 4$.*

(iii) $p = 2$, *H is abelian of exponent 2^e, $e \geq 3$, and $h^g = h^{1+2^{e-1}}$ for all $h \in H$, where $\mathrm{Z}(G) = \mathrm{C}_H(g) = \Omega_{e-1}(H)$.*

(iv) $p > 2$, *H is abelian of exponent p^e, $e \geq 2$, and $h^g = h^{1+p^{e-1}}$ for all $h \in H$, where $\mathrm{Z}(G) = \mathrm{C}_H(g) = \Omega_{e-1}(H)$.*

Theorem 223.1 (Janko). *Let G be a non-Dedekindian p-group in which the normal closure of any nonnormal cyclic subgroup is nonabelian. Then $p = 2$, G has an abelian maximal subgroup A of exponent 2^e, $e \geq 3$, and for an element $v \in G - A$ and for all $h \in A$ we have either $h^v = h^{-1}$ or $h^v = h^{-1+2^{e-1}}$.*

Conversely, let G be a 2-group just defined. Then each subgroup of A is G-invariant and for each $v \in G - A$, $o(v) \leq 4$, $\langle v \rangle$ is nonnormal in G and $\langle v \rangle^G = [A, \langle v \rangle]\langle v \rangle$, where v inverts each element of $[A, \langle v \rangle] = G'$ (of exponent $2^{e-1} \geq 4$) so that $\langle v \rangle^G$ is either *quasidihedral (in case $o(v) = 2$) or quasi-generalized quaternion (in case $o(v) = 4$) and so in any case $\langle v\rangle^G$ is nonabelian.*

Proof. Suppose that G is a title p-group. Let $A < G$ be a maximal normal abelian subgroup of G. Since each cyclic subgroup in A is normal in G, it follows that each subgroup in A is G-invariant.

Suppose $p > 2$. Let B/A be a normal subgroup of order p in G/A. Applying Theorem 125.1 on the subgroup B, we get $\exp(A) = p^e$, $e \geq 2$, and there is $g \in B - A$ such that for all $h \in A$, $h^g = h^{1+p^{e-1}}$. Since B is nonabelian, there is $b \in B - A$ such that $\langle b\rangle$ is not normal in B (and so $\langle b\rangle$ is not normal in G) so that (replacing b with a suitable power b^i, $i \not\equiv 0(\mathrm{mod}\, p)$, if necessary) we have for all $h \in H$, $h^b = h^{1+p^{e-1}}$ and $M = \langle b\rangle^G$ is nonabelian with $M \leq B$. Also, $B' = [A, \langle b\rangle]$ is elementary abelian and $B' \leq Z(G)$ because each subgroup of A is G-invariant. There is $k \in G$ such that $[b, b^k] \neq 1$, where $b^k \in M - A$. Note that $\langle b^p\rangle = \langle b\rangle \cap A$ and so $\langle b^p\rangle \trianglelefteq G$ and therefore $\langle b^k\rangle \cap \langle b\rangle = \langle b^p\rangle$. There is $b' \in \langle b^k\rangle - A$ such that $(b')^p = b^{-p}$ and $[b, b'] \neq 1$. We compute

$$(bb')^p = b^p(b')^p[b', b]^{\binom{p}{2}} = 1\,,$$

and so $o(bb') = p$. Set $bb' = s$ and assume $s \in Z(G)$. But then

$$[b, b'] = [b, b^{-1}s] = 1\,,$$

a contradiction. Thus, s is an element of order p in $B - A$ and $\langle s\rangle$ is not normal in G. By our basic assumption, $\langle s\rangle^G \leq B$ is nonabelian and so there is $l \in G$ such that setting $s' = s^l \in B - A$, we have $[s, s'] \neq 1$. But $[s, s'] = z$ is an element of order p in $Z(G)$ (noting that $B' \leq Z(G)$ and B' is elementary abelian). It follows that $K = \langle s, s'\rangle \cong S(p^3)$ (the nonabelian group of order p^3 and exponent p). Since $K \leq B$, we have $K \cap A \cong E_{p^2}$. On the other hand, $K \cap A \leq Z(G)$ and $|K : (K \cap A)| = p$ so that K is abelian, a contradiction. We have proved that we must have $p = 2$.

Since each subgroup of A is G-invariant and $C_G(A) = A$, we have $\exp(A) \geq 4$. Now assume, by way of contradiction, that $\exp(A) = 4$. Let B/A be any subgroup of order 2 in G/A. By Theorem 125.1, each element $b \in B - A$ inverts each element in A. This implies that G/A has only one subgroup of order 2 and so G/A is either cyclic or generalized quaternion. Suppose $|G/A| > 2$. Then there is $g \in G - A$ such that $(A\langle g\rangle)/A \cong C_4$, where g^2 inverts each element in A. Let $y \in A$ be an element of order 4. Since $\langle y\rangle \trianglelefteq G$, the element g normalizes $\langle y\rangle \cong C_4$ and g^2 inverts $\langle y\rangle$, a contradiction. Hence $|G/A| = 2$ and so $\exp(G) = 4$. Indeed, since each element $v \in G - A$ inverts each element in A, we get $o(v) \leq 4$. Because (by our assumption) G is not Dedekindian, there is $v \in G - A$ such that $\langle v\rangle$ is not normal in G, where $o(v) \leq 4$. We have $[A, \langle v\rangle] = G'$ is elementary abelian and $G' \leq Z(G)$. Set $R = \langle v\rangle^G$ so that our basic assumption implies that R is nonabelian. On the other hand,

$$[A, \langle v\rangle]\langle v\rangle \leq R \text{ and } [A, \langle v\rangle]\langle v\rangle = G'\langle v\rangle \trianglelefteq G$$

so that

$$R \leq [A, \langle v\rangle]\langle v\rangle \text{ and } R = \langle v\rangle^G = [A, \langle v\rangle]\langle v\rangle = G'\langle v\rangle\,.$$

But $G' \leq Z(G)$ implies that v centralizes G' and so R is abelian, a contradiction. We have proved that $\exp(A) = 2^e$ with $e \geq 3$.

Suppose, by way of contradiction, that G/A possesses a subgroup B/A of order 2 such that for all $b \in B - A$ and $h \in A$, we have $h^b = h^{1+2^{e-1}}$. In this case $B' = [A, \langle b \rangle]$ is elementary abelian and $B' \leq Z(G)$. Also, $Z(B) = C_A(b) = \Omega_{e-1}(A)$. If B is not normal in G, then there is $x \in G$ such that $b^x \in G - B$ for some $b \in G - A$. But then $bb^x \in G - A$ and bb^x centralizes A (since b^x acts on A the same way as b does), a contradiction. Hence $B \trianglelefteq G$. Because B is nonabelian and $\exp(B) \geq 8$, it follows that B is not Dedekindian. Hence there is $b \in B - A$ such that $\langle b \rangle$ is not normal in B. By our basic assumption, $\langle b \rangle^G \leq B$ is nonabelian. There is $g \in G$ such that $b^g \in B - A$ and $[b, b^g] \neq 1$. On the other hand, $\langle b^2 \rangle \trianglelefteq G$ and so $\langle b \rangle \cap \langle b^g \rangle = \langle b^2 \rangle$. There is $c \in \langle b^g \rangle - A$ such that $c^2 = b^{-2}$ and $[b, c] \neq 1$. We compute $(bc)^2 = b^2 c^2 [c, b] = [c, b] \neq 1$. Since $[c, b]$ is of order 2, it follows that $d = bc$ is an element of order 4 and $d \in A$. Hence $d \in Z(B)$ and so we get

$$[b, c] = [b, b^{-1}d] = [b, b^{-1}][b, d] = 1 ,$$

a contradiction. We have proved that such a group B/A of order 2 in G/A does not exist. Then using again Theorem 125.1, we see that G/A has exactly one subgroup C/A of order 2 such that for all $c \in C - A$ and $h \in A$, we have either $h^c = h^{-1}$ or $h^c = h^{-1+2^{e-1}}$. In any case $o(c) \leq 4$.

Assume that $|G/A| > 2$. Then there is $g \in G - A$ such that $g^2 \in C - A$ so that if h is an element of order 4 in A, then g normalizes $\langle h \rangle \trianglelefteq G$ and g^2 inverts $\langle h \rangle$, a contradiction. We have proved that $|G/A| = 2$, all elements in $G - A$ are of order ≤ 4, $\exp(A) = 2^e$, $e \geq 3$, and for each $v \in G - A$ and $h \in A$, either $h^v = h^{-1}$ or $h^v = h^{-1+2^{e-1}}$. The structure of our group G is determined.

Conversely, let G be a 2-group defined above and let v be any element in $G - A$. Then $o(v) \leq 4$ and the way in which v acts on A insures that each subgroup of A is G-invariant. Set $R = \langle v \rangle^G \trianglelefteq G$ and then $[A, \langle v \rangle]\langle v \rangle \leq R$. On the other hand, $[A, \langle v \rangle] = G'$ and therefore $[A, \langle v \rangle]\langle v \rangle \trianglelefteq G$ so that $R \leq [A, \langle v \rangle]\langle v \rangle$. We have proved that $[A, \langle v \rangle]\langle v \rangle = R$. But $\exp([A, \langle v \rangle]) = 2^{e-1} \geq 4$ and so v inverts each element of $[A, \langle v \rangle] = G'$ so that $\langle v \rangle^G$ is either quasidihedral (if $o(v) = 2$) or quasi-generalized quaternion (if $o(v) = 4$). Hence in any case $\langle v \rangle^G$ is nonabelian. The theorem is proved. □

Proposition 223.2. *Let M be a quasidihedral or a quasi-generalized quaternion 2-group. If M is minimal nonabelian, then $M \cong D_8$, Q_8 or $\mathcal{H}_2$, where*

$$\mathcal{H}_2 = \langle x, y \mid x^4 = y^4 = 1,\ x^y = x^{-1} \rangle .$$

Proof. Suppose that a 2-group M has an abelian maximal subgroup H of exponent ≥ 4 and there is an element in $M - H$ which inverts each element of H. Suppose in addition that M is minimal nonabelian.

If M is quasidihedral, then there is an involution $t \in M - H$ and let $h \in H$ with $o(h) = 4$. Then $\langle h, t\rangle \cong D_8$ is minimal nonabelian and so $M = \langle h, t\rangle \cong D_8$.

Assume that M is quasi-generalized quaternion. Since $|M'| = 2$, we get

$$H = \langle h\rangle \times V \text{ with } o(h) = 4 \text{ and } \exp(V) \leq 2\,.$$

Let $v \in M - H$, where $o(v) = 4$. If $v^2 = h^2$, then $\langle h, v\rangle \cong Q_8$ is minimal nonabelian and so $M = \langle h, v\rangle \cong Q_8$. If $\langle v\rangle \cap \langle h\rangle = \{1\}$, then $\langle h, v\rangle \cong \mathcal{H}_2$ is minimal nonabelian and so $M = \langle h, v\rangle \cong \mathcal{H}_2$. □

Proposition 223.3. *Let M be quasidihedral or quasi-generalized quaternion. If M is an A_2-group (see § 71), then M is isomorphic to one of the following groups:*
(1) D_{16} *or* Q_{16}*;*
(2) $\langle a, v \mid a^8 = v^4 = 1,\ a^v = a^{-1}\rangle$*;*
(3) $D_8 \times C_2$ *or* $Q_8 \times C_2$*;*
(4) $\mathcal{H}_2 \times C_2$.

Proof. Suppose that a 2-group M has an abelian maximal subgroup H of exponent ≥ 4 and there is an element in $M - H$ which inverts each element of H. Assume in addition that M is an A_2-group. By results of § 71, we know that an A_2-group M which has an abelian maximal subgroup H has the property $1 \neq |M'| \leq 4$. This implies that $\exp(H) = 4$ or 8.

First suppose that M is quasidihedral so that there is an involution $t \in M - H$. If $\exp(H) = 8$, then let $a \in H$ be an element of order 8. In this case $\langle a, t\rangle \cong D_{16}$ is an A_2-group and so $M = \langle a, t\rangle \cong D_{16}$. If $\exp(H) = 4$, then let $a \in H$ be an element of order 4. In this case $\langle a, t\rangle \cong D_8$ is a maximal subgroup of M and therefore $\langle a\rangle$ is a maximal subgroup of H so that $H = \langle a\rangle \times \langle u\rangle$ with $\langle u\rangle \cong C_2$. But then $\langle a, u\rangle\langle t\rangle \cong D_8 \times C_2$ is an A_2-group and so $M = \langle a, u\rangle\langle t\rangle \cong D_8 \times C_2$.

Now assume that M is quasi-generalized quaternion and let $v \in M - H$ with $o(v) = 4$. First we consider the case $\exp(H) = 8$ and let $a \in H$ with $\langle a\rangle \cong C_8$. If $\langle v^2\rangle = \Omega_1(\langle a\rangle)$, then $\langle a, v\rangle \cong Q_{16}$ is an A_2-group and so $M = \langle a, v\rangle \cong Q_{16}$. If $\langle v\rangle \cap \langle a\rangle = \{1\}$, then

$$M = \langle a, v \mid a^8 = v^4 = 1,\ a^v = a^{-1}\rangle$$

is an A_2-group from Proposition 71.2 (b).

It remains to consider the case $\exp(H) = 4$ and let $a \in H$ be an element of order 4. If $v^2 = a^2$, then $\langle a, v\rangle \cong Q_8$ is a maximal subgroup of M so that $\langle a\rangle$ is a maximal subgroup of H. In this case $H = \langle a\rangle \times \langle u\rangle$, where u is an involution and we get $M = \langle a, v\rangle \times \langle u\rangle \cong Q_8 \times C_2$. Suppose that $\langle v\rangle \cap \langle a\rangle = \{1\}$ so that $\langle a, v\rangle \cong \mathcal{H}_2$ is minimal nonabelian. Hence $\langle a, v\rangle$ is a maximal subgroup of M, which implies that $\langle a\rangle \times \langle v^2\rangle \cong C_4 \times C_2$ is a maximal subgroup of H. Assume that H is of type $(4, 4)$. Then there is $h \in H$ such that $h^2 = v^2$ and so $\langle h, v\rangle \cong Q_8$. But then $|M : \langle h, v\rangle| = 4$, contrary to our assumption that M is an A_2-group. Hence H is of type $(4, 2, 2)$ and so $M \cong \mathcal{H}_2 \times C_2$. The proposition is proved. □

Corollary 223.4 (= Problem 3246 (i)). *Let G be a non-Dedekindian p-group in which the normal closure of any nonnormal cyclic subgroup is minimal nonabelian. Then $p = 2$, $G = H \times V$ with $\exp(V) \leq 2$ and for H we have one of the following possibilities:*
(a) $H \cong D_{16}, Q_{16}$ *or* SD_{16}*;*
(b) $H = \langle a, v \mid a^8 = v^4 = 1,\ a^v = a^{-1+4\epsilon}\rangle, \epsilon = 0, 1$*;*
(c) $H = \langle a, b, v \mid a^8 = b^4 = v^4 = [a, b] = 1,\ b^2 = v^2,\ a^v = a^{-1+4\epsilon},\ b^v = b^{-1}\rangle$, $\epsilon = 0, 1$.
Conversely, all the above 2-groups satisfy the assumptions of the corollary.

Proof. Let G be a p-group satisfying the assumptions of our corollary. Then we use the results and the notation of Theorem 223.1. It follows that $p = 2$ and G has an abelian maximal subgroup A of exponent 2^e, $e \geq 3$, and for all $v \in G - A$ and $h \in A$, one has either $h^v = h^{-1}$ or $h^v = h^{-1+2^{e-1}}$. For each $v \in G - A$, $o(v) \leq 4$, $\langle v\rangle$ is nonnormal in G and $\langle v\rangle^G = [A, \langle v\rangle]\langle v\rangle = G'\langle v\rangle$ is isomorphic to one of the groups in Proposition 223.2. This implies $e = 3$ and let a be an element of order 8 in A.

First assume that $G - A$ contains an involution t such that $D_8 \cong \langle t\rangle^G = \langle a^2, t\rangle$. We get $G = \langle a, t\rangle \times V$, where $\langle a, t\rangle \cong D_{16}$ or SD_{16} and $\exp(V) \leq 2$.

In what follows assume that all elements in $G - A$ are of order 4. In this case, for each $v \in G - A$, $\langle v\rangle^G \cong Q_8$ or $\mathcal{H}_2$.

Assume that $\langle v\rangle^G \cong Q_8$. In this case,

$$v^2 = z = a^4 \text{ and } a^v = a^{-1} \text{ so that } G = \langle a, v\rangle \times V \text{ with } \langle a, v\rangle \cong Q_{16} \text{ and } \exp(V) \leq 2\,.$$

Indeed, if $a^v = a^{-1}z$, then

$$(va)^2 = vava = v^2a^va = za^{-1}za = 1\,,$$

and so va is an involution in $G - A$, contrary to our assumption.

Finally, suppose $\langle v\rangle^G \cong \mathcal{H}_2$. Then we get either the case (b) of our corollary or

$$A = \langle a\rangle \times \langle b\rangle \times V \text{ with } o(a) = 2^3\,, \quad o(b) = 4, \text{ and } \exp(V) \leq 2\,.$$

In the second case, replacing b with a^2b (if necessary), we may assume that $b^2 = v^2$ and so we obtain the case (c) of our corollary.

Conversely, it is easy to see that 2-groups defined in our corollary satisfy the assumptions of that corollary. □

Corollary 223.5 (= Problem 3246 (ii)). *Let G be a non-Dedekindian p-group in which the normal closure of any nonnormal cyclic subgroup is an A_2-group (see § 71). Then $p = 2$, $G = H \times V$ with $\exp(V) \leq 2$ and for H we have one of the following possibilities:*
(a) $H \cong D_{32}, Q_{32}$ *or* SD_{32}*;*
(b) $H = \langle a, v \mid a^{16} = v^4 = 1,\ a^v = a^{-1+8\epsilon}\rangle, \epsilon = 0, 1$*;*
(c) $H = \langle a, b, t \mid a^8 = b^4 = t^2 = [a, b] = 1,\ a^t = a^{-1+4\epsilon},\ b^t = b^{-1}\rangle, \epsilon = 0, 1$*;*
(d) $H = \langle a, b, v \mid a^8 = b^4 = v^4 = [a, b] = 1,\ v^2 = a^4,\ a^v = a^{-1},\ b^v = b^{-1}\rangle$*;*

(e) $H = \langle a, b, v \mid a^8 = b^4 = v^4 = [a, b] = 1,\ a^v = a^{-1+4\epsilon},\ b^v = b^{-1}\rangle, \epsilon = 0, 1$;
(f) $H = \langle a, b, c, v \mid a^8 = b^4 = c^4 = v^4 = [a, b] = [a, c] = [b, c] = 1,\ c^2 = v^2,\ a^v = a^{-1+4\epsilon},\ b^v = b^{-1},\ c^v = c^{-1}\rangle, \epsilon = 0, 1$.

Conversely, all the above 2-groups satisfy the assumptions of the corollary.

Proof. Let G be a p-group satisfying the assumptions of our corollary. Then we use the results and the notation of Theorem 223.1. It follows that $p = 2$ and G has an abelian maximal subgroup A of exponent 2^e, $e \geq 3$, and for all $v \in G - A$ and $h \in A$, one has either $h^v = h^{-1}$ or $h^v = h^{-1+2^{e-1}}$. For each $v \in G - A$, $o(v) \leq 4$, $\langle v\rangle$ is nonnormal in G and

$$\langle v\rangle^G = [A, \langle v\rangle]\langle v\rangle = G'\langle v\rangle$$

is isomorphic to one of the groups in Proposition 223.3. This implies that $e = 3$ or $e = 4$.

It is now easy to prove our corollary. In what follows V denotes an arbitrary elementary abelian subgroup (of order ≥ 1).

First assume that there is an involution $t \in G - A$. According to Proposition 223.3,

$$\langle t\rangle^G = [A, t]\langle t\rangle \cong \mathrm{D}_{16} \text{ or } \mathrm{D}_8 \times \mathrm{C}_2\,.$$

In the first case, $A = \langle a\rangle \times V$ with $\langle a\rangle \cong \mathrm{C}_{16}$ so that $G \cong \mathrm{D}_{32} \times V$ or $G \cong \mathrm{SD}_{32} \times V$. In the second case,

$$A = \langle a\rangle \times \langle b\rangle \times V \text{ with } \langle a\rangle \cong \mathrm{C}_8\,, \quad \langle b\rangle \cong \mathrm{C}_4$$

so that G is a group of part (c) of the corollary.

Now assume that there is no involution in $G - A$ so that all elements in $G - A$ are of order 4 and if v is one of them, then $\langle v\rangle^G = [A, \langle v\rangle]\langle v\rangle$.

If $\langle v\rangle^G \cong \mathrm{Q}_{16}$, then

$$A = \langle a\rangle \times V \text{ with } \langle a\rangle \cong \mathrm{C}_{16}$$

so that $v^2 = a^8$, $a^v = a^{-1}$ and we get $G \cong \mathrm{Q}_{32} \times V$. Indeed, if $a^v = a^{-1}a^8$, then

$$(va)^2 = vava = v^2a^va = a^8a^{-1}a^8a = 1\,,$$

and so va would be an involution in $G - A$, contrary to our assumption.

If $\langle v\rangle^G \cong \mathrm{Q}_8 \times \mathrm{C}_2$, then $A = \langle a\rangle \times \langle b\rangle \times V$ with

$$\langle a\rangle \cong \mathrm{C}_8\,, \quad \langle b\rangle \cong \mathrm{C}_4\,, \quad a^4 = v^2 \text{ and } a^v = a^{-1}\,,$$

because in case $a^v = a^{-1}a^4$, va would be an involution in $G - A$, contrary to our assumption. We have obtained the groups from part (d) of our corollary.

Suppose that $\langle v\rangle^G$ is isomorphic to a group from (2) in Proposition 223.3. Then

$$A = \langle a\rangle \times V \text{ with } \langle a\rangle \cong \mathrm{C}_{16} \text{ and } \langle a\rangle \cap \langle v\rangle = \{1\}$$

and we have obtained the groups from part (b) of the corollary.

Finally, assume that $\langle v\rangle^G \cong \mathcal{H}_2 \times C_2$ so that either

$$A = \langle a\rangle \times \langle b\rangle \times V \quad \text{with } \langle a\rangle \cong C_8,\ \langle b\rangle \cong C_4, \text{ and } \langle v\rangle \cap \langle a, b\rangle = \{1\}$$

or

$$A = \langle a\rangle \times \langle b\rangle \times \langle c\rangle \times V \quad \text{with } \langle a\rangle \cong C_8,\ \langle b\rangle \cong \langle c\rangle \cong C_4, \text{ and } v^2 = c^2 .$$

This gives the groups from parts (e) and (f) of our corollary.

Conversely, it is easy to see that all 2-groups G defined in our corollary satisfy the assumptions of that corollary. Indeed, let A be the unique abelian maximal subgroup of G. Then for each $x \in G-A$ and $h \in A$, we have either $h^x = h^{-1}$ or $h^x = h^{-1+2^{e-1}}$, where $2^e = \exp(A)$ and $e = 3$ or $e = 4$. This implies that each subgroup of A is G-invariant. Also we have $\langle x\rangle^G = [A, \langle x\rangle]\langle x\rangle$, where the subgroup $[A, \langle x\rangle] = G'$ is independent on the choice of $x \in G - A$.

In case of groups in part (a) of the corollary, $\langle x\rangle^G \cong D_{16}$ or Q_{16}. In case of groups in part (b) of our corollary, $\langle x\rangle^G$ is isomorphic to the group (2) in Proposition 223.3. In case of groups in part (c) of our corollary, $\langle x\rangle^G \cong D_8 \times C_2$. In case of groups in part (d) of our corollary, $\langle x\rangle^G \cong Q_8 \times C_2$. Finally, in case of groups in parts (e) and (f) of our corollary, $\langle x\rangle^G \cong \mathcal{H}_2 \times C_2$. The proof is complete. □

Problem 1. Classify the non-Dedekindian p-groups in which the normal closure of each nonnormal cyclic subgroup is either abelian or minimal nonabelian.

Problem 2. Classify the non-Dedekindian p-groups in which the normal closure of each nonnormal cyclic subgroup is either abelian or an $\mathcal{A}_2$-group.

Problem 3. Classify the non-Dedekindian p-groups G in which the normal closure of each cyclic subgroup not contained in $\Phi(G)$ is abelian.

Problem 4. Classify the non-Dedekindian p-groups G in which the normal closure of each maximal cyclic subgroup is metacyclic.

Problem 5. Classify the non-Dedekindian p-groups, $p > 2$, in which the normal closure of each absolutely regular subgroup is absolutely regular.

Problem 6. Classify the non-Dedekindian p-groups G, $p > 2$, in which the normal closure of each regular subgroup is regular. (This will be $\Longleftrightarrow$ all maximal regular subgroups of our group are normal.)

Problem 7. Classify the non-Dedekindian p-groups in which all maximal subgroups of exponent p are normal.

§ 224 p-groups in which the normal closure of any cyclic subgroup is abelian

In § 223, we have determined all non-Dedekindian p-groups in which the normal closure of any nonnormal cyclic subgroup is nonabelian. Here we characterize p-groups G in which the normal closure of any cyclic subgroup is abelian (Problem 3589). This also solves Problem 3238 (Theorem 224.2).

Theorem 224.1. *The normal closure of each cyclic subgroup in a p-group G is abelian if and only if each two-generator subgroup of G is of class ≤ 2.*

If each two-generator subgroup of a p-group G is of class ≤ 2, then either G is of class ≤ 2 or $p = 3$ and G is of class 3 and in any case G is metabelian.

Proof. Suppose that the normal closure of each cyclic subgroup in a p-group G is abelian. Let $a, b \in G$ so that $\langle a \rangle^G$ and $\langle b \rangle^G$ are abelian normal subgroups. By the Fitting's lemma (Introduction, Theorem 21), $\langle a \rangle^G \langle b \rangle^G$ is of class ≤ 2 and so $\langle a, b \rangle$ is of class ≤ 2.

Now assume that each two-generator subgroup of a p-group G is of class ≤ 2. Let $x, g \in G$. Then

$$x^g = g^{-1}xg = (xx^{-1})(g^{-1}xg) = x[x, g] ,$$

and so $\mathrm{cl}(\langle x, g \rangle) \leq 2$ implies that $[x, g]$ commutes with x and therefore x^g commutes with x. Hence $\langle x \rangle^G$ is abelian.

Finally, suppose that each two-generator subgroup in a p-group G is of class ≤ 2. Then for each $a, b \in G$, $[a, b]$ commutes with b and so $[a, b, b] = 1$. By [Hup1, Satz III.6.101], G is either of class ≤ 2 or $p = 3$ and $\mathrm{cl}(G) = 3$. In the last case we have

$$[G, G', G] = [G', G, G] = \{1\} ,$$

and so by the Three Subgroups Lemma (see Exercise 1.13 (b)),

$$[G, G, G'] = [G', G'] = \{1\} .$$

Hence in any case G is metabelian and we are done. □

Theorem 224.2. *Let G be a non-Dedekindian p-group all of whose nonnormal abelian subgroups have abelian normalizers (Problem 3238).Then the normal closure of each cyclic subgroup in G is abelian.*

Proof. Let X be any cyclic subgroup in G and let A be a maximal abelian subgroup of G containing X. Then $N_G(A) > A$ is nonabelian and so A is normal in G. But then $X^G \leq A$ is abelian and we are done. □

Problem 1. Classify the p-groups in which the normal closure of any cyclic subgroup (i) Dedekindian, (ii) has the cyclic derived subgroup, in particular, metacyclic, (iii) is of class ≤ 2, (iv) is metabelian.

Problem 2. Classify the p-groups in which the normal closure of any (i) elementary abelian subgroup is elementary abelian, (ii) subgroup of exponent p has exponent p.

Problem 3. Study the p-groups G such that $\exp(H^G) = \exp(H)$ for any $H \leq G$.

Problem 4. Study the irregular p-groups G such that the normal closure of any nonnormal subgroup is irregular.

Problem 5. Classify the non-Dedekindian p-groups all of whose nonnormal cyclic (metacyclic) subgroups have metacyclic normalizers.

Problem 6. Classify the non-Dedekindian p-groups all of whose nonnormal cyclic subgroups have Dedekindian normalizers.

Problem 7. Classify the non-Dedekindian p-groups in which any nonnormal subgroup with cyclic derived subgroup has normalizer with cyclic derived subgroup.

Problem 8. Classify the non-Dedekindian p-groups $G = \Omega_1(G)$ in which the normalizer of any nonnormal elementary abelian (of exponent p) subgroup is elementary abelian (has exponent p).

Problem 9. Study the non-Dedekindian p-groups G such that $\exp(N_G(H)) = \exp(H)$ for any nonnormal $H < G$.

§ 225 Nonabelian p-groups in which any s (a fixed $s \in \{3, \dots, p+1\}$) pairwise noncommuting elements generate a group of maximal class

The nonabelian 2-groups G all of whose two-generator subgroups are of maximal class, have classified. Indeed, if $M \le G$ is minimal nonabelian, then it is of maximal class so of order 8. Now the classification of such 2-groups follows from Theorem 90.1. For an independent proof, see Appendix 96. For $p > 2$, see [GMS1].

Any nonabelian p-group G contains $p + 1 \ge 3$ pairwise noncommuting elements. Indeed, let $M \le G$ be minimal nonabelian and let $L_1, \dots, L_{p+1}$ be all maximal subgroups of M. Take, for each i, $x_i \in L_i - \Phi(G)$. Then the elements $x_1, \dots, x_{p+1}$ are pairwise noncommuting since $\langle x_i, x_j\rangle = M$ is nonabelian for $i \ne j$.

Let $s \in \{3, \dots, p+1\}$ be fixed. In this section we classify the nonabelian p-groups in which any s pairwise noncommuting elements generate a p-group of maximal class.

Remark 1. Let G be a nonabelian p-group. Assume that $\{x_1, x_2, \dots, x_s\}$ be a set of pairwise noncommuting elements of G and $1 < s < p + 1$. Then $\bigcup_{i=1}^{s} C_G(x_i) \ne G$, by Lemma 116.3 (a). Take $y \in G - \bigcup_{i=1}^{s} C_G(x_i)$. Then $\{y, x_1, x_2, \dots, x_s\}$ is the set of $s + 1$ pairwise noncommuting elements. It follows that any set of $s < p + 1$ of pairwise noncommuting elements of a (nonabelian) p-group G is a subset of a set of $p + 1$ pairwise noncommuting elements of G.

Theorem 225.1. *Let s be a fixed member of the set $\{3, \dots, p + 1\}$ and $p > 2$. If any s pairwise noncommuting elements of a nonabelian p-group G generate a p-group of maximal class, then G is also of maximal class with abelian subgroup of index p.*

Proof. (a) We claim that G is of maximal class. Let $A \le G$ be a minimal nonabelian subgroup of the least order. If $A = G$, then any s pairwise noncommuting elements of G generate G (see the second paragraph of the section) so G is of maximal class. Next we assume that $A < G$. As $d(A) = 2$ and $s > 2$, there are pairwise noncommuting $a_1, \dots, a_{s-1} \in A$ that generate A. Let $A < B \le G$, where $|B : A| = p$. Then $d(B) \le d(A) + 1 = 3$. By Remark 1, there is $a_s \in B - A$ such that elements $a_1, \dots, a_{s-1}, a_s$ are pairwise noncommuting and these elements generate B. By hypothesis, B is of maximal class. Thus, all subgroups of G containing A as a subgroup of index p, are of maximal class. In that case, by Exercise 10.10, G is of maximal class, as required.

(b) It remains to prove that G has an abelian subgroup of index p. One may assume that $|G| > p^4$. Assume that G_1, the fundamental subgroup of G (see Theorem 9.6), is nonabelian. By definition, G_1 is not of maximal class. Then, by (a), the subgroup G_1 contains s pairwise noncommuting elements that do not generate the subgroup of maximal class, which is a contradiction. Thus, the group G has an abelian subgroup G_1 of index p. □

Assume that the p-group G of maximal class has an abelian subgroup A of index p. Let us prove that any nonabelian $H < G$ is of maximal class. Let $x \in H - A \subset G - A$. Then $C_A(x) = Z(G)$ so $C_G(x) = C_H(x)$ is of order p^2. By Proposition 1.8, H is of maximal class. In that case any ≥ 2 pairwise noncommuting elements of G generate a subgroup of maximal class.

If any nonabelian two-generator subgroup of a nonabelian p-group G is of maximal class, then G need not necessary is of maximal class. Indeed, the group $G = D \times C$, where D is nonabelian of order p^3 and $C \cong C_p$ is such that all its nonabelian two-generator subgroups $\cong D$ so of maximal class.

Recall that a p-group G is said to be $\mathcal{A}_n$-group if all its subgroups of index p^n are abelian but there is in G a nonabelian subgroup of index p^{n-1}.

Proposition 225.2. *An $\mathcal{A}_n$-group G, $n > 2$, which is a p-group, all of whose $\mathcal{A}_k$-subgroups ($k \in \{2, \ldots, n-1\}$ is fixed) of minimal order are of maximal class, is itself of maximal class.*

Proof. Let $A < B < G$, where A is an $\mathcal{A}_1$-subgroup of minimal possible order (in that case, $|G : A| = p^{n-1} \geq p^2$) and $|B : A| = p^{k-1}$. Then B, being an $\mathcal{A}_k$-subgroup of minimal order, is of maximal class, by hypothesis. Let $A < C < B$, where C is maximal in B. Then any subgroup of G containing C as a subgroup of index p, is of maximal class, by hypothesis. In that case, the group G, by Exercise 10.10, is of maximal class. □

Exercise 1. If any two-generator subgroup of a p-group G, $p > 2$, is metacyclic, then $\Omega_1(G)$ is elementary abelian.

Problem. Study the p-groups in which any two-generator subgroup is metacyclic.

§ 226 Noncyclic p-groups containing only one proper normal subgroup of a given order

A p-group contains $\equiv 1 \pmod p$ normal subgroups of a given order (Sylow). In what follows, G is a noncyclic p-group of order $> p^n$ containing only one normal subgroup N of order p^n and $\mathrm{Z}_n(G)$ is the n-th term of the upper central series of G. If H is a unique normal subgroup of order $> p$ in a p-group G, then it is proved in [GMS2] that, provided $p > 2$, then $H = \mathrm{Z}_n(G)$ for some n (such H is called a *waist*).

If $|\mathrm{Z}_n(G)| = p^n$, then $\mathrm{Z}_n(G)$ is the unique normal subgroup of G of order p^n, which follows immediately from the following easy

Lemma 226.1. *Let H be a normal subgroup of a p-group G and n a natural number. If $|H| \ge p^n$, then $|H \cap \mathrm{Z}_n(G)| \ge p^n$.*

Proof. We proceed by induction on n. We have $H \cap \mathrm{Z}(G) > \{1\}$ so the result holds for $n = 1$. Now let $n > 1$. If $i < n$, then $|H \cap \mathrm{Z}_i(G)| \ge p^i$ by induction. If $H \cap \mathrm{Z}_i(G) < H$, then $H \cap \mathrm{Z}_i(G) < H \cap \mathrm{Z}_{i+1}(G)$ since $\mathrm{Z}(H/Z_i(G)) > \{1\}$. It follows that $|H \cap \mathrm{Z}_{i+1}(G)| \ge p^{i+1}$, and the assertion follows by induction. □

Proposition 226.2. *Let N be the unique proper normal subgroup of order p^n in a noncyclic p-group G. Then*
(a) *$N \le \Phi(G)$.*
(b) *If $n > 1$ and $\mathrm{Z}(N)$ is cyclic, then G is a 2-group of maximal class. In particular, N is cyclic.*

Proof. Clearly, $|G : N| > p$ since $|\Gamma_1| \ge p + 1 > 1$.

(a) Let $H \in \Gamma_1$. Since $|H| > p^n$, there is in H a G-invariant subgroup M of order p^n. By hypothesis, $M = N$. Thus, N is contained in all maximal subgroups of G, and therefore $N \le \Phi(G)$.

(b) Let $R \le G$ be a G-invariant subgroup of order p^2. By hypothesis, $R \le N \le \Phi(G)$. It follows from $|G : \mathrm{C}_G(R)| \le p$ that $R \le \mathrm{Z}(\Phi(G))$ so that $R \le \mathrm{Z}(N)$. By hypothesis, R is cyclic. Thus, any normal subgroup of G of order p^2 is cyclic so that G is a 2-group of maximal class (Lemma 1.4). It follows that N is cyclic since $N \le \Phi(G)$ and $\Phi(G)$ is cyclic. □

Proposition 226.3. *Suppose that N is the unique normal subgroup of order p^2 in a noncyclic p-group G. Then one of the following holds:*
(a) *G is a 2-group of maximal class (in that case, $N = \mathrm{Z}_2(G)$ and $N \cong \mathrm{C}_4$).*
(b) *$N \in \{\mathrm{Z}(G), \mathrm{Z}_2(G)\}$. In that case, if G is not a 2-group of maximal class, then $N \cong \mathrm{E}_{p^2}$.*

Proof. Let $L \le N \cap Z(G)$ be of order p. Then N/L is a unique normal subgroup of order p in G. It follows that the subgroup $H/L = Z(G/L)$ is cyclic; then $H \le Z_2(G)$ and $N \le H$. Assume that H is cyclic. Then, since $|H| \ge p^2$, the group G has no normal subgroup $\cong E_{p^2}$, by hypothesis, so G is a 2-group of maximal class (Lemma 1.4). In what follows we assume that H is noncyclic. Then H is abelian with cyclic subgroup of index p.

Suppose that $|H| = p^2$; then $H = N$ and $N \cong E_{p^2}$. In that case, $Z(G) \le N$ (otherwise, $Z(G)$ contains a subgroup of order p^2 which is $\ne N$, a contradiction). If $Z(G) = N$, we are done. If $Z(G) < N$, then $Z(G) = L$ so that $N = H = Z_2(G)$, and the proof is complete in this case.

Next we assume that $|H| > p^2$. Then $\Omega_1(\Phi(H))$ is of order p, and we conclude that $N = L \times \Omega_1(\Phi(G)) \le Z(G)$. If $N < Z(G)$, then $Z(G)$, being noncyclic, contains two distinct G-invariant subgroups of order p^2, contrary to the hypothesis. Thus, $N = Z(G) \cong E_{p^2}$. □

It follows that if the waist N has order p^2 (i.e., the unique G-invariant subgroup of order p^2 in G), then $N \in \{Z(G), Z_2(G)\}$. It is proved in [GMS2] that if N is a waist of order $> p$ of a noncyclic p-group, $p > 2$, then N is a member of the upper central series of G.

Proposition 226.4. *Let N be a waist in a nonabelian p-group G. If $|N| > p$, then $Z(G) \le N$.*

Proof. Assume that $Z(G) \nleq N$; then $|Z(G)| > p$. Set $L = N \cap Z(G)$ $(> \{1\})$ and let $L < M \le Z(G)$, where $|M : L| = p$; then $|M| \le |N|$, a contradiction since M and N are nonincident. □

If G is a p-group such that $Z_n(G)$ is of maximal class, then $|Z_n(G)| \in \{p^n, p^{n+1}\}$.

Proposition 226.5. *Suppose that G is a p-group such that $Z_{p+1}(G)$ is of maximal class and order p^{p+2}. Then $G = Z_{p+1}(G)$.*

Proof. Assume that $Z_{p+1}(G) < G$. As $|Z_{p+1}(G)| = p^{p+2} > p^{p+1}$ and $Z_{p+1}(G)$ is of maximal class, it follows from Theorem 9.6 (c) that $Z_{p+1}(G)$ has no normal subgroup of order p^p and exponent p. Since $Z_{p+1}(G)$ contains all normal in G subgroups of order p^p (Lemma 226.1), it follows that G has no normal subgroup of order p^p and exponent p. The group G is irregular since $Z_{p+1}(G)$ is (Theorem 9.5). Therefore, Theorem 12.1 (a) implies that G is a p-group of maximal class. As $Z_{p+1}(G)$ is a proper irregular normal subgroup of a p-group of maximal class, it follows that $Z_{p+1}(G) \in \Gamma_1$ (Theorem 9.6). In that case, $|Z_{p+1}(G)|| = p^{p+1} < p^{p+2}$, a contradiction. Thus, $Z_{p+1}(G) = G$. □

Exercise 1. If $n > p$ and $Z_n(G)$ is of maximal class so is G. (**Hint.** Mimic the proof of Theorem 226.6 and use Theorem 13.5.)

Problem 1. Study the p-groups G such that $|Z_n(G)| = p^n$ for a fixed large n.

Problem 2. Study the p-groups G, $p > 2$, such that $G/Z_p(G)$ is of maximal class. Moreover, study the p-groups G containing only one normal subgroup T of its order such that G/T is of maximal class.

Problem 3. Study the p-groups containing $\leq p + 1$ normal subgroups of any given order.

Problem 4. Classify the p-groups in which all members of their upper central series are waists.

§ 227 *p*-groups all of whose minimal nonabelian subgroups have cyclic centralizers

In Theorem 227.1 the title p-groups are characterized. This theorem is just a simple consequence of the very useful in many situations Lemma 57.1.

Suppose that a nonabelian p-group G has an abelian subgroup of index p and the center of order p. We prove, using induction on $|G|$, that G is of maximal class. In that case, $|G : G'| = p|Z(G)| = p^2$ (Lemma 1.1). If $G/Z(G)$ is abelian, it is of order p^2, and so

$$|G| = |G : Z(G)|\,|Z(G)| = p^2|Z(G)| = p^3\,,$$

hence G is of maximal class. If $\bar{G} = G/Z(G)$ is nonabelian, it is of maximal class, by induction, since $|\bar{G} : \bar{G}'| = p^2$ and $Z(G) < G'$. Then G is of maximal class since $|Z(G)| = p$.

Suppose that G is of maximal class with abelian subgroup A of index p and let $M \leq G$ be nonabelian. Then $M = \langle x\rangle(M \cap A)$ for any $x \in M - A$. As $C_G(x) = \langle x, Z(G)\rangle$ is of order p^2, it follows that $C_M(x) = C_G(x)$ so, by Proposition 1.8, M is of maximal class.

Theorem 227.1 (Janko). *If G is a nonabelian p-group, then all its minimal nonabelian subgroups have cyclic centralizers $\iff$ G has an abelian subgroup of index p and $Z(G)$ is cyclic.*

Proof. As the centralizers of all minimal nonabelian subgroups of G are cyclic, the subgroup $Z(G)$ is cyclic. We have to show that G has an abelian subgroup of index p. One may assume that G is not a 2-group of maximal class. Assume that G has no abelian subgroup of index p. Let $E_{p^2} \cong U \lhd G$ (Lemma 1.4) and let $A < G$ be a maximal normal abelian subgroup containing U. Then $C_G(U) \in \Gamma_1$ since U is not a subgroup of the cyclic $Z(G)$. It follows that $C_G(U)$ is nonabelian, by assumption, hence $C_G(U) > A$. Take $g \in C_G(U) - A$. By Lemma 57.1, there is $a \in A$ such that $M = \langle g, a\rangle$ is minimal nonabelian. But then M, the minimal nonabelian subgroup of $C_G(U)$, centralizes the noncyclic subgroup U, a contradiction. Thus, $|G : A| = p$.

Conversely, let G be a nonabelian p-group with an abelian subgroup A of index p and let $Z(G)$ be cyclic. Let $M \leq G$ be minimal nonabelian. We have to prove that $C_G(M)$ is cyclic. Moreover, we shall prove that $C_G(M) = Z(G)$. One may assume that $M < G$ (otherwise, it is nothing to prove further). Assume that there is $y \in C_G(M) - Z(G)$. Then $C_G(y)$ is nonabelian since $M \leq C_G(y)$. If $y \in A - Z(G)$, then $C_G(y) = A$ is abelian, a contradiction. Thus, $y \notin A$. Then $C_G(y) = \langle y, Z(G)\rangle$, a final contradiction. Thus, $C_G(M) = Z(G)$ is cyclic, as desired. □

Problem 1. Classify the nonabelian p-groups all of whose nonabelian subgroups have cyclic centers.

Problem 2. Classify the primary $\mathcal{A}_n$-groups, $n > 1$, all of whose $\mathcal{A}_2$-subgroups have cyclic centralizers.

Problem 3. Study the nonabelian p-groups G such that, whenever $A \leq G$ is minimal nonabelian, then $\mathrm{Z}(\mathrm{C}_G(A))$ is cyclic.

§ 228 Properties of metahamiltonian p-groups

A nonabelian p-group G is called *metahamiltonian* if all nonabelian subgroups of G are normal. This is a natural generalization of Hamiltonian p-groups (i.e., nonabelian Dedekindian p-groups). This is also an essential generalization of the main result of § 16 classifying the p-groups all of whose noncyclic subgroups are normal. The $\mathcal{A}_2$-groups are metahamiltonian.

Metahamiltonian p-groups have been classified by X. Fang and L. An in [FA]. Here we shall present main properties of such p-groups (Problem 768).

Theorem 228.1. *Let G be a metahamiltonian p-group. Then the normal closure of each cyclic subgroup in G is either abelian or minimal nonabelian.*

Proof. Let $x \in G$ and assume that $\langle x\rangle^G$ is nonabelian. Then there is $g \in G$ such that, setting $y = x^g$, we have $[x, y] \neq 1$. Since $K = \langle x, y\rangle$ is nonabelian , it follows $K \trianglelefteq G$ and so $K = \langle x\rangle^G$.

Consider the proper subgroup $L = \langle x, [x, y]\rangle$ of K so that L is not normal in G and therefore L is abelian and this implies $[[x, y], x] = 1$. Similarly, considering $M = \langle y, [x, y]\rangle < K$, we get $[[x, y], y] = 1$. Hence $K' = \langle [x, y]\rangle \leq Z(K)$ and $\mathrm{cl}(K) \leq 2$.

Finally, we consider $N = \langle x, y^p\rangle < K$ so that N is not normal in G and therefore N is abelian. This gives

$$1 = [x, y^p] = [x, y]^p$$

and so $K' = \langle [x, y]\rangle$ is of order p. By Lemma 65.2 (a), $K = \langle x\rangle^G = \langle x, y\rangle$ is minimal nonabelian and we are done. □

Theorem 228.2. *Let G be a metahamiltonian p-group. Then G is of class at most 3 and so G' is abelian.*

Proof. First, we note that each nonabelian section of G is metahamiltonian.

By Theorem 228.1, for each $x \in G$, $K = \langle x\rangle^G$ is either abelian or minimal nonabelian. Since $K' \trianglelefteq G$ and $|K'| \leq p$, we get $K' \leq Z(G)$. Set $\bar{G} = G/Z(G)$ and we use the bar convention. Then for all $\bar{x} \in \bar{G}$, $\langle \bar{x}\rangle^{\bar{G}}$ is abelian. By Theorem 224.1, $\bar{G}$ is either of class ≤ 2 or $p = 3$ and $\bar{G}$ is of class 3. Hence G is either of class ≤ 3 or $p = 3$ and G is of class ≤ 4.

We claim that G is of class ≤ 3. If not, then $p = 3$ and G is of class 4. Let G be a minimal counterexample to Theorem 228.2. Then $p = 3$, $\mathrm{cl}(G) = 4$, $|\mathrm{K}_4(G)| = 3$, and the class of each proper section of G is of class ≤ 3. Also, we may set $G = \langle a, b, c, d\rangle$, where $\langle [a, b, c, d]\rangle = \mathrm{K}_4(G)$.

Set $x = [a, b, c]$. Then Lemma 65.2 (a) implies that $N = \langle x, d\rangle$ is minimal nonabelian. By hypothesis, every subgroup which contains N is normal in G. It follows that G/N is Dedekindian. But $p = 3$ and so G/N is abelian and consequently $G' \leq N$.

By Lemma 57.1, G is generated by its minimal nonabelian subgroups and so there is a minimal nonabelian subgroup M of G distinct from N. Since all subgroups of G

containing M are normal in G and $p = 3$, it follows that $G' \le M$. But then $G' \le M \cap N < N$ and so G' is abelian. This implies $[[c, d], [a, b]] = 1$.

Since $[a, b] \in G' < N$ and $d \in N$, we get

$$[d, [a, b]] \in N' \le Z(G) \quad \text{and so} \quad [d, [a, b], c] = 1 .$$

Now we use the Hall–Witt's formula (see Introduction, Exercise 13 (a)) and get

$$[c, d, [a, b]]^{d^{-1}} [d, [a, b], c]^{[a,b]^{-1}} [[a, b], c, d]^{c^{-1}} = 1 ,$$

which together with $[[c, d], [a, b]] = 1$ and $[d, [a, b], c] = 1$ gives $[a, b, c, d] = 1$, a contradiction.

We have proved that G is of class at most 3 and so G' is abelian. □

Lemma 228.3. *If G is a two-generator group of order 2^4 with a quotient group isomorphic to Q_8, then*

$$G \cong \mathcal{H}_{2,2} = \langle x, y \mid x^4 = y^4 = 1,\ [x, y] = x^2 \rangle .$$

Proof. Let G_0 be a normal subgroup of order 2 in G such that $G/G_0 \cong Q_8$. If $|G'| = 4$, then a result of O. Taussky shows that G is of maximal class. But then $G/G_0 \cong D_8$, a contradiction. Hence $|G'| = 2$ and $G' \neq G_0$ are two distinct minimal normal subgroups in G. By Lemma 65.2 (a), G is minimal nonabelian. If G is of exponent 8, then $G \cong M_{16}$. But M_{16} has only one minimal normal subgroup, a contradiction. It follows $\exp(G) = 4$. Since $G/G_0 \cong Q_8$, G does not possess an elementary abelian subgroup of order 8. Hence G is metacyclic (see Lemma 65.1) and so $G \cong \mathcal{H}_{2,2}$. □

Theorem 228.4. *A nonabelian p-group G is metahamiltonian if and only if G' is contained in every nonabelian subgroup of G.*

Proof. Suppose that a nonabelian p-group G has the property that G' is contained in each nonabelian subgroup of G. Then each nonabelian subgroup of G is normal and so G is metahamiltonian.

Conversely, assume that a p-group G is metahamiltonian. Suppose that G is metahamiltonian of the smallest possible order such that G possesses a minimal nonabelian subgroup $N = \langle a, b \rangle$ such that $G' \nleq N$. Since each subgroup containing N is normal in G, we have $p = 2$ and G/N is Hamiltonian. By the minimality of $|G|$, $G/N \cong Q_8$. Set $G/N = \langle xN, yN \rangle$ and $H = \langle x, y \rangle$. Then

$$G = HN , \quad H/(H \cap N) \cong Q_8 \quad \text{and} \quad z = [x, y] \notin N .$$

Since $d(H) = 2$ and $H/((H \cap N)\langle z \rangle) \cong E_4$, we have $\Phi(H) = (H \cap N)\langle z \rangle$. By Theorem 228.2, H' is abelian. Also,

$$\Phi(H) = \langle x^2, y^2, H' \rangle , \quad \text{where } x^2, y^2 \in \Phi(H) - N .$$

Both normal closures $\langle x \rangle^H$ and $\langle y \rangle^H$ contain $z = [x, y]$ since $x^y = x[x, y]$ and $y^x = y[y, x] = y[x, y]^{-1}$. Also, by Theorem 228.1, both normal closures $\langle x \rangle^H$ and $\langle y \rangle^H$

are either abelian or minimal nonabelian. This gives $[x^2, z] = [y^2, z] = 1$ and therefore $z \in H'$ commutes with each element in H' and with x^2 and y^2 so that $z \in Z(\Phi(H))$. In particular, $[H \cap N, z] = \{1\}$.

Because H is nonabelian, we get $H \trianglelefteq G$ and so $H \cap N \trianglelefteq G$. Since H' covers $\Phi(H)/(H \cap N) \cong C_2$, we have

$$K_3(H) = [H, H'] \leq H' \cap N \quad \text{and} \quad K_3(H) = \langle [z, x], [z, y] \rangle \leq Z(H) .$$

But H is a two-generator group and so $H'/K_3(H)$ is cyclic and $\langle z \rangle$ covers $H'/K_3(H)$. We compute

$$1 = [z, x^2] = [z, x][z, x]^x = [z, x]^2 \quad \text{and} \quad 1 = [z, y^2] = [z, y][z, y]^y = [z, y]^2 ,$$

so that $K_3(H) \leq Z(H)$ is elementary abelian of order ≤ 4. Since

$$z^x = z[z, x] \quad \text{and} \quad z^y = z[z, y] \Rightarrow H' = \langle z \rangle^H .$$

We claim that $H \cap N = \langle x^4, x^2y^2, x^2z \rangle^H$. Indeed, it is clear that x^4, x^2y^2, and x^2z are contained in $H \cap N$ and so

$$H_0 = \langle x^4, x^2y^2, x^2z \rangle^H \leq H \cap N \quad \text{and} \quad H_0 \trianglelefteq H .$$

We compute

$$(x^2z)^x = x^2z^x = x^2z[z, x] \quad \text{and} \quad (x^2)^y = (x^y)^2 = (xz)^2 = x^2z[z, x]z = x^2z^2[z, x] ,$$

so that

$$(x^2z)^y = (x^2)^y z^y = x^2z^2[z, x] \cdot z[z, y] = x^2z^2z[z, x][z, y] .$$

Noting that $x^4 \in H_0$ and so we see that if $H_0 < H \cap N$, then $\langle x^2 \rangle H_0 < \Phi(H)$. However, by the above, H_0 also contains $x^2z[z, x]$ and $x^2z^2z[z, x][z, y]$ and so $\langle x^2 \rangle H_0$ contains $x^2, y^2, z, [z, x], [z, y]$, where $\langle z, [z, x], [z, y] \rangle = H'$. Hence

$$\langle x^2 \rangle H_0 \geq \langle x^2, y^2, H' \rangle = \Phi(H) .$$

This gives the desired equality $H_0 = H \cap N$.

(i) Assume $H \cap N = N$ so that $G = H = \langle x, y \rangle$.

In this case $[N, z] = \{1\}$. Let $M = \langle za, b \rangle$ so that Lemma 65.2 (a) implies that M is minimal nonabelian. Hence $M \trianglelefteq G$ and G/M is also Dedekindian. Since $z \notin M$, G/M is nonabelian. By the minimality of $|G|$, $Q/M \cong Q_8$. It follows

$$M = \langle x^4, x^2y^2, x^2z \rangle^H = N = \langle a, b \rangle ,$$

a contradiction.

In what follows we suppose $H \cap N < N$ so that $H \cap N$ is abelian. Since

$$H \cap N = \langle x^4, x^2y^2, x^2z \rangle^H \text{ and } H \cap N \text{ is abelian} ,$$

it follows $[x^2y^2, x^2z] = 1$ and so $[x^2, y^2] = 1$. We compute

$$1 = [x^2, y^2] = [x^2, y][x^2, y]^y = [x, y]^x[x, y][x, y]^{xy}[x, y]^y = z^x zz^{xy}z^y .$$

On the other hand,

$$z^x = z[z, x] , \quad z^{xy} = z[z, xy] = z[z, y][z, x]^y = z[z, x][z, y] \quad \text{and} \quad z^y = z[z, y] ,$$

and so we get

$$1 = z^4[z, x]^2[z, y]^2 = z^4 .$$

If $z^2 \neq 1$, then $H' = \langle z\rangle K_3(H)$, being abelian, implies $\langle z^2\rangle = \mho_1(H') \leq Z(G)$.

(ii) Suppose $H \cap N \nleq \Phi(N)$.

In this case we may choose $a \in (H \cap N) - \Phi(N)$ and $b \in N - ((H \cap N)\Phi(N))$, where $N = \langle a, b\rangle$. Then $[z, a] = 1$ and $1 = [z^2, b] = [z, b]^z[z, b] = [z, b]^2$.

(ii1) First, assume $[z, b] \neq [a, b]$. Consider the subgroup $M = \langle za, b\rangle$. Because $[z, b] \in K_3(G) \leq Z(G)$, we get

$$[za, b] = [z, b]^a[a, b] = [z, b][a, b] \neq 1 \quad \text{and} \quad [z, b][a, b] \in Z(G) ,$$

and so Lemma 65.2 (a) implies that M is minimal nonabelian. Hence G/M is Dedekindian and since $z \notin M$, G/M is nonabelian. By the minimality of $|G|$, we have $G/M \cong Q_8$. We have

$$G = HN = \langle x, y\rangle\langle a, b\rangle = \langle x, y, a, b\rangle .$$

On the other hand, $z = [x, y]$ and so

$$HM = \langle x, y\rangle\langle za, b\rangle = \langle x, y, a, b\rangle = G ,$$

where $H/(H \cap M) \cong Q_8$. It follows that

$$H \cap M = \langle x^4, x^2y^2, x^2[x, y]\rangle^H = H \cap N$$

and hence $a \in H \cap N = H \cap M \leq M$. Thus $z = (za)a^{-1} \in M$, a contradiction.

(ii2) Now we assume $[z, b] = [a, b]$. Set $L = \langle z, b\rangle \cap N$, where $\langle z, b\rangle$ is minimal nonabelian since $\langle [a, b]\rangle = N' \leq Z(G)$. Hence $\langle z, b\rangle \trianglelefteq G$ and then also $L \trianglelefteq G$, where we note that L is abelian. Let K be a maximal subgroup of N which contains L such that $K \trianglelefteq G$. Since $b \in K$ and $\langle a, b\rangle = N$, we have $a \in N - K$. On the other hand, $|N/K| = 2$ and $a \in H$ and therefore $H = \langle x, y\rangle$ covers $\bar{G} = G/K = \langle xK, yK\rangle = \langle \bar{x}, \bar{y}\rangle$. Hence $\bar{G}$ is a two-generator group of order 2^4 with a quotient group isomorphic to Q_8. By Lemma 228.3, we may set

$$\bar{G} = \langle \bar{x}, \bar{y} \mid \bar{x}^4 = \bar{y}^4 = 1,\ [\bar{x}, \bar{y}] = \bar{x}^2\rangle .$$

Then $\langle \bar{y}\rangle$ and $\langle \bar{x}\bar{y}\rangle$ are not normal in $\bar{G}$. It follows that their complete inverse images are also not normal in G and so they are abelian. This gives $[y, K] = [xy, K] = \{1\}$ and so $\langle x, y\rangle = H$ centralizes K. But then $1 = [z, b] = [a, b]$, where $z \in H$ and $b \in K$, a contradiction.

(iii) We suppose $H \cap N < \Phi(N)$.

We note that a direct product $P = Q_1 \times Q_2$ of quaternion groups Q_1 and Q_2 is not a metahamiltonian group since P contains a subgroup $Q \cong \mathrm{Q}_8$ with $Q \cap Q_1 = \{1\}$, $Q \cap Q_2 = \{1\}$ and Q is not normal in P.

We claim that $H \cap N \neq \{1\}$. Indeed, assume $H \cap N = \{1\}$ so that $G = H \times N$. Since $N \cong G/H$ is Dedekindian, we have $N \cong H \cong \mathrm{Q}_8$, a contradiction. We also have $N' \leq H \cap N$. Otherwise, G/H is Hamiltonian and so the minimality of $|G|$ gives $G/H \cong N/(H \cap N) \cong \mathrm{Q}_8$. But then $G/(H \cap N) \cong \mathrm{Q}_8 \times \mathrm{Q}_8$, a contradiction.

Consider $\bar{G} = G/(H \cap N) = \bar{H} \times \bar{N}$, where $\bar{H} = \langle \bar{x}, \bar{y}\rangle \cong \mathrm{Q}_8$ and $\bar{N}$ is abelian of rank 2 and exponent ≥ 4 since $\Phi(\bar{N}) \neq \{\bar{1}\}$. It follows that we may choose the generators a, b of N so that $\bar{N} = \langle \bar{a}\rangle \times \langle \bar{b}\rangle$ with $\mathrm{o}(\bar{a}) \geq 4$. Set $\bar{K} = \langle \bar{x}\bar{a}\rangle \times \langle \bar{b}\rangle$ so that $\bar{K}$ is not normal in $\bar{G}$ because $[\bar{x}\bar{a}, \bar{y}] = [\bar{x}, \bar{y}] \notin \bar{K}$. Hence the complete inverse image K of $\bar{K}$ is not normal in G. It follows that K is abelian and so $[xa, b] = 1$ giving $[x, b]^a[a, b] = 1$ and so $[x, b] = [a, b]$. Similarly, considering $\langle \bar{y}\bar{a}\rangle \times \langle \bar{b}\rangle$, we get $[y, b] = [a, b]$ and considering $\langle \bar{x}\bar{y}\bar{a}\rangle \times \langle \bar{b}\rangle$, we also get $[xy, b] = [a, b]$. On the other hand, we get

$$[a, b] = [xy, b] = [x, b]^y[y, b] = [a, b]^y[a, b] = [a, b]^2 = 1,$$

a contradiction.

(iv) Now assume $H \cap N = \Phi(N) = N'$.

In this case

$$|N| = 2^3, \quad |H| = 2^4, \quad |G| = 2^6 \quad \text{and} \quad G/N' = (H/N') \times \langle aN'\rangle \times \langle bN'\rangle .$$

Set $A = \langle a, N'\rangle$, $B = \langle b, N'\rangle$ so that $|A| = |B| = 4$ and both A and B are normal in G. Also, $C_G(A)$ and $C_G(B)$ are distinct maximal subgroups of G and so $K = C_G(A) \cap C_G(B)$ is of order 2^4 with $K \cap N = N'$. Since K covers G/N, we have $K/N' \cong \mathrm{Q}_8$. Without loss of generality we may assume $H = K$ and so we can suppose from the start $[H, N] = \{1\}$. Since H is a two-generator group of order 2^4 with $H/N' \cong \mathrm{Q}_8$, Lemma 228.3 implies that we may set

$$H = \langle x, y \mid x^4 = y^4 = 1,\ [x, y] = x^2\rangle \quad \text{and then } N' = \langle x^2y^2\rangle ,$$

because $x^2y^2 = z$ is the only involution in H which is not a square in H and only for such an involution z we have $H/\langle z\rangle \cong \mathrm{Q}_8$. Also, without loss of generality we may choose the generator a of N so that $\mathrm{o}(a) = 4$ and then we have $a^2 = x^2y^2$. We compute

$$(ay)^2 = a^2y^2 = (x^2y^2)y^2 = x^2, \quad [x, ay] = [x, y][x, a]^y = [x, y] = x^2 ,$$

and so $S = \langle x, ay\rangle \cong \mathrm{Q}_8$. On the other hand,

$$[ay, b] = [a, b]^y[y, b] = [a, b] = x^2y^2 \notin S ,$$

and so S is neither abelian nor normal in G, a contradiction.

(v) Finally, suppose that $H \cap N = \Phi(N) > N'$.

Let K be a normal subgroup in G such that $N' \le K < H \cap N$ with $|H \cap N : K| = 2$. Then we consider the factor-group $\bar{G} = G/K$ and we use the bar convention. By Lemma 228.3, we may set

$$\bar{H} = \langle \bar{x}, \bar{y} \mid \bar{x}^4 = \bar{y}^4 = 1,\ [\bar{x}, \bar{y}] = \bar{x}^2 \rangle\,, \quad \text{where } \overline{\Phi(N)} = \overline{H \cap N} = \langle \bar{x}^2 \bar{y}^2 \rangle\,.$$

Also, we may choose the generators $a, b \in N$ so that $\bar{N} = \langle \bar{a} \rangle \times \langle \bar{b} \rangle$ with $o(\bar{a}) = 4$, $o(\bar{b}) = 2$ and then $\bar{a}^2 = \bar{x}^2 \bar{y}^2$.

If $\bar{a} \in Z(\bar{G})$, then $\bar{H}_1 = \langle \bar{x}, \bar{a}\bar{y} \rangle \cong Q_8$ since

$$(\bar{a}\bar{y})^2 = \bar{a}^2 \bar{y}^2 = (\bar{x}^2 \bar{y}^2)\bar{y}^2 = \bar{x}^2 \quad \text{and} \quad [\bar{x}, \bar{a}\bar{y}] = [\bar{x}, \bar{y}] = \bar{x}^2\,.$$

It follows $\bar{H}_1 \trianglelefteq \bar{G}$ and $\bar{G} = \bar{H}_1 \times \bar{N}$. The complete inverse image $H_1 = \langle x, ay \rangle$ of $\bar{H}_1$ covers G/N so that we get $NH_1 = G$ and $H_1 \cap N \le K < \Phi(N)$. Replacing $H = \langle x, y \rangle$ with $H_1 = \langle x, ay \rangle$, we see that we have case (iii), which was already ruled out. It follows that we must have $\bar{a} \notin Z(\bar{G})$.

If $[\bar{a}, \bar{x}] = \bar{1}$, then $[\bar{a}, \bar{y}] = \bar{x}^2 \bar{y}^2$ and hence $\langle \bar{a}\bar{x}, \bar{y} \rangle$ is neither abelian nor normal in $\bar{G}$, a contradiction. Indeed, we have $(\bar{a}\bar{x})^2 = \bar{y}^2$ and $[\bar{a}\bar{x}, \bar{y}] = \bar{y}^2$ and so $\langle \bar{a}\bar{x}, \bar{y} \rangle \cong Q_8$ and $[\bar{a}, \bar{y}] = \bar{x}^2 \bar{y}^2$ shows that $\langle \bar{a}\bar{x}, \bar{y} \rangle$ is not normal in $\bar{G}$. We have proved that we must have $[\bar{a}, \bar{x}] = \bar{x}^2 \bar{y}^2$.

Replacing $\bar{a}$ with $\bar{a}\bar{b}$, we get with the same argument as in the previous paragraph, $[\bar{a}\bar{b}, \bar{x}] = \bar{x}^2 \bar{y}^2$. This relation together with $[\bar{a}, \bar{x}] = \bar{x}^2 \bar{y}^2$ implies $[\bar{b}, \bar{x}] = \bar{1}$. In this case it is easy to see that the abelian groups $\langle \bar{x}, \bar{b} \rangle$ and $\langle \bar{a}\bar{x}, \bar{b} \rangle$ (both of type $(4, 2)$) are not normal in $\bar{G}$. Indeed, we have

$$[\bar{a}, \bar{x}] = \bar{x}^2 \bar{y}^2 \notin \langle \bar{x}, \bar{b} \rangle \quad \text{and} \quad [\bar{a}, \bar{a}\bar{x}] = \bar{x}^2 \bar{y}^2 \notin \langle \bar{a}\bar{x}, \bar{b} \rangle\,.$$

Hence the complete inverse images of $\langle \bar{x}, \bar{b} \rangle$ and $\langle \bar{a}\bar{x}, \bar{b} \rangle$ are not normal in G and therefore they are abelian. We get $[x, b] = [ax, b] = 1$ and so

$$1 = [ax, b] = [a, b]^x [x, b] = [a, b]^x\,,$$

which implies $[a, b] = 1$, a final contradiction. □

Problem 1. Classify the p-groups in which the normal closure of each cyclic subgroup is either abelian or minimal nonabelian. (See Theorem 228.1.)

Problem 2. Classify the p-groups G in which the normal closure of each cyclic subgroup of G is Dedekindian (of class 2).

Problem 3. Classify the p-groups in which the normal closure of each absolutely regular subgroup is regular.

Problem 4. Classify the p-groups all of whose maximal regular subgroups are normal.

Problem 5. Study the p-groups in which the quasinormal closure of each cyclic subgroup is modular.

Problem 6. Classify the p-groups in which the normal closure of each $\mathcal{A}_1$-subgroup is an $\mathcal{A}_2$-subgroup.

Problem 7. Classify the p-groups in which the normal closure of each elementary abelian subgroup is abelian.

Problem 8. Classify the p-groups in which the normal closure of each subgroup of exponent p is regular.

§ 229 p-groups all of whose cyclic subgroups of order $\geq p^3$ are normal

In Theorem 63.4, all p-groups are characterized in which any cyclic subgroup of order $\geq p^2$ is normal in G. As a next step, it is of interest to classify p-groups G all of whose cyclic subgroups of order $\geq p^3$ are normal in G (Theorems 229.1 and 229.3) because this is a first step in studying p-groups G in which the intersection of any two distinct conjugate cyclic subgroups has order $\leq p$.

Theorem 229.1. *Let G be a nonabelian p-group of exponent $\geq p^3$, $p > 2$, all of whose cyclic subgroups of order $\geq p^3$ are normal in G. Then G is generated by its elements of order $\geq p^3$, $|G'| = p$, and $G'\mho_2(G)$ is cyclic.*

Conversely, if G is a p-group of exponent $\geq p^3$, $p > 2$, such that $|G'| = p$ and $G'\mho_2(G)$ is cyclic, then each cyclic subgroup of order $\geq p^3$ is normal in G.

Proof. Set $\exp(G) = p^e$, where $e \geq 3$. Let G_0 be the subgroup of G generated by all elements of orders $\geq p^3$. Assume $G_0 < G$ and note that all elements in $G - G_0$ are of order $\leq p^2$. Let $a \in G_0$ with $o(a) = p^e$ so that $\langle a\rangle \trianglelefteq G$ and let $g \in G - G_0$. We consider the subgroup $H = \langle a, g\rangle = \langle a\rangle\langle g\rangle$, where $\langle a\rangle \cap \langle g\rangle \leq \langle a^p\rangle$. Since H' is cyclic, H is regular (see Theorem 7.1 (c)). Then $\Omega_{e-1}(H) = \langle a^p, g\rangle$ is of exponent p^{e-1} (see Theorem 7.2 (b)) and $\Omega_{e-1}(H)$ is a maximal subgroup of H so that all elements in $H - \Omega_{e-1}(H)$ are of order p^e. But then $o(ag) = p^e$ and $ag \in G - G_0$, a contradiction. We have proved that G is generated by its elements of order $\geq p^3$.

For any $x \in G$ of order $\geq p^3$, $\langle x\rangle \trianglelefteq G$ and $\mathrm{Aut}(\langle x\rangle)$ is abelian so that G' centralizes $\langle x\rangle$. It follows that $G' \leq Z(G)$ and so G is of class 2.

As a next step, we show that G' is contained in each cyclic subgroup of order $\geq p^3$ and $|G'| \leq p^2$. Indeed, let $Z < G$ be any cyclic subgroup of order p^3 in G. Then $Z \trianglelefteq G$ and set $Z_0 = \Omega_2(Z) \cong C_{p^2}$. Let $K/Z_0 \neq \{1\}$ be a cyclic subgroup in G/Z_0. If $K \geq Z$, then K is cyclic of order $\geq p^3$ and then $K \trianglelefteq G$. Suppose that $K \ngeq Z$ so that $K \cap Z = Z_0$ and set $L = KZ$. Since $L/Z \cong K/Z_0$ is cyclic, $L' \leq Z$ is cyclic and so L is regular. If $|K/Z_0| \geq p^3$, then a cyclic subgroup in K of order $\geq p^3$ covers K/Z_0 and so $K \trianglelefteq G$. We may assume that $p^3 \leq |K| \leq p^4$ and so if K is cyclic, then $K \trianglelefteq G$. It remains to consider the case that K is noncyclic. If $\exp(K) = p^3$, then $|K| = p^4$ and K is generated by its cyclic subgroups of order p^3 and so we get again $K \trianglelefteq G$. Therefore we may assume that $\exp(K) = p^2$. Then $\Omega_2(L) = K$ (see Theorem 7.2 (b)) and so all elements in $L - K$ are of order p^3 and they generate L. Hence, $L \trianglelefteq G$ and then $K = \Omega_2(L) \trianglelefteq G$. We have proved that G/Z_0 is Dedekindian and so G/Z_0 is abelian. Hence, $G' \leq Z_0$ and so $|G'| \leq p^2$.

Suppose, by way of contradiction, that $G' \cong C_{p^2}$. Assume for a moment that $e = 3$ and let $a_0 \in G$ with $o(a_0) = p^3$ so that $\langle a_0\rangle \trianglelefteq G$, $G' = \langle a_0^p\rangle \leq Z(G)$. For any $x \in G$,

$$[a_0, x^p] = [a_0^p, x] = 1 \text{ and so } x^p \in Z(G)$$

because x^p centralizes each cyclic subgroup of order p^3 in G. Hence for any $x, y \in G$,

$$[x, y]^p = [x^p, y] = 1 \text{ and so } |G'| \leq p\,,$$

a contradiction. We have proved that we must have $e \geq 4$. We note that G is regular (since G' is cyclic) and so $\exp(\Omega_{e-1}(G)) = p^{e-1}$ so that all elements in $G - \Omega_{e-1}(G)$ are of order p^e and they generate G. In addition, $G' \leq \mho_1(G)$ implies that G is powerful.

If $\mho_1(G) \leq Z(G)$, then for any $x, y \in G$, $[x, y]^p = [x^p, y] = 1$ and so $|G'| \leq p$, a contradiction. Hence, $\mho_1(G) \nleq Z(G)$ and so there is $b \in \mho_1(G)$ such that $b \notin Z(G)$. It follows that we can find $a \in G$ of order p^e such that $[a, b] \neq 1$. By Proposition 26.10, there is $c \in G$ such that $c^p = b$. Hence, $\langle c\rangle$ induces on $\langle a\rangle$ an automorphism of order p^2. We may set

$$a^c = av \text{ with } v \in \langle a^p\rangle \text{ and } o(v) = p^2 \text{ so that } \langle v\rangle = G'\,.$$

From $[a, c] = v$, we get

$$[a, b] = [a, c^p] = [a, c]^p = v^p = z, \text{ where } \langle z\rangle = \Omega_1(\langle a\rangle)\,.$$

We consider the metacyclic subgroup $H = \langle a, c\rangle$, where $H' = G' = \langle v\rangle \cong C_{p^2}$ and $H/\langle a\rangle$ is cyclic of order $\geq p^2$. Since $H' \leq \mho_1(H)$, H is powerful. By Theorem 137.7 (b), each maximal subgroup of H has its derived subgroup of order $\leq p$ and the maximal subgroup $\langle a, b\rangle$ of H is nonabelian. By Lemma 65.2 (a), each maximal subgroup of H is either abelian or minimal nonabelian. Assume that $|H : \langle a\rangle| \geq p^3$. Then H is a metacyclic A_2-group of order $\geq p^{e+3}$. By Proposition 71.2 (a),

$$H = \langle r, s \mid r^{p^m} = 1,\ m \geq 3,\ s^{p^n} = r^{\epsilon p^{m-1}},\ n \geq 2,\ \epsilon \in \{0, 1\},\ r^s = r^{1+p^{m-2}}\rangle\,.$$

If $o(s) \geq p^3$, then by the above, $H' = \langle r^{p^{m-2}}\rangle \leq \langle s\rangle$, a contradiction. Hence, $o(s) = p^2$ and so $\epsilon = 0$, $n = 2$ and $|H| = p^{m+2}$. But then $p^{m+2} \geq p^{e+3}$ and $m \geq e + 1$, contrary to $\exp(H) = p^e$. We have proved that $H/\langle a\rangle \cong C_{p^2}$.

Since $\Phi(H) = \mho_1(H)H' = \mho_1(H) \geq \langle a^p, b\rangle$ and $d(H) = 2$, we get

$$\mho_1(H) = \langle a^p, b\rangle \text{ with } \mho_1(H) \cap \langle a\rangle = \langle a^p\rangle\,.$$

If $\mho_1(H)$ is cyclic (of order p^e), then $o(c) = p^{e+1}$, a contradiction. Hence, $\mho_1(H)$ is noncyclic so that there is an element $i \in \mho_1(H) - \langle a\rangle$ of order p. By Proposition 26.10, there is $d \in H - \langle a\rangle$ such that $d^p = i$ and so

$$H = \langle a\rangle\langle d\rangle \text{ with } \langle d\rangle \cong C_{p^2} \text{ and } \langle a\rangle \cap \langle d\rangle = \{1\}\,.$$

Since $[a, d] \in \langle v\rangle$, we get

$$[a^p, d] = [a, d]^p \in \langle v^p\rangle = \langle z\rangle\,.$$

Let f be an element of order p^3 in $\langle a\rangle$ with $f^{p^2} = z$. Hence, $f \in \langle a^p\rangle$ because $e \geq 4$ and therefore $[f, d] \in \langle z\rangle$. We get

$$(df)^{p^2} = d^{p^2} f^{p^2} [f, d]^{\binom{p^2}{2}} = z \quad \text{and so } o(df) = p^3\,.$$

Since $\langle df\rangle$ covers $H/\langle a\rangle \cong C_{p^2}$, it follows $\langle df\rangle \cap \langle a\rangle = \langle z\rangle$. But then $G' = \langle v\rangle$ is not contained in $\langle df\rangle$, a contradiction. We have proved that $|G'| = p$.

It remains to be proved that $\mho_2(G)$ is cyclic. Let $a \in G$ be an element of order p^e and $x \in G$. We consider the metacyclic subgroup $M = \langle a, x\rangle$. If M is abelian, then M splits over $\langle a\rangle$ and so $M = \langle a\rangle \times \langle b\rangle$ with $o(b) \leq p^2$ since each cyclic subgroup of order $\geq p^3$ contains $G' \cong C_p$. This gives that $x^{p^2} \in \langle a\rangle$. Now suppose that M is nonabelian. Since $M' = G' \leq \langle a\rangle$, it follows that M is minimal nonabelian. By Lemma 65.1, we have

$$M = \langle r, s \mid r^{p^m} = s^{p^n} = 1,\ r^s = r^{1+p^{m-1}}\rangle, \quad m \geq 2, \quad n \geq 1 .$$

Because $M' = \langle r^{p^{m-1}}\rangle \leq \langle r\rangle$, we get $n \leq 2$. On the other hand, $\exp(M) = p^e$ and so $m = e$ and $|M| \leq p^{e+2}$. But then $|M : \langle a\rangle| \leq p^2$ and so again $x^{p^2} \in \langle a\rangle$. We have proved that $\mho_2(G) \leq \langle a\rangle$ and so $\mho_2(G)$ is cyclic.

Since $G' \cong C_p$ and G' is contained in each cyclic subgroup of order $\geq p^3$, it follows that $G' \leq \mho_2(G)$ and so $G'\mho_2(G) = \mho_2(G)$ is cyclic. Our theorem is proved because the converse is clear. □

Proposition 229.2. *Let G be a nonabelian 2-group of exponent ≥ 8 all of whose cyclic subgroups of order ≥ 8 are normal in G. If $|G'| \leq 4$, then G' is contained in each cyclic subgroup of order ≥ 8 and $\mho_2(G)$ is cyclic. In particular, G' is cyclic.*

Proof. Let $C = \langle c\rangle$ be any cyclic subgroup of order 8 in G and set $\langle z\rangle = \Omega_1(C)$ so that $C \trianglelefteq G$ and $z \in Z(G)$. Assume, by way of contradiction, that $G' \nleq C$ so that G/C is nonabelian. In this case $G' \cap C \leq \langle z\rangle$ and so

$$[G, C] \leq G' \cap C \leq \langle z\rangle .$$

Thus, G centralizes $C/\langle z\rangle$.

First we show that G/C is Dedekindian. Indeed, let X/C be any cyclic subgroup in G/C. If $|X/C| = 2$, then the fact that X centralizes C/Z implies that X is either abelian or $X \cong M_{2^4}$. Hence, we have either $X \cong C_{2^4}$ or X is generated by its cyclic subgroups of order 8 and so in any case $X \trianglelefteq G$. Suppose that $X/C \cong C_4$. Let $d \in X - C$ be such that $\langle d\rangle$ covers X/C. If $o(d) \geq 8$, then $\langle d\rangle \trianglelefteq G$ and so $X \trianglelefteq G$. Suppose that $o(d) = 4$. Since $X' = [C, \langle d\rangle] \leq \langle z\rangle$, it follows that $\mathrm{cl}(X) \leq 2$ and so

$$(cd)^4 = c^4 d^4 [d, c]^{\binom{4}{2}} = c^4 = z .$$

Hence

$$\langle cd\rangle \cong C_8 , \quad \langle cd\rangle \trianglelefteq G \ \text{ and so } X = \langle c\rangle\langle cd\rangle \trianglelefteq G .$$

Finally, if $|X/C| \geq 8$, then a cyclic subgroup in X of order ≥ 8 covers X/C and so $X \trianglelefteq G$.

Since G/C is nonabelian and Dedekindian, it follows that G/C is Hamiltonian and so G/C has a subgroup $Q/C \cong Q_8$ and $|Q| = 2^6$. Suppose that Q has a cyclic subgroup $\langle x\rangle$ of order 2^4. Since $Q/\langle x\rangle$ is abelian, $Q' \leq \langle x\rangle$. Also, $\exp(Q/C) = 4$ and so $\mho_2(\langle x\rangle) \leq C$ and therefore $|\langle x\rangle \cap C| \geq 4$. But $Q' \nleq C$ and so $Q' \nleq \langle x\rangle \cap C$ and therefore $|Q'| \geq 8$, a contradiction. We have proved that $\exp(Q) = 8$. Set

$$D = Q'C, \quad \text{where } Q' \cap C \leq \langle z \rangle, \ |D : C| = 2,$$

and D is abelian of type $(8, 2)$ because $|Q : C_Q(C)| \leq 2$. Hence, $\Omega_2(D) = \langle c^2, Q' \rangle$ is abelian of type $(4, 2)$. For each $x \in Q$, $x^2 \in \Omega_2(D)$ because $o(x) \leq 8$. Hence, $\Phi(Q) = \Omega_2(D)$ and so there is a maximal subgroup M of Q such that $M \cap C = \langle c^2 \rangle$ and $M/\langle c^2 \rangle \cong Q_8$. Since $[Q, C] \leq \langle z \rangle$, we have $\langle c^2 \rangle \leq Z(Q)$. All three maximal subgroup of M which contain $\langle c^2 \rangle$ are abelian and so $M' = \langle u \rangle \cong C_2$. In particular, Q' is elementary abelian of order ≤ 4 and $Q' \leq Z(Q)$. Let $q_1, q_2 \in Q - D$ be such that $\langle q_1, q_2 \rangle$ covers $Q/D \cong E_4$. If $o(q_1) \leq 4$, then

$$(cq_1)^4 = c^4 q_1^4 [q_1, c]^{\binom{4}{2}} = c^4 = z \quad \text{and so } o(cq_1) = 8.$$

Hence replacing q_1 with cq_1 (if necessary), we may assume from the start $o(q_1) = 8$. Then we have $\langle q_1 \rangle \trianglelefteq Q$, $\langle q_1 \rangle \cap C = \langle z \rangle$, and $[q_1, q_2] = uz^{\epsilon}$, where $\epsilon \in \{0, 1\}$. But $uz^{\epsilon} \notin \langle q_1 \rangle$, contrary to $\langle q_1 \rangle \trianglelefteq Q$.

We have proved that G' is contained in each cyclic subgroup of order ≥ 8 and so, in particular, G' is cyclic. It remains to prove that $\mho_2(G)$ is cyclic. Let $a \in G$ be a fixed element of order 2^e, $e \geq 3$, and let x be any element in G. We consider the metacyclic subgroup $H = \langle a, x \rangle$. If H is abelian, then there is a complement $\langle b \rangle$ of $\langle a \rangle$ in H. By the above, $\{1\} \neq G' \leq \langle a \rangle$ and so $o(b) \leq 4$. This implies $x^4 \in \langle a \rangle$. Next, suppose that $|H'| = 2$ so that Lemma 65.2 (a) implies that H is minimal nonabelian. Since $\exp(H) \geq 8$, we get by Lemma 65.1,

$$H = \langle r, s \mid r^{2^m} = s^{2^n} = 1, \ r^s = r^{1+2^{m-1}} \rangle, \quad m \geq 2, \quad n \geq 1.$$

Since $H' \leq \langle r \rangle$, we have $n \leq 2$ and H is of exponent 2^m. But then $m = e$, $|H| = 2^{e+n}$ and so $x^4 \in \langle a \rangle$. Finally, assume that $H' \cong C_4$. Since H is metacyclic and $|H'| = 4$, Theorem 137.7 implies that each maximal subgroup H_i of H has the property $|H_i'| \leq 2$ and so Lemma 65.2 (a) gives that H_i is either abelian or minimal nonabelian. But $|H'| = 4$ and so there are minimal nonabelian maximal subgroups. Thus, H is an A_2-group. If $|H : \langle a \rangle| \leq 4$, then again $x^4 \in \langle a \rangle$ and we are done. Therefore we may assume that $|H : \langle a \rangle| \geq 8$ so that $|H| \geq 2^{e+3} \geq 2^6$. Then Proposition 71.2 gives two possibilities for the structure of H. If

$$H = \langle r, s \mid r^8 = 1, \ s^{2^n} = r^{4\epsilon}, \ r^s = r^{-1+4\eta} \rangle, \quad n \geq 1, \quad \epsilon, \eta \in \{0, 1\},$$

then $H' = \langle r^2 \rangle$ and so $n \leq 2$ and $|H| \leq 2^5$, a contradiction. If

$$H = \langle r, s \mid r^{2^m} = 1, \ s^{2^n} = r^{\epsilon 2^{m-1}}, \ r^s = r^{1+2^{m-2}} \rangle, \quad m \geq 4, \quad n \geq 2, \quad \epsilon \in \{0, 1\},$$

then $H' = \langle r^{2^{m-2}} \rangle$ and so $n = 2$ and $\epsilon = 0$. But then $|H| = 2^{m+2} \geq 2^{e+3}$ and so $m \geq e+1$, contrary to $\exp(H) = 2^e$. We have proved that for each $x \in G$, $x^4 \in \langle a \rangle$ and so $\mho_2(G)$ is cyclic. □

Theorem 229.3. *Let G be a nonabelian 2-group of exponent 2^e, $e \geq 3$, all of whose cyclic subgroups of order ≥ 8 are normal in G. Let G_0 be the subgroup of G generated by all elements of order ≥ 8. Then $|G : G_0| \leq 2$ and we have one of the following possibilities.*

(a) *If $|G : G_0| = 2$, then one of the following holds:*
  (a1) *$|G_0'| = 2$, $G_0'\mho_2(G_0)$ is cyclic and each $g \in G - G_0$ inverts G_0/G_0' and $g^2 \in Z(G)$.*
  (a2) *G_0 is abelian and either $\mho_2(G_0)$ is cyclic and $g \in G - G_0$ inverts $G_0/\mho_{e-1}(G_0)$ or $\mho_2(G_0)$ is noncyclic and then for all $g \in G - G_0$ and $a \in G_0$, $a^g = a^{-1+\epsilon 2^{e-1}}$, where $\epsilon \in \{0, 1\}$.*

(b) *If $G = G_0$, then $|G'| = 2$ and $G'\mho_2(G)$ is cyclic.*

Conversely, all the above groups stated in (a) and (b) satisfy the assumptions of the theorem.

Proof. Note that in case $G_0 \neq G$, all elements in $G - G_0$ are of order ≤ 4. For each $x \in G_0$ with $o(x) \geq 8$, $C_G(x) \leq G_0$, $\langle x \rangle \trianglelefteq G$ and since $\mathrm{Aut}(\langle x \rangle)$ is abelian, G' centralizes $\langle x \rangle$. Thus G' centralizes G_0 and so $G' \leq G_0$ and $G' \leq Z(G_0)$. In particular, $\mathrm{cl}(G_0) \leq 2$ and G/G_0 is abelian.

Let a be any element of order 8 in G_0 so that $\langle a \rangle \trianglelefteq G$. Since $C_G(a) \leq G_0$ and $\mathrm{Aut}(\langle a \rangle) \cong E_4$, it follows that G/G_0 is elementary abelian of order ≤ 4. Suppose that there is $g \in G - G_0$ such that $a^g = az$, where $z = a^4$. But then $\langle a, g \rangle' = \langle z \rangle$ and

$$(ga)^4 = g^4 a^4 [a, g]^{\binom{4}{2}} = z \text{ and so } o(ga) = 8\,,$$

a contradiction. Hence if $G_0 < G$, then $|G : G_0| = 2$ and for each $g \in G - G_0$, $a^g = a^{-1+4\epsilon}$, $\epsilon \in \{0, 1\}$, where a is any element of order 8 in G_0. This implies that for any $b \in G_0$ of order ≥ 8, $b^g = b^{-1}v$, where $v \in \langle b^4 \rangle$ and $o(v) \leq 4$ since $o(g) \leq 4$.

(i) Assume $|G : G_0| = 2$ and let $g \in G - G_0$.
Let $a \in G_0$ with $o(a) \geq 8$. Then by the above,

$$a^g = a^{-1}v \text{ with } v \in \langle a^4 \rangle \text{ and so } [a, g] = a^{-1}a^g = a^{-2}v\,,$$
$$\text{where } \langle a^{-2}v \rangle = \langle a^2 \rangle \leq Z(G_0) \text{ since } G' \leq Z(G_0)\,.$$

For any $x \in G_0$, $1 = [a^2, x] = [a, x^2]$ and so x^2 centralizes each element of order ≥ 8 in G_0. Hence, $x^2 \in Z(G_0)$ and so $\mho_1(G_0) \leq Z(G_0)$. For any $x, y \in G_0$, $[x, y]^2 = [x^2, y] = 1$ and so G_0' is elementary abelian.

Suppose that $g^2 = i \notin Z(G_0)$. Then there is $a \in G_0$ of order ≥ 8 such that $[a, i] \neq 1$ and so $\langle g \rangle \cong C_4$ acts faithfully on $\langle a \rangle$, $o(a) \geq 2^4$, and $\langle a \rangle \cap \langle g \rangle = \{1\}$. We have $a^g = a^{-1}v$ with $v \in \langle a^4 \rangle$ and $o(v) = 4$ so that $o(iv) = 4$ because $v \in \mho_1(G_0) \leq Z(G_0)$. But then

$$(ga)^2 = gaga = g^2 a^g a = ia^{-1}va = iv\,,$$

and so $ga \in G - G_0$ and $o(ga) = 8$, a contradiction. We have proved that for each $g \in G - G_0$, $g^2 \in Z(G)$ and so g induces on any $a \in G_0$ of order ≥ 8 an involutory automorphism given by $a^g = a^{-1}z^\epsilon$, $\epsilon \in \{0, 1\}$, where $\langle z \rangle = \Omega_1(\langle a \rangle)$.

(i1) Suppose that G_0 is nonabelian.

Let $\langle a\rangle$ be a fixed cyclic subgroup of order 8 in G_0. We show that $G_0/\langle a\rangle$ is Dedekindian. Indeed, let $S/\langle a\rangle$ be a cyclic subgroup of order ≤ 4 in $G_0/\langle a\rangle$. If $S/\langle a\rangle \cong C_2$, then the fact that G_0' is elementary abelian implies that S is either abelian of type (2^4) or $(2^3, 2)$ or $S \cong M_{16}$. Hence, we have either $S \cong C_{2^4}$ and then $S \trianglelefteq G$ or S is generated by its cyclic subgroups of order 8 and then again $S \trianglelefteq G$. Suppose that $S/\langle a\rangle \cong C_4$ and let $y \in S - \langle a\rangle$ be such that $\langle y\rangle$ covers $S/\langle a\rangle$. If $o(y) \geq 8$, then $\langle y\rangle \trianglelefteq G$ and so $S = \langle a\rangle\langle y\rangle \trianglelefteq G$. Hence, we may assume that $o(y) = 4$. Then we compute

$$(ay)^4 = a^4 y^4 [y, a]^{\binom{4}{2}} = a^4 \text{ and so } o(ay) = 8\,.$$

We get $\langle ay\rangle \trianglelefteq G$ and so $S = \langle a\rangle\langle ay\rangle \trianglelefteq G$.

We have proved that $G_0/\langle a\rangle$ is Dedekindian and so $|(G_0/\langle a\rangle)'| \leq 2$. On the other hand, G_0' is elementary abelian and so $G_0' \cap \langle a\rangle \leq \langle a^4\rangle$. It follows $|G_0'| \leq 4$. By Proposition 229.2, G_0' is cyclic and so $|G_0'| = 2$. Also, $\mho_2(G_0)$ is cyclic and G_0' is contained in each cyclic subgroup of order ≥ 8 in G_0 and so $G_0' \leq \mho_2(G_0)$ and therefore $G_0'\mho_2(G_0)$ is also cyclic. Set $G_0' = \langle z\rangle$.

For each $g \in G - G_0$ and $x \in G_0$ with $o(x) = 2^f \geq 8$, we have

$$g^2 \in Z(G)\,, \quad \langle z\rangle \leq \langle x\rangle \quad \text{and} \quad x^g = x^{-1}z^{\epsilon}\,,\ \epsilon \in \{0, 1\}\,.$$

Let $v \in G_0$ with $o(v) \leq 4$. Then

$$(xv)^{2^{f-1}} = x^{2^{f-1}} v^{2^{f-1}} [v, x]^{\binom{2^{f-1}}{2}} = x^{2^{f-1}} = z, \text{ and so } o(xv) = 2^f\,.$$

Hence

$$(xv)^g = (xv)^{-1}z^{\eta} = v^{-1}x^{-1}z^{\eta} = x^g v^g = x^{-1}z^{\epsilon}v^g\,, \quad \eta \in \{0, 1\}\,.$$

It follows

$$v^{-1}x^{-1} = x^{-1}v^{-1}z^{\zeta}\,, \quad \zeta \in \{0, 1\}\,, \quad \text{and so } v^g = v^{-1}z^{\delta}\,,\ \delta \in \{0, 1\}\,.$$

We have proved that g inverts $G_0/\langle z\rangle$ and so we have obtained the groups stated in part (a1) of our theorem.

(i2) Assume now that G_0 is abelian.

Let $g \in G - G_0$ and $a_1 \in G_0$ with $o(a_1) = 2^e$. Then

$$a_1^g = a_1^{-1}z_1^{\epsilon_1}\,, \quad \text{where } z_1 = a_1^{2^{e-1}} \text{ and } \epsilon_1 \in \{0, 1\}\,.$$

Let C be a complement of $\langle a_1\rangle$ in G_0.

(i2a) Suppose that $\mho_2(G_0)$ is cyclic so that $\exp(C) \leq 4$ and $\mho_{e-1}(G_0) = \langle z_1\rangle$. Let $c \in C$. Then $o(a_1c) = 2^e$ with $\Omega_1(\langle a_1c\rangle) = \langle z_1\rangle$ so that

$$(a_1c)^g = (a_1c)^{-1}z_1^{\eta} = a_1^{-1}c^{-1}z_1^{\eta} = a_1^g c^g = a_1^{-1}z_1^{\epsilon_1}c^g\,, \quad \eta \in \{0, 1\}\,,$$

and we obtain $c^g = c^{-1}z_1^{\epsilon}$, $\epsilon \in \{0, 1\}$. Hence, g inverts $G_0/\mho_{e-1}(G_0)$.

(i2b) Now assume that $\mho_2(G_0)$ is noncyclic and set $\exp(C) = 2^f$, where $e \geq f \geq 3$.

Note that we have $a_1^g = a_1^{-1+\epsilon_1 2^{e-1}}$, $\epsilon_1 \in \{0, 1\}$. Let $a_2 \in C$ be an element of order 2^f so that

$$a_2^g = a_2^{-1} z_2^{\epsilon_2}, \text{ where } \langle z_2 \rangle = \Omega_1(\langle a_2 \rangle) \text{ and } \epsilon_2 \in \{0, 1\}.$$

Also we have

$$o(a_1 a_2) = 2^e \text{ and } (a_1 a_2)^{2^{e-1}} = z_1 a_2^{2^{e-1}},$$

and so we get

$$(a_1 a_2)^g = a_1^g a_2^g = a_1^{-1} z_1^{\epsilon_1} \cdot a_2^{-1} z_2^{\epsilon_2} = (a_1 a_2)^{-1} (z_1 a_2^{2^{e-1}})^{\epsilon}, \quad \epsilon \in \{0, 1\},$$

and this implies $z_1^{\epsilon_1} z_2^{\epsilon_2} = z_1^{\epsilon} a_2^{\epsilon 2^{e-1}}$. If $a_2^{2^{e-1}} = z_2$, then $\epsilon_1 = \epsilon_2 = \epsilon$ and $a_2^g = a_2^{-1+\epsilon_1 2^{e-1}}$. If $a_2^{2^{e-1}} = 1$, then $\epsilon_2 = 0$ and $a_2^g = a_2^{-1} = a_2^{-1+\epsilon_1 2^{e-1}}$. Hence in any case we get $a_2^g = a_2^{-1+\epsilon_1 2^{e-1}}$.

Let D be a complement of $\langle a_2 \rangle$ in C and let $d \in D$. If $o(d) \geq 8$, then we get as in the previous paragraph, $d^g = d^{-1+\epsilon_1 2^{e-1}}$. Now assume $o(d) \leq 4$ so that

$$o(a_1 d) = 2^e, \quad o(a_2 d) = 2^f, \quad \Omega_1(\langle a_1 d \rangle) = \langle z_1 \rangle, \text{ and } \Omega_1(\langle a_2 d \rangle) = \langle z_2 \rangle.$$

This gives

$$\begin{aligned} (a_1 d)^g &= a_1^{-1} z_1^{\epsilon_1} d^g = (a_1 d)^{-1} z_1^{\eta_1} \text{ and so } d^g = d^{-1} z_1^{\eta_1 - \epsilon_1}, \\ (a_2 d)^g &= a_2^{-1+\epsilon_1 2^{e-1}} d^g = (a_2 d)^{-1} z_2^{\eta_2} \text{ and so } d^g = d^{-1} a_2^{-\epsilon_1 2^{e-1}} z_2^{\eta_2}, \\ &\text{where } \eta_1, \eta_2 \in \{0, 1\}. \end{aligned}$$

It follows that $z_1^{\eta_1 - \epsilon_1} = 1$ and so $d^g = d^{-1} = d^{-1+\epsilon_1 2^{e-1}}$. It was proved that for all $a \in G_0$, we get

$$a^g = a^{-1+\epsilon_1 2^{e-1}}, \quad \text{where } \epsilon_1 \in \{0, 1\}.$$

We have obtained all the groups stated in part (a2) of the theorem.

(ii) Finally assume $G = G_0$.

We know that G is of class 2. Let $Z = \langle a \rangle \trianglelefteq G$ be a cyclic subgroup of order 8 in G and set $\Omega_1(Z) = \langle z \rangle$ and $\Omega_2(Z) = Z_0$. We show that G/Z_0 is Dedekindian.

Let K_1/Z_0 be a cyclic subgroup of order 2 in G/Z_0. If K_1 is cyclic (of order 8), then $K_1 \trianglelefteq G$. Assume that K_1 is noncyclic so that $K_1 \cap Z = Z_0$ and we set $H_1 = ZK_1$. Since $\mathrm{cl}(G) = 2$, H_1 cannot be of maximal class (and order 2^4) and so H_1 is either abelian of type $(8, 2)$ or $H_1 \cong M_{2^4}$. In any case, H_1 is generated by its cyclic subgroups of order 8 and so $H_1 \trianglelefteq G$. But $K_1 = \Omega_2(H_1)$ and so $K_1 \trianglelefteq G$.

Let K_2/Z_0 be a cyclic subgroup of order 4 in G/Z_0. If $K_2 > Z$, then K_2 is cyclic (of order 2^4) and $K_2 \trianglelefteq G$. We may assume that K_2 is noncyclic and then $K_2 \cap Z = Z_0$ and we set $H_2 = ZK_2$. Let $v \in K_2 - Z_0$ be such that $\langle v \rangle$ covers K_2/Z_0. If $o(v) \geq 8$, then $K_2 \trianglelefteq G$. Hence, we may assume that

$$K_2 = Z_0 \langle v \rangle, \quad Z_0 \cap \langle v \rangle = \{1\} \quad \text{and} \quad \exp(K_2) = 4.$$

In this case, v^2 centralizes Z and so v induces on Z an automorphism of order ≤ 2. If

$$a^v = a^{-1}z^\epsilon, \quad \epsilon \in \{0,1\}, \text{ then } (a^2)^v = a^{-2}, \text{ where } \langle a^2\rangle = H_2',$$

contrary to the fact that $G' \leq Z(G)$. Hence, we have $a^v = az^\epsilon$, $\epsilon \in \{0,1\}$, and so $H_2' \leq \langle z\rangle$. For each $k \in K_2$,

$$(ak)^4 = a^4k^4[k,a]^{\binom{4}{2}} = z,$$

and so all elements in $H_2 - K_2$ are of order 8 and since they generate H_2, we get $H_2 \trianglelefteq G$ and $K_2 = \Omega_2(H_2) \trianglelefteq G$.

We have proved that G/Z_0 is Dedekindian. Hence, $|G'| \leq 8$ and we assume, by way of contradiction, that $|G'| = 8$. Then G/Z_0 is Hamiltonian and let Q/Z_0 be a quaternion subgroup in G/Z_0. Since $Z_0 \leq G'$, we have $G' \leq Q$ and $G'/Z_0 = (Q/Z_0)'$. Also note that $G' \leq Z(G)$. Hence, all three maximal subgroups of Q containing Z_0 are abelian and so $Q' = \langle u\rangle \cong C_2$. It follows that $G' = Z_0 \times \langle u\rangle$ is abelian of type $(4,2)$ and $Q \cap Z = Z_0$. We set $H = QZ$ and so $|H| = 2^6$. Because $Z_0 \leq Z(H)$, we have $[Z,H] \leq \langle z\rangle$. Let $x_1, x_2 \in Q - Z_0$ be such that $\langle x_1, x_2\rangle$ covers Q/G'. Then $[x_1,x_2] = u$. If $o(x_1) \geq 8$, then $\langle x_1\rangle \trianglelefteq G$ and $\langle x_1\rangle \cap Z_0 > \langle z\rangle$. This is a contradiction since $[x_1,x_2] = u \notin \langle x_1\rangle$. Hence, $o(x_1) = 4$ and we get

$$(ax_1)^4 = a^4x_1^4[x_1,a]^{\binom{4}{2}} = z \quad \text{since } [x_1,a] \in \langle z\rangle.$$

Hence, $o(ax_1) = 8$, $\langle ax_1\rangle \trianglelefteq G$ and $\langle ax_1\rangle \cap Z = \langle z\rangle$. But

$$[ax_1, x_2] = [a,x_2][x_1,x_2] = z^\epsilon u, \text{ where } \epsilon \in \{0,1\} \text{ and } z^\epsilon u \notin \langle ax_1\rangle,$$

a contradiction.

We have proved that $|G'| \leq 4$. By Proposition 229.2, G' is cyclic, $\mho_2(G)$ is cyclic, and G' is contained in each cyclic subgroup of order ≥ 8.

Assume, by way of contradiction, that $G' \cong C_4$. If $e = 3$, then each $x \in G$ induces an automorphism of order ≤ 2 on each cyclic subgroup $\langle a\rangle$ of order 8 in G and so x^2 centralizes $\langle a\rangle$. But all elements of order 8 in G generate G and so $x^2 \in Z(G)$. Hence, $\mho_1(G) \leq Z(G)$ and so for each $x,y \in G$, $[x,y]^2 = [x^2,y] = 1$, a contradiction. We have proved that $e \geq 4$ and so $G' \leq \mho_2(G)$ implying that G is powerful. By Proposition 26.23, each element in $\mho_1(G)$ is a square of an element in G. If $\mho_1(G) \leq Z(G)$, then for any $x,y \in G$, $[x,y]^2 = [x^2,y] = 1$, a contradiction. Thus, there is an element $b \in \mho_1(G)$ with $b \notin Z(G)$. Note that $e \geq 4$ also implies that $\exp(\Omega_{e-1}(G)) = 2^{e-1}$ and so G is generated by all elements in $G - \Omega_{e-1}(G)$ of order 2^e. Hence, there is $a \in G$ of order 2^e such that $[a,b] \neq 1$. Let $c \in G$ be such that $c^2 = b$. Since $[c^4,a] = [c,a]^4 = 1$, c induces on $\langle a\rangle$ an automorphism of order 4. Because $G' \cong C_4$, we may set

$$a^c = av \quad \text{with } o(v) = 4,\ v \in \langle a\rangle \text{ and } \langle v\rangle = G'.$$

Then we get

$$a^b = a^{c^2} = (av)^c = a^cv = av^2 = az, \quad \text{where } z = v^2.$$

Set $H = \langle a, c\rangle$ and since $\mho_2(H)$ is cyclic, we have $\mho_2(H) = \langle a^4\rangle$ and so $H/\langle a\rangle \cong C_4$ and H is powerful because $H' = \langle v\rangle \le \mho_2(H)$. We have $\Phi(H) = \mho_1(H) \ge \langle a^2, b\rangle$ and since $d(H) = 2$, we get $\mho_1(H) = \langle a^2, b\rangle$ and $\mho_1(H) \cap \langle a\rangle = \langle a^2\rangle$ with $b^2 \in \langle a^4\rangle$. Hence, $\mho_1(H)$ is noncyclic and since

$$a^b = az, \text{ we get } (a^2)^b = a^2$$

and so $\mho_1(H)$ is abelian. It follows that there is an involution $i \in \mho_1(H) - \langle a\rangle$. But H is powerful and so there is $d \in H - \langle a\rangle$ such that $d^2 = i$ and so

$$H = \langle a\rangle\langle d\rangle \text{ with } \langle d\rangle \cong C_4 \text{ and } \langle a\rangle \cap \langle d\rangle = \{1\}\,.$$

Let f be an element of order 8 in $\langle a\rangle$ so that $f \in \langle a^2\rangle$ and $f^4 = z$. Then

$$(a^2)^c = (a^c)^2 = (av)^2 = a^2 z \text{ implies } [f, d] \in \langle z\rangle\,,$$

and so we get

$$(df)^4 = d^4 f^4 [f, d]^{\binom{4}{2}} = z\,.$$

Hence, $o(df) = 8$ and $\langle df\rangle$ covers $H/\langle a\rangle \cong C_4$ so that $\langle df\rangle \cap \langle a\rangle = \langle z\rangle$. But then $G' = \langle v\rangle$ is not contained in $\langle df\rangle$, a contradiction.

We have proved that $|G'| = 2$ and since G' is contained in each cyclic subgroup of order ≥ 8, it follows that $G' \le \mho_2(G)$ and so $G'\mho_2(G) = \mho_2(G)$ is cyclic. We have obtained the groups stated in part (b) of our theorem.

Conversely, it is clear that all groups stated in parts (a) and (b) of our theorem satisfy the assumptions of the theorem. □

§ 230 Nonabelian p-groups of exponent p^e all of whose cyclic subgroups of order p^e are normal

Here we generalize the results from §§ 63 and 229. It is surprising that the title groups can be classified with elementary methods (Theorems 230.1 and 230.2) in spite of the fact that our assumptions are quite general. In view of the results in §§ 63 and 229, we may assume that the title 2-groups are of exponent 2^e with $e \geq 4$.

Theorem 230.1 (Janko). *Let G be a nonabelian p-group of exponent p^e, $p > 2$, all of whose cyclic subgroups of order p^e are normal in G. Then $e \geq 2$, G is generated by its elements of order p^e, G is of class 2, G' is cyclic of order p^s, where $1 \leq s \leq \frac{e}{2}$, and G' is contained in each cyclic subgroup of order p^e.*

Conversely, if G is a nonabelian p-group of exponent p^e, $p > 2$, such that G' is contained in each cyclic subgroup of order p^e, then obviously all cyclic subgroups of order p^e are normal in G.

Proof. Let G be a nonabelian p-group of exponent p^e, $p > 2$, all of whose cyclic subgroups of order p^e are normal in G. Then $e \geq 2$.

Let G_0 be the subgroup of G generated by all elements of order p^e. Assume $G_0 < G$. In that case all elements in $G - G_0$ are of orders $\leq p^{e-1}$. Let $g \in G - G_0$ with $g^p \in G_0$ and $a \in G_0$ with $o(a) = p^e$. Set $H = \langle a, g \rangle$, where $\langle a \rangle \cap \langle g \rangle \leq \langle a^p \rangle$ since $o(g^p) \leq p^{e-2}$. Because $H' \leq \langle a \rangle$ is cyclic, Theorem 7.1 (c) implies that H is regular. Then by Theorem 7.2 (b), $\exp(\Omega_{e-1}(H)) = p^{e-1}$ and $\Omega_{e-1}(H) = \langle a^p, g \rangle$ is a maximal subgroup in H. But then $o(ag) = p^e$ and $ag \in G - G_0$, a contradiction. We have proved that $G_0 = G$.

Since G' centralizes each cyclic subgroup of order p^e in G, we have $G' \leq Z(G)$ and so G is of class 2.

Let $\langle a \rangle$ be any cyclic subgroup of order p^e in G so that $\langle a \rangle \trianglelefteq G$. We claim that $G/\langle a \rangle$ is Dedekindian. Indeed, let $X/\langle a \rangle \neq \{1\}$ be any cyclic subgroup in $G/\langle a \rangle$ and let $x \in X$ be such that $\langle x \rangle$ covers $X/\langle a \rangle$. If $o(x) = p^e$, then $\langle x \rangle \trianglelefteq G$ and $X = \langle a \rangle \langle x \rangle \trianglelefteq G$. Suppose that $o(x) \leq p^{e-1}$. Note that $X' \leq \langle a \rangle$ is cyclic and so X is regular. Then $\Omega_{e-1}(X) = \langle a^p \rangle \langle x \rangle$ is of exponent p^{e-1} and all elements in $X - (\langle a^p \rangle \langle x \rangle)$ are of order p^e and they generate X. Hence $X \trianglelefteq G$. We have proved that $G/\langle a \rangle$ is abelian and so G' is cyclic and is contained in each cyclic subgroup of order p^e.

Since G is generated by all elements of order p^e and G' is cyclic, there are elements $a, b \in G$ of orders p^e such that $[a, b] = v$ with $\langle v \rangle = G'$. We set $o(v) = p^s$, where $1 \leq s \leq e - 1$ since $\langle v \rangle \leq \langle a \rangle \cap \langle b \rangle < \langle a \rangle$. On the other hand,

$$[a, b^{p^{s-1}}] = [a, b]^{p^{s-1}} = v^{p^{s-1}} \neq 1 \text{ and so } |\langle b \rangle / \langle a \rangle \cap \langle b \rangle| \geq p^s$$

and this implies that $p^{2s} \leq p^e$ and so $s \leq \frac{e}{2}$. Our theorem is proved. □

Theorem 230.2. *Let G be a nonabelian 2-group of exponent 2^e, $e \geq 4$, all of whose cyclic subgroups of order 2^e are normal in G. Let G_0 be the subgroup of G generated by all elements of order 2^e. Then $|G : G_0| \leq 2$ and we have one of the following possibilities.*
(a) *If $|G : G_0| = 2$, then all elements in $G - G_0$ are of order ≤ 8 and one of the following holds:*
  (a1) *$G_0' = \mho_{e-1}(G_0) \cong C_2$ and for each $g \in G - G_0$ and $a \in G_0$ of order 2^e, $a^g = a^{-1+n2^{e-2}}$, where n is an integer (mod 4).*
  (a2) *G_0 is abelian and either $\mho_{e-1}(G_0) \cong C_2$ is cyclic and each $g \in G - G_0$ inverts $G_0/\mho_{e-1}(G_0)$ or $\mho_{e-1}(G_0)$ is noncyclic and then for all $g \in G - G_0$ and $a \in G_0$, $a^g = a^{-1+\epsilon 2^{e-1}}$, where $\epsilon \in \{0, 1\}$.*
(b) *If $G = G_0$, then G is of class 2, G' is cyclic of order 2^f, where $1 \leq f \leq \frac{e}{2}$, G' is contained in each cyclic subgroup of order 2^e and G is powerful.*
Conversely, all the above groups stated in (a) and (b) satisfy the assumptions of the theorem.

Proof. Let G be a nonabelian 2-group of exponent 2^e, $e \geq 4$, all of whose cyclic subgroups of order 2^e are normal in G. Let G_0 be the subgroup of G generated by all elements of order 2^e. Then all elements in $G - G_0$ are of orders $\leq 2^{e-1}$. If $a \in G_0$ with $o(a) = 2^e$, then $C_G(a) \leq G_0$ and $\langle a \rangle \trianglelefteq G$. This implies that $C_G(G_0) \leq G_0$ and G/G_0 is abelian. Since $G/C_G(a)$ is abelian for each cyclic subgroup $\langle a \rangle$ of order 2^e in G_0, it follows that G' centralizes $\langle a \rangle$ and so $G' \leq Z(G_0)$. In particular, $\mathrm{cl}(G_0) \leq 2$.

Assume $G_0 < G$ and suppose in addition that there are $g \in G - G_0$ and $a \in G_0$ with $o(a) = 2^e$ so that $a^g = av$, where $v \in \langle a^4 \rangle$. Replacing g with some power g^{2^n}, $n \geq 0$, we may assume that $g \in G - G_0$ but $g^2 \in G_0$. We consider the metacyclic subgroups $H = \langle a, g \rangle$ and $H_1 = \langle a, g^2 \rangle$, where H_1 is of class ≤ 2. From $a^g = av$, we get

$$(a^2)^g = (a^g)^2 = (av)^2 = a^2v^2, \text{ where } v^2 \in \langle a^8 \rangle.$$

Thus,

$$a^{-2}(a^2)^g = [a^2, g] \in \langle a^8 \rangle \text{ and so } o([a^2, g]) \leq 2^{e-3},$$

and since

$$[a^2, g^2] = [a^2, g][a^2, g]^g = v^2(v^2)^g, \text{ we get } o([a^2, g^2]) \leq 2^{e-3}.$$

Then we compute

$$(ga)^2 = gaga = g^2a^ga = g^2ava = g^2(a^2v) \text{ and so}$$
$$(ga)^{2^{e-1}} = ((ga)^2)^{2^{e-2}} = (g^2(a^2v))^{2^{e-2}} = g^{2^{e-1}}(a^2v)^{2^{e-2}}[a^2v, g^2]^{\binom{2^{e-2}}{2}}.$$

But

$$g^{2^{e-1}} = 1, \quad (a^2v)^{2^{e-2}} = a^{2^{e-1}} = z, \text{ where } \langle z \rangle = \Omega_1(\langle a \rangle),$$

and $[a^2v, g^2] = [a^2, g^2]$ since $[v, g^2] = 1$ and so $[a^2v, g^2]^{\binom{2^{e-2}}{2}} = 1$. Hence we have obtained $(ga)^{2^{e-1}} = z$ and so $o(ga) = 2^e$ and $ga \in G - G_0$, a contradiction. We have proved that there is no such $g \in G - G_0$ chosen above, which implies that in case $G_0 < G$, we must have $|G : G_0| = 2$ and for each $g \in G - G_0$ and $a \in G_0$ with $o(a) = 2^e$, we have $a^g = a^{-1}v$, where $v \in \langle a^4 \rangle$. It follows $[a, g] = a^{-2}v$ and so (since $v \in \langle a^4 \rangle$) $\langle a^{-2}v \rangle = \langle a^2 \rangle \le G' \le Z(G_0)$. Hence for each $x \in G_0$, $1 = [a^2, x] = [a, x^2]$ and so x^2 centralizes each element of order 2^e in G_0 and therefore $x^2 \in Z(G_0)$ and $\mho_1(G_0) \le Z(G_0)$. For any $x, y \in G_0$, $[x, y]^2 = [x^2, y] = 1$ and so G_0' is elementary abelian. Since $v \in \langle a^4 \rangle$, we have $o(v) \le 2^{e-2}$ and so

$$(a^{2^{e-2}})^g = (a^{-1}v)^{2^{e-2}} = (a^{2^{e-2}})^{-1}, \text{ implying that } C_{\langle a \rangle}(g) = \langle z \rangle = \Omega_1(\langle a \rangle) .$$

From $a^g = a^{-1}v$, we get

$$a^{g^2} = (a^{-1}v)^g = av^{-1}v^g \text{ and so } [a, g^2] = v^{-1}v^g .$$

It follows

$$1 = [a^2, g^2] = [a, g^2]^2 = (v^{-1}v^g)^2 = v^{-2}(v^2)^g, \ (v^2)^g = v^2 \text{ giving } v^2 \in \langle z \rangle \text{ and } o(v) \le 4 .$$

In particular, $(a^4)^g = (a^{-1}v)^4 = a^{-4}$ and so g inverts $\langle a^4 \rangle$ and $v^g = v^{-1}$.

(i) Assume $G_0 < G$. Then we know that $|G : G_0| = 2$, $\mho_1(G_0) \le Z(G_0)$, G_0' is elementary abelian and for each $g \in G - G_0$ and $a \in G_0$ with $o(a) = 2^e$, $a^g = a^{-1}v^n$, where $v = a^{2^{e-2}}$, $z = v^2$ and n is an integer (mod4). Also, g inverts $\langle a^4 \rangle$ and so $v^g = v^{-1}$. Set $i = g^2$ so that $o(i) \le 2^{e-2}$. Then

$$a^i = a^{g^2} = (a^{-1}v^n)^g = (av^{-n})v^{-n} = av^{-2n} = az^n \text{ and } (ai)^{2^{e-2}} = v[i, a]^{\binom{2^{e-2}}{2}} = v .$$

We get

$$(ai)^g = (ai)^{-1}v^m = a^{-1}i^{-1}z^r v^m = a^g i = a^{-1}v^n i, \text{ where } m, r \text{ are some integers} .$$

Noting that i centralizes $\langle v \rangle$, the above relation implies that $i^2 \in \langle v \rangle$. But g inverts v and so $\langle v \rangle \nleq \langle g \rangle$ and therefore $\langle i^2 \rangle = \langle v \rangle$ is not possible. Hence $i^2 \in \langle z \rangle$ and so $o(g) \le 8$. We have proved that all elements in $G - G_0$ are of order ≤ 8.

(i1) Suppose that G_0 is nonabelian. Let $\langle a \rangle \trianglelefteq G$ be a cyclic subgroup of order 2^e, $e \ge 4$, in G_0. Set $\langle z \rangle = \Omega_1(\langle a \rangle)$ so that $z \in Z(G)$. Since G_0' is elementary abelian, we have $[G_0, \langle a \rangle] \le \langle z \rangle$ and so G_0 centralizes $\langle a \rangle / \langle z \rangle$.

Let $X/\langle a \rangle \ne \{1\}$ be any cyclic subgroup in $G_0/\langle a \rangle$ and let $d \in X - \langle a \rangle$ be such that $\langle d \rangle$ covers $X/\langle a \rangle$. If $o(d) = 2^e$, then $\langle d \rangle \trianglelefteq G$ and so $X = \langle a \rangle \langle d \rangle \trianglelefteq G$. We assume $o(d) \le 2^{e-1}$ and note that $X' = [\langle a \rangle, \langle d \rangle] \le \langle z \rangle$. We have

$$(da)^{2^{e-1}} = d^{2^{e-1}} a^{2^{e-1}} [a, d]^{\binom{2^{e-1}}{2}} = z \text{ and so } o(da) = 2^e .$$

Hence $\langle da \rangle \trianglelefteq G$ and so $X = \langle a \rangle \langle da \rangle \trianglelefteq G$. We have proved that $G/\langle a \rangle$ is Dedekindian.

Suppose that $G_0/\langle a\rangle$ is nonabelian and so $G_0/\langle a\rangle$ is Hamiltonian. Let $Q/\langle a\rangle$ be a quaternion subgroup in $G_0/\langle a\rangle$. Set $D = \langle a\rangle Q'$, where $Q' \le Z(G_0)$ so that D is abelian of type $(2^e, 2)$. Let $q_1, q_2 \in Q-D$ be such that $\langle q_1, q_2\rangle$ covers $Q/D \cong E_4$. Then $[q_1, q_2] \in Q' - \langle a\rangle$ and $[q_1, q_2]$ is an involution since Q' is elementary abelian (of order ≤ 4). If $o(q_1) \le 2^{e-1}$, then

$$(q_1a)^{2^{e-1}} = q_1^{2^{e-1}} z[a, q_1]^{\binom{2^{e-1}}{2}} = z \text{ and so } o(q_1a) = 2^e\,.$$

Replacing q_1 with q_1a (if necessary), we may assume that $o(q_1) = 2^e$ so that $\langle q_1\rangle \trianglelefteq Q$ and $\Omega_1(\langle q_1\rangle) = \langle z\rangle$. But then $[q_1, q_2] \notin \langle q_1\rangle$, contrary to $\langle q_1\rangle \trianglelefteq Q$. It follows that $G_0/\langle a\rangle$ is abelian and so $G_0' \le \langle a\rangle$. But G_0' is also elementary abelian and so $G_0' = \langle z\rangle \cong C_2$. We have proved that $G_0' \cong C_2$ and G_0' is contained in each cyclic subgroup of order 2^e.

We claim that $G_0' = \mho_{e-1}(G_0) \cong C_2$. Indeed, let $a \in G_0$ with $o(a) = 2^e$ and $x \in G_0$. Then we consider the metacyclic subgroup $\langle a, x\rangle$. If $\langle a, x\rangle/\langle a\rangle$ is cyclic of order $\ge 2^e$, then $o(x) = 2^e$ and $\langle a\rangle \cap \langle x\rangle = \{1\}$. But then, by the above, $G_0' \le \langle a\rangle \cap \langle x\rangle = \{1\}$, a contradiction. Thus $\langle a, x\rangle/\langle a\rangle$ is cyclic of order $\le 2^{e-1}$ and so $x^{2^{e-1}} \in \langle a\rangle$ and we are done.

(i2) Assume that G_0 is abelian. Then each $g \in G - G_0$ induces on $\langle a_1\rangle \le G_0$ with $o(a_1) = 2^e$ an involutory automorphism induced by

$$a_1^g = a_1^{-1+\epsilon 2^{e-1}} = a_1^{-1} z_1^{\epsilon}, \text{ where } z_1 = a_1^{2^{e-1}} \text{ and } \epsilon \in \{0, 1\}\,.$$

(i2a) Suppose that $\mho_{e-1}(G_0) = \langle z_1\rangle$ is cyclic and let C be a complement of $\langle a_1\rangle$ in G_0 so that $\exp(C) \le 2^{e-1}$. Let $c \in C$ so that $o(a_1c) = 2^e$ and $\Omega_1(\langle a_1c\rangle) = \langle z_1\rangle$. Then

$$(a_1c)^g = a_1^{-1}c^{-1}z_1^{\epsilon_1} = a_1^g c^g = a_1^{-1} z_1^{\epsilon} c^g \text{ and so } c^g = c^{-1} z_1^{\epsilon_1-\epsilon}, \text{ where } \epsilon_1 \in \{0, 1\}\,.$$

We have proved that g inverts $G_0/\mho_{e-1}(G_0)$.

(i2b) Assume that $\mho_{e-1}(G_0)$ is noncyclic. In that case there is $a_2 \in G_0$ with $o(a_2) = 2^e$ and $\langle a_1\rangle \cap \langle a_2\rangle = \{1\}$ and let D be a complement of $\langle a_1, a_2\rangle$ in G_0. We have $a_1^g = a_1^{-1+\epsilon 2^{e-1}} = a_1^{-1}z_1^{\epsilon}$ and $a_2^g = a_2^{-1}z_2^{\zeta}$ with $z_2 = a_2^{2^{e-1}}$ and $\zeta \in \{0, 1\}$. Then $o(a_1a_2) = 2^e$ and $\Omega_1(\langle a_1a_2\rangle) = \langle z_1z_2\rangle$ and so

$$(a_1a_2)^g = a_1^{-1}a_2^{-1}(z_1z_2)^{\delta} = a_1^{-1}z_1^{\epsilon}a_2^{-1}z_2^{\zeta}, \quad \delta \in \{0, 1\}, \text{ implying}$$
$$\delta = \epsilon = \zeta \text{ and } a_2^g = a_2^{-1}z_2^{\epsilon} = a_2^{-1+\epsilon 2^{e-1}}\,.$$

Let $d \in D$. If $o(d) = 2^e$ with $\Omega_1(\langle d\rangle) = \langle z_3\rangle$, then we get as above

$$d^g = d^{-1}z_3^{\epsilon} = d^{-1+\epsilon 2^{e-1}}\,.$$

Now assume $o(d) \le 2^{e-1}$. Then

$$o(a_1d) = o(a_2d) = 2^e \text{ and } \Omega_1(\langle a_1d\rangle) = \langle z_1\rangle, \quad \Omega_1(\langle a_2d\rangle) = \langle z_2\rangle\,.$$

Hence

$$(a_1d)^g = a_1^{-1}d^{-1}z_1^{\delta_1} = a_1^g d^g = a_1^{-1}z_1^{\epsilon}d^g\,, \quad \delta_1 \in \{0,1\}\,,$$
$$(a_2d)^g = a_2^{-1}d^{-1}z_2^{\delta_2} = a_2^g d^g = a_2^{-1}z_2^{\epsilon}d^g\,, \quad \delta_2 \in \{0,1\}\,.$$

These relations imply

$$d^g = d^{-1}z_1^{\delta_1-\epsilon} = d^{-1}z_2^{\delta_2-\epsilon} \text{ and then } \delta_1-\epsilon = \delta_2-\epsilon = 0\,.$$

This gives $d^g = d^{-1} = d^{-1+\epsilon 2^{e-1}}$ and so we have proved that in this case for all $a \in G_0$, $a^g = a^{-1+\epsilon 2^{e-1}}$.

(ii) Suppose that $G_0 = G$. In this case $\mathrm{cl}(G) = 2$. Let $Z = \langle a \rangle$ be any cyclic subgroup of order 2^e in G, where $2^e = \exp(G)$, $e \geq 4$. Set $\langle z \rangle = \Omega_1(Z)$ and $Z_0 = \Omega_{e-1}(Z)$. We show that G/Z_0 is Dedekindian.

Let $X/Z_0 \neq \{1\}$ be any nontrivial cyclic subgroup in G/Z_0. If $X \geq Z$, then X is cyclic and so $X = Z \trianglelefteq G$. Hence we may assume that $X \ngeq Z$ and so $X \cap Z = Z_0$ and then we set $K = XZ$. Let $x \in X - Z_0$ be such that $\langle x \rangle$ covers X/Z_0. If $\mathrm{o}(x) = 2^e$, then $X = Z_0\langle x \rangle \trianglelefteq G$ and we are done. Thus we may assume that $\mathrm{o}(x) \leq 2^{e-1}$. We have $K = \langle a \rangle\langle x \rangle$ and $K' \leq Z \cap X = Z_0$, where $K' = \langle [a,x] \rangle$. Suppose for a moment that $K' = Z_0$. Then $Z_0 \leq \mathrm{Z}(K)$ and so

$$1 = [a^2, x] = [a, x^2]\,,$$

which implies that $\langle x \rangle$ induces on $Z = \langle a \rangle$ an involutory automorphism induced by $a^x = a^{-1}z^{\epsilon}$, $\epsilon \in \{0,1\}$. But then

$$(a^2)^x = (a^x)^2 = (a^{-1}z^{\epsilon})^2 = a^{-2}\,,$$

and so x inverts $\langle a^2 \rangle$, where $\langle a^2 \rangle = Z_0$, contrary to $Z_0 \leq \mathrm{Z}(K)$. We have proved that $K' = \langle [a,x] \rangle$ is a proper subgroup of Z_0 and so K' is cyclic of order $\leq 2^{e-2}$. Let $x_1, x_2 \in X$ be any elements of orders $\leq 2^{e-1}$ and then

$$(x_1x_2)^{2^{e-1}} = x_1^{2^{e-1}}x_2^{2^{e-1}}[x_2,x_1]^{\binom{2^{e-1}}{2}} = 1 \text{ and so } \mathrm{o}(x_1x_2) \leq 2^{e-1}\,.$$

Hence $\exp(X) = 2^{e-1}$. Let ay with $y \in X$ be any element in $K - X$. Then we compute

$$(ay)^{2^{e-1}} = a^{2^{e-1}}y^{2^{e-1}}[y,a]^{\binom{2^{e-1}}{2}} = a^{2^{e-1}} = z \text{ so that } \mathrm{o}(ay) \leq 2^e\,.$$

Hence all elements in $K - X$ are of order 2^e and they generate K. This implies that $K \trianglelefteq G$ and then $\Omega_{e-1}(K) = X \trianglelefteq G$. We have proved that G/Z_0 is Dedekindian.

Assume, by way of contradiction, that G/Z_0 is nonabelian. Then G/Z_0 is Hamiltonian and so G/Z_0 possesses a subgroup $Q/Z_0 \cong \mathrm{Q}_8$. If $Q \geq Z$, then for an element $q \in Q - Z$, $q^2 \in Z - Z_0$ and so $\mathrm{o}(q) = 2^{e+1}$, a contradiction. It follows that $Z \cap Q = Z_0$ and we set $H = ZQ$, where $|H| = 2^{e+3} \geq 2^7$. Let $D = Z_0Q'$, where $Q' \leq \mathrm{Z}(G)$ and $|D : Z_0| = 2$. Hence $ZD = \langle a \rangle Q'$ is abelian of type $(2^e, 2)$. If D is cyclic (of order 2^e), then for an element $q \in Q - D$, we have $q^2 \in D - Z_0$ and so $o(q) = 2^{e+1}$, a contradiction. Hence D is abelian of type $(2^{e-1}, 2)$. Since $H/(ZD) \cong Q/D \cong E_4$, each element $h \in H-(ZD)$ induces on Z an automorphism of order ≤ 2. Suppose that there is $h \in H$ such that

$$a^h = a^{-1}z^\epsilon, \quad \epsilon \in \{0, 1\}, \text{ and then } \langle a^2 \rangle \leq H' .$$

But $(a^2)^h = a^{-2}$, contrary to $H' \leq Z(G)$. Hence for each $h \in H$,

$$a^h = az^\eta, \quad \eta \in \{0, 1\}, \text{ and so } (a^2)^h = a^2 \text{ and therefore } \langle a^2 \rangle = Z_0 \leq Z(H) .$$

All three maximal subgroups of Q containing Z_0 are abelian and so $Q' = \langle u \rangle \cong C_2$, where $u \in D - Z_0$. Let $q_1, q_2 \in Q - D$ be such that $\langle q_1, q_2 \rangle$ covers $Q/D \cong E_4$ so that $[q_1, q_2] = u$. If $o(q_1) = 2^e$, then $\langle q_1 \rangle \trianglelefteq H$ and $\langle q_1 \rangle \cap Z_0 \geq \langle z \rangle$. But $[q_1, q_2] = u \notin \langle q_1 \rangle$, contrary to $\langle q_1 \rangle \trianglelefteq H$. Hence $o(q_1) \leq 2^{e-1}$ and we compute

$$(aq_1)^{2^{e-1}} = a^{2^{e-1}} q_1^{2^{e-1}} [q_1, a]^{\binom{2^{e-1}}{2}} = z \text{ and so } o(aq_1) = 2^e \text{ with}$$
$$\langle aq_1 \rangle \trianglelefteq H \text{ and } \langle aq_1 \rangle \cap Z_0 > \langle z \rangle .$$

On the other hand,

$$[aq_1, q_2] = [a, q_2][q_1, q_2] = z^\zeta u, \quad \zeta \in \{0, 1\} .$$

But then $z^\zeta u \notin \langle aq_1 \rangle$, contrary to $\langle aq_1 \rangle \trianglelefteq H$. We have proved that G/Z_0 is abelian and so $G' \leq Z_0$. Hence G' is cyclic of order $\leq 2^{e-1}$ and G' is contained in each cyclic subgroup of order 2^e.

We set $|G'| = 2^f$ and so if $a \in G$ with $o(a) = 2^e$, then

$$G' \leq \langle a \rangle \text{ implies } a^{2^{e-f}} \in G' \leq Z(G) .$$

Then for each $x \in G$,

$$1 = [a^{2^{e-f}}, x] = [a, x^{2^{e-f}}] .$$

It follows that $x^{2^{e-f}}$ commutes with each element of order 2^e in G and so $x^{2^{e-f}} \in Z(G)$ and $\mho_{e-f}(G) \leq Z(G)$. Then for any $x, y \in G$,

$$[x, y]^{2^{e-f}} = [x^{2^{e-f}}, y] = 1, \text{ and so } |G'| = 2^f \leq 2^{e-f} .$$

Hence $f \leq e - f$ and this implies $1 \leq f \leq \frac{e}{2}$. In particular, $f < e - 1$ and this gives $G' \leq \mho_2(G)$ and so G is powerful. Our theorem is proved.

□

Problem. Classify the p-groups of exponent p^e, all of whose cyclic subgroups of order p^e are quasinormal.

§ 231 p-groups which are not generated by their nonnormal subgroups

The purpose of this section is to classify p-groups which are not generated by their nonnormal subgroups (Problem 1668 (i)). In fact, the only known such non-Dedekindian p-groups are groups of Theorem 1.25 classifying the non-Dedekindian p-groups all of whose nonnormal subgroups have order p. We prove here the following result.

Theorem 231.1. *Let G be a non-Dedekindian p-group and let G_0 be the subgroup generated by all nonnormal subgroups of G, where we assume $G_0 < G$. Then G is of class 2, G/G_0 is cyclic and for each $g \in G - G_0$, $\{1\} \neq \langle g\rangle \cap G_0 \trianglelefteq G$ and $G/(\langle g\rangle \cap G_0)$ is abelian so that G' is cyclic.*

Proof. Since our group G has at least p (nonnormal) conjugate cyclic subgroups, it follows that the subgroup G_0 is noncyclic. Let $x \in G - G_0$. Then $\langle x\rangle \trianglelefteq G$, by hypothesis, and so G' centralizes $\langle x\rangle$. It follows from $\langle G - G_0\rangle = G$ that $G' \le Z(G)$ and so $\mathrm{cl}(G) = 2$.

Let $g \in G - G_0$. Then $Z = \langle g\rangle \triangleleft G$. Write $Z_0 = Z \cap G_0$; then Z_0, being the intersection of two G-invariant subgroups, is G-invariant. We claim that G/Z_0 is Dedekindian. Indeed, let X/Z_0 be any proper subgroup in G/Z_0. We have to show that $X \triangleleft G$. If $X \nleq G_0$, then $X \trianglelefteq G$. Now assume that $X < G_0$ (the subgroup G_0 is G-invariant). Then $XZ = ZX$ is normal in G since $XZ \nleq G_0$. By the product formula, one has

$$|XZG_0| = |ZG_0| = \frac{|Z|\,|G_0|}{|Z_0|}\,.$$

On the other hand,

$$|XZG_0| = \frac{|XZ|\,|G_0|}{|XZ \cap G_0|} = \frac{|X|\,|Z|}{|Z_0|}\cdot\frac{|G_0|}{|XZ \cap Z_0|} = |XZG_0|\cdot\frac{|X|}{|XZ \cap G_0|}$$

which implies $X = XZ \cap G_0 \triangleleft G$, and we are done. We have proved that G/Z_0 is Dedekindian. In particular, $Z_0 \neq \{1\}$ since G is non-Dedekindian, by hypothesis. If $p > 2$, then G/Z_0 is abelian and so $G' \le Z_0$ and G' is cyclic. If $p = 2$, then G/Z_0 is either abelian or Hamiltonian (= nonabelian Dedekindian).

It follows from the above that $\Omega_1(G) \le G_0$.

Now assume that $p > 2$. Amongst all elements in the set $G - G_0$, we choose an element a of the smallest possible order. Then $a^p \in G_0$ and $G' \le \langle a^p\rangle$ (see the previous paragraph). We set $|G'| = p^d$, $d \ge 1$. Suppose that G/G_0 is not cyclic. Then there is $b \in G - (G_0\langle a\rangle)$ such that $b^p \in G_0$. We have $\langle a\rangle \cap \langle b\rangle \ge G'$ and $\mathrm{o}(b) \ge \mathrm{o}(a)$ by the minimality of $\mathrm{o}(a)$. Set

$$|\langle a\rangle/(\langle a\rangle \cap \langle b\rangle)| = p^s, \text{ where } s \ge 1 \text{ and } \mathrm{o}(a) \ge p^{d+s}\,.$$

Hence there is $b' \in \langle b\rangle - \langle a\rangle$ such that $a^{p^s} = (b')^{-p^s}$. In that case, since $\mathrm{cl}(G) = 2$, one obtains

$$(ab')^{p^s} = a^{p^s}(b')^{p^s}[b', a]^{\binom{p^s}{2}} = [b', a]^{\binom{p^s}{2}}\,,$$

where $s \geq 1$, $o(a) \geq p^{d+s}$ and $\langle [b', a]^{\binom{p^s}{2}} \rangle < G'$ so that $o([b', a]^{\binom{p^s}{2}}) < p^d$. It follows that

$$o(ab') < p^{d+s} \text{ and so } o(ab') < o(a) .$$

If $b' \in \langle b^p \rangle \leq G_0$, then $ab' \in G - G_0$. If $\langle b' \rangle = \langle b \rangle$, then $ab' \in G - (G_0\langle a \rangle)$ and so again $ab' \in G - G_0$. But this contradicts the minimality of $o(a)$. We have proved that in case $p > 2$, G/G_0 is cyclic.

Suppose $p = 2$ and G/G_0 is nonabelian. Then for each $g \in G - G_0$, $G/(\langle g \rangle \cap G_0)$ is Hamiltonian (i.e., Dedekindian nonabelian). Let Q/G_0 be a subgroup of G/G_0 which is isomorphic to Q_8 and let R/G_0 be a unique subgroup of order 2 in Q/G_0. Then for each $x \in Q - R$, $x^2 \in R - G_0$. Let $a, b \in Q - R$ be such that $\langle a, b \rangle$ covers $Q/R \cong E_4$. Note that $\langle a \rangle \trianglelefteq G$, $\langle b \rangle \trianglelefteq G$ and since $\langle a \rangle \cap G_0 \neq \{1\}$ and $\langle b \rangle \cap G_0 \neq \{1\}$, we get $o(a) = 2^s$, $s \geq 3$, and $o(b) \geq 2^3$. Because

$$[a, b] \in R - G_0 \text{ and } [a, b] \in \langle a \rangle \cap \langle b \rangle ,$$

we have

$$\langle a \rangle \cap \langle b \rangle = \langle a^2 \rangle = \langle b^2 \rangle = \langle [a, b] \rangle .$$

But then $C = \langle a, b \rangle$ is a 2-group of maximal class and order 2^{s+1}, $s \geq 3$, and in this case $\langle a \rangle$ is a unique cyclic subgroup of order 2^s in C, contrary to the fact that $o(b) = 2^s$. We have proved that in case $p = 2$, G/G_0 must be abelian and so $G' \leq G_0$.

Suppose that G' is noncyclic. By the above, $p = 2$ and for each $g \in G - G_0$, $\{1\} \neq \langle g \rangle \cap G_0 \trianglelefteq G$, where $G/(\langle g \rangle \cap G_0)$ is Hamiltonian (= nonabelian Dedekindian). Set $D = \langle g \rangle \cap G_0$ and $R/D = (G/D)' \cong C_2$, where $R = G'D$. We know that $G' \leq G_0$ (since G/G_0 is abelian) and so $R \leq G_0$ and G/R is elementary abelian. In particular, $G/G_0 \neq \{1\}$ is elementary abelian and $\langle g^2 \rangle = D$. Note that all quaternion subgroups in a Hamiltonian 2-group X generate X. Hence there is a quaternion subgroup $K/D \cong Q_8$ in the Hamiltonian group G/D such that $K \nleq G_0$. We have $K > R$ and $K/R \cong E_4$ so that for each $x \in K - R$, $x^2 \in R - D$. We may choose some elements $a, b \in K - G_0$ such that $Q = \langle a, b \rangle$ covers K/R and so Q also covers K/D. Note that $\langle a \rangle \trianglelefteq G$, $\langle b \rangle \trianglelefteq G$ and $[a, b] \in R - D$. Also,

$$[a, b] \in \langle a \rangle \cap \langle b \rangle \text{ and so } \langle [a, b] \rangle = \langle a^2 \rangle = \langle b^2 \rangle = \langle a \rangle \cap \langle b \rangle .$$

This gives $|Q : Q'| = 4$ and so (by a result of O. Taussky) Q is a 2-group of maximal class with two distinct cyclic subgroups $\langle a \rangle$ and $\langle b \rangle$ of index 2. By inspection of 2-groups of maximal class (and noting that G is of class 2), we get $o(a) = o(b) = 4$ and $Q \cong Q_8$ with $Q' = \langle a^2 \rangle = \langle b^2 \rangle$. Hence $K = Q \times D$ since $Q \trianglelefteq G$ and Q covers $K/D \cong Q_8$. Also, $\langle g \rangle \trianglelefteq G$ and $Q \cap \langle g \rangle = \{1\}$ and so Q centralizes $\langle g \rangle$. The factor-group $G/\langle a^2 \rangle$ is Hamiltonian and so

$$o(g) = 4 , \quad D = \langle g^2 \rangle \cong C_2, \text{ and } G' = \langle a^2, g^2 \rangle \cong E_4$$

since G' covers $\langle a^2, g^2 \rangle / \langle a^2 \rangle$ and G' is noncyclic. For each $x \in G$,

$$x^4 \in \langle a^2 \rangle \cap \langle g^2 \rangle = \{1\} \text{ and so } \exp(G) = 4 .$$

Let $K_1/\langle a^2\rangle \cong Q_8$ with $K_1 \nleq G_0$. Then choose $a_1, b_1 \in K_1 - G_0$ such that $\langle a_1, b_1\rangle$ covers $K_1/\langle a^2\rangle$. We get

$$Q_1 = \langle a_1, b_1\rangle \cong Q_8 \text{ with } Q \cap Q_1 = \{1\} \text{ and } Q_1' = \langle a_1^2\rangle = \langle b_1^2\rangle, \quad \text{so } \langle Q, Q_1\rangle = Q \times Q_1 .$$

Set $a^2 = t$, $a_1^2 = t_1$, and let $x \in Q - \langle t\rangle$, $x_1 \in Q_1 - \langle t_1\rangle$ so that xx_1 is one of 36 elements of order 4 with $(xx_1)^2 = x^2x_1^2 = tt_1$. We claim that $\langle xx_1\rangle$ is not normal in $Q \times Q_1$ and so $xx_1 \in G_0$. Indeed, let $y \in Q - \langle x\rangle$ so that

$$(xx_1)^y = x^{-1}x_1 = (xx_1)t, \quad \text{where } (xx_1)t \notin \langle xx_1\rangle .$$

But all these 36 elements of order 4 generate $Q \times Q_1$ (of order 64) and so $Q \times Q_1 \leq G_0$, a contradiction. We have proved that also in case $p = 2$, G' is cyclic.

In the following five paragraphs, we assume that G/G_0 is noncyclic. By the above, $p = 2$ and G/G_0 is abelian.

Assume that there are $a_1, a_2 \in G - G_0$ such that $\langle a_1\rangle \cap \langle a_2\rangle = \{1\}$. We know that $G/(\langle a_1\rangle \cap G_0)$ and $G/(\langle a_2\rangle \cap G_0)$ are Dedekindian and $[a_1, a_2] \in \langle a_1\rangle \cap \langle a_2\rangle = \{1\}$ and so $\langle a_1, a_2\rangle$ is abelian. If both $G/(\langle a_1\rangle \cap G_0)$ and $G/(\langle a_2\rangle \cap G_0)$ are abelian, then

$$G' \leq (\langle a_1\rangle \cap G_0) \cap (\langle a_2\rangle \cap G_0) = \{1\} ,$$

a contradiction. Assume for a moment that both $G/(\langle a_1\rangle \cap G_0)$ and $G/(\langle a_2\rangle \cap G_0)$ are Hamiltonian. Then for each $x \in G$,

$$x^4 \in (\langle a_1\rangle \cap G_0) \cap (\langle a_2\rangle \cap G_0) = \{1\} \text{ and so } \exp(G) = 4 .$$

In particular,

$$o(a_1) = o(a_2) = 4, \quad \langle a_1^2, a_2^2\rangle \cong E_4 \text{ with } \langle a_1^2, a_2^2\rangle \leq Z(G) .$$

We have

$$G' \leq \langle a_1^2, a_2^2\rangle, \quad G' \text{ covers } \langle a_1^2, a_2^2\rangle/\langle a_1^2\rangle \text{ and } \langle a_1^2, a_2^2\rangle/\langle a_2^2\rangle \text{ and } G' \text{ is cyclic}$$

and so $G' = \langle a_1^2a_2^2\rangle$. For each

$$x \in G, \quad [a_2, x] \in \langle a_2\rangle \cap G' = \{1\} \text{ and so } a_2 \leq Z(G) .$$

But then in the Hamiltonian 2-group $G/\langle a_1^2\rangle$ the element $(\langle a_2\rangle\langle a_1^2\rangle)/\langle a_1^2\rangle \cong C_4$ of order 4 lies in its center, a contradiction. We have proved that if $a_1, a_2 \in G - G_0$ are such that $\langle a_1\rangle \cap \langle a_2\rangle = \{1\}$, then one of $G/(\langle a_1\rangle \cap G_0)$ and $G/(\langle a_2\rangle \cap G_0)$ is abelian and the other one is Hamiltonian.

Assume in addition that $(G_0\langle a_1, a_2\rangle)/G_0$ is noncyclic. Set $\Omega_1(\langle a_1\rangle) = \langle t_1\rangle$ and $\Omega_1(\langle a_2\rangle) = \langle t_2\rangle$ so that $\langle t_1, t_2\rangle \cong E_4$ and $\langle t_1, t_2\rangle \leq Z(G)$. Without loss of generality we may suppose that $G/(\langle a_1\rangle \cap G_0)$ is abelian and $G/(\langle a_2\rangle \cap G_0)$ is Hamiltonian. Since G/G_0 is elementary abelian, we obtain

$$o(a_1) = 4, \quad G' = \langle a_1^2\rangle \cong C_2 \text{ and } 1 \neq a_2^2 \in G_0 .$$

It follows that $(G_0\langle a_1, a_2\rangle)/G_0 \cong E_4$. Let a_2' be an element of order 4 in $\langle a_2\rangle$ so that

$$(a_1a_2')^2 = a_1^2(a_2')^2 = t_1t_2 \text{ and } a_1a_2' \in G - G_0\,.$$

But then $\langle a_1\rangle$, $\langle a_2\rangle$, $\langle a_1a_2'\rangle$ are three cyclic subgroups in G which are not contained in G_0 and they have pairwise a trivial intersection. By the previous paragraph, this is not possible. We have proved that whenever $a_1, a_2 \in G-G_0$ are such that $(\langle a_1, a_2\rangle G_0)/G_0$ is noncyclic, then $\langle a_1\rangle \cap \langle a_2\rangle \neq \{1\}$.

Let E/G_0 be a four-subgroup in the noncyclic abelian group G/G_0. Among all elements in $E - G_0$ choose an element a of the smallest possible order 2^s. We have $s \geq 2$ since $a^2 \neq 1$. Set $F = G_0\langle a\rangle$ and let b be any element in $E - F$ so that $o(b) \geq 2^s$. By the above, $D = \langle a\rangle \cap \langle b\rangle \neq \{1\}$. Let b' be an element of order 2^s in $\langle b\rangle$ such that

$$a^{2^n} = (b')^{-2^n}\,, \quad \text{where } |\langle a\rangle : D| = |\langle b'\rangle : D| = 2^n,\ n \geq 1, \text{ and } D = \langle a^{2^n}\rangle = \langle (b')^{2^n}\rangle\,.$$

We compute

$$(ab')^{2^n} = a^{2^n}(b')^{2^n}[b', a]^{\binom{2^n}{2}} = [b', a]^{2^{n-1}(2^n-1)}\,, \quad \text{where } ab' \in E-G_0 \text{ and } [b', a] \in D.$$

Since a was an element of the smallest possible order in the set of all elements in $E - G_0$, we get

$$n = 1\,, \quad a^2 \in D\,, \quad \text{and } \langle [b', a]\rangle = D \neq \{1\}\,.$$

On the other hand,

$$[b, a]^2 = [b, a^2] = 1 \text{ and so } [b^2, a] = [b, a]^2 = 1\,.$$

Hence, if $b' \in \langle b^2\rangle$ (in case $o(b) > o(a) = 2^s$), we get $[b', a] = 1$ and so $D = \{1\}$, a contradiction. It follows that

$$o(a) = o(b) = 2^s \text{ and } \langle [b, a]\rangle = D \cong C_2\,, \quad \text{where } D = \langle a^2\rangle = \langle b^2\rangle\,.$$

Hence

$$s = 2\,, \quad o(a) = o(b) = 4\,, \quad \text{and } Q = \langle a, b\rangle \cong Q_8\,.$$

We have proved that all elements in $E-F$ are of order 4 and each such element has the same square a^2. We know that G' is cyclic, $G' \leq G_0$, $G' \leq Z(G)$ and $G' \geq \langle a^2\rangle = \langle a, b\rangle'$. Suppose that $G' > \langle a^2\rangle$ and let $x \in G' - \langle a^2\rangle$ be such that $x^2 = a^2$, where $[x, a] = 1$. But then xa is an involution in $E - G_0$, a contradiction. Hence $G' = \langle a^2\rangle \cong C_2$. Since all elements in $E - F$ are of order 4 and they generate E and $E' = \langle a^2\rangle \cong C_2$, we get $\exp(G) = 4$. In particular, all elements in $F-G_0$ are of order 4 and let $y \in F-G_0$. Then y is also of the smallest possible order 4 in $E-G_0$. By repeating the above argument with the element y (instead of a), we get that for each $b \in E - F$, $b^2 = y^2$ and so $y^2 = a^2$. We have proved that for each $x \in E - G_0$, $x^2 = a^2$. For any $x, y \in G$,

$$[x^2, y] = [x, y]^2 = 1 \text{ since } G' = \langle a^2\rangle \cong C_2\,.$$

Hence $\mho_1(G) \le Z(G)$.

Let c be an element of order 4 in G_0. Then

$$ac \in E - G_0 \text{ and so } a^2 = (ac)^2 = a^2c^2[c, a] \text{ implying } c^2 = [a, c] \in \langle a^2\rangle \text{ and } c^2 = a^2 .$$

But then $\langle c\rangle \trianglelefteq G$ and so there is $b \in E - G_0$ which centralizes $\langle c\rangle$. It follows that bc is an involution in $E - G_0$, a contradiction. We have proved that G_0 is elementary abelian. If $G_0 \nleq Z(E)$, then there are $t \in G_0 - \langle a^2\rangle$ and $x \in E - G_0$ such that $[t, x] = a^2 = x^2$. But then $\langle t, x\rangle \cong D_8$ and so there are involutions in $\langle t, x\rangle - G_0$, a contradiction. We have proved that E is Hamiltonian and so $E \neq G$ because G is not Dedekindian.

Let $v \in G - E$ be such that $v^2 \in E$. Since $\mho_1(G) \le Z(G)$, we get $1 \neq v^2 \in Z(E) = G_0$. Then, by the above, $\langle v\rangle \cap \langle a\rangle \neq \{1\}$ and so $v^2 = a^2$. Let $a, b \in E - G_0$ be such that $\langle a, b\rangle$ covers E/G_0. Because there are no involutions in $G - G_0$, we have

$$[v, a] = [v, b] = [a, b] = a^2 \quad \text{and} \quad [v, ab] = [v, a] = [v, b] = a^2a^2 = a^4 = 1 .$$

But then $(ab)^2 = v^2 = a^2$ implies that $(ab)v$ is an involution in $G - G_0$, a final contradiction. We have proved that also in case $p = 2$, G/G_0 is cyclic.

Suppose that $p = 2$ and there is $g \in G - G_0$ such that $G/(\langle g\rangle \cap G_0)$ is Hamiltonian. We set $D = \langle g\rangle \cap G_0 \neq \{1\}$ and note that $G' \le G_0$ implies that G/G_0 is elementary abelian. But G/G_0 is also cyclic and so $|G : G_0| = 2$. We obtain $g^2 \in G_0$ and so $D = \langle g^2\rangle \neq \{1\}$. Since the Hamiltonian group G/D is generated by its quaternion subgroups, there is a quaternion subgroup K/D in G/D such that $K \nleq G_0$. Let $a, b \in K - G_0$ be such that $Q = \langle a, b\rangle$ covers K/D, where $ab \in G_0$. Let R/D be a unique subgroup of order 2 in K/D so that $R \le G_0$ and G' covers R/D. We have

$$a^2 \in R - D, \quad b^2 \in R - D, \quad (ab)^2 \in R - D, \quad \text{and} \quad [a, b] \in R - D .$$

On the other hand,

$$[a, b] \in \langle a\rangle \cap \langle b\rangle \text{ and so } \langle a\rangle \cap \langle b\rangle = \langle a^2\rangle = \langle b^2\rangle = \langle [a, b]\rangle .$$

Since $Q/Q' \cong E_4$, Q is of maximal class (by O. Taussky) and since Q has two distinct cyclic subgroups $\langle a\rangle$ and $\langle b\rangle$ of index 2, we get

$$Q \cong Q_8 , \quad o(a) = o(b) = 4 , \quad \langle [a, b]\rangle \cong C_2 , \quad Q \cap \langle g^2\rangle = \{1\}$$
$$\text{and so } \langle Q, \langle g\rangle\rangle = Q \times \langle g\rangle .$$

Also,

$$G' \le R \text{ and } G' \ge \langle [a, b]\rangle \cong C_2 ,$$

and so the fact that G' is cyclic implies $G' \cap \langle g^2\rangle = \{1\}$. It follows

$$G' = \langle [a, b]\rangle = \langle a^2\rangle \cong C_2$$

and for any $x, y \in G$,

$$[x^2, y] = [x, y]^2 = 1 \text{ implying } \mho_1 G) \le Z(G) .$$

Since $G' \cap \langle g \rangle = \{1\}$, we have $\langle g \rangle \leq Z(G)$ and so $G = G_0 * \langle g \rangle$ gives that G_0 is nonabelian. We have $ab \in G_0$ and so $abg \notin G_0$ which implies $\langle abg \rangle \trianglelefteq G$. We compute

$$(abg)^2 = (ab)^2 g^2 = a^2 g^2 .$$

If $g^4 \neq 1$, then

$$(abg)^4 = g^4 \neq 1 \text{ and so } G' = \langle a^2 \rangle \nleq \langle abg \rangle .$$

If $g^4 = 1$,then $a^2 g^2$ is an involution distinct from a^2 and so again $G' \nleq \langle abg \rangle$. It follows that in any case $G' \nleq \langle abg \rangle$ and so $\langle abg \rangle \leq Z(G)$. But then

$$ab \in Z(G) \text{ giving } C_4 \cong \langle ab \rangle \leq Z(Q) ,$$

a contradiction. We have proved that for each $g \in G - G_0$, $G/(\langle g \rangle \cap G_0)$ is abelian. Our theorem is proved. □

Let G be a nonabelian p-group containing an abelian subgroup A of index p and covered by minimal nonabelian subgroups. Then $\mathrm{cl}(G) = 2$ and G' is elementary abelian.

Proof. Let $a \in A$ and $g \in G - A$ be such that $[a, g] \neq 1$. We shall prove that $[a, g]$ is of order p and $[a, g] \in Z(G)$. Indeed, by hypothesis, there is a minimal nonabelian subgroup M in G such that $a \in M$. Let $g_1 \in M - A$ so that $g = g_1 a_1$ for some $a_1 \in A$. Since $a \notin Z(G)$, we have $[a, g_1] \neq 1$ and so $M' = \langle [a, g_1] \rangle$, where $[a, g_1]$ is of order p and $[a, g_1] \in Z(G)$. We obtain

$$[a, g] = [a, g_1 a_1] = [a, g_1][a, a_1]^{g_1} = [a, g_1]$$

and we are done. □

Problem 1. Classify the non-Dedekindian p-groups that are not generated by nonnormal nonabelian (minimal nonabelian, noncyclic) subgroups.

Problem 2. Study the nonmodular p-groups that are not generated by non-quasinormal subgroups (cyclic subgroups, abelian subgroups).

Problem 3. Classify the non-Dedekindian p-groups that are not generated by nonnormal abelian subgroups (noncyclic abelian subgroups).

§ 232 Nonabelian p-groups in which any nonabelian subgroup contains its centralizer

The purpose of this section is to initiate a classification of the title p-groups (see Problem 9). If $p > 2$, then it is known that a title p-group G is either of maximal class or $\Omega_1(G)$ is elementary abelian (Corollary A.74.3). However if $p = 2$, then one can prove a stronger result.

Examples of the title groups:

(1) Minimal nonabelian p-groups.

(2) $\mathcal{A}_2$-groups all of whose maximal subgroups are nonabelian.

(3) p-groups of maximal class with abelian subgroup of index p.

(4) The minimal nonmetacyclic group G of order 2^5. If $L < G$ is of order 2. then G/L contains a nonabelian subgroup H/L such that $C_{G/L}(H/L) \nleq H/L$.

Theorem 232.1 (Janko). *Let G be a nonabelian 2-group in which $C_G(H) < H$ for each nonabelian subgroup H in G. Then one of the following holds:*

(a) *G is of maximal class.*

(b) *$E = \Omega_1(G)$ is elementary abelian of order 4 or 8. If $|E| = 4$, then such groups G are classified in § 82. If $|E| = 8$, then $\Omega_2(C_G(E))$ is of exponent 4.*

Proof. Let G be a nonabelian 2-group in which $C_G(H) < H$ for each nonabelian subgroup H in G. Assume, in addition, that G is not of maximal class. Then Proposition 10.17 implies that $E = \Omega_1(G)$ is elementary abelian. Indeed, G has no subgroup $\cong D_8$.[1] Also, we have $|E| \geq 4$. Indeed, if $|E| = 2$, then G is generalized quaternion and so G is of maximal class (Proposition 1.3), a contradiction.

(i) Assume that $E \nleq Z(G)$. Let $g \in G - C_G(E)$ be such that $g^2 \in C_G(E)$. Then

$$o(g) \geq 4 \quad \text{and} \quad \langle z \rangle = \langle g \rangle \cap E \cong C_2 .$$

Also, g induces on E an involutory automorphism. Then it is well known that[2]

$$|C_E(g)| \geq |E : C_E(g)| \quad \text{and} \quad [\langle g \rangle, E] \leq C_E(g) .$$

(i1) First suppose that there is $t \in E - \langle z \rangle$ such that $[g, t] \neq 1$ but $[g, t] \in \langle g^2 \rangle$. Then

$$[g, t] \in \langle g^2 \rangle \cap E = \langle z \rangle \quad \text{and so } [g, t] = z .$$

1 If $H \cong D_8$ is a subgroup of G, then $C_G(H) < H$ so G is of maximal class, by Proposition 10.17, contrary to the assumption.

2 This is easy to deduce from the following fact: if $a \in E$, then $aa^g \in Z(\langle g, E \rangle)$.

In this case, $\langle g, t\rangle' = \langle z\rangle$ and then Lemma 65.2 (a) implies that $\langle g, t\rangle$ is minimal nonabelian. If $o(g) = 4$, then $\langle g, t\rangle \cong D_8$. But then our hypothesis together with Proposition 10.17 implies that G is of maximal class, a contradiction. Hence we get $\langle g, t\rangle \cong M_{2^{n+1}}$, $n \geq 3$, where $o(g) = 2^n$ and $Z(\langle g, t\rangle) = \langle g^2\rangle \cong C_{2^{n-1}}$. But $C_E(g)$ centralizes $\langle g, t\rangle$ and so $C_E(g) = \langle z\rangle$ implying $E = \langle t, z\rangle \cong E_4$. Our group G has exactly three involutions and such groups are classified in § 82.

(i2) Now assume that there is $t \in E - \langle z\rangle$ such that $[g, t] \neq 1$ but $[g, t] = z_0 \notin \langle g^2\rangle$.

Then $\langle g, t\rangle/\langle g^2\rangle \cong D_{2^n}$, $n \geq 3$. Note that

$$[\langle g\rangle, E] \leq C_E(g) \quad \text{and so } z_0 \in C_E(g) \text{ implying } z_0 \in Z(\langle g, t\rangle)\,.$$

On the other hand,

$$Z(\langle g, t\rangle) \geq \langle g^2\rangle \text{ and so } Z(\langle g, t\rangle) = \langle g^2\rangle \times \langle z_0\rangle \text{ since } \langle g, t\rangle/\langle g^2\rangle \cong D_{2^n}\,.$$

This gives that $C_E(g)$ (centralizing $\langle g, t\rangle$) is equal $\langle z, z_0\rangle \cong E_4$ and so $E \cong E_8$ or E_{16}. However, if $E \cong E_{16}$, then $C_E(g) = [\langle g\rangle, E] \cong E_4$ and so there is $t' \in E$ such that $[g, t'] = z \in \langle g^2\rangle$. But then by (i1) $E \cong E_4$, a contradiction. Hence we have $E \cong E_8$.

(ii) Now suppose that $E \leq Z(G)$.

Let M be any minimal nonabelian subgroup in G. Then our hypothesis implies that $E \leq M$ and this gives (by Lemma 65.1) that $|E| \leq 8$. Again, if $|E| = 4$, then such groups G are classified in § 82.

Finally, assume that $|E| = 8$ and set $G_0 = C_G(E)$. We show here that G_0 has the property (X) in the sense of Definition 1 in § 153. Indeed, let $x, y \in G_0$ be such that $x^2 = y^2$. We must show that in this case $[x, y] = 1$. Suppose $[x, y] \neq 1$ so that $\langle x, y\rangle$ is nonabelian and $\langle x^2\rangle \leq Z(\langle x, y\rangle)$.

If $[x, y] \in \langle x^2\rangle$, then $\langle x, y\rangle/\langle x^2\rangle \cong E_4$ and $\langle x^2\rangle = Z(\langle x, y\rangle)$. But our hypothesis implies that $E \cong E_8$ must be contained in $Z(\langle x, y\rangle)$, a contradiction.

If $[x, y] \notin \langle x^2\rangle$, then $\langle x, y\rangle/\langle x^2\rangle \cong D_{2^n}$, $n \geq 3$, and so $\langle x^2\rangle$ is a subgroup of index ≤ 2 in $Z(\langle x, y\rangle)$. Again, E is not contained in $Z(\langle x, y\rangle)$, contrary to our hypothesis. We have proved that $[x, y] = 1$, and so any two elements in G_0 with the same square commute. By Theorem 153.4, G_0/E also has the property (X). Set $F/E = \Omega_1(G_0/E)$ so that, by Theorem 153.4, F/E is elementary abelian. Let a, b be any elements of order ≤ 4 in G_0. Then $a^2, b^2 \in E$ and so $a, b \in F$. But then $ab \in F$ and so $(ab)^2 \in E$, implying $o(ab) \leq 4$. We have proved that $\Omega_2(G_0)$ is of exponent 4 and our theorem is proved. □

If G is a nonabelian p-group such that $C_G(A) < A$ for any minimal nonabelian $A \leq G$ then this containment holds for any nonabelian $A \leq G$. In view of this, we state the following

Exercise 1. Let G be a nonabelian p-group such that $C_G(A) < A$ for any nonabelian $A \leq G$. Then one of the following holds:
(a) $\Omega_1(G)$ is elementary abelian.
(b) G is of maximal class.

Solution. Let $p = 2$ and assume that $\Omega_1(G)$ is nonabelian. Then there are in G two noncommuting involutions x, y. In that case, $\langle x, y \rangle$ contains a subgroup $D \cong D_8$. As $C_G(D) < D$, Proposition 10.17 implies that G is of maximal class.

Next we assume that $p > 2$. Let $E < G$ be a normal elementary abelian subgroup of maximal order. Assume that $E < \Omega_1(G)$. Then there is $x \in \Omega_1(G) - E$ of order p. By Theorem 10.1, the subgroup $H = \langle x, E \rangle$ is nonabelian. Then the subgroup $L = \langle a, x \rangle$ is minimal nonabelian for some $a \in E$ (Lemma 57.1). It follows from $\Omega_1(L) = L$ that $L \cong S(p^3)$. We have $C_G(L) < L$ so G is of maximal class (Proposition 10.17).

If G is a nonabelian p-group such that $C_G(A) < A$ for any minimal nonabelian $A \leq G$ then this containment holds for any nonabelian $A \leq G$. In view of this we state the following

Problem 1. Classify the $\mathcal{A}_n$-groups G, $n > 2$, of prime power order such that $C_G(H) < H$ for all $\mathcal{A}_2$-subgroups $H < G$.

Problem 2. Classify the nonabelian p-groups G such that $C_G(H) < H$ for all $\mathcal{A}_1$-subgroups $H < G$ of minimal order.

Problem 3. Study the nonabelian nonmetacyclic p-groups G such that $C_G(H) \leq H$ for all maximal metacyclic $H < G$.

Problem 4. Study the nonabelian p-groups of exponent $> p$ all of whose maximal elementary abelian subgroups are maximal abelian.

§ 233 On monotone p-groups

A (finite) group G is monotone if, for every pair H, K of subgroups of G, $H < K$ implies $d(H) \leq d(K)$. Subgroups of monotone groups are monotone. Monotone p-groups have been introduced and investigated by A. Mann, who determined their structure if $p > 2$ (see [Man35] and [Man36]). It turns out that if $p > 3$, they are (apart from a small number of exceptions) modular, whereas for $p = 3$, there are also some monotone 3-groups of maximal class and some other related groups. See Theorems 8, 9, 10 in [Man35] and the corrigendum [Man36] for a full account.

Eleonora Crestani and Federico Menegazzo gave in [CrM] a complete classification of monotone 2-groups. From that classification (Theorem 1.1 in [CrM]), we see in particular that in all monotone 2-groups G we have that G'' is abelian and so the derived length of G is 3. In that classification, an important role has the following nice characterization of monotone 2-groups.

Theorem 233.1 (A. Mann). *A 2-group G is monotone if and only if every two-generator subgroup of G is metacyclic.*

Proof. If G is a monotone 2-group, then every maximal subgroup of a two-generator subgroup is two-generator and so by Corollary 36.6, every two-generator subgroup is metacyclic.

We prove the converse by contradiction. Suppose that there exists a 2-group G such that every two-generator subgroup of G is metacyclic but G is not monotone. Let G be a minimal counterexample and let H and K be such $H < K \leq G$ but $d(H) > d(K)$. By the minimality of $|G|$, we get $K = G$ and we set $d(G) = r$. Among all proper subgroups H of G with $d(H) > r$, we choose one of the maximal possible order. Let $G_0 \leq G$ be such that $H < G_0$ with $|G_0 : H| = 2$. Then $d(G_0) \leq r$ and so the minimality of $|G|$ gives $G_0 = G$ and therefore H is maximal in G. Consider $G/\Phi(H)$ so that $\Phi(H) \leq \Phi(G)$ gives $d(G/\Phi(H)) = r$ and $d(H/\Phi(H)) = d(H) > r$. The minimality of $|G|$ gives $\Phi(H) = \{1\}$ and so H is elementary abelian and $\exp(G) \leq 4$. But G is not monotone and so G is not abelian and therefore $\exp(G) = 4$.

Let $g \in G - H$ with $o(G) = 4$. For any $h \in H$, $M = \langle g, h\rangle$ is metacyclic and $M = \langle g\rangle(M \cap H)$. Suppose that M is nonabelian. Then

$$h \notin \langle g^2\rangle \text{ and so } M \cap H = \langle h, g^2\rangle \cong E_4 \text{ and } M \cong D_8\,.$$

In any case h normalizes $\langle g\rangle$ and so $\langle g\rangle \trianglelefteq G$. We have $G/\langle g\rangle \cong H/\langle g^2\rangle$ is elementary abelian and so

$$\mho_1(G) \leq \langle g\rangle \cap H = \langle g^2\rangle \text{ giving } \Phi(G) = \mho_1(G) = \langle g^2\rangle\,.$$

But then $d(H) = d(G)$, a contradiction. □

Problem 1. Study the p-groups G such that, whenever $A < B \leq G$, where A is abelian, then $d(A) \leq d(B)$.

Problem 2. Study the irregular p-groups all of whose regular subgroups are monotone.

Problem 3. Study the irregular p-groups all of whose minimal irregular subgroups are monotone.

§ 234 p-groups all of whose maximal nonnormal abelian subgroups are conjugate

Here we solve problem 2820 by showing that the non-Dedekindian title p-groups are actually metahamiltonian (see § 228).

Theorem 234.1. *Let G be a non-Dedekindian p-group all of whose maximal nonnormal abelian subgroups are conjugate. Then G is metahamiltonian (i.e., all nonabelian subgroups are G-invariant) of class 2.*

Moreover, if H is a maximal nonnormal abelian subgroup of G, then $G_0 = H^G < G$ is abelian, $|G_0 : H| = p$ and G_0 is generated by all nonnormal subgroups of G (see § 231). Also, both G/G_0 and G' are cyclic.

Proof. Let G be a non-Dedekindian p-group all of whose maximal nonnormal abelian subgroups are conjugate. Let H be a maximal nonnormal abelian subgroup of G. Set $K = H^G$ so that $H < K < G$. Let $x \in G - K$ and assume that $\langle x \rangle$ is not normal in G. Let H_1 be a maximal nonnormal abelian subgroup of G containing $\langle x \rangle$. Since $H_1 \nleq K$, H_1 is not conjugate to H, a contradiction. Thus, for each $x \in G - K$, $\langle x \rangle \trianglelefteq G$. It follows that each subgroup X of G with $X \nleq K$ is normal in G. Hence $G_0 = K = H^G$ is the subgroup of G generated by all nonnormal subgroups of G. Since $G_0 < G$, Theorem 231.1 implies that G is of class 2. In particular, each maximal abelian subgroup is normal in G.

Let L be a maximal abelian subgroup in G containing H. Then $L \trianglelefteq G$ and so $H < L$. We get $G_0 = H^G \leq L$ and so G_0 is abelian. Since each nonnormal subgroup of G is contained in G_0, it follows that G is metahamiltonian.

Let $H < U \leq G_0$ be such that $|U : H| = p$. By the maximality of H, we get $U \trianglelefteq G$ and so $H^G = U = G_0$. We have proved that $|G_0 : H| = p$.

The fact that $G_0 < G$ together with Theorem 231.1 implies that both G/G_0 and G' are cyclic and we are done. □

§ 235 On normal subgroups of capable 2-groups

Recall that a p-group H is called "capable" if there exists a p-group G such that $G/Z(G) \cong H$. We show here that the structure of capable 2-groups is somewhat restricted since they cannot have normal subgroups of certain type.

Theorem 235.1 (S. Shahriari [Sh]). *Let H be a capable 2-group. Then H does not have a normal subgroup which is generalized quaternion or semidihedral.*

In fact, this theorem is equivalent with the following result.

Theorem 235.2. *Let G be a 2-group. Then $G/Z(G)$ does not have a normal subgroup which is generalized quaternion or semidihedral.*

Proof (Janko). Let $H/Z(G)$ be a normal subgroup in $G/Z(G)$ such that $H/Z(G) \cong Q_{2^n}$, $n \geq 3$, or $H/Z(G) \cong SD_{2^n}$, $n \geq 4$.

Let $n = 3$; then $H/Z(G) \cong Q_8$. In this case, each maximal subgroup of $H/Z(G)$ is cyclic of order 4 and so H has three pairwise distinct abelian maximal subgroups and this implies $|H'| = 2$ and $H' \nleq Z(G)$ (indeed, if $H' \leq Z(G)$, then $H/Z(G)$ as an epimorphic image of H/H', is abelian so $\ncong Q_8$). But H' of order 2 is G-invariant, and so $H' \leq Z(G)$, a contradiction. We have proved that $n \geq 4$ and so $H/Z(G)$ has a unique cyclic subgroup $H_0/Z(G) \cong C_{2^{n-1}}$ of index 2 and H_0 is abelian.

Let $A/Z(G) < H/Z(G)$ be cyclic of index 2 and let $B/Z(G) < H/Z(G)$ be cyclic of order 4 such that $B \nleq A$. Then A, B are nonincident abelian subgroups of the group $H = AB$ and $|A \cap B| = 2|Z(G)|$. But $A \cap B = Z(H)$ (this follows from the fact that the center of $H/Z(G)$ has order 2), and we conclude that $|Z(H) : Z(G)| = 2$. By Lemma 1.1,

$$|H| = 2|Z(H)|\,|H'| = 4|Z(G)|\,|H'| \Rightarrow 2^n = |H|/|Z(G)| = 4|H'| \Rightarrow |H'| = 2^{n-2}\,.$$

Since $|(H/Z(G))'| = 2^{n-2} = |H'|$, it follows that $H' \cap Z(G) = \{1\}$ which is impossible since $\{1\} < H' \lhd G$. □

Exercise 1. Suppose that G is a capable p-group. Then G has no normal subgroup $\cong M_{p^n}$.

Problem 1. Does there exist a capable p-group containing a normal nonabelian metacyclic subgroup of order p^4 and exponent p^2?

Problem 2. Classify the abelian capable p-groups.

§ 236 Non-Dedekindian p-groups in which the normal closure of any cyclic subgroup has a cyclic center

We solve here Problem 2744 (ii) by proving the following result.

Theorem 236.1 (Janko). *Let G be a non-Dedekindian p-group in which the normal closure of any cyclic subgroup has a cyclic center. Then $p = 2$ and $G = H \times E$, where H is of maximal class and* $\exp(E) \leq 2$.

Proof. Let G be a non-Dedekindian p-group in which the normal closure of any cyclic subgroup has a cyclic center. Let Z be a nonnormal cyclic subgroup in G. Suppose that the normal closure Z^G is abelian. By our hypothesis, Z^G is cyclic. But then $Z \trianglelefteq G$, a contradiction. Hence Z^G is nonabelian and so we may use Theorem 223.1. It follows that $p = 2$ and G has an abelian maximal subgroup A of exponent 2^e, $e \geq 3$, where for an element $v \in G - A$ and all $h \in A$, we have either

$$h^v = h^{-1} \quad \text{or} \quad h^v = h^{-1+2^{e-1}} .$$

This gives $o(v) \leq 4$ and $\langle v \rangle$ is not normal in G. Also, we have

$$\langle v \rangle^G = [A, \langle v \rangle]\langle v \rangle ,$$

v inverts each element of $[A, \langle v \rangle] = G'$ so that $\langle v \rangle^G$ is either quasidihedral (in case of $o(v) = 2$) or quasi-generalized quaternion (in case $o(v) = 4$) (see Definition 1 in § 223).

By our hypothesis, $Z(\langle v \rangle^G)$ is cyclic and this implies that $[A, \langle v \rangle]$ is cyclic and $\langle v^2 \rangle \leq [A, \langle v \rangle]$ so that $\langle v \rangle^G$ is dihedral or generalized quaternion and $A = Z \times E$, where $Z \cong C_{2^e}$ and E is elementary abelian with $\langle v^2 \rangle \leq Z$. It follows that $G = H \times E$, where H is of maximal class and $\exp(E) \leq 2$. Our theorem is proved. □

For a partial case, see Theorem 221.1, where it has been proven that if any nonnormal abelian subgroup of a non-Dedekindian p-group has cyclic center, then G is a generalized quaternion group. That result follows immediately from Theorem 236.1.

§ 237 Noncyclic p-groups all of whose nonnormal maximal cyclic subgroups are self-centralizing

Here we generalize Problem 2917 and prove the following result.

Theorem 237.1 (Janko). *Let G be a non-Dedekindian p-group such that every maximal nonnormal cyclic subgroup in G is self-centralizing. Then $p = 2$ and G is generalized quaternion of order > 8.*

Proof. Suppose that G is a non-Dedekindian p-group such that every maximal nonnormal cyclic subgroup in G is self-centralizing. Clearly, $Z(G)$, being a subgroup of all maximal nonnormal cyclic subgroups of G, must be cyclic. Let $X = \Omega_1(Z(G))$; then $|X| = p$ and X is the unique normal subgroup of order p in G. If X is the unique subgroup of order p in G, then G is a generalized quaternion group Q_{2^n}, $n > 3$ (since G is non-Dedekindian), by Proposition 1.3, and we are done in this case. Now assume that there is in G a subgroup $Y \neq X$ of order p; then Y is not normal in G, by what has just been said. Let $Y < U < G$, where U is maximal cyclic in G. As $Y = \Omega_1 U)$ is nonnormal in G so U is also non-G-invariant. However, $Z(G) < U$ so that $X < U$, and we conclude that $X = Y$, which is a contradiction. Thus, Y does not exist so, as we noted above, G is a generalized quaternion group of order > 8. □

Exercise 1. Classify the non-Dedekindian p-groups in which the centralizer of any maximal nonnormal cyclic subgroup has no normal abelian subgroup of type (p, p).

Solution. Let A be a nonnormal maximal cyclic subgroup of G (such A exists). Set $C = C_G(A)$. Assume that G is not a generalized quaternion group of order > 8. In that case, by Theorem 237.1, one may assume that for some choice of A, C is noncyclic. Then, by Lemma 1.4, C is a 2-group of maximal class, which is impossible since, otherwise, $|A| > 2 = |Z(C)|$. Thus, G is a generalized quaternion group of order > 8.

Problem. Does there exist a nonabelian p-group G, $p > 2$, in which the centralizer of any nonnormal maximal cyclic subgroup has no normal subgroup $\cong E_{p^3}$?

§ 238 Nonabelian p-groups all of whose nonabelian subgroups have a cyclic center

Here we solve a partial case of Problem 2369 and prove the following result.

Theorem 238.1 (Janko). *Let G be a nonabelian p-group all of whose nonabelian subgroups have a cyclic center. Then G has an abelian subgroup H of index p and the intersection of any two distinct maximal abelian subgroups is equal $Z(G)$.*

If $p = 2$, then G is isomorphic to one of the following 2-groups:

(a) *Group of maximal class.*

(b) M_{2^n}.

(c) $G = D * C$ *(central product), where* $D \cong D_{2^n}$, $C \cong C_{2^m}$, $m \geq 2$, *is cyclic of order* 2^m *and* $D \cap C = Z(D)$.

(d) $G = \langle x, t \mid (xt)^2 = a,\ a^{2^m} = t^2 = 1,\ m \geq 2,\ x^2 = ab,\ b^{2^{n-1}} = 1,\ n \geq 3,\ b^t = b^{-1},\ [a, x] = [a, t] = 1,\ t^x = tb,\ a^{2^{m-1}} = b^{2^{n-2}} \rangle$, *where* $|G| = 2^{m+n}$, $m \geq 2$, $n \geq 3$, $Z(G) = \langle a \rangle \cong C_{2^m}$, $G' = \langle b \rangle \cong C_{2^{n-1}}$, *and* $M = \langle x, a \rangle$ *is a unique abelian maximal subgroup of G. We have* $C_G(t) = \langle t \rangle \times \langle a \rangle \cong C_2 \times C_{2^m}$ *and* $\langle b, t \rangle \cong D_{2^n}$.

(e) $G = \langle g, h \mid g^{2^n} = h^{2^m} = 1,\ m \geq 3,\ n \geq 3,\ g^{2^{n-1}} = h^{2^{m-1}},\ h^g = h^{-1} \rangle$, *where G is metacyclic,* $|G| = 2^{m+n-1}$ *since* $\langle g \rangle \cap \langle h \rangle = \langle g^{2^{n-1}} \rangle \cong C_2$. *Also,* $Z(G) = \langle g^2 \rangle \cong C_{2^{n-1}}$, $G' = \langle h^2 \rangle \cong C_{2^{m-1}}$ *and* $M = \langle h, g^2 \rangle$ *is a unique abelian maximal subgroup of G.*

If $p > 2$, then there is $g \in G - H$ so that $Z = C_H(g)$ is cyclic and any such group satisfies the hypothesis.

Proof. Let $A \neq B$ be any two distinct maximal abelian subgroups in G. Then $A \cap B = Z(\langle A, B \rangle)$, where $\langle A, B \rangle$ is nonabelian. By our hypothesis, $A \cap B$ is cyclic. Hence any two distinct maximal abelian subgroups in G have a cyclic intersection.

If $p = 2$, then such 2-groups are determined up to isomorphism in Theorem 91.1.

From now on, we assume in addition $p > 2$. By Lemma 1.4, G possesses a normal abelian subgroup U of type (p, p). Since $Z(G)$ is cyclic, $H = C_G(U)$ is a subgroup of index p in G. By our hypothesis, H is abelian. Let $g \in G - H$ and set $Z = C_H(g)$. It follows $g^p \in Z$ and $Z = Z(G)$ so that Z is cyclic. Maximal abelian subgroups of G are H and $Z\langle k \rangle$ for any $k \in G - H$, where $k^p \in Z$. We see that any two distinct maximal abelian subgroups of G intersect in $Z = Z(G)$.

Conversely, let G be a nonabelian p-group with an abelian maximal subgroup H and an element $g \in G - H$ such that $Z = C_H(g)$ is cyclic. Let K be any nonabelian subgroup in G. Then $K \nleq H$ and $Z(K) \leq Z(G) = Z$ and so $Z(K)$ is cyclic. □

Problem 1. Study the nonabelian p-groups G such that $A \cap B$ is cyclic for any distinct maximal abelian subgroups $A, B < G$. (For a partial case, see Theorem 238.1.)

Problem 2. Classify the primary $\mathcal{A}_n$-groups, $n > 1$, all of whose $\mathcal{A}_1$- and $\mathcal{A}_2$-subgroups have cyclic centers.

Problem 3. Study the nonmetacyclic p-groups all of whose maximal metacyclic subgroups have cyclic centers.

Problem 4. Study the p-groups G in which the center of the centralizer of any noncyclic abelian subgroup A coincides with A. (Examples: D_{2^n}, SD_{2^n}.)

Problem 5. Classify the primary $\mathcal{A}_n$-groups, $n > 1$, all of whose proper nonabelian (non-Dedekindian) subgroups have cyclic center.

Problem 6. Study the irregular p-groups all of whose maximal regular subgroups have cyclic centers.

Problem 7. (A partial case of some unsolved problems.) Study the nonabelian p-groups all of whose nonabelian two-generator subgroups have cyclic centers.

§ 239 p-groups G all of whose cyclic subgroups are either contained in $Z(G)$ or avoid $Z(G)$

Here we initiate a study of the title groups (Problem 236) and prove the following result.

Theorem 239.1. *Let G be a nonabelian p-group such that, whenever $C < G$ is cyclic, then either $C \leq Z(G)$ or $C \cap Z(G) = \{1\}$. Then $Z(G)$ is elementary abelian and if A is any normal abelian subgroup in G, then A is elementary abelian. Moreover, if $p = 2$, then $\Omega_1(G) < G$.*

Proof. Let G be a nonabelian p-group such that either $C \leq Z(G)$ or $C \cap Z(G) = \{1\}$ for each cyclic subgroup C in G. Let $X/Z(G)$ be a subgroup of order p in $G/Z(G)$; then X is abelian. Then all elements in $X - Z(G)$ are of order p and this implies that $Z(G)$ is elementary abelian. Indeed, if $x \in Z(G)$ and $o(x) \geq p^2$, then for any $y \in X - Z(G)$, $o(xy) \geq p^2$ and $xy \in X - Z(G)$, a contradiction.[1]

Let A be a maximal normal abelian subgroup in G and suppose that A is not elementary abelian. Set $Y = \Omega_1(\mho_1(A)) \neq \{1\}$, where each element in Y is a pth power of an element in A. This implies that $Y \cap Z(G) = \{1\}$. On the other hand, Y is characteristic in A and so $Y \triangleleft G$ so that $Y \cap Z(G) \neq \{1\}$, a contradiction. Thus, any maximal normal abelian subgroup of G has exponent p.

Assume, in addition, that $p = 2$. Let $B/Z(G)$ be a normal subgroup of order 2 in $G/Z(G)$ so that B is abelian. By the above, B is elementary abelian. Let $b \in B - Z(G)$ and let $g \in G - C_G(b)$ be an involution. By Lemma 200.2, $\langle b, g\rangle$ is minimal nonabelian. Hence $\langle b, g\rangle \cong D_8$ (since both b and g are involutions) and $1 \neq [b, g] \in Z(G)$, by the choice of g. We get that $[b, g]$ is a square in G, a contradiction, We have proved that there is no involution in $G - C_G(b)$ and so $\Omega_1(G) \leq C_G(b) < G$ (recall that $b \notin Z(G)$), and we are done. □

Problem 1. Study the p-groups G such that any of their cyclic subgroups of order p^2 is either contained in $Z(G)$ or avoids $Z(G)$.

Problem 2. Let N be a nontrivial normal subgroup of a p-group G. Study the structure of G provided any its cyclic subgroup (cyclic subgroup of order p^2) is either contained in N or avoids N.

1 Another proof: $X = \langle X - Z(G)\rangle \leq \Omega_1(X)$ so X is elementary abelian.

§ 240 p-groups G all of whose nonnormal maximal cyclic subgroups are conjugate

Here we classify the title groups (Problem 2210) and prove the following result.

Theorem 240.1. *Let G be a non-Dedekindian p-group all of whose nonnormal maximal cyclic subgroups are conjugate. Then G is of class 2 and if C is a nonnormal maximal cyclic subgroup in G, then $H = C^G < G$ is also the subgroup of G generated by all nonnormal subgroups of G. The subgroup H is abelian of rank 2, $\Omega_1(G) \cong E_{p^2}$ and G is metahamiltonian (i.e., all nonabelian subgroups are normal in G; see § 228). Moreover, G/H is cyclic and we may choose $g \in G-H$ so that $\langle g\rangle$ covers G/H, $\langle g\rangle \trianglelefteq G$ and $G = \langle g\rangle C$ is splitting metacyclic with $\langle g\rangle \cap C = \{1\}$.*

Proof. Let G be a non-Dedekindian p-group all of whose nonnormal maximal cyclic subgroups are conjugate. Let C be a nonnormal maximal cyclic subgroup in G. Set $H = C^G$ so that $C < H < G$.

Let X be a subgroup in G with $X \nleq H$. We claim that $X \trianglelefteq G$. Suppose that this is false. Then there is $x \in X-H$ such that $\langle x\rangle$ is nonnormal in G. Let D be a maximal cyclic subgroup of G containing $\langle x\rangle$. It follows that D is nonnormal in G. By our hypothesis, D must be conjugate to C in G, a contradiction. We have proved that each subgroup of G which is not contained in H is normal in G and so H is the subgroup of G generated by all nonnormal subgroups of G.

Since $H < G$, we may use Theorem 231.1. It follows that G is of class 2 and for each $g \in G - H$, $\{1\} \neq \langle g\rangle \cap H \trianglelefteq G$ and $G/(\langle g\rangle \cap H)$ is abelian and so G' is cyclic. Moreover, G/H is cyclic.

Since $\mathrm{cl}(G) = 2$, each maximal abelian subgroup of G is normal in G. Let K be a maximal abelian subgroup of G containing C. Since $K \trianglelefteq G$, it follows that $H = C^G \leq K$ is abelian. In particular, each nonabelian subgroup of G is normal in G and so G is metahamiltonian (see § 228). By Theorem 228.4, G' is contained in each nonabelian subgroup of G.

Set $M = \mathrm{N}_G(C) < G$ so that $H \leq M$ and let $g \in G-M$. We have $\langle g\rangle \trianglelefteq G$ and $C < H$ and therefore H is of rank ≥ 2. We know that $G/(\langle g\rangle \cap H)$ is abelian and so $(\langle g\rangle \cap H)C \trianglelefteq G$. But then the normal closure $H = C^G$ is contained in $(\langle g\rangle \cap H)C \leq H$ and this implies $H = (\langle g\rangle \cap H)C$. Thus, H is of rank 2. Since there are no elements of order p in $G - H$, we get $\Omega_1(G) \cong E_{p^2}$. Because G/H is cyclic, we may choose $g \in G-H$ so that $\langle g\rangle$ covers G/H and therefore $G = \langle g\rangle C$ is metacyclic.

Let G_1 be a maximal subgroup of G containing H. Since G is nonabelian and G is generated by its minimal nonabelian subgroups, there is a minimal nonabelian subgroup $K \nleq G_1$. Then for each $g \in K-G_1$, $\langle g\rangle$ covers G/H and so K covers G/H. It is easy to see that K is splitting metacyclic. Indeed, if this is not the case, then Lemma 65.1 implies that $p = 2$ and $K \cong Q_8$. If in this case $\mathrm{C}_G(K) \nleq K$, then $K\mathrm{C}_G(K)$ is nonmetacyclic, a contradiction. Thus $\mathrm{C}_G(K) \leq K$ and then Theorem 10.17 gives that G is a 2-group of maximal nonnormal class. But G is of class 2 and so $G = K \cong Q_8$, contrary to our hypothesis that G is not Dedekindian. We have proved that K is splitting metacyclic. We may set

$$K = \langle k\rangle\langle l\rangle \text{ with } \langle k\rangle \trianglelefteq K, \quad \langle k\rangle \cap \langle l\rangle = \{1\} \text{ and } \langle [k, l]\rangle = \Omega_1(\langle k\rangle) = K'.$$

Since $\langle l\rangle$ is nonnormal in K, we get $\langle l\rangle \leq H$. But K covers G/H and so $\langle k\rangle$ covers G/H. Let C_1 be a maximal cyclic subgroup in G containing $\langle l\rangle$. Then C_1 is nonnormal in G implying $C_1 \leq H$ and $|C_1| = |C|$. It follows that

$$G = \langle k\rangle C_1 \text{ with } \langle k\rangle \cap C_1 \leq \langle k\rangle \cap \langle l\rangle = \{1\}.$$

Hence $\langle k\rangle \cap C_1 = \{1\}$ and so our group G is splitting metacyclic. Our theorem is proved. □

Problem 1. Study the nonmetacyclic p-groups all of whose maximal nonnormal metacyclic subgroups are conjugate.

Problem 2. Study the nonabelian p-groups all of whose nonnormal $\mathcal{A}_1$-subgroups are conjugate.

Remark 1. Smallest examples of p-groups G satisfying the assumptions of our Theorem 240.1 are M_{16} and M_{p^3} (p odd). In these cases, the subgroup H generated by all nonnormal subgroups of G is isomorphic to E_{p^2} and G/H is cyclic.

§ 241 Non-Dedekindian p-groups with a normal intersection of any two nonincident subgroups

Here we solve Problem 320 by proving the following result.

Theorem 241.1. *Let G be a non-Dedekindian p-group such that, whenever $A, B < G$ are nonincident, then $A \cap B \trianglelefteq G$. Then for each $H \leq G$, we have $|H : H_G| \leq p$ and G' centralizes $\mho_1(G)$ so that in case $p = 2$, G is metabelian. Moreover, G has no elementary abelian subgroups of order p^3 (and such p-groups have been considered in § 50 and Theorem 13.7).*

Proof. Let G be a non-Dedekindian p-group such that, whenever $A, B < G$ are nonincident, then $A \cap B \trianglelefteq G$.

First we show that for each $x \in G$, $\langle x^p \rangle \trianglelefteq G$. Indeed, let $\langle a \rangle$ be a maximal cyclic subgroup in G containing $\langle x \rangle$. Let H be a subgroup in G containing $\langle a \rangle$ such that $|H : \langle a \rangle| = p$. If $a \in \Phi(H)$, then H is cyclic, a contradiction. Hence $a \notin \Phi(H)$ and so there is a maximal subgroup M of H such that $a \notin M$. Then $M \cap \langle a \rangle = \langle a^p \rangle$ and so our hypothesis implies $\langle a^p \rangle \trianglelefteq G$. Since $x \in \langle a \rangle$, we get $x^p \in \langle a^p \rangle$ and so $\langle x^p \rangle \trianglelefteq G$.

Since $\langle x^p \rangle \trianglelefteq G$ and $\mathrm{Aut}(\langle x^p \rangle)$ is abelian, G' centralizes $\langle x^p \rangle$ for each $x \in G$ and so G' centralizes $\mho_1(G)$. If $p = 2$, then $\Phi(G) = \mho_1(G)$ and so $G' \leq \mho_1(G)$ implies that G' is abelian.

For each subgroup $H \leq G$, we have $|H : H_G| \leq p$. Indeed, we may assume that H is noncyclic (because if $H = \langle x \rangle$ is cyclic, then by the above, $\langle x^p \rangle \trianglelefteq G$). Also, we may suppose that $H < G$ and let $K \leq G$ be such that $H < K$ and $|K : H| = p$. Then $\Phi(K) \leq H$ and if $\Phi(K) = H$, then K is cyclic and so H is cyclic, a contradiction. Thus $\Phi(K) < H$ and so there is a maximal subgroup K_1 of K such that $H \neq K_1$. But then H and K_1 are nonincident with $|H : (H \cap K_1)| = p$ and $H \cap K_1 \trianglelefteq G$ and we are done.

Let $E \cong \mathrm{E}_{p^3}$ be an elementary abelian subgroup of order p^3 in G. Let $x \neq 1$ be an element in E. Choose $y, z \in E$ such that $\langle x, y, z \rangle = E$. Then

$$\langle x, y \rangle \cap \langle x, z \rangle = \langle x \rangle \trianglelefteq G$$

and so $x \in \mathrm{Z}(G)$. We have proved that $E \leq \mathrm{Z}(G)$. Let $\langle a \rangle$ be any cyclic subgroup in G. Then $|\langle a \rangle \cap E| \leq p$ and so there are $v, w \in E$ such that

$$\langle v, w \rangle \cong \mathrm{E}_{p^2} \text{ and } \langle v, w \rangle \cap (\langle a \rangle \cap E) = \{1\}\,.$$

It follows that $\langle a, v \rangle$ and $\langle a, w \rangle$ are nonincident and $\langle a, v \rangle \cap \langle a, w \rangle = \langle a \rangle$, implying that $\langle a \rangle \trianglelefteq G$. But then G is Dedekindian, a contradiction. □

Problem 1. Study the non-Dedekindian p-groups G in which the intersection of any two nonincident nonnormal cyclic subgroups is G-invariant.

Problem 2. Study the non-Dedekindian p-groups G in which the intersection of any two distinct conjugate subgroups (cyclic subgroups, abelian subgroups) is G-invariant.

Problem 3. Let H be a maximal subgroup of a p-group G. Study the structure of G provided $H \cap A \lhd G$ for any $A < G$ nonincident with H.

Problem 4. Study the non-Dedekindian p-groups G in which the intersection $A \cap B \lhd G$ for any non-incident $A \lhd G$ and $B < G$. Consider in detail the case when $\exp(G) = p$.

§ 242 Non-Dedekindian p-groups in which the normal closures of all nonnormal subgroups coincide

Here we solve Problem 3296 by proving the following result.

Theorem 242.1. *Let G be a non-Dedekindian p-group. Then the normal closures of all nonnormal subgroups in G coincide if and only if G is isomorphic to a modular group M_{p^r}, $r \geq 3$, (where in case $r = 3$, we have $p > 2$).*

Proof. Let G be a non-Dedekindian p-group in which the normal closures of all nonnormal subgroups coincide with a subgroup $G_0 < G$. Then there is a maximal cyclic subgroup Z which is nonnormal in G so that $G_0 = Z^G$. By our hypothesis, each nonnormal subgroup in G is contained in G_0 and G_0 is generated by its nonnormal subgroups. By Theorem 231.1, G is of class 2. This implies that each maximal abelian subgroup is normal in G since it contains $\mathrm{Z}(G)$ and $G' \leq \mathrm{Z}(G)$. Let H be a maximal abelian subgroup of G containing Z so that H is normal in G. Then $G_0 = Z^G \leq H$ and therefore G_0 is abelian. But each nonnormal subgroup X in G is contained in G_0 and so X is abelian and therefore G is metahamiltonian (i.e., all nonabelian subgroups are normal in G).

Let $H_0 < G_0$ be such that $H_0 \trianglelefteq G$ and $|G_0 : H_0| = p$. Then our hypothesis implies that each subgroup of H_0 is normal in G. Let X be a nonnormal subgroup of G. Then $X \leq G_0$ but $X \nleq H_0$ so that $X \cap H_0 \trianglelefteq G$ and therefore $X_G = X \cap H_0$ and $|X : X_G| = p$.

By Theorem 231.1, G/G_0 is cyclic and if $g \in G - G_0$ is such that $\langle g\rangle$ covers G/G_0, then $\langle g\rangle \trianglelefteq G$ and $G/(\langle g\rangle \cap G_0)$ is abelian so that $G' \leq \langle g\rangle \cap G_0$ is cyclic. We have $(\langle g\rangle \cap G_0)Z \trianglelefteq G$ and so $G_0 = Z^G \leq (\langle g\rangle \cap G_0)Z$ implying $G_0 = (\langle g\rangle \cap G_0)Z$ and $G = \langle g\rangle Z$ is metacyclic.

Let G_1 be a maximal subgroup of G containing G_0. Since G is nonabelian and G is generated by its minimal nonabelian subgroups, there is a minimal nonabelian subgroup $K \nleq G_1$. Then for each $g \in K - G_1$, $\langle g\rangle$ covers G/G_0 and so K covers G/G_0. It is easy to see that K is splitting metacyclic. Indeed, if this is not the case, then Lemma 65.1 implies that $p = 2$ and $K \cong \mathrm{Q}_8$. If in this case $\mathrm{C}_G(K) \nleq K$, then $K\mathrm{C}_G(K)$ is nonmetacyclic, a contradiction. Thus $\mathrm{C}_G(K) \leq K$ and then Theorem 10.17 gives that G is a 2-group of maximal class. But G is of class 2 and so $G = K \cong \mathrm{Q}_8$, contrary to our hypothesis that G is not Dedekindian. We have proved that K is splitting metacyclic. We may set

$$K = \langle k\rangle\langle l\rangle \quad \text{with } \langle k\rangle \trianglelefteq K,\ \langle k\rangle \cap \langle l\rangle = \{1\},\ \text{and } \langle [k, l]\rangle = \Omega_1(\langle k\rangle) = K'\,.$$

Since $\langle l\rangle$ is nonnormal in K, we get $\langle l\rangle \leq G_0$. But K covers G/G_0 and so $\langle k\rangle$ covers G/G_0. On the other hand,

$$G_0 = \langle l\rangle^G = \langle l\rangle^{G_0\langle k\rangle} = \langle l\rangle^{\langle k\rangle} \text{ and so } G = G_0\langle k\rangle = \langle l\rangle^{\langle k\rangle}\langle k\rangle = \langle l\rangle\langle k\rangle = K\,.$$

We know that $\langle l^p \rangle \trianglelefteq G$ (since, by the above, $|\langle l \rangle : \langle l \rangle_G| = p$) and so $\langle l^p \rangle$ centralizes $\langle k \rangle$. We get

$$[k, l]^p = [k, l^p] = 1 \text{ and so } K' = G' = \langle [k, l] \rangle \cong C_p .$$

By Lemma 65 (a), $G = K$ is minimal nonabelian. We may set

$$G = \langle k, l \mid k^{p^{m+1}} = l^{p^n} = 1, \quad [k, l] = k^{p^m} = z \rangle , \quad \text{where } m \geq 1 , \quad n \geq 1 .$$

We note that $\langle l \rangle$ is nonnormal in G and so $l \in G_0$ and $G_0 = \langle l \rangle^G = \langle l \rangle \times \langle z \rangle$. If $\langle kl \rangle$ is nonnormal in G, then $kl \in G_0$. But then $G = \langle k, l \rangle = G_0$, a contradiction. Hence we must have $\langle kl \rangle \trianglelefteq G$. We compute

$$(kl)^l = k^l l = (kl)z \text{ and so } z \in \langle kl \rangle \text{ and } (kl)^{p^m} = k^{p^m} l^{p^m} [l, k]^{\binom{p^m}{2}} = z^{1-\binom{p^m}{2}} l^{p^m} .$$

Also,

$$(kl)^{p^{m+1}} = k^{p^{m+1}} l^{p^{m+1}} [l, k]^{\binom{p^{m+1}}{2}} = l^{p^{m+1}}$$

and so if $l^{p^{m+1}} \neq 1$, then $z \notin \langle kl \rangle$, a contradiction. Hence $l^{p^{m+1}} = 1$ and so $m + 1 \geq n$. But if $m + 1 = n$, then

$$(kl)^{p^m} = z^{1-\binom{p^m}{2}} l^{p^{n-1}}$$

is an element of order p in $G - \langle k \rangle$ implying $z \notin \langle kl \rangle$, a contradiction. Hence

$$m + 1 > n \text{ and } (kl)^{p^m} = z^{1-\binom{p^m}{2}} .$$

It follows that $z \in \langle kl \rangle$ unless $p = 2$, $m = 1$, and $n = 1$, in which case $G \cong D_8$, a contradiction. Indeed, D_8 has two conjugate classes of nonnormal subgroups of order 2 and they generate D_8.

Suppose that $n > 1$. Let k' be an element of order p^n in $\langle k \rangle$. Since $n < m + 1$, $\langle k' \rangle \leq Z(G)$ and we consider the subgroup $\langle k' l \rangle \cong C_{p^n}$. We have

$$[k' l, k] = [k', k][l, k] = [l, k] = z^{-1} ,$$

and so $\langle k' l \rangle$ is nonnormal in G because

$$\Omega_1(\langle k' l \rangle) \nleq \langle k \rangle \text{ and so } \Omega_1(\langle k' l \rangle) \neq \langle z \rangle .$$

Hence we get $\langle k' l \rangle \leq G_0$. On the other hand,

$$G_0 = \langle l \rangle^G = \langle l \rangle \times \langle z \rangle \text{ and so } \langle k' l \rangle \nleq G_0 ,$$

a contradiction. It follows that $n = 1$ and therefore $G \cong M_{p^r}$, $r \geq 3$, where in case $r = 3$, we have $p > 2$.

Conversely, all the above modular p-groups G satisfy the assumptions of our theorem since in this case $G_0 = \Omega_1(G) \cong E_{p^2}$ and all subgroups of order $> p$ are normal in G. Our theorem is proved. □

Problem 1. Classify the non-Dedekindian p-groups in which the normal closures of all nonnormal abelian subgroups coincide.

Problem 2. Study the non-Dedekindian p-groups in which the normal closures of all nonnormal minimal nonabelian subgroups have the same order.

Problem 3. Study the non-Dedekindian p-groups in which the normal closures of all nonnormal cyclic subgroups are either abelian or $\mathcal{A}_k$-groups, $k \leq 2$.

§ 243 Nonabelian p-groups G with $\Phi(H) = H'$ for all nonabelian $H \le G$

Here we initiate a study of Problem 2775 by proving the following result.

Theorem 243.1. *Let G be a nonabelian p-group in which $\Phi(H) = H'$ for all nonabelian $H \le G$. If A is a maximal normal abelian subgroup in G, then $\exp(G/A) = p$, each element in $G - A$ is of order $\le p^2$ and each minimal nonabelian subgroup of G is of order p^3. If $p = 2$ and $\exp(A) > 4$, then $|G : A| = 2$.*

Proof. Let G be a title p-group. Then our hypothesis implies that G/G' is elementary abelian. If H is a minimal nonabelian subgroup in G, then $\Phi(H) = H' \cong C_p$ together with $d(H) = 2$ implies $|H| = p^3$.

Let A be a maximal normal abelian subgroup in G and let $g \in G-A$. By Lemma 57.1, there is $a \in A$ such that $\langle g, a\rangle$ is minimal nonabelian. We have

$$\langle [g, a]\rangle \cong C_p, \ [g, a] \in A, \ \text{and } \Phi(\langle g, a\rangle) = \langle g, a\rangle' = \langle [g, a]\rangle .$$

This implies $g^p \in \langle [g, a]\rangle \le A$ and so $\exp(G/A) = p$ and each element in $G - A$ is of order $\le p^2$.

Suppose in addition that $p = 2$. If $\exp(A) > 4$, then Lemma 57.2 gives $|G : A| = 2$ and our theorem is proved. □

§ 244 p-groups in which any two distinct maximal nonnormal subgroups intersect in a subgroup of order $\leq p$

Here we initiate a study of a problem posed by the first author by proving the following result.

Theorem 244.1. *Let G be a non-Dedekindian p-group in which any two distinct maximal nonnormal subgroups which are conjugate in G intersect in a subgroup of order $\leq p$. Then all subgroups of order $\geq p^3$ are normal in G and such groups are classified in Theorems 229.1 and 229.3.*

Proof. Let G be a title p-group. Let M be a maximal nonnormal subgroup in G and let $M < H$ with $|H : M| = p$. Then $H \triangleleft G$. Let $x \in G$ be such that $M \neq M^x$. It follows that $H = MM^x$. By our hypothesis,

$$|M \cap M^x| \leq p \textit{ and so } |M| = |M^x| \leq p^2 .$$

We have proved that all subgroups of order $\geq p^3$ are normal in G and such groups are classified in Theorems 229.1 and 229.3. □

§ 245 On 2-groups saturated by nonabelian Dedekindian subgroups

1^0 *Introduction.* In what follows G is a prime power group, as a rule, a 2-group. Next, m, n, s, k, t are positive integers, p is a prime (as a rule, $p = 2$).

A group is said to be Dedekindian if all its subgroups are normal. Abelian groups are Dedekindian. By $\delta(G)$ we denote the number of nonabelian Dedekindian members in the set Γ_1 of maximal subgroups of a group G.

The order structure of a p-group G is given by the string $(\mathrm{N}_1(G), \mathrm{N}_2(G), \dots)$, where $\mathrm{N}_i(G)$ is the number of elements of order p^i in G. For example, the order structure of D_8 is $(5, 2)$. If $\exp(G) = p^e$, then $|G| = 1 + \sum_{i=1}^{e} \mathrm{N}_i(G)$.

Definition 1. Given n, a 2-group G is said to be a $\mathcal{D}_n$-group, if it satisfies the following conditions:

($1\mathcal{D}_n$) All subgroups of index 2^n in G are Dedekindian.
($2\mathcal{D}_n$) There is in G a non-Dedekindian subgroup of index 2^{n-1}.
($3\mathcal{D}_n$) G is not an $\mathcal{A}_n$-group.[1]

Condition ($3\mathcal{D}_n$) in the above definition can be replaced by the following equivalent condition:
($3'\mathcal{D}_n$) There is in G a nonabelian Dedekindian subgroup of index 2^n.

In particular, a 2-group G is a $\mathcal{D}_1$-group $\iff$ it is non-Dedekindian and non-$\mathcal{A}_1$-group but all its maximal subgroups are Dedekindian. Any maximal subgroup of a $\mathcal{D}_1$-group is either abelian or nonabelian Dedekindian. Thus, $\mathcal{D}_1$-groups are those minimal non-Dedekindian 2-groups that are not $\mathcal{A}_1$-groups (of course, $\mathcal{A}_1$-groups $\not\cong$ Q_8 are minimal non-Dedekindian).

It follows from ($2\mathcal{D}_n$) that if $H \le G$ is non-Dedekindian, then $|G : H| \le 2^{n-1}$. If a 2-group satisfies ($1\mathcal{D}_n$) and ($2\mathcal{D}_n$), then it is either $\mathcal{D}_n$- or $\mathcal{A}_n$-group.

If a 2-group G is a $\mathcal{D}_n$-group and C is of order 2, then $G \times C$ is a $\mathcal{D}_{n+1}$-group. For any $m > 2$, there is an $\mathcal{A}_1$-group of order 2^m (see Lemma 65.1). As Theorem 245.A shows, there is only one $\mathcal{D}_1$-group, namely Q_{2^4}, the generalized quaternion group of order 2^4. It follows that the set of 2-groups that are $\mathcal{A}_n$-groups is infinite. By the above, the set of $\mathcal{D}_n$-groups is finite for any n. By Theorem 245.B, a $\mathcal{D}_n$-group is an $\mathcal{A}_k$-group, where $k \in \{n + 1, n + 2\}$, but the converse is not true (this is not obvious from the start).

If a p-group G is an $\mathcal{A}_n$-group, then $|G| \ge p^{n+2}$. This is the best possible estimate since any p-group of maximal class of order p^{n+2} is an $\mathcal{A}_n$-group since it contains a nonabelian subgroup of order p^3 (Theorem 9.6 (f)). If G is a $\mathcal{D}_n$-group, then $|G| \ge 2^{n+3}$;

1 Recall that a p-group is said to be an $\mathcal{A}_n$-group if all its subgroups of index p^n are abelian but it contains a nonabelian subgroup of index p^{n-1}.

the generalized quaternion group of order 2^{n+3} is a $\mathcal{D}_n$-group, and this estimate is best possible (indeed, $Q_{2^{n+3}}$ is a $\mathcal{D}_n$-group).

In general, non-Dedekindian proper epimorphic images of $\mathcal{D}_n$-groups need not necessarily are $\mathcal{D}_s$-groups for all s. For example, if $G = Q_{2^{n+3}}$, then $G/Z(G)$ is not a $\mathcal{D}_s$-group for all s (see Lemma J(d), below). Note that any nonabelian p-group is an $\mathcal{A}_n$-group for an appropriate n.

If G is an $\mathcal{A}_n$-group, $n > 1$, then H is an $\mathcal{A}_{n-1}$-group for some $H \in \Gamma_1$. It is not true that a $\mathcal{D}_n$-group, $n > 1$, contains a proper $\mathcal{D}_s$-subgroup of index 2 for some s. As follows from Theorems 245.A and 245.B if a $\mathcal{D}_s$-group H is a maximal subgroup of a $\mathcal{D}_n$-group, $|G| = 2^{n+3}$, $n > 1$, then $s = n - 1$.

The $\mathcal{A}_n$-groups are classified for $n \leq 3$ (see Lemma 65.1, § 71, and [ZZLS]). A classification of so narrow class of $\mathcal{D}_3$-groups without tables of groups of order 2^7 is fairly difficult.

In 3^0 the $\mathcal{D}_2$-groups are investigated in some detail. In 4^0 a number of related problems is stated.

Some known results which are used in what follows are collected in the following lemma.

Lemma J. *Let G be a nonabelian 2-group.*

(a) *(Lemma 65.1) If G is an $\mathcal{A}_1$-group, then $d(G) = 2$, $Z(G) = \Phi(G)$, $|\Omega_1(G)| \leq 2^3$ and, if $|\Omega_1(G)| < 2^3$, then G is metacyclic. If G is metacyclic and has no cyclic subgroup of index 2, then $Z(G)$ is noncyclic since it contains $\Omega_1(G) \in \{E_4, E_8\}$.*

(b) *(Theorem 1.20) If G is a nonabelian Dedekindian group, then $G = Q \times E$, where $Q \cong Q_8$ and $\exp(E) \leq 2$. If $A < G$ is nonidentity abelian of exponent > 2, then $|A : \Omega_1(A)| = 2$. The set Γ_1 has exactly three abelian members.*

(c) *(Theorem 10.28) The group G is generated by $\mathcal{A}_1$-subgroups.*

(d) *If $G > \{1\}$ has no subgroup $\cong D_8$, then $\exp(\Omega_1(G)) = 2$.*

(e) *(Lemma 57.1) Let $A < G$ be a maximal normal abelian subgroup. Then for any $x \in G - A$ there is $a \in A$ such that $\langle a, x\rangle$ is an $\mathcal{A}_1$-subgroup.*

(f) *Any $\mathcal{D}_n$-group contains a subgroup $\cong Q_8$.*

(g) *(Exercise 1.6) The number of abelian members in the set Γ_1 is either 0, or 1, or 3.*

(h) *(Lemma 1.1.) If $A \in \Gamma_1$ is abelian, then $|G| = 2|G'||Z(G)|$. (The similar equality holds for arbitrary nonabelian prime power groups.)*

(i) *(Schreier; see [Suz1, p. 185] Let $H < G$ be of index j and $d(G) = d$. Then $d(H) \leq 1 + j(d - 1)$. In particular, if $d = j = 2$, then $d(H) \leq 3$. (This is true for arbitrary groups.)*

(j) *(Corollary A.17.3, Theorem 90.1) If all $\mathcal{A}_1$-subgroups of G are $\cong Q_8$, then $G = Q \times E$, where $Q \cong Q_{2^n}$ and $\exp(E) \leq 2$.*

(k) *All maximal abelian subgroups of an extraspecial 2-group G of order 2^{2m+1} have order 2^{m+1}.*

(l) *An extraspecial group of order 2^{2m+1}, $m > 1$, has a subgroup $\cong Q_8$, $\Omega_1(G) = G$, $\Omega_2^*(G) = G$.*

(m) *(Lemma 65.2 (a)) If $G' \le \Omega_1(Z(G))$ and $d(G) = 2$, then G is an $\mathcal{A}_1$-group.*
(n) *(Proposition 1.6) If a 2-group G is not of maximal class, then $|G : G'| > 4$.*
(o) *(Lemma 4.3) If $|G'| = 2$ and $A \le G$ is minimal nonabelian, then $G = AC_G(A)$. In particular, if G is extraspecial, then $G = S_1 * \cdots * S_n$, where S are nonabelian of order 8.*
(p) *Let $G = Q_8 * C_4$ be of order 16. Then $G = D_8 * C_4$, G has only one subgroup $\cong Q_8$ and exactly three subgroups $\cong D_8$. Next, $c_1(G) = 7$ and $c_2(G) = 4$, $\Omega_1(G) = G$.*

Most results listed in Lemma J have analogs for nonabelian p-groups with $p > 2$.

Exercise 1. Prove Lemma J(d).

Proof of Lemma J(f). Assume that a $\mathcal{D}_n$-group G has no subgroup $\cong Q_8$. Then all subgroups of index 2^n in G are abelian (Lemma J(b)). By ($2\mathcal{D}_n$), G has a non-Dedekindian so nonabelian subgroup of index 2^{n-1}, and we conclude that G is an $\mathcal{A}_n$-group, contrary to ($3\mathcal{D}_n$). □

Proof of Lemma J(j). We deduce this result from Theorem 90.1. By that theorem, if all minimal nonabelian subgroups of a 2-group G are isomorphic to Q_8, then $G = HZ(G)$, where $\mho_1(Z(G)) \le Z(H)$ and either $H \cong Q_{2^n}$ or extraspecial. (i) Let $H \cong Q_{2^n}$ and let $C \le Z(G)$ be cyclic of order 4. Let $Q_8 \cong Q \le H$. Then $Q * C$ (of order 16) contains a subgroup $\cong D_8$ (Lemma J(p)), a contradiction. Thus, $\exp(Z(G)) = 2$. In that case, $Z(G) = Z(H) \times E$ and hence $G = H \times E$. (ii) Now let H be extraspecial. We claim that then $H \cong Q_8$. Let $A \le H$ be an $\mathcal{A}_1$-subgroup and $C < Z(G)$ be cyclic of order 4. As $A' = H'$ and $\exp(H/H') = 2$, it follows that $|A| = 8$. By hypothesis, $A \cong Q_8$. As above, $A * C$ contains a subgroup $\cong D_8$, contrary to the hypothesis. Thus, $\exp(Z(G)) = 2$. It remains to show that $|H| = 8$. Assume that this is false. By Lemma J(o), $G = A * C_G(A)$. Assume that $|C_G(A)'| = 2$. Let $B \le C_G(A)$ be minimal nonabelian and $Z < A$ cyclic of order 4. As above, $B * Z$ has a subgroup $\cong D_8$, a contradiction. Thus, $C_G(A) = Z(G)$ is of exponent 2, and, as above, G is Dedekindian. □

Exercise 2. If G is extraspecial group of order 2^{2m+1}, $m > 1$, then $\Omega_1(G) = G$. (**Hint.** Mimic, with small changes, the proof of Lemma J(j).)

Proof of Lemma J(l). Assume that G has no subgroup $\cong Q_8$. Then, by Lemma J(o), $G = D_1 * \cdots * D_m$, where all $D_i \cong D_8$, $m > 1$. Let $H = D_1 * C$, where $C \cong C_4$ is a subgroup of D_2. Then H has a subgroup $\cong Q_8$ (Appendix 16). Let us prove that $\Omega_1(G) = G$. Let again G admits the above factorization with all $|D_i| = 8$. One may assume that $D_1 \cong Q_8$. Let $C < D_2$ be cyclic of order 4. Then $D_1 * C = U * C$, where $U \cong D_8$, and $\Omega_1(U * C) = U * C = D_1 * C$ so that $D_1 \le \Omega_1(G)$. Thus, all factors $\cong Q_8$ are contained in $\Omega_1(G)$. It follows that $\Omega_1(G) = G$. If $D_2 \cong D_8$ and $Z < D_1$ is cyclic of order 4, then $D_2 * Z = V * Z$, where $V \cong Q_8$, and hence $\Omega_2^*(V * Z) = V * Z = D_2 * Z$. Thus, all factors $\cong D_8$ are contained in $\Omega_2^*(G)$ so that $\Omega_2^*(G) = G$. □

In this section the following three theorems are proved (Theorems 245.B and 245.C are due to the first author).

Theorem 245.A (G. A. Miller [Mil9]). *If G is a $\mathcal{D}_1$-group, then $G \cong Q_{2^4}$.*

Theorem 245.B. *If G is a $\mathcal{D}_n$-group, then $G \in \{2^{n+3}, 2^{n+4}\}$.*[2]

Theorem 245.C. *If G is a $\mathcal{D}_2$-group, then $|G| = 2^5$.*

We have not read [Mil9] and believe that Miller has proved Theorem 245.A correctly.

2⁰ Proofs of Theorems 245.A–245.C. Our theorems will follow from a number of auxiliary results.

Lemma 245.1. *Suppose that a nonabelian p-group G has two distinct abelian subgroups E_1 and E_2 of index p. Then $G = SZ(G)$, where S is minimal nonabelian. If, in addition, $\exp(E_i) = p$, $i = 1, 2$, then $G = S \times E$, where $\exp(E) \le p$ and one of the following holds:*
(a) *If $p = 2$, then $S \cong D_8$*
(b) *If $p > 2$, then $G \cong S(p^3)$.*

Proof. One has $E_1 \cap E_2 = Z(G)$ hence $|G : Z(G)| = p^2$. Let $S \le G$ be minimal nonabelian. As $|S : (S \cap Z(G))| = p^2 = |S : Z(S)|$, it follows that $S \cap Z(G) = Z(S)$ and $G = SZ(G)$, by the product formula. Let, in addition, $\exp(E_i) = p$, $i = 1, 2$. In that case, $S \cap E_1$ and $S \cap E_2$ be two distinct elementary abelian subgroups of index p in S. It follows from Lemma 65.1 that $S \in \{D_8, S(p^3)\}$. In that case, $S \cap Z(G) = Z(S)$ and $Z(G) = Z(S) \times E$, where $\exp(E) \le p$ since $\exp(Z(G)) = p$. In that case, $G = S \times E$. □

We assume in what follows that $p = 2$.

Lemma 245.2 (= Lemma 171.1). *If a 2-group G has three pairwise distinct elementary abelian subgroups U, V, W of index 2, then* $\exp(G) = 2$.

Proof. Assume that G is nonabelian. Then $T = U \cap V = V \cap W = W \cap U = Z(G)$ and $G/T \cong E_4$. It follows that $U \cup V \cup W = G$, and therefore $\exp(G) = 2$. □

Lemma 245.3. *Suppose that a 2-group G contains two distinct nonabelian Dedekindian subgroups B_1 and B_2 of index 2 (i.e., $\delta(G) \ge 2$). Then $B_1' = B_2'$ and G/B_1' is either elementary abelian or $\cong D_8 \times E$, where $\exp(E) \le 2$.*

Proof. Assume that $B_1' \ne B_2'$. One has $B_1' = \Phi(B_1) \le \Phi(G) < B_2$, and so B_2/B_1' is a nonabelian Dedekindian subgroup of index 2 in G/B_1' (note that $B_1' \triangleleft G$) and $\exp(B_1/B_1') = 2$. Therefore the subgroup $(B_1/B_1') \cap (B_2/B_1')$ is elementary abelian of

2 If, in addition, the group G from Theorem 245.B has no $\mathcal{A}_1$-subgroup of order 2^5, then $|G| = 2^{n+3}$. This follows from $(2\mathcal{D}_2)$ immediately.

index 2 in a nonabelian Dedekindian group B_2/B_1', which is impossible (see Lemma J(b)). Thus, $B_1' = B_2'$, and so the group G/B_1' has two distinct elementary abelian subgroups B_1/B_1' and B_2/B_1' of index 2. Now the result follows from Lemma 245.1 applied to G/B_1'. □

Lemma 245.4. *Suppose that a 2-group G contains three pairwise distinct nonabelian Dedekindian subgroups A, B, C of index 2. If G has no subgroup $\cong \mathrm{D}_8$, then it is Dedekindian.*

Proof. By Lemma 245.3, $A' = B' = C'(\triangleleft\, G)$ so G/A' has three distinct elementary abelian subgroups A/A', B/A' and C/A' of index 2. Then, by Lemma 245.2, G/A' is itself elementary abelian hence $A' = G'$. If $M < G$ is an $\mathcal{A}_1$-subgroup, then $M' = G'$ and $\exp(M/G') = \exp(G/G') = 2$, and we conclude that $|M| = 8$ (Lemma J(a)). By hypothesis, $M \cong \mathrm{Q}_8$ since D_8 and Q_8 are unique nonabelian groups of order 8. By Lemma J(o), $G = (B_1 * \cdots * B_s)\mathrm{Z}(G)$, where $B_i \cong \mathrm{Q}_8$ for all i. All cyclic subgroups of G of order 4 contain G' since $\exp(G/G') = 2$. By Lemma J(p), G has no subgroup $\cong \mathrm{Q}_8 * \mathrm{C}_4$ of order 16 since this group contains a subgroup $\cong \mathrm{D}_8$. It follows that $s = 1$ and $\exp(\mathrm{Z}(G)) = 2$. If $\mathrm{Z}(G) = \mathrm{Z}(B_1) \times E$, where $E < \mathrm{Z}(G)$ is of index 2. Then $G = B_1 \times E$ is Dedekindian since $B_1 \cong \mathrm{Q}_8$, by hypothesis. □

Remark 1. Let us find $\delta(G)$, the number of nonabelian Dedekindian members in the set Γ_1, for the group $G = Q * D$ of order 2^5, where $Q \cong \mathrm{Q}_8$ and $D \cong \mathrm{D}_8$. The group G is extraspecial so it has only one nonlinear irreducible character, say χ, and $\chi(1) = 4(=\sqrt{|G| - |G/G'|})$. By the Frobenius–Schur formula for the number of involutions (see [BZ, § 4.6]), one has $\mathrm{c}_1(G) = |G/G'| \pm \chi(1) - 1 = 15 \pm 4 \in \{11, 19\}$. Let $D = \langle a, b\rangle$, where a, b are two noncommuting involutions. Then $\mathrm{c}_2(Q \times \langle a\rangle) = 6 = \mathrm{c}_2(Q \times \langle b\rangle)$ so that

$$\mathrm{c}_2(G) \ge \mathrm{c}_2(Q \times \langle a\rangle) + \mathrm{c}_2(Q \times \langle b\rangle) - \mathrm{c}_2(Q) = 6 + 6 - 3 = 9\,.$$

Thus, G has at least $2 \cdot 9$ elements of order 4 so that $\mathrm{c}_1(G) \le |G| - 18 - 1 = 13$, and we conclude that $\mathrm{c}_1(G) = 11$. Clearly, G has no subgroup $\cong \mathrm{E}_8$ so it has $\frac{11-1}{2} = 5$ subgroups $\cong \mathrm{E}_4$. Any $\cong \mathrm{E}_4$ subgroup of G is normal.[3] If $\mathrm{E}_4 \cong U < G$, then $U \triangleleft G$ and so $|G : \mathrm{C}_G(U)| = 2$ since $|\mathrm{Z}(G)| = 2$. By Lemma J(h), G has no abelian subgroup of index 2 so that $\mathrm{C}_G(U)$ is nonabelian. Next, $\mathrm{C}_G(U)$ is not minimal nonabelian since it is not generated by two elements (recall that $\mathrm{Z}(G) = 4$). Thus, $\mathrm{C}_G(U)$ has a proper minimal nonabelian subgroup, say A. If $L \nleq A$ is of order 2 and $L < \mathrm{Z}(\mathrm{C}_G(U))$, then $\mathrm{C}_G(U) = A \times L$, by the above, has no subgroup $\cong \mathrm{E}_8$, and this implies that $A \cong \mathrm{Q}_8$. Thus, $\mathrm{C}_G(U)$ is Dedekindian. It follows that $\delta(G) = 5$ (= the number of subgroups $\cong \mathrm{E}_4$ in G). It will be shown in the sequel that if G is a 2-group, then $\delta(G) \in \{0, 1, 2, 5\}$.

Exercise 3. Let $G = S * T$ be extraspecial of order 2^5, where $S \cong \mathrm{Q}_8 \cong T$. Then $\delta(G) = 0$.

---

3 Indeed, $\mathrm{Z}(G) = G'$ has order 2 and, if $G > R \cong \mathrm{E}_4$, then, since G has no subgroup $\cong \mathrm{E}_8$, one has $G'R = R \triangleleft G$.

Solution. One has $G = S_1 * T_1$, where $S_1 \cong D_8 \cong T_1$ (see Appendix 16). It follows from the second factorization that G has a subgroup $E \cong E_8$. Assume that there is in G a subgroup $H = Q \times C$, where $C \cong C_2$. Then $HE = G$ implies $H \cap E = Z(H)$ and so $C_G(H \cap E) \geq HE = G$, i.e., $|Z(G)| > 2$, a contradiction. Thus, $\delta(G) = 0$.

Proof of Theorem 245.A. As G has no subgroup $\cong D_8$, then $\delta(G) < 3$ (Lemma 245.4). By Lemma J(f), G has a subgroup $\cong Q_8$ so that $\delta(G) \geq 1$ since all nonabelian members of the set Γ_1 are Dedekindian. It follows from Lemma J(g), that $\delta(G) = 2$ since the set Γ_1 has at least two nonabelian members; then $d(G) = 2$ since $|\Gamma_1| \leq \delta(G) + 3 = 5$ (recall that a nonabelian 2-group has at most three abelian subgroups of index 2 and $|\Gamma_1| = 1 + 2 + \cdots + 2^{d(G)-1}$).

Let $B_1, B_2 \in \Gamma_1$ be distinct nonabelian Dedekindian. By Lemma 245.3, $B_1' = B_2' (\triangleleft\, G)$ is of order 2 and G/B_1' is either elementary abelian or $\cong D \times E$, where $D \cong D_8$ and $\exp(E) \leq 2$. As $d(G) = 2$, in the first case we have $|G| = 8$ so G is an $\mathcal{A}_1$-group, contrary to $(3\mathcal{D}_2)$. In the second case, $G/B_1' \cong D_8$ so that $|G : G'| = 4$, and hence G is of maximal class, by Lemma J(n). In that case, $G \cong Q_{16}$ since $\exp(\Omega_1(G)) = 2$, by Lemma J(d). □

Corollary 245.5. *An arbitrary group G is Dedekindian $\iff$ all its two-generator subgroups are Dedekindian.*

Proof. All Dedekindian groups are nilpotent. Assume that G is minimal counterexample. Then G is minimal non-Dedekindian.

Let G be nonnilpotent; then it is minimal nonnilpotent. In that case, G is two-generator so Dedekindian hence nilpotent, a contradiction.

Now let G be nilpotent. Then $G = Q \times S$, where $Q \in \mathrm{Syl}_2(G)$ is nonabelian and S is abelian. In that case, if $S > \{1\}$, then Q is Dedekindian, by hypothesis, so G is Dedekindian, contrary to the assumption. Thus, $S = \{1\}$ so G is a $\mathcal{D}_1$-group. By Theorem 245.A, $G \cong Q_{16}$ is two-generator and non-Dedekindian, a contradiction. □

Remark 2. Let G be nonnilpotent minimal non-Dedekindian group, which is not minimal nonabelian. Then $G = PQ$, where $P \cong \mathrm{Syl}_p(G)$ is cyclic and $Q = G' \in \mathrm{Syl}_q(G)$ (Appendix 22). As G is not minimal nonabelian, Q is nonabelian Dedekindian. It follows that a special subgroup $Q \cong Q_8$. As $\mathrm{Aut}(Q_8) \cong S_4$ is an $\{2, 3\}$-group, it follows that $p = 3$. Any such group is minimal non-Dedekindian. It is asserted in [Mil9] that nonnilpotent minimal non-Dedekindian group which is not minimal nonabelian, is $\cong SL(2, 3)$. This is a noncorrect assertion as the above result shows.

Exercise 4 (this is a consequence of Lemma J(j)). Suppose that all minimal nonabelian subgroups of a nonabelian 2-group G are $\cong Q_8$. Prove that $\Omega_1(G) \leq Z(G)$.

Solution. By Lemma J(d), $\exp(\Omega_1(G)) = 2$. Assume that $\Omega_1(G) \nleq Z(G)$. As G is generated by its minimal nonabelian subgroups (Lemma J(c)) which are $\cong Q_8$, one has $\Omega_2^*(G) = G$ and there is in G a minimal nonabelian subgroup $Q (\cong Q_8)$ that does not centralize $\Omega_1(G)$. In that case, there is $C = \langle a \rangle < Q$ of order 4 that does not centralize

$\Omega_1(G)$. In that case, the subgroup $H = C\Omega_1(G)$ is nonabelian. Note that $|H| = 2|\Omega_1(G)|$ and $\exp(H) = 4$. Then, by Lemma J(e), there is $x \in \Omega_1(G)$ such that $U = \langle a, x\rangle$ is an $\mathcal{A}_1$-subgroup. As $c_1(U) > 1$, we get $U \ncong Q_8$, contrary to the hypothesis. Thus, $\Omega_1(G) \le Z(\Omega_2(G))$. As, by the above, $\Omega_2(G) = G$ (Lemma J(c)), we are done.

Exercise 5. If $\mathcal{D}_n$-group has order $> 2^{n+3}$, then $\Omega_1(G) \le Z(\Omega_2(G))$.

Solution. Assume that the assertion is false. Then there is $C = \langle x\rangle < G$ of order 4 such that the subgroup $U = \langle x, \Omega_1(G)\rangle$ is nonabelian (note that $|U : \Omega_1(G)| = 2$). By Lemma J(d), $\exp(\Omega_1(G)) = 2$. By Lemma J(e), there is $a \in \Omega_1(G)$ such that $V = \langle a, x\rangle$ is an $\mathcal{A}_1$-subgroup. As the abelian group V/V' is generated by elements of orders 2 and ≤ 4 (note that $a \notin \langle x\rangle$), we get $|V/V'| \le 8$ so that $|V| = 16$. By Lemma J(b), the subgroup V is non-Dedekindian so that, by $(2\mathcal{D}_n)$, $|G| = 2^{n-1} \cdot 2^4 = 2^{n+3}$, contrary to the hypothesis.

Proof of Theorem 245.B. We proceed by induction on $|G|$. If $n = 1$, then $G \cong Q_{16}$ (Theorem A), so the result holds since $|G| = 2^{1+3} < 2^{1+4}$. Now let $n > 1$. Then, by $(2\mathcal{D}_n)$, there is a non-Dedekindian subgroup M of index 2^{n-1}. All maximal subgroups of M are Dedekindian so, by Theorem 245.A, either $M \cong Q_{2^4}$ or M is an $\mathcal{A}_1$-subgroup. In the first case, $|G| = 2^{n-1} \cdot 2^4 = 2^{n+3}$. Next we assume that M is an $\mathcal{A}_1$-subgroup. Then, by $(2\mathcal{D}_n)$, $|G : M| \le 2^{n-1}$. Therefore, if $|M| \le 2^5$, then $|G| \le 2^{n-1} \cdot 2^5 = 2^{n+4}$ and we are done. Next we assume that M is an $\mathcal{A}_1$-subgroup of order $\ge 2^6$.

By Lemma J(f), there is in G a subgroup $Q \cong Q_8$. Let $\mathcal{M}$ be the set of those non-Dedekindian subgroups of G that contain a subgroup $\cong Q_8$. As $G \in \mathcal{M}$, the set $\mathcal{M}$ is nonempty. If $K \in \mathcal{M}$, then $|G| \le 2^{n-1}|K|$ so that, if $|K| < 2^6$, then $|G| \le 2^{n+4}$, and we are done. Next we assume that any member of the set $\mathcal{M}$ has order $\ge 2^6$.

Let $H \in \mathcal{M}$ be of minimal possible order. If $Q_8 \cong Q < H$ and $Q < D < H$ (recall that $|H : Q| \ge 2^3$), where $|H : D| = 2$, then D is nonabelian Dedekindian, by the choice of H. If all minimal nonabelian subgroups of H are $\cong Q_8$, then, by Lemma J(j), $H = T \times E$, where $T \cong Q_{2^t}$ and $\exp(E) \le 2$ ($t > 3$ since H is non-Dedekindian). In fact, $t = 4$, since D is nonabelian Dedekindian of index 2 in H, and, since T is non-Dedekindian, one obtains $E = \{1\}$. Then $|G| = 2^{n-1}|T| = 2^{n+3}$, and we are done in this case. Now assume that H has a minimal nonabelian subgroup $B \ncong Q_8$. Assume, by way of contradiction, that $|G| > 2^{n+4}$. Then B, being non-Dedekindin (Lemma J(b)), has order $\ge 2^6$ (see the last sentence in the first paragraph of the proof). In that case, $|H| \ge 2|B| \ge 2^7$ (one has $B < H$ since B has no subgroup $\cong Q_8$) and $|B| = \frac{1}{2}|H| \ge 2^6$. One has (Lemma J(b)) $D = Q \times E$, where $Q \cong Q_8$ and $E \cong E_{2^k}$, $k \ge 3$ (see the first sentence of this paragraph). Then $d(D) = d(Q) + k \ge 5$. Note that $D' \lhd H$ and $D' = \Phi(D) \le \Phi(H)$. Set $U = B \cap D$; then the subgroup U is abelian of index 2 in B, by the product formula. As any abelian subgroup of D has an elementary abelian subgroup of index 2, then the subgroup U of order $\ge 2^5$ has an elementary abelian subgroup of order 2^4, which is, in the same time, a subgroup of the $\mathcal{A}_1$-subgroup B. Thus, $|\Omega_1(B)| \ge 2^4$, contrary to Lemma J(b). The proof is complete. □

Exercise 6. If $G = Q \times C$, where $Q \cong Q_8$ and $C \cong C_4$, then $\delta(G) = 1$.

Solution. Clearly, $Q \times \Omega_1(C) \in \Gamma_1$ is nonabelian Dedekindian, so that $\delta(G) \geq 1$. Let $L < \Omega_1(G)$ be of order 2, $L \notin \{Z(Q), \Omega_1(C)\}$. Set $\bar{G} = G/L$. Then $\bar{G} = \bar{Q} * \bar{C}$ is of order 2^4, where $\bar{Q} \cong Q$ and $\bar{C} \cong C$. In that case $\bar{G}$ contains three distinct subgroups $\bar{A} \cong \bar{B} \cong \bar{C} \cong D_8$. The subgroups $A, B, C \in \Gamma_1$ are non-Dedekindian (Lemma J(b)). Next, the set Γ_1 has three pairwise distinct abelian members. As $|\Gamma_1| = 7$, it follows that $\Omega_1(C) \times Q$ is the unique nonabelian Dedekindian member of the set Γ_1, as was to be shown.

Exercise 7. If $G = Q_8 * Q_8$ is extraspecial of order 2^5, then $\delta(G) = 0$.

Solution. One has $G \cong D_8 * D_8$. Then G contains a subgroup $\cong D_8 \times C_2$ so there is in G a subgroup $E \cong E_8$. Assume that $H \in \Gamma_1$ is nonabelian Dedekindian; then $H \cap E = Z(H) \cong E_4$. In that case, $C_G(H \cap E) \geq HE = G$ so that $H \cap E \cong E_4$ is contained in $Z(G)$. As $|Z(G)| = 2$, this is a contradiction. Thus, $\delta(G) = 0$.

Suppose that X is a non-Dedekindian 2-group and E of exponent 2 and $G = X \times E$. We claim that $\delta(G) = \delta(X)$. Indeed, if $M < X$ is nonabelian Dedekindian of index 2, then $M \times E \in \Gamma_1$ is nonabelian Dedekindian, and we conclude that $\delta(G) \geq \delta(X)$. Now let $H \in \Gamma_1$ be nonabelian Dedekindian. Assume that $E \nleq H$. Then there is in E a subgroup C of order 2 that is not contained in H. Then $G = H \times C$ is Dedekindian, a contradiction since $X < G$ is non-Dedekindian.

Exercise 8. If G is an extraspecial 2-group of order $\geq 2^5$ and $\delta(G) > 0$, then $G = Q_8 * D_8$ is of order 2^5.

Solution. Let $|G| = 2^{2m+1}$; then $m > 1$. Let $D \in \Gamma_1$ be nonabelian Dedekindian. Then D of order 2^{2m} contains an abelian subgroup of order 2^{2m-1}. As all maximal abelian subgroups of G have order 2^{m+1} (Lemma J(k)), one has $2m - 1 \leq m + 1$, and so $m = 2$; then $|G| = 2^5$. By Exercise 7 and Remark 1, $G = Q_8 * D_8$ since for $X = Q_8 * Q_8 \cong D_8 * D_8$ one has $\delta(X) = 0$.

Lemma 245.6. *If G is a non-Dedekindian 2-group, then $\delta(G) \in \{0, 1, 2, 5\}$. If $\delta(G) = 5$, then $G = (D * Q) \times E$, where $D \cong D_8$, $Q \cong Q_8$ and $\exp(E) \leq 2$ (in this case, by Remark 1, $\delta(G) = 5$).*

Proof. In view of Exercise 8 and Remark 1, one may assume that G is not extraspecial. Suppose that $\delta(G) > 2$ and prove that $\delta(G) = 5$. By Lemma 245.4, $|G'| = 2$ and G/G' is elementary abelian, i.e., $G' = \Phi(G)$. We proceed by induction on $|G|$.

Suppose that $Z(G)$ is noncyclic; then $Z(G) \nleq \Phi(G)$. Therefore, there is $L < Z(G)$ of order 2 such that $L \neq G' = \Phi(G)$. In that case, $G = L \times M$ for some $M \in \Gamma_1$; then M is non-Dedekindian (otherwise, G is Dedekindian). By paragraph, preceding Exercise 8, $\delta(G) = \delta(M)$. By induction, $\delta(M) = 5$ and $M = (D_1 * Q_1) \times E_1$, where $D_1 \cong D_8$, $Q_1 \cong Q_8$ and $\exp(E_1) \leq 2$ (see also Remark 1), and we conclude that $\delta(G) = 5$. We are done in this case.

Now let $Z(G) = C$ be cyclic; then $C \cong C_4$ since $\exp(G) = 4$ and G is not extraspecial. Assume that $|G| = 16$. In that case, $G = Q * Z(G)$, where $Q \cong Q_8$. By Lemma J(p), $\delta(G) = 1$, contrary to the assumption. Thus, $|G| > 16$. If $A \in \Gamma_1$ is nonabelian Dedekindian, then $C \nleq A$ since $\exp(Z(A)) = 2$. In that case, $G = A * C$. There is in A an abelian subgroup B of index 2. Then the abelian subgroup $U = C * B \in \Gamma_1$, by the product formula. By Lemma J(h), $|G : Z(G)| = 2|G'| = 4$ so that $|G| = 4|Z(G)| = 16$, a final contradiction. □

Proof of Theorem 245.C. We have to prove that the order of a $\mathcal{D}_2$-group G is equal to 2^5. If $K < G$ is a $\mathcal{D}_1$-group, then $K \cong Q_{2^4}$ (Theorem 245.A) so that, by ($2\mathcal{D}_2$), one has $|G| = 2|K| = 2^5$, and we are done in this case. The same argument shows that $|G| = 2^5$ if G has a non-Dedekindian subgroup of order 2^4.

Assume the theorem is false. Then, by Theorem 245.B, $|G| = 2^{2+4} = 2^6$. In that case all non-Dedekindian subgroups of G have orders $> 2^4$ (see the previous paragraph). Therefore, G has no $\mathcal{D}_1$-subgroup. Let $M \in \Gamma_1$ be non-Dedekindian. Then, by ($1\mathcal{D}_2$), all maximal subgroups of M are Dedekindian. As M is not a $\mathcal{D}_1$-group, it follows that it is an $\mathcal{A}_1$-subgroup as M is nonabelian. Thus, any member of the set Γ_1 is either Dedekindian or an $\mathcal{A}_1$-subgroup. By Theorem 245.A, at least one of members of the set Γ_1 is a non-Dedekindian $\mathcal{A}_1$-subgroup (otherwise, G is a $\mathcal{D}_1$-group).

Let $H, D \in \Gamma_1$, where H is an $\mathcal{A}_1$-subgroup and D is nonabelian Dedekindian (it is important that, by ($3\mathcal{D}_n$), G is not an $\mathcal{A}_2$-group so D exists). Let $D = Q \times E$, where $Q \cong Q_8$ and $E \cong E_4$ (recall that $|D| = \frac{1}{2}|G| = 2^5$). Then $d(D) = d(Q) + d(E) = 2 + 2 = 4$; therefore, by Lemma J(i), $d(G) > 2$. It follows from $d(G) \le d(H) + 1 = 3$ that $d(G) = 3$. By Lemma J(a), $|\Phi(H)| = 2^3 = |\Phi(G)|$, and so $\Phi(G) = \Phi(H)$ since $\Phi(H) \le \Phi(G)$. Next, $\Phi(G) = \Phi(H) = Z(H)$ for any minimal nonabelian $H \in \Gamma_1$. By Lemma 245.6 and Lemma J(d), $\delta(G) < 3$. By Lemma J(g), the set Γ_1 has at most $2 + 1 = 3$ abelian members so it has at least $|\Gamma_1| - 3 - \delta(G) = 7 - 3 - 2 = 2$ minimal nonabelian members; let $F, H \in \Gamma_1$ be distinct minimal nonabelian. Then, by the above, $C_G(\Phi(G)) \ge FH = G$, so that $\Phi(G) \le Z(G)$. As $\Phi(G) \le Z(D) \cong E_8$, it follows that $\Phi(G) \cong E_8$.

One has $H' \lhd G$ and H/H' is abelian metacyclic (recall that H is an $\mathcal{A}_1$-subgroup so $d(H) = 2$, by Lemma J(a)), $D/D' \cong E_{16}$, and we conclude that $H' \ne D'$ for any minimal nonabelian $H \in \Gamma_1$. Indeed, otherwise, $H' = D'$ so $(H/H') \cap (D/H') \cong E_8$, as a maximal subgroup of $D/H' \cong E_{2^4}$, and this is a contradiction since H/H' is abelian of rank 2. Then $|G'| \ge |H' \times D'| = 4$ and therefore, by Lemma J(h), the set Γ_1 has at most one abelian member. The number of abelian and minimal nonabelian members in the set Γ_1 is $\ge |\Gamma_1| - \delta(G) - 1 \ge 7 - 2 - 1 = 4$. One has $|G : Z(G)| > 4$ since the set Γ_1 has at most one abelian member, and therefore $\Phi(G) \cong E_8 \cong Z(G)$ (note that $Z(G) \le Z(D) \cong E_8$) which implies $\Phi(G) = Z(G)$. It follows from the above that, if $M \in \Gamma_1$, then $\Omega_1(M) = \Phi(G) = Z(G) \cong E_8$ (here we use Lemma J(a,b)). By Lemma J(d), $\exp(\Omega_1(G)) = 2$. Assume that $\Omega_1(G) > Z(G)$; then $\Omega_1(G) \cong E_{16}$ (note that $\exp(G) \ge \exp(D) = 4$). If $\Omega_1(G) < A \in \Gamma_1$, then A is abelian since $|\Omega_1(A)| > 8$ (indeed, for all nonabelian $X \in \Gamma_1$ one has $|\Omega_1(X)| = 8$). As $G/\Omega_1(G) \cong E_4$ (indeed, $G/\Omega_1(G)$ is an epimorphic image of order 4 of $G/\Phi(G) \cong E_8$), any member of the set Γ_1 containing $\Omega_1(G)$, must be abelian, a contradiction since the set Γ_1, by the above, has at most one abelian member in view of that $|G'| > 2$. Thus, $\Omega_1(G) = Z(G)(\cong E_8)$. By the above, $L \in \Gamma_1 \Rightarrow |L'| \le 2$.

Assume that $A \in \Gamma_1$ is abelian. Then, by Lemma J(h), $|G'| = \frac{1}{2}|G : Z(G)| = 4$. Since $G' < \Phi(G)$, we get $G' \cong E_4$. Thus, in any case, $|G'| \in \{4, 8\}$ (see Lemma J(n)).

(i) Assume that $|G'| = 8$. Then $G' = \Phi(G) = Z(G) \cong E_8$ so that G is a special group. By the previous paragraph, G has no abelian subgroup of index 2. Let $M, N \in \Gamma_1$ be distinct and assume that $M' = N'$. Then M/M' and N/M' are distinct abelian subgroups of index 2 in a nonabelian group G/M' of order 2^5. In that case, $|(G/M') : Z(G/M')| = 4$, and application of Lemma J(h) shows that $|G'/M'| = |(G/M')'| = 2$. In that case, $|G'| = 4$, contrary to the assumption. Thus, distinct maximal subgroups of the group G have distinct derived subgroups (of order 2). As $c_1(G') = 7 = |\Gamma_1|$, it follows that all elements of G' are commutators.

Next we use atlas [HS]. As $Z(G) = G'$, G is a stem and its rank is equal to $\log_2(|G|) = 6$, G belongs to the family Γ_9 and its order structure is $(7, 56)$. There is in the family Γ_9 only one group with the above proved properties (in particular, with such order structure), namely, the group with number 152. Using MAGMA, Qinhai Zhang (Shanxi Normal University) has found that the set Γ_1 has five members isomorphic to $M_2(2, 2, 1)$ and two members isomorphic to $M_2(2, 2) \times C_2$, and therefore G is not a $\mathcal{D}_2$-group.

(ii) Thus, $|G'| = 4$. In that case, G has the rank (in sense of [HS]) $\log_2(|G/Z(G)|) = 3$. It follows from [HS] that G belongs to the family Γ_2 and so $|G'| = 2$, a final contradiction. We have proved that any $\mathcal{D}_2$-group has order 2^5, as desired. □

For $n = 1, 2$, the order of $\mathcal{D}_n$-group is equal to 2^{n+3}, We do not know if there is $n > 2$ such that there is a $\mathcal{D}_n$ group of order 2^{n+4}.

In the following remark we prove an additional property of extraspecial 2-groups.

Remark 3. (See Lemma J(l).) Let us prove that the extraspecial 2-group G is generated by subgroups $\cong D_8$, unless $G \cong Q_8$. One may assume that $|G| = 2^{2m+1}$, $m > 1$. By Lemma J(o), $G = (Q_1 * \cdots * Q_r) * (D_1 * \cdots * D_s)$, where $r+s = m$, $Q_i \cong Q_8$, $D_j \cong D_8$. One may assume that $r > 0$. Let D be a subgroup of G generated by all its subgroups $\cong D_8$. It suffices to show that $Q_i \le D$ for all $i \in \{1, \dots, r\}$. Let $H = Q_1 * C$, where C is a cyclic subgroup of order 4 in $C_G(Q_1)$. Then H (of order 16) has exactly three subgroups $\cong D_8$ (Lemma J(p)) and therefore $Q_1 < H < D$. Similarly, $Q_i < D$ for all $i \in \{1, \dots, r\}$, and so $D = G$.

We suggest to readers to prove that if an extraspecial 2-group $G \ncong D_8$, then D is generated by subgroups $\cong Q_8$ and so $\Omega_2^*(G) = G$.

Exercise 9. Any $\mathcal{D}_n$-group of order 2^{n+3} contains a non-Dedekindian subgroup of order 2^4 which is either an $\mathcal{A}_1$-group or $\cong Q_{2^4}$.

Hint. Indeed, by definition, G possesses a non-Dedekindian subgroup H of order 2^4. If H is not an $\mathcal{A}_1$-group, it is a $\mathcal{D}_1$-group so $\cong Q_{2^4}$, by Theorem 245.A.

Exercise 10. Any elementary abelian p-group of an $\mathcal{A}_n$-group G has order $\le p^{n+2}$.

Hint. Proceed by induction using Lemma J(a) taking into account that the assertion holds for $n = 1$.

Exercise 11. Classify the $\mathcal{D}_n$-groups all of whose nonabelian Dedekindian subgroups have orders $\le 2^4$.

Exercise 12. Classify the non-Dedekindian 2-groups all of whose maximal subgroups, except one, are Dedekindian. (**Hint.** Use Lemma 245.6.)

3° Some additional information on the subgroup structure of $\mathcal{D}_2$*-groups.* In this subsection G is a $\mathcal{D}_2$-group; then $|G| = 2^5$ (Theorem 245.C). By Theorem 245.A, there is $H \in \Gamma_1$ (of order 2^4) which is non-Dedekindian. By definition, any maximal subgroup of H is either abelian or $\cong Q_8$. If H has no subgroup $\cong Q_8$, it is an $\mathcal{A}_1$-subgroup.

(a) Assume that the (non-Dedekindian) subgroup H has a subgroup $Q \cong Q_8$. Then, as H has no subgroup $\cong D_8$, it follows that H is a $\mathcal{D}_1$-group so that $H \cong Q_{2^4}$ (Theorem 245.C; this also follows easily from Proposition 10.17 and Lemma J(p)). If G has no normal abelian subgroup $\cong E_4$, it is of maximal class (Lemma 1.4), and so, by Lemma J(d) and Theorem 1.2, $G \cong Q_{2^5}$. Now assume that G has a normal subgroup $R \cong E_4$. Then $G = HR$, $H \cap R = Z(H)$, $G/R \cong D_8$ and $G/Z(H) \cong D_8 \times C_2$; then $d(G) = 3$. Let $U/R < G/R$ be nonnormal of order 2. Then, by classification of the groups of order 8 and Lemma J(d), the subgroup U is abelian. Since the nonnormal subgroups of G/R of order 2 generate G/R, it follows that their inverse images are contained in $C_G(R)$ so that $R \le Z(G)$. If $L < R$ is of order 2 and $L \ne Z(H)$, then $G = H \times L$.

Next one assumes that G is not of maximal class. Then $|G'| \le 4$ (Lemma J(n)).

(b) Assume that $|G'| = 2$. Then, by Lemma J(o), $G = Q * C_G(Q)$, where $Q \cong Q_8$. If $C_G(Q)$ contains a cyclic subgroup $Z \cong C_4$, then $Q * C$ has a subgroup $\cong D_8$ (Lemma J(p)), contrary to Lemma J(d). Thus, $\exp(C_G(Q)) = 2$. Then $C_G(Q) = Z(G) = Z(Q) \times E$, where $E < Z(G)$ is of order 4. In that case, $G = Q \times E$ is Dedekindian, a contradiction. Thus, $|G'| = 4$. It follows from Lemma J(h) that $|G : Z(G)| > 4$.

(c) Now assume that all non-Dedekindian members of the set Γ_1 have no subgroups $\cong Q_8$. Then these members are $\mathcal{A}_1$-subgroups. Let $H \in \Gamma_1$ be an $\mathcal{A}_1$-subgroup and let $D \in \Gamma_1$ be nonabelian Dedekindian (D exists since, by ($2\mathcal{D}_2$), G is not an $\mathcal{A}_2$-group). One has $D = Q \times C$, where $Q \cong Q_8$ and $|C| = 2$. As $|\Phi(H)| = 4$, it follows that $|\Phi(G)| \ge 4$. The intersection $U = H \cap D$ is maximal abelian subgroup of $D \cong Q_8 \times C_2$ so $Z(D) \le U$ and U is abelian of type $(4, 2)$. Assume that H is nonmetacyclic; then $E = \Omega_1(H) \cong E_8$ (Lemma J(a)), and then $C_G(\Omega_1(U)) \ge DE = G$, i.e, $\Omega_1(D) = \Omega_1(U) \le Z(G)$. As $|G'| > 2$, one has $Z(G) = \Omega_1(D) = Z(D) \cong E_4$.

(d) Assume that $\Phi(G) = G'(\cong E_4)$; then $d(G) = 3$ since $|G/\Phi(G)| = 2^3$. In that case, $\Phi(G) = \Phi(H) = Z(H)$. If $M \in \Gamma_1$ is an $\mathcal{A}_1$-subgroup, then $\Phi(G) = Z(M)$ and, if $M \neq H$, then $\Phi(G) = Z(G)$. By Lemma 245.6, $\delta(G) \leq 2$. Therefore, since $|\Gamma_1| = 7$ and the set Γ_1 has at most one abelian member, one may assume that $M \neq H$, and then $\Phi(G) = Z(G)$.

Next we assume that $d(G) = 2$; then $Z(G) \leq \Phi(G)$. As G has no cyclic subgroup of index 2, then $\Phi(G) = \mho_1(G)$ is abelian of order 8 (Lemma 1.4) but noncyclic. It follows from Lemma J(m,n) that $|G'| = 4$. It follows that $G/Z(G)$ is nonabelian, and hence $Z(G) \neq G'$. The quotient group G/D' is minimal nonabelian (Lemma J(m)) and G/G' is abelian of type $(4, 2)$.

(e) Suppose that $\Gamma_1 = \{H, A, D\}$, where H is an $\mathcal{A}_1$-subgroup, D is a nonabelian non-Dedekindian subgroup and A is abelian. By Lemma J(m,g), one has $|G'| = 4$ and G/G' is abelian of type $(4, 2)$. As $H \cap A$, $D \cap A$ are maximal abelian in H, D, respectively, one has

$$Z(H) < A \cap H < A, Z(D) < D \cap A \Rightarrow C_G(Z(H)) \geq AH = G\,,$$

$$C_G(Z(D)) \geq AD = G \Rightarrow Z(H) = Z(G) = Z(D) \cong E_4\,.$$

Since $H \neq D$ and $d(G) = 2$, one has

$$H/Z(G) \cong E_4 \cong D/Z(G) \Rightarrow G/Z(G) \cong D_8\,,$$

since that nonabelian quotient group of order 8 has two distinct subgroups $\cong E_4$ (indeed, $Z(G)$ and G' are different). Next,

$$D > G' \neq Z(G) = Z(D) = \Omega_1(D) \Rightarrow G' \cong C_4\,.$$

As $D/D' \cong E_8$ and G/D' is minimal nonabelian (Lemma J(m)), it follows that $D/G' = \Omega_1(G/G')$ (Lemma J(a)). In that case, G/G' is abelian of type $(4, 2)$ so that $A/G' \cong C_4 \cong H/H'$ hence A and H are metacyclic since G' is cyclic. As $G' \cong C_4$ and $|H'| = |D'| = 2$, it follows that $H' = \Omega_1(G') = D'$. As the two-generator 2-group $G/\Omega_1(G')$ has two distinct abelian maximal subgroups $H/\Omega_1(G') \cong C_4 \times C_2$ and $D/\Omega_1(G') \cong E_8$, it follows that $G/\Omega_1(G') \cong M_2(2, 1, 1)$ is an $\mathcal{A}_1$-group and hence $A/\Omega_1(G')$ is abelian of type $(4, 2)$.

(f) Let $\Gamma_1 = \{H, F, D\}$, where H is an $\mathcal{A}_1$-subgroup, D is nonabelian non-Dedekindian and F is nonabelian. As G is not an $\mathcal{A}_1$-group, we get $Z(G) < \Phi(G)$ so that $|Z(G)| \leq 4$ and $|G'| = 4$ (Lemma J(m,g)).
Assume that $|Z(G)| = 4$. Then $Z(G) = Z(H) = Z(D) \cong E_4$ and $\exp(G/Z(G)) = 2$ (otherwise, the set Γ_1 has an abelian member), and this implies $d(G) > 2$, a contradiction.
Thus, $Z(G) \cong C_2$. It follows that G has no subgroup $\cong E_8$ (otherwise, the intersection of that subgroup with D is contained in $Z(G)$ and has order $4 > 2 = |Z(G)|$, a contradiction). It follows that $\Omega_1(H) \cong E_4$ so that H is metacyclic (Lemma J(a)).

The intersection $H \cap D$, being a maximal abelian subgroup of D, is abelian of type $(4, 2)$. As $Z(G)$ is of order 2 and $(E_4 \cong)\Omega_1(H \cap D) \leq Z(D)$, we get $\Omega_1(H \cap D) \nleq Z(H)$. It follows that $H \cong M_{2^4}$ (indeed, if a metacyclic minimal nonabelian 2-group X of order $> 2^3$ is such that $\Omega_1(X) \nleq Z(X)$, then X has a cyclic subgroup of index 2). If F is (nonabelian) Dedekindian, then $Z(D) = \Omega_1(D) = \Omega_1(F) = Z(F)$ (here we use the fact that $\Omega_1(D) = \Omega_1(H) = \Omega_1(F)$) so that $Z(D) \leq Z(G)$, a contradiction since $|Z(D)| = 4 > 2 = |Z(G)|$. Thus, F is an $\mathcal{A}_1$-subgroup and, as H, it is $\cong M_{2^4}$. It follows that $\Omega_1(G) < F \cap H = \Phi(G)$. One has $H/\Omega_1(G) \cong C_4 \cong F/\Omega_1(G)$ and $D/\Omega_1(G) \cong E_4$. It follows from $|G/\Omega_1(G)| = 8$ that $G/\Omega_1(G)$ is abelian of type $(4, 2)$, and we conclude that $\Omega_1(G) = G' \cong E_4$. As $Z(G) \cong C_2$, it follows that $Z(F) \neq Z(H)$. It follows that $Z(G) = D' = H' = F'$. As $D/D' \cong E_8$, it follows that $G/Z(G) \cong M_2(2, 1, 1)$ is nonmetacyclic. Thus, in the case under consideration, all members of the set Γ_1 are described up to isomorphism.

4^0 Problems.

Problem 1. Study the $\mathcal{D}_n$-groups with abelian subgroup of index 2.

Problem 2. Classify the $\mathcal{D}_n$-groups G such that $|G : Z(G)| = 2^3$.

Problem 3. Classify the $\mathcal{D}_n$-groups with cyclic G'.

Problem 4. Study the $\mathcal{D}_n$-groups containing a $\mathcal{D}_1$-subgroup.

Problem 5. Study the $\mathcal{D}_n$-groups generated by subgroups $\cong Q_8$.

Problem 6. Classify the subgroup structure of the 2-groups containing an elementary abelian subgroup of index ≤ 4.

Problem 7. Does there exist a $\mathcal{D}_n$-group of order 2^{n+4}?

Problem 8. Study the 2-groups all of whose nonabelian maximal subgroups are of the form $M \times E$, where M is of maximal class and $\exp(E) \leq 2$.

Problem 9. Study the 2-groups G containing a subgroup $\cong Q_8$ such that, whenever $Q \cong Q_8$ and $Q \leq M \in \Gamma_1$, then M is Dedekindian.

Problem 10. Study the 2-groups all of whose nonabelian maximal subgroups are of the form $D \times A$, where D is nonabelian of order 8 and A is abelian.

Problem 11. Study the 2-groups all of whose nonabelian maximal subgroups are direct products of groups of maximal class.

Problem 12. Study the $\mathcal{D}_n$-groups, $n > 1$, (i) all of whose maximal subgroups are $\mathcal{D}_{n-1}$-groups, (ii) such that, whenever $H \in \Gamma_1$, then H contains a subgroup $\cong Q_8$.

Problem 13. Study the $\mathcal{D}_n$-groups that have no proper $\mathcal{D}_s$-subgroups for all $s < n$.

Problem 14. Classify the $\mathcal{D}_n$-groups G containing a generalized quaternion subgroup of index 4 (in that case, $|G| = 2^{n+3}$).

Problem 15. Study the two-generator $\mathcal{D}_n$-groups.

Problem 16. Classify the $\mathcal{D}_n$-groups, $n \in \{2, 3\}$.

Problem 17. Classify the $\mathcal{D}_n$-groups containing a metacyclic subgroup of index 2.

Problem 18. Study the 2-groups that are lattice isomorphic to $\mathcal{D}_n$-groups.

§ 246 Non-Dedekindian p-groups with many normal subgroups

Here we initiate a study of p-groups described in Problem 2303 and prove the following result.

Theorem 246.1. *Let G be a non-Dedekindian p-group with $H \in \Gamma_1$ such that $\langle h, x\rangle$ is G-invariant for all $h \in H - \{1\}$ and $x \in G - H$. Then G is of class 2 and we have the following two possibilities.*

(i) *If there is an element $g \in G - H$ of order p, then $\mathrm{C}_G(g) = \langle g\rangle \times C$, where $C \neq \{1\}$ is cyclic, $\mathrm{C}_G(g)$ is an abelian maximal subgroup of G and in case $p = 2$ such groups are described in § 48.*

(ii) *If there is no an element of order p in $G - H$, then each nonnormal subgroup of G is contained in H and such groups are described in § 231.*

Proof. Let G be a non-Dedekindian p-group with $H \in \Gamma_1$ such that $\langle h, x\rangle$ is G-invariant for all $h \in H - \{1\}$ and $x \in G - H$. It is easy to see that G is of class 2. Indeed, let $x \in G - H$. If $\mathrm{o}(x) > p$, then $\langle x, x^p\rangle = \langle x\rangle$ is normal in G and so G' centralizes $\langle x\rangle$. If $\mathrm{o}(x) = p$, then take an element $h \in H$ of order p such that $h \in \mathrm{Z}(G)$. It follows that $\mathrm{E}_{p^2} \cong \langle x, h\rangle$ is normal in G and so again G' centralizes $\langle x, h\rangle$. We have proved that G' centralizes each element in $G - H$ and so G' centralizes G, $G' \leq \mathrm{Z}(G)$ and G is of class 2.

Assume that there is an element $g \in G - H$ of order p. If $\langle g\rangle$ is normal in G, then $G = H \times \langle g\rangle$ and for any $1 \neq h \in H$, $\langle h, g\rangle = \langle h\rangle \times \langle g\rangle$ is normal in G so that $\langle h\rangle = \langle h, g\rangle \cap H$ is normal in G and therefore H is Dedekindian. But then $G = H \times \langle g\rangle$ is also Dedekindian, a contradiction. We have proved that $\langle g\rangle$ is nonnormal in G. Let $h \in H$ be an element of order p such that $h \in \mathrm{Z}(G)$. Then $V = \langle h, g\rangle = \langle h\rangle \times \langle g\rangle$ is normal in G and $\langle g\rangle^G = V$. We see that $G_1 = \mathrm{C}_G(V)$ is of index p in G and $\mathrm{C}_H(V) = \mathrm{C}_H(g) = G_1 \cap H$. If $\langle h_1\rangle \neq \langle h\rangle$ is another subgroup of order p in $\mathrm{C}_H(g)$, then $\langle h_1, g\rangle$ is normal in G and so $\langle h_1, g\rangle \cap \langle h, g\rangle = \langle g\rangle$ is normal in G, a contradiction. Thus, $\mathrm{C}_H(g)$ has only one subgroup of order p. It follows that $\mathrm{C}_H(g)$ is either cyclic or $p = 2$ and $\mathrm{C}_H(g)$ is generalized quaternion. But G is of class 2 and so if $\mathrm{C}_H(g)$ is generalized quaternion, then $\mathrm{C}_H(g) \cong \mathrm{Q}_8$. Using results of § 49 (Theorem 49.1 (A_1)), we have in the last case that G has a maximal subgroup isomorphic to Q_{2^4} or SD_{2^4}, contrary to the fact that G is of class 2. We have proved that $\mathrm{C}_G(g) = \langle g\rangle \times \mathrm{C}_H(g)$, where $\mathrm{C}_H(g) \neq \{1\}$ is cyclic and in case $p = 2$ such groups are described in § 48.

Suppose that there is no element of order p in $G - H$. Then for each $x \in G - H$, $\langle x\rangle$ is normal in G. Let X be any nonnormal subgroup of G. If $X \nleq H$, then for each $x \in X - H$, $\langle x\rangle$ is normal in G and so X is normal in G, a contradiction. Hence each nonnormal subgroup of G is contained in H and so G is not generated by its nonnormal subgroups. Such groups are described in § 231. Our theorem is proved. □

§ 247 Nonabelian p-groups all of whose metacyclic sections are abelian

We initiate a study of nonabelian p-group all of whose metacyclic sections are abelian (Problem 2846) and prove the following result.

Theorem 247.1. *Let G be a nonabelian p-group all of whose metacyclic sections are abelian. Then we must have $p > 2$, G is nonmodular, and each modular section of G is abelian. Each minimal nonmodular section of G is a nonmetacyclic minimal nonabelian group.*

Proof. Let G be a nonabelian p-group all of whose metacyclic sections are abelian.

Suppose that $p = 2$. Since G is D_8-free and Q_8-free, it follows by the results of § 73 that G is a non-Hamiltonian modular 2-group. By Iwasawa's theorem 73.15, G has a proper abelian normal subgroup N and an element $t \in G - N$ such that $G = \langle N, t\rangle$ and there is a positive integer $s > 2$ such that $a^t = a^{1+2^s}$ for all $a \in N$. Also, there is $a_0 \in N$ such that $a_0^t = a_0^{1+2^s} \neq a_0$ (since G is nonabelian) so that $\langle a_0, t\rangle$ is a nonabelian metacyclic group, a contradiction. We have proved that we must have $p > 2$.

Suppose that in case $p > 2$, G has a nonabelian modular section X. By Theorem 73.15, X has a nonabelian metacyclic subgroup, a contradiction. It follows that G is nonmodular and each modular section of G is abelian.

Let Y be a minimal nonmodular section of G. But then each proper subgroup of Y is modular and so is abelian. It follows that Y is minimal nonabelian and (by our assumption) Y is nonmetacyclic. Our theorem is proved. □

§ 248 Non-Dedekindian p-groups G such that $H^G = HZ(G)$ for all nonnormal $H < G$

Here we begin a study of p-groups from Problem 3121 and prove the following result.

Theorem 248.1. *Let G be a non-Dedekindian p-group such that $H^G = HZ(G)$ for all nonnormal $H < G$. Then G is of class 2, G' is elementary abelian and $\mho_1(G) \leq Z(G)$.*

Proof. Let G be a non-Dedekindian p-group such that $H^G = HZ(G)$ for all nonnormal $H < G$. We first show that $G/Z(G)$ is Dedekindian. Indeed, let $L/Z(G)$ be a nonnormal cyclic subgroup of $G/Z(G)$. Let K be a cyclic subgroup of L such that K covers $L/Z(G)$. Then K is nonnormal in G (otherwise, $L/Z(G)$ is normal in $G/Z(G)$) and (by our assumption) $K^G = KZ(G) = L$ is normal in G, a contradiction.

Assume that $G/Z(G)$ is Hamiltonian. Then $p = 2$ and there is a normal subgroup $H/Z(G)$ in $G/Z(G)$ such that $H/Z(G) \cong Q_8$. Let $M_i/Z(G)$, $i = 1, 2, 3$, be one of three maximal subgroups in $H/Z(G)$. Then $M_i/Z(G) \cong C_4$ and so each M_i, $i = 1, 2, 3$, is abelian. It follows

$$H' \cong C_2 \text{ and } (H'Z(G))/Z(G) = (H/Z(G))' \cong C_2\,, \quad \text{so that } H' \nleq Z(G)\,.$$

But

$$H' \text{ char } H \trianglelefteq G \text{ and therefore } H' \trianglelefteq G \text{ and so } H' \leq Z(G)\,,$$

a contradiction.

We have proved that $G/Z(G)$ (being Dedekindian) is abelian and so $G' \leq Z(G)$ and G is of class 2.

Let X be a nonnormal subgroup in G. By our assumption, $X^G = XZ(G)$ and so X^G contains G'. We may use Theorem 191.5 which implies that G' is elementary abelian and $\mho_1(G) \leq Z(G)$. Our theorem is proved. □

Problem. Study the non-Dedekindian p-groups G such that $A^G \leq AZ(G)$ for any nonnormal subgroup (abelian subgroup) A of G.

§ 249 Nonabelian p-groups G with $A \cap B = Z(G)$ for any two distinct maximal abelian subgroups A and B

We solve here a problem posed by the first author by proving the following result.

Theorem 249.1. *A nonabelian p-group G has the property $C_G(x) = A$ for each maximal abelian subgroup A and each $x \in A - Z(G)$ if and only if $A \cap B = Z(G)$ for any two distinct maximal abelian subgroups A and B of G. Such groups have been described in Theorems 91.2 and 91.3.*

Proof. Suppose that a nonabelian p-group G has the property $C_G(x) = A$ for each maximal abelian subgroup A and each $x \in A - Z(G)$. Let $A \neq B$ be two distinct maximal abelian subgroups of G so that $Z(G) \leq A \cap B$. Assume that $A \cap B > Z(G)$ and let $x \in (A \cap B) - Z(G)$. Then $C_G(x) \geq \langle A, B \rangle > A$, a contradiction. We have proved that $A \cap B = Z(G)$ for any two distinct maximal abelian subgroups A and B of G.

Conversely, suppose that we have $A \cap B = Z(G)$ for any two distinct maximal abelian subgroups A and B of a nonabelian p-group G. Let A be a maximal abelian subgroup of G and assume that for an element $a \in A - Z(G)$ we have $C_G(a) > A$. Let $b \in C_G(a) - A$ be such that $b^p \in A$. Let B be a maximal abelian subgroup in G containing the abelian subgroup $\langle a, b \rangle$. Then $A \neq B$ and since $a \in A \cap B$ with $a \notin Z(G)$, we get a contradiction. Our theorem is proved. □

Problem 1. Study the primary $\mathcal{A}_n$-groups G, $n > 1$, such that $A \cap B \leq Z(G)$ for any two distinct $\mathcal{A}_1$-subgroups $A, B < G$.

Problem 2. Study the primary $\mathcal{A}_n$-groups G, $n > 1$, such that $A \cap B \leq Z(G)$ for any $A, B < G$, where A is maximal abelian and B is minimal nonabelian.

Problem 3. Study the primary $\mathcal{A}_n$-groups G, $n > 1$, such that $A_G = Z(G)$ for any minimal nonabelian $A < G$.

Problem 4. Study the primary $\mathcal{A}_n$-groups G, $n > 1$, such that $Z(G) < A^G$ for any minimal nonabelian $A < G$.

§ 250 On the number of minimal nonabelian subgroups in a nonabelian p-group

As we know (see § 76), the structure of a p-group G depends on the invariant $\alpha_1(G)$, the number of minimal nonabelian subgroups of G. For example, if $\alpha_1(G) \le 1+p+p^2$, then all subgroups of G of index p^3 are abelian, i.e., provided G is nonabelian, it is $\mathcal{A}_n$-group for $n \le 3$ (see § 76). Therefore, it is important to estimate the number $\alpha_1(G)$ in general case.

It is fairly difficult to find the number $\alpha_1(G)$ for complicated p-groups G (e.g., try to find $\alpha_1(\Sigma_{p^n})$). In some cases, it is possible to estimate that number.

In this section we obtain a lower bound of $\alpha_1(G)$. The result will follow from Lemma 57.1. We also study in some detail the groups in which the obtained estimate is attained. As it will be shown, such p-groups have small order and very special structure.

Denote by $\mathrm{c}(G)$ the number of nonidentity cyclic subgroups of a group G. One has $\mathrm{c}(G) = \mathrm{c}_1(G) + \mathrm{c}_2(G) + \cdots$, where $\mathrm{c}_k(G)$ is the number of cyclic subgroups of order p^k in a p-group G. For example, using Hall's enumeration principle and induction, one obtains

$$\mathrm{c}(\mathcal{H}_{2,2}) = 3 \cdot 5 - 2 \cdot 3 = 9\,, \quad \mathrm{c}(\mathrm{M}_{2^{n+1}}) = 2n + 1 = \mathrm{c}(\mathrm{C}_{2^n} \times \mathrm{C}_2)\,,$$
$$\mathrm{c}(\mathrm{Q}_{2^{n+1}}) = 2 \cdot 2^{n-2} + n = 2^{n-1} + n\,, \quad \mathrm{c}(\mathrm{D}_{2^{n+1}}) = 2^n + n\,,$$
$$\mathrm{c}(\mathrm{SD}_{2^{n+1}}) = (2^{n-2} + n - 1) + (2^{n-1} + n - 1) + n - 2(n-1) = 3 \cdot 2^{n-2} + n\,.$$

In particular, $\mathrm{c}(\mathrm{Q}_{2^4}) = 2^2 + 3 = 7$. Applying Hall's enumeration principle, one obtains $\mathrm{c}(\mathrm{D}_{2^4}) = 2 \cdot 6 + 3 - 2 \cdot 2 = 11$, $\mathrm{c}(\mathrm{SD}_{2^4}) = 4 + 6 + 3 - 2 \cdot 2 = 9$. If $G = Q * C$, where $Q \cong \mathrm{Q}_8$, $C \cong \mathrm{C}_4$ and $|G| = 2^4$, then $\mathrm{c}(G) = \mathrm{c}_1(G) + \mathrm{c}_2(G) = 7 + 4 = 11$. If $G = \mathrm{D}_8 \times \mathrm{C}_2$, then $\mathrm{c}(G) = 13$. It is easy to show that if G is noncyclic of order 2^4, then $7 \le \mathrm{c}(G) \le 15$.

If G is a noncyclic p-group of order p^n, there is $T \triangleleft G$ such that $G/T \cong \mathrm{E}_{p^2}$. Let $H_1/T, \dots, H_{p+1}/T$ be all subgroups of order p in G/T. Then $H_1, \dots, H_{p+1} \in \Gamma_1$. If $C < G$ is cyclic, then $C \le H_i$ for some $i \le p+1$. In that case, the following identity holds:

$$\mathrm{c}(G) = \mathrm{c}(H_1) + \cdots + \mathrm{c}(H_{p+1}) - p\mathrm{c}(T). \qquad (*)$$

It is possible to use $(*)$ for counting $\mathrm{c}_k(G)$ instead of $\mathrm{c}(G)$. If $G = \mathrm{D}_8 \times \mathrm{C}_2$ and $T = Z(G)$, then $\mathrm{c}(G) = 7 + 7 + 5 - 2 \cdot 3 = 13$.

The following basic result, due to the second author, plays in what follows a crucial role (this result is used many times in our book).

Lemma 250.1 (= Lemma 57.1). *Let A be a maximal normal abelian subgroup of a nonabelian p-group G. Then for any $x \in G - A$ there is in A an element a such that the subgroup $\langle a, x\rangle$ is minimal nonabelian (= $\mathcal{A}_1$-subgroup).*

Proof. We offer a detailed proof of that lemma. Bearing in mind our aim, one may assume without loss of generality that $G = \langle x, A\rangle$. Clearly, $C_A(x) \le Z(G)$, and the subgroup $U = \langle x\rangle C_A(x)$ is maximal abelian in G. Let $U < B \le G$ be such that $|B : U| = p$ so that $U \triangleleft B$ and B is nonabelian. By the modular law, $B = \langle x\rangle(A \cap B)$, and we get $A \cap B > C_A(x)$ (otherwise, $B = U$) so that the set $(A \cap B) - C_A(x)$ is nonempty; moreover, $|(A \cap B) : C_A(x)| = p$. Take $a \in (A \cap B) - C_A(x)$; then $A \cap B = \langle a, C_A(x)\rangle$ and $a^p \in C_A(x)$. Set $M = \langle a, x\rangle$; then M is nonabelian, by the choice of a. We claim that M is minimal nonabelian. One has $A \cap B \triangleleft B$ since $A \triangleleft G$. Next,

$$B = \langle U, a\rangle = \langle x, C_A(x), a\rangle = \langle x, A \cap B\rangle .$$

Since U is maximal abelian in B, it follows that $C_A(x) \le Z(B) < U$. Next, $C_A(x) = U \cap (A \cap B)$. Set $\bar{B} = B/C_A(x)$. In that case, $\langle \bar{a}\rangle = \overline{A \cap B} \triangleleft \bar{B}$ (moreover, $\bar{a} \in Z(\bar{B})$ since $o(\bar{a}) = p$) and this implies that $\bar{B} = \langle \bar{x}\rangle \times \langle \bar{a}\rangle$ is abelian of type $(o(\bar{x}), p)$ with cyclic subgroup $\langle \bar{x}\rangle$ of index p. Then $\bar{B}$ has two cyclic subgroups $\bar{S} = \langle \bar{x}\rangle$ and, say $\bar{T}$, both of index p so that S and T are abelian subgroups of index p in B. It follows that $S \cap T = Z(B)$ has index p^2 in B, and Lemma 1.1 implies that $|B'| = p$. Thus, $|M'| = p$ since $M \le B$ is nonabelian. As $d(M) = 2$, it follows from Lemma 65.2(a) that M is minimal nonabelian. □

Theorem 250.2. *Let A be a maximal normal abelian subgroup of a nonabelian p-group G. Then $\alpha_1(G) \ge c(G/A)$. Next we assume, in addition, that $\alpha_1(G) = c(G/A)$ and $|G/A| > p$.*[1] *Then*

(a) $p > 2$.

(b) $A = \Phi(G) = \Omega_1(G) \cong E_{p^3}$, $d(G) = 2$, $\alpha_1(G) = p + 1$, $|G| = p^5$ *and G is an $\mathcal{A}_2$-group.*

Proof. Write $\bar{G} = G/A$. Let $\bar{H}_i = \langle \bar{x}_i\rangle$, $i = 1, \dots, k = c(\bar{G})$, be all pairwise distinct nonidentity cyclic subgroups of $\bar{G}$. Then the subgroups $H_i = \langle x_i\rangle A$, $i = 1, \dots, k$, are pairwise distinct nonabelian. By Lemma 57.1, for every $i \le k$, there is $a_i \in A$ such that $B_i = \langle a_i, x_i\rangle$ is minimal nonabelian. One has $B_iA = H_i \ne H_j = B_jA$ so that $B_i \ne B_j$ for $i \ne j$. Thus, $B_1, \dots, B_k$ are pairwise distinct $k = c(G/A)$ minimal nonabelian subgroups of the group G. We see that to each cyclic subgroup of $\bar{G}$ corresponds an $\mathcal{A}_1$-subgroup of G and so obtained $\mathcal{A}_1$-subgroups are pairwise distinct. It follows that $\alpha_1(G) \ge c(G/A)$.

Next we assume that $\alpha_1(G) = c(G)$ and $|G/A| > p$ (if $|G/A| = p$, then G is an $\mathcal{A}_1$-group).

(i) Take $A < H \le G$, where $|H| = p|A|$. In that case, $H = \langle x, A\rangle$ for some $x \in H - A$; then $x^p \in A$, by the choice of H. As $\alpha_1(G) = c(G/A)$, it follows that $\alpha_1(H) = 1$. Indeed, assume that H is not an $\mathcal{A}_1$-subgroup. Let $F = \langle a, x\rangle < H$ be an $\mathcal{A}_1$-subgroup for some

1 In that case, if $B/A < G/A$ is of order p, then B is an $\mathcal{A}_1$-subgroup and such B exhaust all $\mathcal{A}_1$-subgroups of G.

$a \in A$ (Lemma 57.1). Note that $FA = H$ since $x \in FA - A$ and $A < \langle x \rangle F = FA$ has index p. Then there is in H an $\mathcal{A}_1$-subgroup $C \neq F$ (Proposition 10.28) and, obviously, $CA = H$. Thus, $\alpha_1(H) > 1$ and we obtain an $\mathcal{A}_1$-subgroup of G which is not contained in the set of $c(G/A)$ minimal nonabelian subgroups above constructed so that $\alpha_1(G) > c(G/A)$, contrary to the assumption. Thus, H is minimal nonabelian hence any subgroup of G containing A as a subgroup of index p is minimal nonabelian. Then, by Lemma 65.1, $d(A) \le 3$.

(ii) If $C/A \le G/A$ is cyclic of order p^2, then we must have $\alpha_1(C) = 2$ (if $\alpha_1(C) > 2$, then $\alpha_1(G) > c(G/A)$). Let $A < B < C$, where $|C : B| = p$. By Lemma 76.5 (a), $\beta_1(C, B)(= \alpha_1(C) - \alpha_1(B)) \ge p - 1$. If $p > 2$, then $\alpha_1(C) \ge \alpha_1(B) + (p - 1) = p > 2$, and this implies $\alpha_1(G) > c(G/A)$, a contradiction. Thus, if $p > 2$, then such C/A does not exist, and therefore $\exp(G/A) = p$. Now let $p = 2$ and let $C/A \le G/A$ be cyclic of order 4. Then, by Lemma 76.5 (a), C contains an abelian subgroup, say U, of index 2. Let us show that this is impossible. The subgroup $Z(C) = A \cap U$ has index 2 in A so index 8 in C. Then $C/Z(C)$, as a noncyclic extension of subgroup A/C of order 2 by a cyclic group of order 4, is abelian of type $(4, 2)$. If $L_i/Z(C)$ are distinct cyclic of index 2 in $C/Z(C)$, $i = 1, 2$, then L_1 and L_2 are abelian of index 2 in C so that $L_1 \cap L_2 = Z(C)$ has index $4 = |C : A|$ in C and so $Z(C)A > A$ is abelian of order $s|A|$, a contradiction. Thus, C/A does not exist so that in this case we have $\exp(G/A) = 2$.

Let us prove that if $|G/A| > p$, then $A \le \Phi(G)$. Assume that this is false. Then there is $T \in \Gamma_1$ such that $A \nleq T$. In that case, T has no $\mathcal{A}_1$-subgroup since, by the above, all $\mathcal{A}_1$-subgroups of G contain A as a subgroup of index p, and hence T is abelian. The number of those members of the set Γ_1 that do not contain A, is a multiple of p (that number is equal to $|\Gamma_1|$ minus the number of maximal subgroups of the quotient group G/A and both these numbers are $\equiv 1 \pmod p$)). As all members of the set Γ_1 containing A are nonabelian (indeed, A is a maximal abelian subgroup of G and $|G : A| > p$), it follows that the number of abelian members in the set Γ_1 is a positive multiple of p, contrary to Exercise 1.6. Thus, T does not exist hence $A \le \Phi(G)$.

(iii) As $C_G(A) = A$ and $|G/A| > p$, it follows that $|A| > p^2$. By (ii), $\exp(G/A) = p$ and, if $S/A \le G/A$, then $\alpha_1(S) = c(S/A)$. Let $S/A \le G/A$ be of order p^2. Then $c(S/A) = p + 1$ since $S/A \cong E_{p^2}$. It follows from Theorem 76.4(c2) that S is an $\mathcal{A}_2$-group of order p^5, $p > 2$.[2] In that case, $A \cong E_{p^3}$ (Lemma J(a)). Since $C_G(A) = A$ and $|\mathrm{Aut}(A)|_p = p^3$, it follows that $|G/A| \le p^3$.

(iv) Assume that $|G/A| = p^3$. Then $G/A \cong S(p^3)$. If $F/A = \Phi(G/A)$, then $F = \Phi(G)$. As $|F/A| = p$, it follows that F is minimal nonabelian, and we conclude that $d(\Phi(G)) = 2$. By Theorem 44.12, $\Phi(G)$ is metacyclic. This is a contradiction since $\Phi(G)$ contains a

2 For further information on S, see § 71 where the primary $\mathcal{A}_2$-groups are classified; see also Proposition 76.4 (a).

subgroup $A \cong E_{p^3}$. Thus, $|G : A| = p^2$ and it follows from $A \le \Phi(G)$ that $A = \Phi(G)$. In that case, $|G| = |A|\,|G : A| = p^5$.

(v) By (iv), $\alpha_1(G) = c(G/A) = p + 1$. Then, by Theorem 76.6 (b), G is an $\mathcal{A}_2$-group. The proof is complete.

□

Applying Hall's enumeration principle to Σ_9, a Sylow 3-subgroup of the symmetric group S_9 of degree 9, one obtains $c(\Sigma_9) = 2\cdot 7 + 2\cdot 13 - 3\cdot 4 = 28$ (note that $2+2 = |\Gamma_1|$). Now let $G = \Sigma_{27} \in \mathrm{Syl}_3(S_{27})$. One has $G = C_3 \operatorname{wr} \Sigma_9$. Then $A \cong E_{3^9}$, the base subgroup of G, is a maximal abelian normal subgroup of G. As $G/A \cong \Sigma_9$, it follows from the above and Theorem 250.1 that $\alpha_1(G) \ge c(G/A) = c(\Sigma_9) = 28$. In fact, here $\alpha_1(G) > c(G/A)$ since $d(A) = 9 > 3$ (see Theorem 250.1).

If G is the holomorph of the cyclic group L of order 2^n, $n > 1$, then $\alpha_1(G) \ge c(G/L) = 2n - 3$.

For more detailed description of G, see § 76.

Exercise 1. Find $c(\Sigma_{p^2})$ (see paragraph containing formula $(*)$).

Hint. Set $G = \Sigma_{p^2}$. One has $\exp(G) = p^2$ so that $c(G) = c_1(G) + c_2(G)$. It is known that the set Γ_1 contains exactly two members of exponent p. Now one can use Hall's enumeration principle. For example,

$$\begin{aligned} c_1(G) &= 2(1 + p + \cdots + p^{p-1}) + (p-1)(1 + p + \cdots + p^{p-2}) - p(1 + p + \cdots + p^{p-2}) \\ &= 1 + p + \cdots + p^{p-2} + 2p^{p-1}\,. \end{aligned}$$

It is easy to check that $c_2(G) = (p-1)\cdot \frac{p^p - p^{p-1}}{p(p-1)} = p^{p-2}(p-1)$.

Exercise 2. Let A be a maximal normal abelian subgroup of a nonabelian p-group G and let $\exp(G/A) = p$, $|G/A| > p$. Study the structure of G if all subgroups of G containing A as a subgroup of index p are minimal nonabelian. Is it true that A is noncyclic (this holds provided $p > 2$ since then $\mathrm{Aut}(A)$ is cyclic)?

Exercise 3. If G is a p-group of maximal class and order p^m with an abelian subgroup of index p, then $\alpha_1(G) = p^{m-3}$.

Hint. It is known that all nonabelian subgroups of G are of maximal class. Use induction on m.

Exercise 4. Let G be a p-group of maximal class and order p^m with the abelian Frattini subgroup. Prove that $\alpha_1(G) > p^{m-3}$, unless the fundamental subgroup G_1 is abelian.

Exercise 5. Let $L_1, \dots, L_k$ be all pairwise distinct nonidentity cyclic subgroups of a p-group $G > \{1\}$. Then

$$|G| - 1 = \varphi(|L_1|) + \cdots + \varphi(|L_k|)\,,$$

where $\varphi(*)$ is Euler's totient function. (**Hint.** The number of generators of L_i is equal to $\varphi(L_1)$.)

Exercise 6. Classify the 2-groups G of order 2^n satisfying $\mathrm{c}(G) \geq 2^n + n$. State a similar problem for the p-groups with $p > 2$.

Exercise 7. Let A be a maximal normal abelian subgroup of a p-group G such that G/A contains an abelian subgroup of type (p, p). Suppose that each subgroup of G containing A as a subgroup of index p is minimal nonabelian. Study the structure of G.

Exercise 8. It is easy to find that $\mathrm{c}(\mathrm{M}_{p^{n+1}}) = pn + 1$.
(a) Is it true that if G is a noncyclic p-group of order p^{n+1}, then $\mathrm{c}(G) \geq pn + 1$?
(b) Classify the p-groups G of order p^{n+1} such that $\mathrm{c}(G) = pn + 1$.

Exercise 9. Classify the p-groups G of order $|M|$ such that $\mathrm{c}(G) = \mathrm{c}(M)$, where M is a 2-group of maximal class.

Exercise 10. Let H be a homocyclic p-group. Study the p-groups G of order $|H|$ such that $\mathrm{c}(G) = \mathrm{c}(H)$.

Exercise 11. Let A be an absolutely regular p-group. Study the p-groups G of order $|A|$ such that $\mathrm{c}(G) = \mathrm{c}(A)$.

Exercise 12. Let R be a regular p-group of exponent p^n and order $p^{n(p-1)}$. Study the p-groups G of order $|R|$ such that $\mathrm{c}(G) = \mathrm{c}(E)$.

Exercise 13. Find $\mathrm{c}(A)$, where A is an abelian group of type $(p^{e_1}, \dots, p^{e_k})$.

Exercise 14. Find $\mathrm{c}(G)$, where G is a regular p-group of exponent p^e provided the numbers $|\Omega_1(G)|$, $|\Omega_2(G)/\Omega_1(G)|, \dots, |\Omega_e(G)/\Omega_{e-1}(G)|$ are known.

Exercise 15. Find $\mathrm{c}(G)$, where G is a minimal nonabelian p-group.

Exercise 16. Find $\mathrm{c}(G)$, where G is an $\mathcal{A}_2$-group.

Exercise 17. Study the p-groups G such that G contains an elementary abelian subgroup A of index p.

Exercise 18. Let a p-group $G = A \times B$ with known $\mathrm{c}(A)$ and $\mathrm{c}(B)$. Find $\mathrm{c}(G)$.

Exercise 19. Find $\mathrm{c}(G)$, where G is minimal nonabelian of order 2^5. Classify all 2-groups W satisfying $\mathrm{c}(W) = \mathrm{c}(G)$.

Exercise 20. Classify the groups G of order p^n with $\mathrm{c}(G) \leq (p + 1)n$.

Exercise 21. Classify the 2-groups G of order $|E|$ satisfying $\mathrm{c}(G) = \mathrm{c}(E)$, where E is an extraspecial 2-group.

Exercise 22. Is it true that whenever G is a group of order p^n and exponent $> p$, then $\mathrm{c}(G) < \mathrm{c}(\mathrm{E}_{p^n})$?

Problem 1. Classify the nonabelian p-groups G such that, whenever $A < G$ is maximal abelian, then $A < B \leq G$, where B is minimal nonabelian.

Problem 2. Let $n > 2$ and G a p-group of class n. Study the p-groups G such that for any $k < n$ and any $A < G$ of class k there is $B \leq G$ of class $k + 1$ such that $A < B$.

Problem 3. Classify the p-groups G satisfying $\alpha_1(G) = \mathrm{c}(G/A) + 1$, where A is a maximal normal abelian subgroup of G.

Problem 4. Find $\mathrm{c}_1(\Sigma_{p^n})$, $\mathrm{c}(\Sigma_{p^n})$, $\mathrm{c}_1(\mathrm{UT}(n, p))$, $\mathrm{c}(\mathrm{UT}(n, p))$.

Problem 5. Estimate $\alpha_1(G)$, where G is a two-generator metabelian group of exponent p.

Problem 6. Classify the p-groups G containing a maximal normal abelian subgroup A such that $\alpha_1(G) = (p - 1) + \mathrm{c}(G/A)$.

§ 251 p-groups all of whose minimal nonabelian subgroups are isolated

The first author came to an excellent idea to introduce the concept of the so-called isolated subgroups in a p-group. In § 239, we have already considered the case of an isolated in G center. A subgroup H of a p-group G is said to be isolated if for each cyclic $C < G$, we have either $C \le H$ or C avoids H, i.e., $H \cap C = \{1\}$.

Here we shall classify nonabelian p-groups G all of whose minimal nonabelian subgroups are isolated in G (Theorem 251.1). Obviously, minimal nonabelian p-groups and nonabelian p-groups with exponent $p > 2$ satisfy this assumption.

Theorem 251.1. *Let G be a nonabelian p-group all of whose minimal nonabelian subgroups are isolated in G. Suppose in addition that G is not minimal nonabelian and $\exp(G) > p$. Then $p > 2$ and G is an irregular p-group with a unique abelian maximal subgroup A. The subgroup A is of exponent $> p$ and rank ≥ 3, $A = \mathrm{H}_p(G)$ and each minimal nonabelian subgroup of G is isomorphic to $\mathrm{S}(p^3)$ (the nonabelian group of order p^3 and exponent p).*

Proof. Let G be a p-group satisfying all the assumptions of Theorem 251.1, i.e., G is a nonabelian p-group all of whose minimal nonabelian subgroups are isolated, G is not minimal nonabelian, and $\exp(G) > p$. In our proof, we shall use freely Theorems 7.1, 7.2 and Lemmas 1.4, 57.1, 65.1, and 65.2.

Let $H < G$ be a minimal nonabelian subgroup in G and let $H < M \le G$ be such that $|M : H| = p$. Then all elements in $M - H$ are of order p. If $p = 2$, then a result of Burnside implies that H is abelian, a contradiction. Hence we must have $p > 2$. By Lemma 1.4, G has a normal subgroup U which is abelian of type (p, p).

(i) In what follows A shall always denote a maximal normal abelian subgroup in G which contains U so that A is of rank ≥ 2. Suppose that $\exp(G/A) > p$ and let B/A be a subgroup of G/A isomorphic to C_{p^2}. Let $g \in B - A$ be such that $\langle g \rangle$ covers B/A and set $C = A\langle g^p \rangle$ so that $|C : A| = p$. By Lemma 57.1, there is $a \in A$ such that $K = \langle g^p, a \rangle$ is minimal nonabelian. But $K \le C$ and $g \notin C$ with $\langle g \rangle \cap K = \langle g^p \rangle > \{1\}$, a contradiction. We have proved that $\exp(G/A) = p$.

(ii) Suppose, by way of contradiction, that there is a minimal nonabelian subgroup H in G containing A. Then $|H : A| = p$ and $H < G$. Let $g \in G - H$ so that (i) implies $g^p \in A$ and therefore $o(g) = p$. Hence all elements in $G - H$ are of order p. If $\exp(H) = p$, then $\exp(G) = p$ and this is a contradiction. Hence $\exp(H) > p$. By Lemma 65.1, $\Omega_1(H)$ is elementary abelian of order p^2 or p^3 and so H is generated by elements of order $> p$ in $H - \Omega_1(H)$. Thus, $\mathrm{H}_p(H) = H$ and so $\mathrm{H}_p(H) = \mathrm{H}_p(G) = H \trianglelefteq G$. For each $g \in G - H$, $C_H(g)$ is elementary abelian since all elements in $(C_H(g)\langle g \rangle) - C_H(g)$ are of order p. In particular, $Z(G) < A$ then implies that $Z(G)$ is elementary abelian.

If $U \nleq Z(G)$, then there is $g \in G - H$ such that g does not centralize U so that $U\langle g\rangle \cong S(p^3)$. If in this case, $U \leq Z(H)$, then $\mho_1(H) \cap U \neq \{1\}$ since

$$Z(H) = \Phi(H) = \mho_1(H)H'$$

and then there is $h \in H$ with

$$1 \neq h^p \in U \leq U\langle g\rangle \cong S(p^3)\,,$$

a contradiction. Hence $U \nleq Z(H)$ and so there is $u \in U$ with $u \notin Z(H)$. All elements in $H - A$ are not of order p (otherwise, $\exp(H) = p$ and $\exp(G) = p$, a contradiction) and so there is $h \in H - A$ with $o(h) > p$ so that $H = \langle h, u\rangle$, $u^h = uz$, where $\langle z\rangle = H'$ and $\Omega_1(\langle h\rangle) \cap U = \{1\}$. We have $\mho_1(H) = \langle h^p\rangle$, $A = \langle h^p\rangle \times U$, and so $\langle h^p\rangle \trianglelefteq G$ since $H \trianglelefteq G$. If $o(h^p) > p$, then there is $h' \in \langle h^p\rangle$ of order p^2 with $\langle h'\rangle \trianglelefteq G$ and so $\langle h', g\rangle \cong M_{p^3}$ (minimal nonabelian) and $\langle h\rangle \cap \langle h', g\rangle \neq \{1\}$, a contradiction. Hence $o(h) = p^2$ and $A = \langle h^p\rangle \times U \cong E_{p^3}$. Since $U \trianglelefteq G$, $U \nleq Z(H)$ and $U\langle g\rangle \cong S(p^3)$, it follows that we may choose $h \in H - A$ so that $k = gh$ (of order p) centralizes U. Then for each $a \in A - U$, we have $\langle a, k\rangle \cong S(p^3)$. In particular (since $h^p \in A - U$)

$$\langle h^p, k\rangle \cong S(p^3) \text{ with } \langle h\rangle \cap \langle h^p, k\rangle = \langle h^p\rangle \neq \{1\}\,,$$

a contradiction. We have proved that $U \leq Z(G)$.

We know that $Z(G) < A$ and $Z(G)$ is elementary abelian and so $|Z(G) : U| \leq p$. Let B be a G-invariant subgroup of A with $Z(G) < B$ and $|B : Z(G)| = p$. Let $x \in B - Z(G)$. If each $g \in G - H$ centralizes x, then $x \in Z(G)$, a contradiction. Hence there is $g \in G - H$ (of order p) so that $[g, x] \neq 1$ and then $[g, x] \in Z(G)$ and $\langle g, x\rangle$ is minimal nonabelian (by Lemma 65.2) because $\langle g, x\rangle$ is of class 2 and so

$$[g, x]^p = [g^p, x] = 1 \quad \text{and} \quad \langle g, x\rangle' = \langle [g, x]\rangle \cong C_p\,.$$

Because $gx \in G - H$, we have $o(gx) = p$ and so

$$1 = (gx)^p = g^p x^p [x, g]^{\binom{p}{2}} = x^p \text{ and } o(x) = p\,.$$

It follows that all elements in $B - Z(G)$ are of order p and so $B = \langle B - Z(G)\rangle$ is elementary abelian. By Lemma 65.1,

$$B = \Omega_1(H) \cong E_{p^3} \text{ and so } U = Z(G) \cong E_{p^2}\,,$$

and $C_H(g) = U$ since $C_H(g)$ is elementary abelian and $C_B(g) = U$. We also have

$$\langle g, x\rangle \cong S(p^3) \text{ with } x \in B - U, g \in G - H, \quad [g, x] \neq 1, \quad \text{and}$$
$$B_1 = \langle g, x\rangle \cap H = \langle [g, x], x\rangle \cong E_{p^2}\,.$$

It follows that no element in B_1 is a pth power of an element in H. If $B \leq Z(H)$, then

$$S = \Omega_1(\mho_1(H)) \cong E_{p^2} \text{ and so } S \cap B_1 \neq \{1\}$$

and each element in $S \cap B_1$ is a pth power of an element in H, a contradiction. We get $B \nleq Z(H)$ and so $\Omega_1(Z(H)) = U$ and therefore for any $h \in H - A$ (of order $> p$) and $x \in B - U$, $\langle h, x\rangle = H$ with $[h, x] \in U$. We have

$$A = \langle h^p \rangle \times B_1 \text{ and } \mho_1(H) = \langle h^p \rangle \trianglelefteq G .$$

If $o(h^p) > p$, then there is an element h' of order p^2 in $\langle h^p\rangle$ so that $\langle h', g\rangle \cong M_{p^3}$ (minimal nonabelian), a contradiction. Hence $o(h^p) = p$ and so $A = B \cong E_{p^3}$. It follows that G stabilizes the chain $A > U > \{1\}$ and so $G/A \cong E_{p^2}$ and $G = H\langle g\rangle$ is of order p^5. It follows that there is $g' \in G - H$ (of order p) such that for an $a \in A - U$, we have $\langle [g', a]\rangle = \langle h^p\rangle \leq U$. But $\langle g', a\rangle \cong S(p^3)$ is minimal nonabelian, a contradiction. We have proved that there is no minimal nonabelian subgroup of G containing A.

(iii) We study in the sequel the general situation, where A is a maximal normal abelian subgroup of rank ≥ 2 in G with $\exp(G/A) = p$ (see (i)) and there is no minimal nonabelian subgroup in G containing A (by (ii)).

Let $g \in G - A$ so that $g^p \in A$. By Lemma 57.1, there is $a \in A$ such that $H = \langle g, a\rangle$ is minimal nonabelian. Set $H_1 = H \cap A$ and we have $H = H_1\langle g\rangle$ with $|H : H_1| = p$ and $H_1 < A$ by (ii). Set

$$A_1/H_1 = \Omega_1(A/H_1) \neq \{1\} \text{ so that for each } x \in A_1 - H_1 ,\ x^p \in H_1 \text{ and so } o(x) = p .$$

We have proved that all elements in $A_1 - H_1$ are of order p and so H_1 is elementary abelian. In fact,

$$A_1 = \Omega_1(A) > H_1 \text{ and since } p^2 \leq |H_1| \leq p^3, \text{ we have } p^3 \leq |H| \leq p^4 , \quad o(g) \leq p^2$$

and A is of rank ≥ 3. Each element $g \in G - A$ is of order $\leq p^2$ and $g^p \in \Omega_1(A)$. Let $H < X \leq A_1\langle g\rangle$ with $|X : H| = p$ so that all elements in $X - H$ are of order p. We have proved that for each $g \in G - A$, there are elements of order $p(A\langle g\rangle) - A$.

For each $g \in G - A$, we consider $C_A(g)$. Set $C = C_{A\langle g\rangle}(g)$, $C_1 = C \cap A = C_A(g)$ and consider $A_1 \leq A\langle g\rangle$ with $A_1 > C$ and $|A_1 : C| = p$. For any $x \in (A_1 \cap A) - C_1$, $\langle g, x\rangle$ is minimal nonabelian (see the proof of Lemma 57.1) and so $o(x) = p$. Hence all elements in $(A_1 \cap A) - C_1$ are of order p and so $C_1 = C_A(g)$ is elementary abelian. Thus, for each $g \in G - A$, $C_A(g)$ is elementary abelian and since $Z(G) \leq C_A(g)$, we get that $Z(G)$ is elementary abelian.

(iv) First we consider the case, where all elements in $G - A$ are of order p. Since $\exp(G) > p$, we get in this case $\exp(A) > p$ and then the Hughes subgroup $H_p(G)$ is equal to A. Let H be any minimal nonabelian subgroup in G. Since H is generated by all elements in $H - (H \cap A)$ and all these elements are of order p, it follows that $\exp(H) = p$ and therefore $H \cong S(p^3)$.

Suppose in addition that $|G/A| > p$. Then we consider a subgroup K/A of G/A with $|K/A| = p^2$ so that K is metabelian and $\mathrm{H}_p(K) = A$. By a result of Hogan–Kappe (see [HogK] in vol. 3), in a metabelian p-group the index of the Hughes subgroup is at most p. This is a contradiction and so $|G/A| = p$. We have obtained the groups stated in our theorem.

(v) Now suppose that there are elements of order p^2 in $G - A$ and let g be one of them. Then $g^p \in A_1 = \Omega_1(A)$ and $o(g^p) = p$. Let $a \in A_1$ be such that $H = \langle a, g\rangle$ is minimal nonabelian (Lemma 57.1) and set $H_1 = H \cap A$ so that $H_1 \cong \mathrm{E}_{p^2}$ or $H_1 \cong \mathrm{E}_{p^3}$ and $\Omega_1(H) = H_1$. We know that there is an element $g_1 \in (A\langle g\rangle) - A$ of order p. We may choose a generator $g \in \langle g\rangle$ so that $g_1 = cg$ for some $c \in A - H_1$ (since there are no elements of order p in $H - H_1$). We have $H' = \langle a'\rangle$, where

$$a' = [g, a] \neq 1 \text{ with } a' \in H_1 .$$

Then

$$[g_1, a] = [cg, a] = [c, a]^g[g, a] = [g, a] = a' \text{ and so } \langle g_1, a\rangle \cong \mathrm{S}(p^3) ,$$

where we note that a' commutes with g and so a' commutes with $g_1 = cg$. Obviously, $g \notin \langle g_1, a\rangle$ and so if $\langle a'\rangle = \langle g^p\rangle$, then $H \cong \mathrm{M}_{p^3}$ and then we get a contradiction. Hence $H_1 \cong \mathrm{E}_{p^3}$ and $\mathrm{Z}(H) = \langle g^p, a'\rangle \cong \mathrm{E}_{p^2}$. We know that $\mathrm{Z}(G)$ is elementary abelian and so $\mathrm{Z}(G) \leq A_1$. Suppose that $z \in \mathrm{Z}(G)$ with $z \in A_1 - H_1$. Then $o(zg) = p^2$ with $(zg)^p = g^p$ and $zg \notin H$, a contradiction. Hence

$$\mathrm{Z}(G) \leq \mathrm{Z}(H) = \langle g^p, a'\rangle \cong \mathrm{E}_{p^2} .$$

Assume in addition that $A_1 < A$, i.e., A is not elementary abelian. Let A_2 be a G-invariant subgroup in A such that $A_1 < A_2$ and $|A_2 : A_1| = p$. Then all elements in $A_2 - A_1$ are of order p^2 and if a_0 is one of them, then

$$\mho_1(A_2) = \langle a_0^p\rangle \cong \mathrm{C}_p \text{ and so } \langle a_0^p\rangle \leq \mathrm{Z}(G) .$$

By the previous paragraph, $a_0^p \in \mathrm{Z}(H)$ but $a_0 \notin H$, a contradiction. We have proved that A is elementary abelian of order $\geq p^4$ (since $H_1 < A$). If an element $b \in A - H_1$ centralizes g, then $bg \notin H$ but $1 \neq (bg)^p = g^p \in H$,a contradiction. We have also proved that $\mathrm{C}_A(g) = \mathrm{Z}(H) \cong \mathrm{E}_{p^2}$. We know that

$$|H| = p^4 , \quad H \cap A = H_1 \cong \mathrm{E}_{p^3} \text{ with } \exp(H) = p^2 .$$

Let $M > H$ with $|M : H| = p$. Since

$$|H : \mho_1(H)| = p^3 , \quad \text{we have } |M : \mho_1(M)| \leq p^4 .$$

If $p \geq 5$, then $|M : \mho_1(M)| < p^p$ and so M is absolutely regular. But then M is regular and since all elements in $M - H$ are of order p, Lemma 7.2 implies that M is of exponent

p, a contradiction. We have proved that $p = 3$ and so G is an irregular 3-group. We know that

$$C_A(g) = Z(H) = \langle g^3, H' \rangle \cong E_9 \text{ and } Z(G) \le Z(H) \text{ with } Z(A\langle g\rangle) = Z(H)\,.$$

Suppose that there is an element $g_1 \in (A\langle g\rangle) - (A \cup H)$ of order 3^2. Since $g_1^3 \in Z(A\langle g\rangle) = Z(H)$, we have a contradiction. Hence all elements in $(A\langle g\rangle) - (A \cup H)$ are of order 3 and so $H_3(A\langle g\rangle) = H$. But $A\langle g\rangle$ is metabelian and so we must have $|(A\langle g\rangle) : H| = 3$ which implies that $A \cong E_{3^4}$.

Assume, by way of contradiction, that $|G/A| > 3$. Let

$$A\langle g\rangle < B \le G \text{ with } |B : (A\langle g\rangle)| = 3\,.$$

If all elements in $B - (A\langle g\rangle)$ are of order 3, then $H_3(B) = H$ and $|B : H| = 9$. But $B' \le A$ and so B is metabelian. By a Kappe–Hogan result, in a metabelian p-group the index of the Hughes subgroup is at most p. This gives a contradiction. We conclude that there is $g_2 \in B - (A\langle g\rangle)$ of order 9 so that $1 \ne g_2^3 \in A - H_1$. We have

$$\langle H', g^3\rangle = Z(H) = Z(A\langle g\rangle) \trianglelefteq B\,,$$

and considering the action of g_2 on A (instead of g), we have $|C_A(g_2)| = 3^2$ and so g_2 acts non trivially on $Z(H) = \langle H', g^3\rangle$. It follows that $L = \langle g_2, Z(H)\rangle$ is minimal nonabelian (of order 3^4). But $g \notin L$ and $1 \ne g^3 \in L$, a contradiction. We have proved that $|G/A| = 3$ and so $G = A\langle g\rangle$ is an irregular 3-group of order 3^5 with $H_3(G) = H$, which is the nonmetacyclic minimal nonabelian group of order 3^4 and exponent 3^2. Also we have

$$Z(G) = Z(H) = \langle H', g^3\rangle \cong E_9$$

and G is of class 3.

By Lemma 1.1, $|Z(G)| = 9$ implies

$$3^5 = |G| = 3|Z(G)||G'| \text{ and so } |G'| = 9\,.$$

On the other hand, $G' \le A \cap H$ and so $G' \cong E_9$ with

$$G' \cap Z(G) = H' \cong C_3 \text{ and } C_G(G') = A\,.$$

Let $g_1 \in G - (A \cup H)$ be an element of order 3 so that $S = G'\langle g_1\rangle \cong S(27)$ and $S \trianglelefteq G$. Since $A \cap S = G'$ and $A \cong E_{3^4}$, it follows that A covers G/S and so $G/S \cong E_9$. On the other hand,

$$g^3 \in Z(G) \text{ and } S \cap Z(G) = H' \text{ so that } \langle g\rangle \cap S = \{1\}\,.$$

But then G is a splitting extension of S with $\langle g\rangle \cong C_9$ and so $G/S \cong C_9$, a final contradiction. Our theorem is proved. □

Problem 1. Study the primary $\mathcal{A}_n$-groups, $n > 2$, all of whose $\mathcal{A}_2$-subgroups are isolated.

Problem 2. Classify the nonmetacyclic p-groups all of whose minimal nonmetacyclic subgroups are isolated.

Problem 3. Study the irregular p-groups all of whose maximal regular subgroups are isolated.

Problem 4. Study the nonmetacyclic p-groups all of whose maximal metacyclic subgroups are isolated.

Problem 5. Classify the nonabelian p-groups all of whose maximal abelian subgroups are isolated.

§ 252 Nonabelian p-groups all of whose maximal abelian subgroups are isolated

We recall that a subgroup H of a p-group G is said to be *isolated* if for each cyclic $C < G$, we have either $C \leq H$ or C avoids H, i.e., $H \cap C = \{1\}$.

Here we shall solve Problem 5 from § 251 stated by the first author. We shall classify nonabelian p-groups all of whose maximal abelian subgroups are isolated. We prove the following somewhat surprising result.

Theorem 252.1. *Let G be a nonabelian p-group all of whose maximal abelian subgroups are isolated. Then $p > 2$ and G is of exponent p.*

Proof. If G is nonabelian of exponent $p > 2$, then G satisfies the hypothesis. Therefore, one may assume in the sequel that $\exp(G) \geq p^2$.

Let G be a nonabelian p-group all of whose maximal abelian subgroups are isolated. Suppose that $p = 2$ and let A be a maximal normal abelian subgroup of G. Let

$$A < B \leq G \text{ with } |B : A| = 2\,.$$

Then all elements in $B - A$ must be of order 2. If $\exp(A) = 2$, then B is elementary abelian, contrary to $C_G(A) = A$. Hence $\exp(A) > 2$, and if $i \in B - A$, then i inverts each element in A. Therefore, one has $C_A(i) = A_1 = \Omega_1(A)$. Let X be a maximal abelian subgroup of G containing (the abelian subgroup) $\langle i \rangle C_A(i)$. Then $X \cap A = A_1$ and so, if a is an element of order 4 in $A - A_1$, then $1 \neq a^2 \in A_1 \leq X$, a contradiction. Thus, $p > 2$. Let A, B be any cyclic subgroups with $A \cap B > \{1\}$ and let M be a maximal abelian subgroup in G containing A. If

$$B \nleq M\,, \quad \text{then } B \cap M \geq B \cap A > \{1\}$$

hence M is not isolated in G, which is a contradiction. Hence $B \leq M$ and so $\langle A, B \rangle$ is abelian. We have proved that any two cyclic subgroups with a non-trivial intersection generate an abelian subgroup. Such groups have been classified in § 160.

If $\Omega_1(G)$ is abelian, then we consider a maximal normal abelian subgroup A of G containing $\Omega_1(G)$. But then for $g \in G - A$, the subgroup $\Omega_1(\langle g \rangle)$ is contained in $\Omega_1(G) \leq A$ so that A is not isolated in G, a contradiction. Hence, the subgroup $\Omega_1(G)$ is nonabelian and hence G is a group of Theorem 160.5.

By Theorem 160.5, G has a proper subgroup G_0 such that $G_0 = H_p(G)$ and G_0 is either abelian or G_0 is a group of Theorem 160.1. In the second case, G_0 is nonabelian but $\Omega_1(G_0)$ is elementary abelian. In that case, let A_1 be a maximal abelian subgroup of G_0 containing $\Omega_1(G_0)$ and let A be a maximal abelian subgroup of G containing A_1. But then for an $x \in G_0 - A_1$, one has

$$\langle x \rangle \cap A_1 = \langle x \rangle \cap A > \{1\}\,,$$

which is a contradiction since $o(x) > p$.

We have proved that the subgroup $G_0 = \mathrm{H}_p(G)$ is abelian so $G_0 < G$, by hypothesis, and $\exp(G_0) > p$, by assumption. Assume, by way of contradiction, that $|G/G_0| > p$. Then we consider a subgroup

$$C \text{ with } G_0 < C \leq G \text{ and } |C/G_0| = p^2 .$$

In this case, C is metabelian and $\mathrm{H}_p(C) = G_0$. By a result of Hogan–Kappe [HogK] (see also Theorem 39.7), the index of a nontrivial Hughes subgroup in a metabelian p-group is at most p. This is a contradiction. Thus, $|G/G_0| = p$.

Let $K \neq G_0$ be a maximal abelian subgroup of G. Then K covers G/G_0, and let $g \in K - G_0$. Set

$$K_0 = \mathrm{C}_{G_0}(g) \text{ so that } K = \langle g\rangle \times K_0 \text{ and } K_0 = \mathrm{Z}(G) \text{ is elementary abelian} .$$

It follows from $\exp(G_0) > p$ that

$$L = \mho_1(G_0) > \{1\} \text{ and } L \cap K_0 = \{1\} ,$$

because each element in L is a pth power of an element in G_0. Being characteristic in $G_0 \lhd G$, the subgroup L is G-invariant and so L contains nontrivial central elements of G, contrary to $K_0 = \mathrm{Z}(G)$ and $L \cap K_0 = \{1\}$. The proof is complete. □

Problem 1. Classify the nonabelian p-groups containing an isolated maximal Abelian subgroup of rank two and exponent $> p$.

Problem 2. Classify the p-groups, $p > 2$, containing a proper isolated subgroup $\cong \mathrm{M}_{p^n}$.

Problem 3. Classify the p-groups G, $p > 2$, containing a proper isolated subgroup M of maximal class and order p^{p+1}.[1]

Problem 4. Study the irregular p-groups, $p > 2$, all of whose maximal regular subgroups are isolated.

Problem 5. Classify the irregular p-groups all of whose maximal absolutely regular subgroups are isolated. (Example: D_{2^n}.)

Problem 6. Study the nonabelian p-groups whose derived subgroup is isolated. (Example: any nonmetacyclic minimal nonabelian p-group.)

Problem 7. Classify the p-groups G satisfying $\Phi(G) = \mathrm{H}_p(G)$ (this will be $\iff$ $\Phi(G)$ is isolated).

---

1 Assume that $|M| > p^{p+1}$ and M is as above. Let $M < H \leq G$, where $|H : M| = p$. As all elements of the set $H - M$ have order p, it follows that $\mathrm{c}_1(H) \equiv \mathrm{c}_1(M) \equiv 1 + p + \cdots + p^{p-2} \pmod{p^p}$ (Lemma 12.3 (a)). It follows from Theorem 13.2 (a) that H is of maximal class. Then, by Exercise 10.10, G is of maximal class. The intersection $M_1 = M \cap G_1$ is the fundamental subgroup of M isolated in G_1. By Exercise 251.2, G_1 is a group of maximal class and order $\geq p^{p+2}$ so irregular, contrary to Theorem 9.6 (e). This explains why in Problem 3 stands p^{p+1}.

§ 253 Maximal abelian subgroups of p-groups, 2

Here we shall solve a special case of Problem 3325 by proving the following result.

Theorem 253.1 (Janko). *Let G be a nonabelian p-group which is not minimal nonabelian such that whenever H is a nonabelian proper subgroup in G, then all maximal abelian subgroups of H, except one, are also maximal abelian in G. Then G has an abelian subgroup of index p and either G is of maximal class or $\Omega_1(G)$ is elementary abelian which is in case $p = 2$ of order ≤ 8.*

Proof. Suppose that G is a nonabelian p-group which is not minimal nonabelian such that whenever H is a nonabelian proper subgroup in G, then all maximal abelian subgroups of H, except one, are also maximal abelian in G.

Let $H \in \Gamma_1$ be nonabelian. Let A_1 be a unique maximal abelian subgroup of H which is not maximal abelian in G. Then $A_1 \trianglelefteq G$ (indeed, the set of maximal abelian subgroups of H is G-invariant) and let A be a maximal abelian subgroup of G containing A so that A covers G/H, $A \cap H = A_1$, and $C_G(A_1) = A$. Let us prove the last equality. One has $HC_G(A_1) \geq HA = G$ and hence

$$p|H| = |G| = \frac{|H|\,|C_G(A_1)|}{C_H(A_1)} = \frac{|H|\,|C_G(A_1)|}{|A_1|} \Rightarrow |C_G(A_1) : A_1| = p \Rightarrow C_G(A_1) = A\,.$$

It follows that $A \trianglelefteq G$ and so A is a maximal normal abelian subgroup of G. One has $A \cap H = A_1$ hence, by the product formula, $|A : A_1| = p$.

Assume, by way of contradiction, that $|G/A| > p$ and let

$$A < B < G \text{ with } |B : A| = p \text{ and set } B_1 = B \cap H\,.$$

Assume that B_1 is abelian. Then $(B \cap H =)B_1 = A_1$ since $A_1 < H$ is minimal nonabelian. In that case,

$$(1) \qquad |G| = |BH| = \frac{|B|\,|H|}{|B \cap H|} = \frac{p|A|\,|H|}{|A_1|} = p^2|H| > |G|,$$

which is a contradiction. Thus, B_1 is nonabelian so that $B_1 > A_1$. As in (1), one has

$$(2) \qquad p|H| = |G| = |BH| = \frac{p|A|\,|H|}{|B_1|} = \frac{p^2|A_1|\,|H|}{|B_1|} \Rightarrow |B_1 : A_1| = p.$$

Also, $B < G$ is nonabelian (by assumption, $|G| \geq p^2|A| = p|B|$). Let $b_1 \in B_1 - A_1$ so that

$$b_1^p \in Z = Z(B_1) = C_{A_1}(b_1) \neq \{1\} \text{ and } Z < A_1\,.$$

Let

$$(\langle b_1 \rangle Z) < X \leq B_1 \text{ with } |X : (\langle b_1 \rangle Z)| = p\,.$$

Then X is nonabelian with maximal abelian subgroups $\langle b_1 \rangle Z$ and $X_1 = X \cap A_1$. Since $X_1 \leq A_1 < A$, X_1 is not a maximal abelian subgroup in G. By our hypothesis, $\langle b_1 \rangle Z$

must be a maximal abelian subgroup in G and so $C_{A_1}(b_1) = C_A(b_1) = Z$. It follows

$$Z(B) = Z(B_1) = Z \text{ and so } |A : Z(B)| \geq p^2 .$$

Now we consider the nonabelian proper subgroup B and note that A is a maximal abelian subgroup of B which is also a maximal abelian subgroup of G. By our hypothesis, there is a maximal abelian subgroup of B which is not a maximal abelian subgroup of G. Hence there is $b \in B - (A \cup B_1)$ so that $b^p \in Z$ and $\langle b \rangle Z$ is a maximal abelian subgroup of B but it is not a maximal abelian subgroup of G. Let

$$(\langle b \rangle Z) < Y \leq B \text{ be such that } |Y : (\langle b \rangle Z)| = p .$$

Then Y is nonabelian with maximal abelian subgroups $\langle b \rangle Z$ and $Y_1 = Y \cap A$. Since $|Y_1| = p|Z|$ we have $Y_1 < A$ and so Y_1 is not a maximal abelian subgroup of G. By our hypothesis, $\langle b \rangle Z$ must be a maximal abelian subgroup of G, a contradiction. We have proved that $|G/A| = p$ and so A is an abelian subgroup of index p in G.

From the hypothesis it follows that each nonabelian subgroup in G contains its centralizer in G. Then, by Exercise 1 in § 232, either G is of maximal class or $\Omega_1(G)$ is elementary abelian and by Theorem 232.1, in case $p = 2$, one has $|\Omega_1(G)| \leq 8$. □

Problem. Study the nonabelian p-groups G such that, whenever $M \leq G$ is minimal nonabelian, the at most one maximal subgroup of M is not maximal abelian in G.

§ 254 On p-groups with many isolated maximal abelian subgroups

Recall (see § 251 and § 252) that a subgroup A of a group G is isolated if, whenever a cyclic $C \leq G$ satisfies $C \cap A > \{1\}$, then $C \leq A$. (See the above sections and appendix on properties of isolated subgroups of a p-group.) In particular, if $A < B \leq G$, where A is isolated and $|B : A| = p$, then $\Omega_1(B) = B$. As B is isolated in G and $M \leq G$, it follows that $B_1 = B \cap M$ is isolated in M.

By Theorem 252.1 (= Theorem 104.1), if all maximal abelian subgroups of a nonabelian p-group G are isolated, then $\exp(G) = p$. In this section we reprove that result using an alternate approach (see Corollary 254.2, below).

Let $\mathcal{A}(G)$ be the set of maximal abelian subgroups in a nonabelian p-group G.

Theorem 254.1. *Let G be a nonabelian p-group, let $A \in \Gamma_1$ be abelian of exponent $p^e > p$ and $B \in \mathcal{A}(G) - \{A\}$. If B is isolated in G, then* $\exp(Z(G)) = p^e$ *and the subgroup A is not isolated in G.*

Proof. We retain the same notation as in the statement. Let G be a counterexample of minimal order. One has $AB = G$ so that $A \cap B = Z(G)$. Write $\mho = \mho_{e-1}(A) \cap Z(G)$. As $\mho \lhd G$ is a nonidentity central subgroup (of exponent p), there is $x \in \mho$ of order p; then $x \in \mho \leq Z(G) = A \cap B$. By Theorem 7.2 (c), there is a cyclic $C \leq A$ of order p^e such that $x \in C$. By Exercise A.101.7, the subgroup $Z(G) = A \cap B$ is isolated in A since B is. One has $x \in C \cap Z(G)$ so that $C \leq Z(G)$ since $Z(G)$ is isolated in A hence $\exp(Z(G)) \geq \exp(C) = p^e$. It follows from $Z(G) < A$ that $\exp(Z(G)) \leq \exp(A) = p^e$. Thus, $\exp(Z(G)) = p^e$ proving the first assertion.

It remains to prove that A is not isolated in G. Assume that this is false. Then all elements of the set $G - A$ have order p. Let $x \in G - A$ and let $y \in Z(G))(< A)$ be of order p^e. Then $o(x) = p$, $xy \notin A$ so that $o(xy) = p$. However, $(xy)^p = x^p y^p = y^p \neq 1$, a final contradiction. □

Corollary 254.2 (= Theorem 252.1 (Janko)). *If all maximal abelian subgroups of a nonabelian p-group G are isolated, then* $\exp(G) = p$.

Proof. Assume that G is a counterexample of minimal order; then $\exp(G) = p^e > p$. Let $A \in \mathcal{A}(G)$ be of exponent p^e and let $A < M \leq G$, where $|M : A| = p$; then M is nonabelian (of exponent p^e). Assume that $B_1 \in \mathcal{A}(M) - \{A\}$ and let $B_1 \leq B \in \mathcal{A}(G)$. As B is isolated in G, it follows that $B_1 = B \cap M$ is isolated in M (Exercise A.101.7). Thus, all maximal abelian subgroups are isolated in a nonabelian p-group $M \leq G$. Therefore, if $M < G$, then, by induction, $\exp(M) = p < \exp(A)$, which is a contradiction since $A < M$ has exponent $p^e > p$. Thus, $M = G$ so that $A \in \Gamma_1$. By Theorem 254.1, the subgroup A, being of exponent $p^e > p$, is not isolated in G, contrary to the hypothesis. Thus, all maximal abelian subgroups of G have exponent p, and this implies that $\exp(G) = p$. □

Exercise 1. Suppose that a nonabelian p-group G contains an isolated maximal normal abelian subgroup A of exponent $p^e > p$. Then $\mathrm{Z}(G)$ is not isolated in A.

Solution. Assume this is false. As we know, $\exp(\mathrm{Z}(G)) = p$. Then any subgroup of $\mathrm{Z}(G)$ of order p is isolated in A. Therefore, if $L \leq \mathrm{Z}(G)$ is of order p, then $L \nleq \mho_1(A) = \Phi(A)$. Then there is in A a subgroup A_1 of index p such that $L \nleq A_1$. It follows that $A = A_1 \times L$, and we conclude that $A = \mathrm{Z}(G) \times B$ for some $B < A$. Obviously, $\mho_{e-1}(A) \leq B$. As the nonidentity subgroup $\mho_{e-1}(A) \lhd G$, we get $\mho_{e-1}(A) \cap \mathrm{Z}(G) > \{1\}$. However, $\mho_{e-1}(A) \cap \mathrm{Z}(G) \leq B \cap \mathrm{Z}(G) = \{1\}$, and this is a contradiction.

§ 255 Maximal abelian subgroups of p-groups, 3

Here we solve Problem 2324 and improve Theorem 91.3 for odd p.

Theorem 255.1. *Let G be a nonabelian p-group satisfying the following condition:*

($*$) Whenever $A < B \le G$ are nonabelian, then $Z(A) \le Z(B)$.

This condition holds $\iff$ the intersection of any two distinct maximal abelian subgroups of G is equal to $Z(G)$. In that case either G has an abelian subgroup of index p or $\mho_1(G) \le Z(G)$.

Proof. Let G be a p-group satisfying condition ($*$). Let $X \neq Y$ be any two distinct maximal abelian subgroups of G. Then $Z(G) \le X \cap Y$ and the subgroup $\langle X, Y\rangle$ is nonabelian with $X \cap Y \le Z(\langle X, Y\rangle)$. By the hypothesis, $Z(\langle X, Y\rangle) \le Z(G)$, and so $\langle X, Y\rangle = Z(G)$. Thus, the intersection of any two distinct maximal abelian subgroups of G coincides with $Z(G)$. By Theorem 91.2, the subgroup $C_G(x)$ is abelian for each $x \in G - Z(G)$. This fact then obviously implies that condition ($*$) holds.[1]

Assume, in addition, that G has no abelian subgroup of index p. If $p = 2$, then Theorem 91.3 implies that $\mho_1(G) \le Z(G)$. Therefore suppose in what follows that $p > 2$. Let $A < G$ be maximal normal abelian subgroup. Take $x \in G - A$. It follows from Lemma 57.1 that there is $a \in A$ such that the subgroup $B = \langle a, x\rangle$ is minimal nonabelian. Since $x^p \in \Phi(B) = Z(B)$ and $Z(B) \le Z(G)$, by hypothesis, it follows that $x^p \in Z(G) < A$, and we conclude that $\exp(G/A) = p$. Moreover, all elements in the set $(G/Z(G)) - (A/Z(G))$ have order p so that $H_p(G/Z(G)) \le A/Z(G)$. By assumption, $|G/A| > p$ so there is in G/A a subgroup H/A of order p^2. The subgroup H and so $H/Z(G)$ are metabelian. Assume that $\exp(A/Z(G)) > p$. Then $H_p(G/Z(G)) = A/Z(G))$. In particular, $H_p(H/Z(G)) = A/Z(G))$. By the theorem of Hogan–Kappe [HogK], we have in that case $|H/Z(G) : H_p(H/Z(G)| = p$. It follows from the above that $|H : A| = p$, contrary to a choice of H. Thus, the group G/A has no subgroup of order p^2 so that $|G/A| = p$. By the above, if G has no abelian subgroup of index p, then $\exp(A/Z(G)) = p$, i.e., $A/Z(G)$ is elementary abelian. Since $x^p \in Z(G)$ for all $x \in G - A$, we get $\exp(G/Z(G) = p$ so that $\mho_1(G) \le Z(G)$, completing the proof. □

1 Indeed, let $A < B \le G$ be nonabelian and assume that $Z(A) \nleq Z(B)$. If $x \in Z(A) - Z(B)$, then $C_G(x) \ge A$ is nonabelian, contrary to what has just been said.

§ 256 A problem of D. R. Hughes for 3-groups

We recall that the Hughes subgroup $\mathrm{H}_p(G)$ (for p prime) of a group G is the (normal) subgroup of G generated by all elements of G which do not have order p. Then the Hughes conjecture asserts that either $\mathrm{H}_p(G) = \{1\}$ or $\mathrm{H}_p(G) = G$ or $|G : \mathrm{H}_p(G)| = p$. This conjecture is not true for $p \geq 5$ (G. E. Wall [Wa1]) but holds for $p = 2$ (Burnside) and here we show that it holds for $p = 3$ (Straus and Szekeres [SS]). This conjecture also holds for all metabelian p-groups (see [HogK]).

To facilitate the proof of the Hughes conjecture for $p = 3$, we first prove two auxiliary lemmas.

Lemma 256.1. *Let $h, g_1, \ldots, g_n$ be elements in a group G and $a_1, \ldots, a_n$ some integers. Then we denote*

$$h^{a_1 g_1 + \cdots + a_n g_n} = g_1^{-1} h^{a_1} g_1 \cdot g_2^{-1} h^{a_2} g_2 \cdot \cdots \cdot g_n^{-1} h^{a_n} g_n .$$

If $h \in \mathrm{H}_p(G)$ and $x \notin \mathrm{H}_p(G)$, then

$$h^{1+x+x^2+\cdots+x^{p-1}} = 1 .$$

Proof. Since $hx^{p-1} \notin \mathrm{H}_p(G)$, we have

$$1 = (hx^{p-1})^p = hx^{p-1} \cdot hx^{p-1} \cdot \cdots \cdot hx^{p-1} = h \cdot x^{-1}hx \cdot x^{-2}hx^2 \cdot \cdots \cdot x^{-(p-1)}hx^{p-1} = \\ = h^{1+x+x^2+\cdots+x^{p-1}} .$$ □

Lemma 256.2. *If $h \in \mathrm{H}_3(G)$ and $x, y \in G - \mathrm{H}_3(G)$ with $x\mathrm{H}_3(G) \neq y\mathrm{H}_3(G)$, then $h^{x+y} = h^{y+x}$ (= commutativity of the "addition" of group elements).*

Proof. By our assumption, $z = x^{-1}y \notin \mathrm{H}_3(G)$ and $z^{-1} \notin \mathrm{H}_3(G)$. Hence by Lemma 256.1, we have for $h_1 \in \mathrm{H}_3(G)$,

$$1 = h_1^{1+z+z^2} = h_1^{1+z^{-1}+z^{-2}} = h_1^{1+z^2+z} .$$

This gives $h_1^{z+z^2} = h_1^{z^2+z}$ and so setting $h_1 = h^{xz^2}$ with $h \in \mathrm{H}_3(G)$, we get

$$h^{xz^2(z+z^2)} = h^{xz^2(z^2+z)}, \quad h^{x(1+z)} = h^{x(z+1)},$$
$$h^{x(1+x^{-1}y)} = h^{x(x^{-1}y+1)}, \quad h^{x+y} = h^{y+x} .$$ □

Theorem 256.3. *The Hughes conjecture holds for 3-groups.*

Proof. Let G be a 3-group with $|G : \mathrm{H}_3(G)| > 3$. We consider a subgroup $H/\mathrm{H}_3(G)$ in $G/\mathrm{H}_3(G)$ of order 9 so that $H/\mathrm{H}_3(G) \cong \mathrm{E}_9$. Let $x, y \in H - \mathrm{H}_3(G)$ be such that $\langle x, y\rangle$ covers $H/\mathrm{H}_3(G)$. We have

$$3 = -(1+x+x^2)y - y^2(1+x+x^2) + (1+y+y^2) + (1+xy+y^2x^2) + (1+x^2y+y^2x) = f(x, y),$$

where we note that

$$(xy)^2 = (xy)^{-1} = y^{-1}x^{-1} = y^2x^2 \text{ and } (x^2y)^2 = (x^2y)^{-1} = y^{-1}x^{-2} = y^2x\,.$$

But according to Lemma 256.2, we have therefore $h^3 = h^{f(x,y)}$ for every $h \in \mathrm{H}_3(G)$. Then Lemma 256.1 implies $h^3 = 1$ and so $\mathrm{H}_3(G) = \{1\}$. The Hughes conjecture for 3-groups is proved. □

Appendix 58 Alternate proof of Passman's Theorem 1.23

In this subsection, the alternate proof of important Passman's Theorem 1.23 due to the second author is offered.

Theorem A.58.1 (= Theorem 1.23 (Passman)). *Let G be a nonabelian p-group such that for every G-invariant subgroup K of index p in G', the quotient group G/K is Dedekindian. Then G is also Dedekindian.*

Proof (Janko). Let K be a G-invariant subgroup of index p in G'. Then $p = 2$ since G/K is nonabelian Dedekindian (Theorem 1.20).

Let G be a minimal counterexample. Let R be a G-invariant subgroup of index 2 in K and assume that $R \neq \{1\}$. As G is a minimal counterexample, it follows that G/R is Dedekindian. But then $(G/R)'$ is of order 4, contrary to Theorem 1.20. Hence $R = \{1\}$ and so $|K| = 2$ and $|G'| = 4$. As G/K is nonabelian Dedekindian, it follows that $G' = \Phi(G)$.

It follows from the description of nonabelian Dedekindian 2-groups (see Theorem 1.20) that there is in G/K a subgroup Q/K such that $Q > G'$ and $Q/K \cong \mathrm{Q}_8$. If $Q' = G'$, then Q is of maximal class, by Taussky's Proposition 1.6, in which case $Q/K \cong \mathrm{D}_8$ (Theorem 1.2), contrary to the choice of Q. Hence $|Q'| = 2$ and since Q' covers G'/K, it follows that $G' = Q' \times K \cong \mathrm{E}_4$ which together with $Q' \lhd G$ (indeed, Q' is characteristic in $Q \lhd G$) implies that $G' \leq Z(G)$.

By Corollary 1.17.3, G possesses a minimal nonabelian subgroup A distinct from Q_8 (otherwise, $G = S \times E$ with S being generalized quaternion, $Z(S) \lhd G$ and $\exp(E) \leq 2$; then it is easily seen that $S \cong \mathrm{Q}_8$ is Dedekindian (otherwise, $S/Z(Q) \cong \mathrm{D}_8$ is non-Dedekindian so $/Z(G)$ is non-Dedekindian, a contradiction). It follows that $A > K$ since G/K is Dedekindian. The fact that $|A : \Phi(A)| = 4$ and $|\Phi(A)| = |G'| = 4$ yields $|A| \in \{8, 16\}$.

Assume that $|A| = 16$. Then $\Phi(A) = \Phi(G) = G'$. There are exactly two such minimal nonabelian groups (see Lemma 65.1):

(i) $A = \langle x, y \mid x^4 = y^4 = 1,\ x^y = x^3\rangle$ with $\Phi(A) = \langle x^2, y^2\rangle = G'$. In that case, $G/\langle y^2\rangle \cong \mathrm{D}_8$ is non-Dedekindian, a contradiction.

(ii) $A = \langle x, y \mid x^4 = y^2 = 1,\ [x, y] = z,\ z^2 = [x, z] = [y, z] = 1\rangle$. In that case, $x^2 \in Z(G)$ and $G/\langle x^2\rangle \cong \mathrm{D}_8$ is non-Dedekindian, a contradiction.

Now let $|A| = 8$; then $A \cong \mathrm{D}_8$ with $A \geq K$. Let $v \in Z(G) - A$. Then $G/\langle v\rangle \cong A \cong \mathrm{D}_8$, contrary to the assumption that $G/\langle v\rangle$ is Dedekindian. □

It follows from Theorem A.60.1 that if the derived subgroup of a p-group G has order $> p$, then there is a proper non-Dedekindian epimorphic image of G of order $2|G/G'|$.

Exercise 1. Suppose that G is a non-Dedekindian p-group. If $|G'| = p^k > p$, then there is in G a nonnormal subgroup of order $\geq p^k$.

Solution. By Theorem A.60.1, there is in G' a G-invariant subgroup K of index p such that G/K is non-Dedekindian. Therefore, there is in G/K a nonnormal subgroup L/K; then L is not normal in G. Since $|L| \geq |L/K|\,|K| \geq p \cdot p^{k-1} = p^k$, we are done.

Exercise 2. Given k, suppose that all subgroups of G of order $\geq p^k$ are normal. Then $|G'| \leq p^k$.

Solution. Assume that $|G'| = p^s > p^k$; then G is non-Dedekindian (Theorem 1.20). Then, by Exercise 1, there is in G a nonnormal subgroup of order $p^s > p^k$, a contradiction.

Appendix 59 Iwasawa's theorem on modular p-groups

In this appendix, we prove Iwasawa's theorem characterizing the modular p-groups. Recall that a p-group G is *modular* if all its subgroups are quasinormal, i.e., permutable with all subgroups of G. All sections of modular p-groups are modular. The groups $S(p^3)$ and D_8 are nonmodular. The modular p-groups are classified in § 73. The minimal nonmodular p-groups are treated in § 78.

Theorem A.59.1 (Iwasawa [Iwa]). *If G is a nonmodular p-group, it has a section $U \in \{S(p^3), D_8\}$.*

Proof. It suffices to assume that G is minimal nonmodular. The groups $S(p^3)$, D_8 are (minimal) nonmodular. The group G contains two nonpermutable cyclic subgroups A and B such that the number $|A|\,|B|$ is minimal possible. One has $G = \langle A, B\rangle$ since $\langle A, B\rangle$ is not modular. Then $d(G) = 2$, $A \cap B \leq Z(G)$ and $G/(A \cap B)$ is not modular since $A/(A \cap B)$ and $B/(A \cap B)$ are not permutable. Therefore, without loss of generality, one may assume that $A \cap B = \{1\}$. As G is minimal nonmodular, $\Phi(A)$ and $\Phi(B)$ are quasinormal in G (e.g., $\Phi(A) \leq \Phi(G)$ so that $\langle \Phi(A), C\rangle < G$, where $C < G$ is cyclic). One has $H = \Phi(A)B = B\Phi(A) < G$. Assume that $AH = HA$. Then

$$AH = A(\Phi(A)B) = (A\Phi(A))B = AB,$$
$$HA = (B\Phi(A))A = B(\Phi(A)A) = BA\,,$$

so that $G = AB = BA$, contrary to the choice of A and B. Thus, A and H are not permutable. Set $F = A\Phi(B)(< G)$. As above, F and H are not permutable. Indeed, if $FH = HF$, then

$$FH = A\Phi(B)B\Phi(A) = AB\Phi(A) = A\Phi(A)B = AB\,,$$
$$HF = B\Phi(A)A\Phi(B) = BA\Phi(B) = B\Phi(B)A = BA$$

so that $AB = BA$, a contradiction. The subgroup $F \cap H = \Phi(A)\Phi(B)$ has index p in F and H. It follows that $F \cap H \lhd \langle F, H\rangle = G$. Since $G/(F \cap H)$ is not modular, one may assume, without loss of generality, that $F \cap H = \{1\}$. Then $F = \langle f \mid f^p = 1\rangle \cong C_p$ and $H = \langle h \mid h^p = 1\rangle \cong C_p$. In that case, $G = \langle f, h\rangle$.

If $p = 2$, then G, generated by two involutions f and h, is dihedral so it contains a subgroup $\cong D_8$, and we are done in this case.

Now let $p > 2$. If G is metacyclic, then $G = \langle f, h\rangle = \Omega_1(G) \cong E_{p^2}$, i.e., modular, a contradiction. Thus, G is not metacyclic. In that case, $|G/\mho_1(G)| > p^2$ (Corollary 36.8). Since G is generated by two elements, $G/\mho_1(G)$ is nonabelian. Let $H_1/\mho_1(G)$ be a minimal nonabelian subgroup of $G/\mho_1(G)$; then $H_1/\mho_1(G) \cong S(p^3)$ (Lemma 65.1), completing the proof. □

We deduce from Theorem A.59.1 the following facts. (1) A p-group G is modular $\iff$ it is $S(p^3)$- and D_8-free. (2) Metacyclic p-groups, $p > 2$, are modular. In particular, if $p > 2$ and any two-generator subgroup of a group G is metacyclic, then G is modular. Assume that this is false. Then G has a section $M/L \cong S(p^3)$. Let $H \le M$ be minimal such that $HL = M$. Then $d(H) = 2$ and H is nonmodular since $H/(H \cap L) \cong M/L$, contrary to the hypothesis.

Exercise 1. Classify the modular minimal nonabelian p-groups. (**Hint.** Nonmetacyclic minimal nonabelian p-groups are nonmodular.)

Exercise 2. Classify the modular $\mathcal{A}_2$-groups of prime power order.

Exercise 3. Prove that any minimal nonmodular p-group G is two-generator.

Solution. As we know, there are $L \trianglelefteq M \le G$ such that $M/L \in \{S(p^3), D_8\}$. Since M is nonmodular, we get $M = G$ so that $L \trianglelefteq G$. Assume that $L \nleq \Phi(G)$. Then there is a minimal $H < G$ such that $G = LH$. It follows from $H/(L \cap H) \cong G/L \in \{S(p^3), D_8\}$ that G has a proper nonmodular two-generator subgroup H, a contradiction. Thus, $L < \Phi(G)$, and we conclude that $d(G) = d(G/L) = 2$.

Appendix 60 On p-groups, containing only one noncyclic subgroup of order p^e, $e \geq 3$

Isaacs [BI] has noted that if G is a p-group of order $> p^e$, $e \geq 3$, and P is the unique noncyclic subgroup of order p^e in G, then P is abelian of type (p^{e-1}, p). In this appendix to classify the p-groups G satisfying the above condition.

By Theorem 1.17 (b) and Sylow's theorem, if a p-group G is neither cyclic nor a 2-group of maximal class, then the number of noncyclic subgroups of order $p^e > p$ in G is $\equiv 1 \pmod{p}$.

Theorem A.60.1. *Let $e \geq 3$ and suppose that a noncyclic p-group G of order $p^n > p^e$ has only one noncyclic subgroup of order p^e. Then one of the following holds:*
(a) $G \cong C_{p^{n-1}} \times C_p$.
(b) $G \cong M_{p^n}$.
(c) $p = 2$, $e = 3$, *G is metacyclic and*

$$G = \langle a, b \mid a^{2^{n-2}} = b^8 = 1,\ a^b = a^{-1},\ a^{2^{n-3}} = b^4,\ n > 4\rangle\,.$$

Here

$$Z(G) = \langle b^2\rangle\,, \quad G' = \langle a^2\rangle\,, \quad G/G' \cong C_4 \times C_2\,,$$
$$\Phi(G) = \langle a^2, b^2\rangle\,, \quad \Omega_2(G) = \langle a^{2^{n-4}}, b^2\rangle\,,$$

i.e., G is a group of Lemma 42.1 (c), $e = 3$.

Proof. If G has a cyclic subgroup of index p, it is as in (a) or (b), by Theorem 1.2 and the basic theorem on abelian p-groups. Next we assume that G has no cyclic subgroup of index p. Then it has a normal subgroup $R \cong E_{p^2}$ (Lemma 1.4). By hypothesis, G/R has only one subgroup of order p^{e-2}. By Proposition 1.3, either G/R is cyclic or a generalized quaternion group with $e - 2 = 1 \Rightarrow e = 3$.

Assume that G/R is cyclic. As G has no cyclic subgroup of index p, we get $\Omega_1(G) \cong E_{p^3}$. If $M/R \leq G/R$ is of order p^{e-1} (then $|M| = p^{e+1}$), then all > 1 maximal subgroups of M are noncyclic (of order p^e), a contradiction.

Now let $G/R \cong Q_{2^{n-2}}$, $e = 3$. Let $S/R < G/R$ be of order 2. Clearly, $S < C_G(R)$, so that S is abelian. Assume that $R < \Omega_1(G)$. Then $\Omega_1(G) = S$ is of order $2^e = 2^3$ and all remaining subgroups of order 2^3 of G are cyclic. If $S < T < G$, where $|T| = 2^4$, then all maximal subgroups (of order $2^3 = 2^e$) of T are noncyclic, a contradiction. Thus, $\Omega_1(G) = R$; then $\Omega_2(G) = S$ has order $2^3 = 2^e = 2^{2+1}$ so G is as in Lemma 42.1 (c). □

In particular, the unique noncyclic subgroup of G of order p^e is abelian of type (p^{e-1}, p), and this agrees with the above stated result.

Exercise 1. Suppose that P is a noncyclic p-group of order $> p^e \geq p^3$. If P has no normal noncyclic subgroup of order p^e, then P is a 2-group of maximal class and order $> 2^{e+1}$.

Solution. If P has a cyclic subgroup of index p and satisfies the hypothesis, it is a 2-group of maximal class of order $> 2^{e+1}$. Next assume that P has no cyclic subgroup of index p; then P is not a 2-group of maximal class. Then by Sylow and Theorem 1.17 (b), the number of noncyclic subgroups of order p^e in P is $\equiv 1 \pmod p$, and so one of these subgroups is P-invariant.

Exercise 2. If a noncyclic p-group G contains at most $p + 1$ subgroups of each order, then $G \in \{Q_8, C_{p^n} \times C_p, M_{p^{n+1}}\}$.

Hint. One may assume that $|G| > p^3$. By hypothesis, G and all its subgroups are two-generator. We use induction on $|G|$. Then all proper noncyclic subgroups of G are such as in the conclusion so metacyclic. It follows that G is either metacyclic or minimal nonmetacyclic and, if $p = 2$, then G is metacyclic (see Theorems 66.1 and 69.1). If G has a cyclic subgroup of index p, then $G \in \{C_{p^n} \times C_p, M_{p^{n+1}}\}$. Next we assume that G has no cyclic subgroup of index p. Let G be metacyclic. In that case, $|G| > p^4$ (if $|G| = p^4$, then $s_2(G) > p + 1$). Then G contains a normal abelian subgroup $R = \Omega_1(G)$ of type (p, p). Take $H/R < G/R$ of order p. Then H is abelian of type (p^2, p) (Proposition 10.19) so $s_2(G) = p+1$). It follows that H/R is the unique subgroup of order p is G/R so that G/R is either cyclic or generalized quaternion. If G/R is cyclic, then $G \in \{C_{p^n} \times C_p, M_{p^{n+1}}\}$. Now let $G/R \cong Q_{2^n}$. Then $\Omega_2(G) = H$ hence G is as in Lemma 42.1 (c) and $n = 5$. However, $s_3(G) = 7 > 3$ (here we use Hall's enumeration principle), a contradiction. Now let G be minimal nonmetacyclic. By Theorem 66.1, $p > 2$ (if $p = 2$, then $d(G) = 3$) and G is of maximal class and order 3^4. However, such group does not satisfy the hypothesis since $c_2(G) = \frac{3^4-3^2}{3(3-1)} = 12 > 3 + 1$.

Exercise 3. Study the p-groups G containing at most $p+1$ proper noncyclic subgroups of each possible order.

Exercise 4. Study the p-groups G containing exactly one subgroup $\cong E_{p^2}$.

Appendix 61 A necessary and sufficient condition for a p-group G to satisfy $\Phi(G) \le Z(G)$

In this appendix, we classify the p-groups in which $M \cap \Phi(G) = \Phi(M)$ for all minimal nonabelian $M \le G$. It appears that this condition is equivalent to inclusion $\Phi(G) \le Z(G)$. Namely, the following assertion holds.

Theorem A.61.1. *For a nonabelian p-group G, the following conditions are equivalent:*
(a) $\Phi(G) \le Z(G)$.
(b) $\Phi(G) \cap M = \Phi(M)$ *for any minimal nonabelian subgroup M of G.*

Proof. (i) Let $\Phi(G) \le Z(G)$. Take in G a minimal nonabelian subgroup M. Then $M/(M \cap \Phi(G)) \cong M\Phi(G)/\Phi(G)$ as a subgroup of $G/\Phi(G)$ is elementary abelian so its order is $\le p^2$ (indeed, $d(M) = 2$) so that $M \cap \Phi(G) \ge Z(M) = \Phi(M)$. It follows that

$$\Phi(M) \le M \cap \Phi(G) \Rightarrow |M : (M \cap \Phi(G))| \le |M : \Phi(M)| = p^2 .$$

As M is nonabelian and $\Phi(G) \le Z(G)$, it follows that $|M : (M \cap \Phi(G))| \ge p^2$, and so $|M : (M \cap \Phi(G))| = p^2$. In that case, $\Phi(G) \cap M = \Phi(M)$ hence (a) $\Rightarrow$ (b).

(ii) Now suppose that $\Phi(G) \cap M = \Phi(M)$ for any minimal nonabelian subgroup M of G. In that case, $\Phi(G)$ has no minimal nonabelian subgroup so it is abelian. Let $H/\Phi(G) < G/\Phi(G)$ be of order p. Assume that H is nonabelian. Then there is in H a minimal nonabelian subgroup M. As $\Phi(G)$ is abelian, $M \not\le \Phi(G)$. Therefore, $M\Phi(G) = H$ so, by the product formula, $|M : (M \cap \Phi(G))| = p$, and this implies that $M \cap \Phi(G) > \Phi(M)$, contrary to the hypothesis. Thus, all such subgroups H are abelian, and, obviously, they generate G. Then

$$C_G(\Phi(G)) \ge \langle H \mid \Phi(G) < H < G,\ |H : \Phi(G)| = p\rangle = G ,$$

so that $\Phi(G) \le Z(G)$, completing the proof. □

Exercise 1. Let $K < G$ be a proper normal subgroup of a nonabelian p-group G such that $\exp(G/K) = p$ and $M \cap K \le \Phi(M)$ for each minimal nonabelian subgroup M of G. Then $K \le Z(G)$.

Solution. As in the theorem, K is abelian. Let $K < L \le G$, where $|L : K| = p$. Assume that L is nonabelian. By hypothesis, L is not minimal nonabelian. Let $M < L$ be minimal nonabelian. Again $K \not\le M$. In that case,

$$KM = L \Rightarrow |M : (M \cap K)| = |L : K| = p \Rightarrow M \cap K \not\le \Phi(M) ,$$

contrary to the hypothesis. Thus, L is abelian for any of its choice. Then $C_G(K)$ contains all such L and these L generate G in view of $\exp(G/K) = p$. It follows that $C_G(K) = G$, and we are done.

Remark 1. It is possible to replace, in Exercise 1, the condition "$\exp(G/K) = p$" by "$\Omega_1(G/K) = G/K$."

Problem. Study the primary $\mathcal{A}_n$-groups G, $n > 2$, such that $M \cap \Phi(G) = \Phi(M)$ for every $\mathcal{A}_2$-subgroup M of G.

Appendix 62 Subgroups of some p-groups

We expose our results in the form of exercises.

Exercise 1. Let

$$G = \langle a, b \mid o(a) = 2^n,\ o(b) = 4,\ a^b = a^{-1+2^{n-2}},\ n \geq 4\rangle$$

be a metacyclic group of order 2^{n+2}. Set $z = a^{2^{n-1}}$. Here

$$\Omega_1(G) = R = \langle b^2, z\rangle \cong \mathrm{E}_4\,, \quad G/R \cong \mathrm{SD}_{2^n}\,, \quad \mathrm{Z}(G) = \langle z\rangle\,.$$

(This G is a group of Theorem 67.2 (c).) Describe the maximal abelian subgroups of G.

Solution. The group G is not of maximal class since $G/\langle a\rangle \cong \mathrm{C}_4$ hence $R = \Omega_1(G) \cong \mathrm{E}_4$ (Proposition 10.19). One has $a^{b^2} = a^{1+2^{n-1}}$ so that $C = \langle a, b^2\rangle \cong \mathrm{M}_{2^{n+1}}$ and $C/R \cong \mathrm{C}_{2^{n-1}}$. By Theorem 1.2, $G/R \cong \mathrm{SD}_{2^n}$. Let

$$\Gamma_1 = \{D, Q, C\}, \text{ where } D/R \cong \mathrm{D}_{2^{n-1}}\,, \quad Q/R \cong \mathrm{Q}_{2^{n-1}}\,, \quad C/R \cong \mathrm{C}_{2^{n-1}}\,.$$

By Theorem 1.2, D and Q has no cyclic subgroups of index 2 so that $\mathrm{c}_n(G) = \mathrm{c}_n(C) = 2$. If $L/R < D/R$ is of order 2, then L is abelian since D is not of maximal class (Proposition 10.19). As D/R is generated by involutions, we get $\mathrm{C}_D(R) = D$ and $R \not\leq \mathrm{Z}(C)$ so that $\mathrm{C}_G(R) = D$, and we get $C \cong \mathrm{M}_{2^{n+1}}$. Thus, all members of the set Γ_1 are nonabelian so all abelian subgroups of G of order 2^n are maximal abelian. If $\Phi/R < D/R$ is cyclic of index 2, then $\Phi = \Phi(G) = \langle R, a^2\rangle$ is maximal abelian in G and it is of type $(2^{n-1}, 2)$. Next, $V = \mathrm{C}_G(b) = \langle b, z\rangle$ is maximal abelian in G; V is abelian of type $(4, 2)$. Write $A = \langle a\rangle$. Let a cyclic subgroup $Z < G$ be maximal abelian; then $|Z| > 4$ since G is not of maximal class (Proposition 1.8). If $AZ = G$, then $A \cap Z = \mathrm{Z}(G) = \langle z\rangle$ so $Z \cong \mathrm{C}_8$. Let $M < G$ be noncyclic maximal abelian of order > 8 and $M \not\leq C$. In that case, M/R, as an abelian subgroup of order > 2 of $G/R \cong \mathrm{SD}_{2^{n-1}}$, is of order 4; then $|M| = 16$. One has $AM = G$ (otherwise, $AM = C$ which is impossible since $M \not\leq C$) so that $\mathrm{Z}(G) \geq A \cap M$, a contradiction since $|A \cap M| = 4 > 2 = |\mathrm{Z}(G)|$. Thus, if $M \not\leq C$, then $|M| = 8$. Note that $G' = \langle a^2\rangle \cong \mathrm{C}_{2^{n-1}}$ and G/G' is abelian of type $(4, 2)$.

Exercise 2. Let

$$G = \langle a, b \mid a^{2^n} = b^4 = 1,\ z = a^{2^{n-1}},\ a^b = a^{-1}z^{\epsilon},\ \epsilon = 0, 1,\ n \geq 3\rangle$$

be a metacyclic group of order 2^{n+2}. Here

$$R = \langle b^2, z\rangle = \mathrm{Z}(G) \cong \mathrm{E}_4\,, \quad G/R \cong \mathrm{D}_{2^n}\,.$$

(This G is a U_2-group of Theorem 67.2 (a).) Describe the maximal abelian and minimal nonabelian subgroups of G.

Hint. Let

$$\Gamma_1 = \{D, D_1, C\}, \text{ where } D/R \cong \mathrm{D}_{2^{n-1}} \cong D_1\,, \quad C/R \cong \mathrm{C}_{2^{n-1}}\,.$$

Then C is abelian of type $(2^n, 2)$, $R = \mathrm{Z}(G)$. Since $C/R = C/\mathrm{Z}(G) \cong G'$ (Lemma 1.1), $G' \cong \mathrm{C}_{2^{n-1}}$; then G/G' is abelian of type $(4, 2)$. Let $M/\mathrm{Z}(G) < G/\mathrm{Z}(G)$ be abelian of type $(2, 2)$. Since $MC = G$, $M \cap C = \Omega_2(C) > R = \mathrm{Z}(G)$ and $|\mathrm{Z}(G)| = 4$, it follows that M is nonabelian so that $M \cong \mathcal{H}_{2,2}$ since $\exp(M) = 4$. Let $U \neq C$ be maximal abelian in G. As $U \cap C = \mathrm{Z}(G)$, it follows that U is of type $(4, 2)$. Thus, all $\neq C$ maximal abelian subgroups of G are of type $(4, 2)$. By the above (see the argument with M), all minimal nonabelian subgroups of G are $\cong \mathcal{H}_{2,2}$.

Exercise 3. Let

$$G = \langle a, b \mid a^{2^n} = b^8 = 1,\ a^b = a^{-1},\ a^{2^{n-1}} = b^4,\ n > 3\rangle$$

be a metacyclic group of order 2^{n+2}. (This G is a group of Lemma 42.1 (c).) Here

$$\mathrm{Z}(G) = \langle b^2\rangle\,, \quad G' = \langle a^2\rangle\,, \quad G/G' \cong \mathrm{C}_4 \times \mathrm{C}_2\,,$$
$$\Phi(G) = \langle a^2, b^2\rangle\,, \quad \Omega_2(G) = \langle a^{2^{n-2}}, b^2\rangle\,.$$

Describe the maximal abelian and minimal nonabelian subgroups of G.

Hint. By Lemma 42.1 (c), $\mathrm{Z}(G) = \langle b^2\rangle \cong \mathrm{C}_4$. The subgroup $C = \langle a, b^2\rangle \in \Gamma_1$ is abelian of type $(2^n, 2)$. Next, $R = \Omega_1(G) \cong \mathrm{E}_4$. Write $\bar{G} = G/R$; then $\bar{G} \cong \mathrm{Q}_{2^n}$ so that $\Omega_2(G)$ is abelian of type $(4, 2)$. Let $A < G$ be maximal abelian, $A \neq C$ and $R < A$. Since $A \cap C = \mathrm{Z}(G)$, then $|A| = 8$ and $R \nleq A$, and we conclude that $A \cong \mathrm{C}_8$. Thus, C is the unique maximal abelian subgroup of G containing R so all $\neq C$ maximal abelian subgroups of G are cyclic of order 8. By Proposition 10.19, G has no nonabelian subgroup of order 8. Let a cyclic $A < G$ be maximal abelian. Then $AR \cong \mathrm{M}_{2^4}$ is minimal nonabelian (this follows from Theorem 1.2 and Proposition 10.19). If $\bar{U} < \bar{G}$ is isomorphic to Q_8, then U is not minimal nonabelian (otherwise, $U' < R$ so $\bar{U}$ must be abelian). Thus, all minimal nonabelian subgroups of G are isomorphic to M_{16}.

Exercise 4. Given $n > 3$, let

$$G_n^1 = \langle a, b \mid o(a) = 2^n,\ o(b) = 2^{n-2},\ a^b = a^5\rangle\,,$$
$$G_n^2 = \langle a, b \mid o(a) = 2^n,\ o(b) = 2^{n-1},\ a^b = a^5\rangle\,,$$
$$G_n^3 = \langle a, b \mid o(a) = 2^n,\ o(b) = 2^{n-1},\ a^{2^{n-1}} = b^{2^{n-2}},\ a^b = a^5\rangle$$

be metacyclic groups of orders 2^{2n-2}, 2^{2n-1}, and 2^{2n-2}, respectively. Describe the maximal abelian and minimal nonabelian subgroups of these groups.

Exercise 5. Classify the minimal nonabelian p-groups G such that $|A : (A \cap B)| = p$ for all distinct cyclic $A, B < G$ of equal order.

Hint. One may assume that $|G| > p^3$. If G is metacyclic, then $G \cong \mathrm{M}_{p^n}$ (Lemma 65.1). Let

$$G = \langle a, b \mid o(a) = p^m,\ o(b) = p^n,\ m \geq n,\ c = [a, b],\ [a, c] = [b, c] = c^p = 1\rangle$$

be nonmetacyclic; then $m > 1$, by assumption. Let $A = \langle a\rangle$, $B = \langle b\rangle$. If $n > 1$, then $\Omega_n(A) \cap B = \{1\}$ has index $p^n > p$ in $\Omega_n(A)$, a contradiction. Thus, $n = 1$.

Exercise 6. Let G be a nonabelian p-group of order $> p^{p+1}$. If the subgroup $\mathrm{Z}_{p+1}(G)$ is of maximal class, then $G = \mathrm{Z}_{p+1}(G)$ is of order p^{p+2}.

Solution. Note that $\mathrm{Z}_{p+1}(G)$ contains all normal subgroups of G of orders $\leq p^{p+1}$. If G is of maximal class, then its order equals p^{p+2} (Theorem 9.6). Assume that G is not of maximal class. Then, by Theorem 12.1 (a), there is $R \lhd G$ of order p^p and exponent p. As $R < \mathrm{Z}_{p+1}(G)$, Theorem 9.6 (c) implies that $|\mathrm{Z}_{p+1}(G)| = p^{p+1}$. By Theorem 13.6, the number of subgroups of maximal class and order p^{p+1} in G is divisible by p^2. Therefore, there is in G a normal subgroup $T \neq \mathrm{Z}_{p+1}(G)$ of maximal class and order p^{p+1}. On the other hand, $T \leq \mathrm{Z}_{p+1}(G)$, and we conclude that $T = \mathrm{Z}_{p+1}(G)$, a final contradiction. Thus, G is of maximal class and order p^{p+2}.

Let $\mathcal{A}_1(G)$ be the set of minimal nonabelian subgroups of a nonabelian p-group G.

Exercise 7. Let a p-group G be an $\mathcal{A}_n$-group, $n > 1$. Suppose that for each $A \in \mathcal{A}_1(G)$ one has $\langle \mathcal{A}_1(G) - \{A\}\rangle < G$. Prove that $p = 2$. (**Hint.** Use Theorem 10.28 and results of §76 on $\beta_1(G, H) = \alpha_1(G) - \alpha_1(H)$ for $H < G$. By those results, $\beta_1(G, H) \geq p - 1$.)

Appendix 63 Intersections of nonnormal cyclic subgroups

The following theorem, a fairly easy consequence of Kazarin's Theorem 63.4, is a partial case of the main result of § 187.

Theorem A.63.1. *Suppose that any two distinct nonnormal cyclic subgroups of equal order of a non-Dedekindian p-group G have intersection* $\{1\}$. *Then one and only one of the following holds:*

(a) $\exp(G) = p$.

(b) $p = 2$ *and the* H_2*-subgroup of G has index 2 in G, i.e.,* $G = \langle x\rangle \cdot A$, *where* $o(x) = 2$ *and* $a^x = a^{-1}$ *(a generalized dihedral group).*

(c) $|G'| = p$, $\exp(G) > p$ *and* $\Phi(G)$ *is cyclic.*

Proof. Obviously, groups (a), (b), and (c) satisfy the hypothesis. For example, if G is as in (b), then all nonnormal cyclic subgroups of G have order 2. Let G be as in (c) and assume that $C < G$ is cyclic of order $> p$. Then $\Omega_1(C) = \Omega_1(\Phi(G)) = G'$ so that $C \trianglelefteq G$. In what follows we assume that $\exp(G) > p$.

Assume that there is in G a nonnormal cyclic subgroup C of order $> p$. As any cyclic subgroup of G containing C is not G-invariant, one may assume without loss of generality that C is a maximal cyclic subgroup of G. Let $C < M \le G$ be such that $|M : C| = p$; then M is noncyclic.

Assume that $C_1 < M$ is cyclic of index p and $C_1 \neq C$. Then $C_1 \cap C > \{1\}$ so, by hypothesis, $C_1 \trianglelefteq G$. It follows that $C \cap C_1 = \mho_1(C) = \mho_1(C_1) \trianglelefteq G$. Let $x \in G - N_G(C) (\neq \emptyset)$. Then $C^x \neq C$ is not normal in G, and $C \cap C^x = \mho_1(C) > \{1\}$, in view of normality of $\mho_1(C)$ in G, contrary to the hypothesis. Thus, C_1 does not exist hence C is the unique cyclic subgroup of index p in M so M, being noncyclic, is a 2-group of maximal class $\not\cong Q_8$ (Theorem 1.17 (b)). In that case, G is a 2-group of maximal class (Exercise 10.10). However, a 2-group of maximal class has no nonnormal cyclic subgroup of order > 2 which has trivial intersection with all its conjugates.

Now assume that all cyclic subgroups of G of order $> p$ are normal. Then G is a $\mathcal{K}_p$-group (see § 63). By Theorem 63.4, G is a group from (b) or (c) since, by assumption, $\exp(G) > p$. □

In § 187 the non-Dedekindian p-groups in which the intersection of any two distinct conjugate cyclic subgroups is trivial are classified (this is a generalization of Theorem A.63.1).

Exercise 1. Classify the non-Dedekindian p-groups in which the intersection of any two distinct conjugate maximal cyclic subgroups is trivial.

Exercise 2. Classify the non-Dedekindian p-groups G containing a nonnormal cyclic subgroup C of order $> p$ such that for any $x \in N_G(C)$ one has $C \cap C^x = \{1\}$.

Appendix 64 Some remarks on p-groups all of whose nonnormal subgroups are abelian

The following theorem contains some remarks on the non-Dedekindian p-groups all of whose nonnormal subgroups are abelian.

Theorem A.64.1 (see Problem 768). *Suppose that G is a non-Dedekindian p-group all of whose nonnormal subgroups are abelian. Let $A < G$ be maximal abelian.*

(a) *If $A \triangleleft G$, then the quotient group G/A is Dedekindian. In particular, if $p > 2$, then G is metabelian.*

(b) *If A is nonnormal in G, then $\mathrm{N}_G(A) \triangleleft G$, $\mathrm{N}_G(A)/A$ is either cyclic or $\cong \mathrm{Q}_8$. If $B/A \leq \mathrm{N}_G(A)/A$ is of order p, then $B \triangleleft G$ and G/B is Dedekindian and $|B : \mathrm{Z}(B)| = p^2$. Next, $|G : \mathrm{N}_G(A)| = p$.*

Proof. All subgroups of G properly containing A, are nonabelian so G-invariant. It follows that $\mathrm{N}_G(A)$ normal in G and $\mathrm{N}_G(A)/A$ is Dedekindian. In particular, if $A \triangleleft G$, then G/A is Dedekindian. The last assertion of (a) follows from Theorem 1.20.

Now let A be not normal in G. By (a), $\mathrm{N}_G(A)/A$ is Dedekindian. Assume that there are in $\mathrm{N}_G(A)/A$ two distinct subgroups U/A, V/A of order p. By the above, $U, V \triangleleft G$ so that $A = U \cap V \triangleleft G$, a contradiction. Thus, $\mathrm{N}_G(A)/A$ has only one subgroup of order p so this quotient group is either cyclic or generalized quaternion (Proposition 1.3). In the second case $\mathrm{N}_G(A)/A \cong \mathrm{Q}_8$. If $A < U \leq G$, where $|U : A| = p$, then G/U is Dedekindian. Since A is not characteristic in $U(\triangleleft G)$ (otherwise, $A \triangleleft G$), there is $\phi \in \mathrm{Aut}(U)$ such that $A^\phi \neq A$. Since U is nonabelian, it follows that $A \cap A^\phi = \mathrm{Z}(U)$ has index p^2 in U. By Exercise 1.6, there are in U exactly $p + 1$ abelian subgroups of index p. It follows that $|G : \mathrm{N}_G(A)| = p$. □

It is interesting to describe the structure of the quotient group $G/\mathrm{Z}(G)$, where G is a group of Theorem A.65.1. Note that the classification of groups of Theorem A.65.1 is given in [FA] (where those groups are called *metahamiltonian*). However, the proof in [FA] is quite involved with great number of computations.

Appendix 65 On p-groups G with $|\Omega_1(G)| = p^n$

In this appendix, the letter of Mann obtained by the first author a few years ago is reproduced.

From Mann's letter. Looked in the problems at the end. In Problem 9 you ask for a bound for $|G/\mho_1(G)|$ if $|\Omega_1(G)| = p^n$. Thought a little about that, and I think that the following is true:

Let G be a p-group, p odd. If $|\Omega_1(G)| = p^n$, then $|G/\mho_1(G)| \leq p^{n^2}$.

Proof. Let $C = C_G(\Omega_1(G))$. Then G/C is an automorphism group of C, and therefore its order is at most $p^{n(n-2)/2}$. In C, all elements of order p are central. Such groups are called p-central, and Thompson proved that $d(C) \leq d(Z(C))$, where $d(X)$ is the minimum generators of X [Hup1, III.12.2] (see also Theorem 15.1). Here $d(Z(C)) \leq n$. Moreover, subgroups of p-central p-groups are p-central, and therefore each subgroup of C can be generated by n elements. Write $H = C/\mho_1(C)$. Let A be a maximal normal abelian subgroup of H. Then A is elementary abelian of order at most p^n (as the inverse image of A in C is generated by n elements), and $H/A = C_H(A)$ is a group of automorphisms of A, so again we have $|H/A| \leq p^{n(n-2)/2}$. Combining these inequalities, we get $G/\mho_1(G)| \leq p^{n^2}$. □

The bound can be improved a little. I just now thought of this argument, and did not look for the best possible result. For $p = 2$ we can get a similar result using elements of order 4, with worse bounds. It is still interesting to find the best bound for $n = p$, for $n < p$ we have the best bounds.

Exercise 1. Find the best possible bound for $|G/\mho_1(G)|$ for $n = 3, 4$.

Appendix 66 Metacyclic p-groups containing an abelian subgroup of index p

The p-groups with an abelian subgroup of index p play an important role in p-group theory. Defining relations of such p-groups were obtained in [NRSB], which is fairly nontrivial; however, some challenging problems have not been solved for those groups.

In this appendix we consider in some detail the partial case when our p-group is metacyclic with an abelian subgroup of index p. We prove the following

Theorem A.66.1. *Let a metacyclic 2-group G be neither abelian nor minimal non-abelian. Suppose that G has an abelian subgroup A of index 2. Then*
(a) *The group $G/Z(G)$ is of maximal class.*
(b) *If $B \neq A$ is a maximal abelian subgroup of G, then $|B| = 2|Z(G)|$.*
(c) *If $B < C \leq G$, where B is as in (b) and $|C : B| = 2$, then C is minimal nonabelian.*

Proof. By hypothesis, $|G'| > 2$ (Lemma 65.2(a)). Since G is not minimal nonabelian, A is the unique abelian subgroup of index 2 in G. Write $\bar{G} = G/Z(G)$. Then $|\bar{G}| > p^2$. By Lemma 1.1, $\bar{A}$ is isomorphic with the cyclic subgroup G'.

Assume that $\bar{T} \neq \bar{A}$ is cyclic of index 2 in $\bar{G}$ ($\bar{T}$ exists if either $\bar{G}$ is abelian or $\bar{G} \cong M_{2^n}$); then $T \neq A$ is the abelian subgroup of index 2 in G, contrary to the result of the previous paragraph. It follows that $\bar{A}$ is the unique cyclic subgroup of index 2 in $\bar{G}$. By Theorem 1.2, $\bar{G}$ is a 2-group of maximal class, completing the proof of (a).

It remains to prove (b) and (c). Let $B \neq A$ be a maximal abelian subgroup of G, Then $AB = G$ so that $A \cap B = Z(G)$ and $|B : (A \cap B)| = 2$, by the product formula, proving (b). Let $B < C < G$, where $|C : B| = 2$ (recall that $|G : B| > 2$). The subgroup C is nonabelian so that $C/Z(G) \cong E_4$. As $d(C) = 2$ (indeed, C is metacyclic as a subgroup of G), it follows that all maximal subgroups of a nonabelian group C are abelian so C is minimal nonabelian, completing the proof of (c) and thereby the theorem. □

For a stronger result, see Theorem A.53.1.

Let G be a nonabelian metacyclic p-group, $p > 2$ and suppose that G has an abelian subgroup A of index p. Then G is minimal nonabelian, by Theorem A.66.1.

Exercise 1. If a nonabelian 2-group has no nonabelian epimorphic image of order 8 and possesses an abelian subgroup of index 2, then it is minimal nonabelian. (**Hint.** Repeat the argument in the second paragraph of the proof of Corollary 148.2(b2).)

Exercise 2. Let G be a nonabelian metacyclic p-group containing an abelian subgroup A of index p. Let $B < C \leq G$, where $B \neq A$ is maximal abelian, $|C : B| = p$ and $d(C) = 2$. Then C is minimal nonabelian. If $A/Z(G)$ is cyclic and $d(G) = 2$, then either G is minimal nonabelian or $G/Z(G)$ is a 2-group of maximal class.

Hint. As $C/Z(G) \cong E_{p^2}$, then C, being two-generator nonabelian, is minimal nonabelian. If $A/Z(G)$ is the unique cyclic subgroup of index p in $\bar{G} = G/Z(G)$, then $\bar{G}$ is a 2-group of maximal class (Theorem 1.2). Otherwise, G is minimal nonabelian since $d(G) = 2$ and the set Γ_1 has two abelian members.

Problem 1. Given n, does there exist a two-generator p-group containing $\geq n$ maximal abelian subgroups of pairwise distinct orders?

Problem 2. Classify the two-generator p-groups containing an abelian subgroup of index p. (See Theorem A.53.1.)

Problem 3. Study the two-generator p-groups G such that $G/Z(G)$ is metacyclic.

Problem 4. Study the automorphism group of the group G which is as in Problem 2.

Problem 5. Study the p-groups G such that $|G'| = p$ and G/G' is homocyclic. Describe $\mathrm{Aut}(G)$.

Problem 6. Study the metacyclic p-groups G containing an abelian subgroup A of index p^2 (in view of Exercise 1.6, one may assume that $A \lhd G$). Consider in detail the metacyclic p-groups with abelian Frattini subgroup.

Problem 7. Study the two-generator 2-groups with cyclic derived subgroup.[1]

Problem 8. Describe a Sylow p-subgroup of the holomorph of a minimal nonabelian p-group. (See [Mal7] and [Kur].)

Problem 9. Given a homocyclic p-group G, find the maximal order of an abelian subgroup of $\mathrm{Aut}(G)$.

Problem 10. Study the nonabelian p-groups satisfying the following two conditions simultaneously: (a) G has an abelian subgroup A of index p. (b) If $B < C \leq G$, where $B \neq A$ is maximal abelian and $|C : B| = p$, then $d(C) = 2$.

Problem 11. Classify the p-groups G with metacyclic $\mathrm{Aut}(G)$.

1 Commentary of Mann: For odd p, it is proved in the paper of Mann and Posnick–Fradkin [MP-F], that a two-generator p-group G has a cyclic derived subgroup $G' \iff G/Z(G)$ is metacyclic.

Appendix 67 p-groups in which the intersection of all their subgroups of order p^2 has order p

The structure of a finite p-group depends essentially on intersections of some of its subgroups. For example, if the intersection of any two distinct maximal subgroups of a nonabelian p-group G is cyclic, then either $|G| = p^3$ or G has a cyclic subgroup of index p. In this appendix, we classify the title p-groups.

The stated property is inherited by noncyclic subgroups of order $> p^2$. If a p-group G is such that the intersection of all its subgroups of order p^2 has order p, then it has no subgroup isomorphic to E_{p^3}. Therefore, a large amount of information yields Theorem 13.7 and the main result of § 50. Note that the extraspecial 2-group $D_8 * Q_8$ of order 2^5 and all its maximal subgroups satisfy the above condition. The following theorem contains some additional information on such p-groups.

Theorem A.67.1. *Let G be a nonabelian 2-group of order $> 2^3$. If the intersection of all subgroups of G of order 4 has order 2, then one and only one of the following holds:*
(a) *There is in G a cyclic subgroup of index 2.*
(b) *There is in G an abelian subgroup A of type $(2^n, 2)$ with $|G : A| = 2$, $Z(G)$ is cyclic.*
(c) *$G = Q \times C$, where Q is a generalized quaternion group and $|C| = 2$.*
(d) *The group G has a maximal subgroup $A = Q \times C$, where Q is a generalized quaternion group, $|C| = 2$ and $Z(G) = Z(Q)$.*

If a p-group G of order $> p^3$ is such that all its subgroups of order p^2 have intersection of order p, then it has no subgroup isomorphic to E_{p^3} since that subgroup has trivial Frattini subgroup.

Proof of Theorem A.67.1. By the previous paragraph, G has no subgroup $\cong E_8$.

If $X \leq G$ is noncyclic, then $I = I(X)$ denotes the intersection of all subgroups of X having order 4. If G is noncyclic abelian, then it has a cyclic subgroup of index 2. Next we assume that G is nonabelian. If G has a cyclic subgroup of index 2, then $I(G) = \Omega_1(Z(G))$ has order 2. In what follows we assume that G has no cyclic subgroup of index 2. Next, G has no metacyclic subgroup of order 16 and exponent 4 (this group contains two distinct cyclic subgroups of order 4 whose intersection is equal to $\{1\}$).

Let $R \triangleleft G$ be abelian of type $(2, 2)$ (Lemma 1.4). Write $A = C_G(R)$; then $|G : A| \leq 2$. We claim that A is either abelian of type $(2^n, 2)$ or $A = Q \times C$, where Q is a generalized quaternion group and $|C| = 2$. If A is abelian, it is of type $(2^n, 2)$ since it has no subgroups of types $(2, 2, 2)$ and $(4, 4)$; then $|G : A| = 2$ since G is nonabelian. If A is nonabelian and $|G : A| = 2$, then $R \nleq Z(G)$.

(i) Let $C_G(R) = A \in \Gamma_1$ be abelian of type $(2^n, 2)$, $n > 1$. Then $Z(G) < A$ so that $Z(G)$ is cyclic. One has $I(G) = I(A) = \Omega_1(\Phi(A))$. Let $L < G$ of order 4 be such that $L \nleq A$.

Then $A \cap L \leq Z(G)$. As $Z(G)$ is cyclic, one obtains $A \cap L = \Omega_1(\Phi(A)) = I(A) = I(G)$ so G satisfies the hypothesis.

(ii) Now assume that $A = C_G(R)$ is nonabelian; then $|A| > 2^3$. In that case A has a minimal nonabelian subgroup M. Since M has no subgroup $\cong E_8$, it is metacyclic (Lemma 65.1). Since M has no subgroups of order 16 and exponent 4, it follows that M has a cyclic subgroup of index 2 (Lemma 65.1 again). It follows that $R \nleq M$ (otherwise, M would be abelian). Since G has no subgroup $\cong E_8$, a consideration of the subgroup RM shows that M has no abelian subgroup of type $(2, 2)$, and we conclude that $M \cong Q_8$. By Corollary A.17.3, $A = Q \times C$, where Q is a generalized quaternion group and $|C| = 2$.[1] If $A = G$, then G satisfies the hypothesis. Now we let $|G : A| = 2$. Then $Z(G) = Z(Q)$ (indeed, $Z(G) < R$ and $Z(Q) \leq Z(G)$). Let $L < G$ be of order 4 and $L \nleq A$. Then $L \cap A = Z(G)$ since $C_G(L \cap A) = AL = G$ and hence $L \cap A = Z(Q) = I(A) = I(G)$ so that our group satisfies the hypothesis. □

Any group from Lemma 42.1 (c) satisfies the hypothesis of Theorem A.67.1; this group is one of the groups from part (b) of that theorem.

The first part of the proof Proposition A.67.2 repeats, word by word, the first part of the proof of Theorem A.67.1.

Theorem A.67.2 (see [ZW], Theorem 18). *Let G be a noncyclic p-group of order $> p^3$, $p > 2$, and suppose that the intersection of all subgroups of G of order p^2 has order p. Then one and only one of the following holds:*

(a) *There is in G a cyclic subgroup of index p, i.e., either $G \cong M_{p^n}$ or abelian of type (p^{n-1}, p).*
(b) *G is a 3-group of maximal class and order 3^4, $G \ncong \Sigma_9 \in Syl_3(S_9)$.*
(c) *$G = C\Omega_1(G)$, where C is cyclic of order $> p$, $\Omega_1(G)$ is nonabelian of order p^3 and exponent p.*

Proof. By the paragraph preceding the proof of Theorem A.67.1, G has no subgroup $\cong E_{p^3}$. Therefore, G is one of the groups of Theorem 13.7. Let $R \lhd G$ be abelian of type (p, p) (Lemma 1.4).

(i) Suppose that G is metacyclic. As a metacyclic group of order p^4 and exponent p^2 does not satisfy the hypothesis, G has no such subgroup. It follows that G/R is cyclic (Lemma 1.4); then G has a cyclic subgroup of index p, i.e., it is as in (a).

(ii) Suppose that G is a 3-group of maximal class. As G has no subgroup $\cong E_{27}$, the fundamental subgroup G_1 of G is metacyclic (Theorems 9.5 and 9.6). By (i), G_1 has a cyclic subgroup of index 3. As G has no normal cyclic subgroup of order 9, it follows that $|G_1| = 3^3$ so that $|G| = 3^4$.

(iii) Suppose that $G = CE$, where $E = \Omega_1(G) \cong S(p^3)$ and C is cyclic of order $> p$. Obviously, G satisfies the hypothesis $\iff$ $\Omega_2(G)$ is. Therefore, one may assume,

1 See the proof of Theorem 134.2 where a similar argument is presented.

without loss of generality, that $|G| = |\Omega_2(G)| = p^4$. It is easy to see that then G satisfies the hypothesis (indeed, all subgroups of G of order p^2 contain G'). □

For related results, see § 168.

Problem 1. Classify the p-groups G in which the order of the intersection of any two distinct subgroups of order p^2 is equal to p.

Problem 2. Given $k > 2$, study the p-groups in which the order of the intersection of any two distinct subgroups of order p^k is $\geq p^{k-2}$.

Problem 3. Study the p-groups of exponent $\geq p^k$ in which the intersection of any two distinct cyclic subgroups of order $p^k > p$ has order p^{k-1} (k is fixed).

Appendix 68 The 2-groups all of whose nonabelian two-generator subgroups are minimal nonabelian

In this appendix we prove the following

Theorem A.68.1 (Janko). *The following conditions for a nonabelian 2-group G are equivalent:*

(a) *Any two noncommuting elements of G generate a minimal nonabelian subgroup.*
(b) $\mathrm{cl}(G) = 2$ *and G' is elementary abelian.*

Proof. (a) $\Rightarrow$ (b): Let $x, y \in G$. As $\mathrm{cl}(\langle x, y\rangle) \le 2$ and $|\langle x, y\rangle'| \le 2$ (Lemma 65.1), one has

$$[x^2, y] = [x, y]^2 = 1 \Rightarrow G' \le \Phi(G) = \mho_1(G) \le Z(G) \Rightarrow \mathrm{cl}(G) = 2\,.$$

Also, $[x, y]^2 = [x^2, y] = 1$, and we conclude that G' is elementary abelian.

(b) $\Rightarrow$ (a): Conversely, let G be a 2-group of class 2 with elementary abelian derived subgroup G'. Then for any two noncommuting elements x, y of G one has

$$Z(G) \ge G' \ge \langle x, y\rangle' \cong C_2\,,$$

and so Lemma 65.2 (a) implies that the two-generator subgroup $\langle x, y\rangle$ is minimal nonabelian. □

Theorem A.68.1 solves a part of Problem 2825.

Exercise 1. Classify the nonabelian 2-groups G, all of whose nonabelian three-generator subgroups are either $\mathcal{A}_1$- or $\mathcal{A}_2$-subgroups.

Solution. Let $x, y \in G$ be noncommuting. Write $H = \langle x, y\rangle$. If $z \in H - \{x, y\}$, then the nonabelian subgroup $\langle x, y, z\rangle = H$ so that H is either $\mathcal{A}_1$- or $\mathcal{A}_2$-subgroup, by hypothesis. If for any such choice of x, y the subgroup H is an $\mathcal{A}_1$-subgroup, then G is a group of Theorem A.68.1. Now Let H be an $\mathcal{A}_2$-subgroup for some $x, y \in G$. Assume that $H < G$. If $z \in G - H$, then nonabelian subgroup $\langle x, y, z\rangle$ is neither $\mathcal{A}_1$- nor $\mathcal{A}_2$-subgroup, a contradiction. Thus, if G is not as in Theorem A.68.1, it is either $\mathcal{A}_1$- or $\mathcal{A}_2$-group.

Problem 1. Study the nonabelian p-groups, $p > 2$, all of whose nonabelian two-generator subgroups are minimal nonabelian.

Problem 2. Study the p-groups, $p > 2$, all of whose nonabelian subgroups of rank $p - 1$ are absolutely regular.

Problem 3. Study the nonabelian p-groups, all of whose nonabelian two-generator subgroups are metacyclic.

Problem 4. Study the nonabelian p-groups, all of whose nonabelian two-generator subgroups are either metacyclic or minimal nonabelian.

Problem 5. Study the nonabelian p-groups, all of whose nonabelian two-generator subgroups are either minimal nonabelian or of maximal class.

Appendix 69 Supplement to Theorem 200.1

Theorem 200.1 asserts that a nonabelian p-group G contains a minimal nonabelian subgroup A such that $A \cap Z(G) > \{1\}$.

The following assertion is obvious.

Exercise 1. If G is a p-group with cyclic derived subgroup $G' > \{1\}$, then $A' \le Z(G)$ for any minimal nonabelian subgroup A of G.

See the proof of Theorem A.69.1.

It appears that the condition on cyclicity of G' in Exercise 1 is superfluous. Corresponding assertion, for any general p-group, is contained in [WZ, Theorem 1.1]. Their proof is based on the following assertion. If $\mathrm{cl}(G) = n$, then there are in G elements $x_1, \ldots, x_n$ such that $o([x_1, \ldots, x_n]) = p$. The same result was proved by the second author before publication of [WZ].

Theorem A.69.1. *If G is a p-group with derived subgroup $G' > \{1\}$, then $A' \le Z(G)$ for some minimal nonabelian subgroup A of G.*

Proof (Janko). Let $x \in Z_2(G) - Z(G)$. Then for any $y \in G - C_G(x)$ one has $z = [x, y] \neq 1$. Set $U = \langle x, y\rangle$ and $Z = \langle z\rangle$. Then $U' = Z \le Z(G)$ (indeed, $U' \le [Z_2(G), G] \le Z(G)$). Let $A \le U$ be minimal nonabelian. Then, as $U' = Z$ is cyclic, one has $A' \le U'$ and, by Theorem 65.1, $|A'| = p$. It follows that $A' = \Omega_1(U') \le Z(G)$. □

Exercise 2. A nonmetacyclic p-group G contains a minimal nonmetacyclic subgroup M such that $M \cap Z(G) > \{1\}$.

Hint. Assume that this is false. Then, by Theorems 66.1. and 69.1, any minimal nonmetacyclic subgroup H of G contains an abelian subgroup R of type (p,p). Let $L \le Z(G)$ be of order p. One may assume that $L \not\le H$. Then $M = R \times L \cong E_{p^3}$ is minimal nonmetacyclic and $\{1\} < L \le M \cap Z(G)$, contrary to the assumption.

Problem 1. Let G be a primary $\mathcal{A}_n$-group, $n > 3$. Is it true that there is in G an $\mathcal{A}_2$-subgroup M such that $M \cap Z(G) > \{1\}$?

Problem 2. Let G be a nonabelian p-group. Assume that for any $x \in Z_2(G) - Z(G)$ and any $y \in G - C_G(x)$, the subgroup $\langle x, y\rangle$ is minimal nonabelian. Study the structure of G.

If there are $x, y \in G$, $x \in Z_2(G) - Z(G)$ and $y \in G - C_G(x)$ such that $\langle x, y\rangle$ is not a minimal nonabelian subgroup, then there is in $\langle x, y\rangle$ an $\mathcal{A}_2$-subgroup M containing $\Omega_1(\langle x, y\rangle') \le Z(G)$ so that $M \cap Z(G) > \{1\}$.

Problem 3. Is it true that an irregular p-group G contains a minimal irregular subgroup H such that $H \cap Z(G) > \{1\}$. (For $p = 2$ this follows from Theorem A.69.1, since regular 2-groups are abelian.)

Appendix 70 Nonabelian p-groups all of whose maximal cyclic subgroups coincide with their centralizers

In this appendix, the title groups will be classified. The following easy theorem, due to the second author, holds:

Theorem A.70.1. *If all maximal cyclic subgroups of a noncyclic p-group G coincide with their centralizers, then $G \cong Q_{2^n}$.*

Proof. Clearly, G is nonabelian. Let $x \in Z(G)$ be of order p. In view of Proposition 1.3, it suffices to show that $\langle x \rangle$ is a unique subgroup of order p in G. Assume that there is in G an element u of order p such that $\langle u \rangle \neq \langle x \rangle$. Let $u \in S < G$, where S is a maximal cyclic subgroup of G. Then $x \in C_G(S)$ and $x \notin S$, contrary to the hypothesis. Thus, u does not exist so that $\langle x \rangle$ is the unique subgroup of order p in G. Therefore, $p = 2$ and G is a generalized quaternion group, as was to be shown. □

This solves Problem 737.

Exercise 1. Suppose that a 2-group G is such that, whenever $x \in G$ is an involution, then there is a generalized quaternion subgroup $Q \leq G$ containing x. Check the following assertion: If any generalized quaternion subgroup of G contains its centralizer, then G is a generalized quaternion group.

Exercise 2. Classify the non-Dedekindian p-groups all of whose nonnormal maximal cyclic subgroups coincide with their centralizers. (**Hint.** Prove that G has only one subgroup of order p.)

Problem 1. Classify the p-groups of exponent $> p$ all of whose maximal cyclic subgroups of order $> p$ coincide with their centralizers.

Problem 2. Classify the p-groups of exponent $> p$ all of whose maximal elementary abelian subgroups coincide with their centralizers.

Problem 3. Classify the p-groups G containing a maximal subgroup H that contains no maximal cyclic subgroups of G.

Problem 4. Classify the non-Dedekindian p-groups of exponent $> p$ all of whose nonnormal maximal cyclic subgroups of order $> p$ coincide with their centralizers.

Problem 5. Study the nonmetacyclic p-groups G such that $C_G(M) \leq M$ for any maximal metacyclic $M < G$.

Problem 6. Study the nonmetacyclic p-groups all of whose maximal metacyclic subgroups are normal. (Let G be a 3-group of maximal class such that $H_p(G) = G_1$, where G_1 is the fundamental subgroup of G. Then G satisfies the hypothesis.)

Problem 7. Study the irregular p-groups all of whose maximal absolutely regular subgroups are normal.

Appendix 71 Finite groups G containing a p-subgroup permutable with all Sylow q-subgroups of G for all $q \neq p$

If A is a subgroup of a group G, then A^G is the normal closure of A in G (= the intersection of all normal subgroups of G containing A). If X is a group, then $\mathrm{S}(X)$ denotes its solvable radical (= the product of all normal solvable subgroups of X). Clearly, $\mathrm{S}(X)$ is characteristic in X so that $N \lhd X \Rightarrow \mathrm{S}(N) \lhd X$. Let $\pi(G)$ be the set of all prime divisors of $|G|$.

In this appendix we prove the following

Theorem A.71.1 (Isaacs [Isa21]). *Suppose that $P > \{1\}$ is a p-subgroup of a group G permutable with all Sylow q-subgroups of G, where q runs over all $\neq p$ prime divisors of $|G|$. Then $P^G \leq \mathrm{S}(G)$.*

In fact, this is a partial case of essentially more general Isaacs' result [Isa21, Theorem A].

The following two exercises are also contained in Isaacs' paper and used in the proofs of Theorems A.71.1 and A.71.3.

Exercise 1. Let $U < H < G$, where U is permutable with all Sylow q-subgroups of a group G for some $q \in \pi(H) - \pi(U)$. Then U is permutable with all Sylow q-subgroups of H for the same prime q.

Solution. Suppose that $\pi(U) \neq \pi(H)$ and let $q \in \pi(H) - \pi(U)$, $Q \in \mathrm{Syl}_q(H)$. Let $Q \leq Q_1 \in \mathrm{Syl}_q(G)$; then $UQ_1 = Q_1U$, by hypothesis. One has $Q_1 \cap H = Q$. By the modular law, $UQ_1 \cap H = U(H \cap Q_1) = UQ$. Then $UQ = UQ_1 \cap H$ is a subgroup of H, and therefore $UQ = QU$.

Exercise 2. Suppose that U is a subgroup of a group G permutable with all Sylow q-subgroups of G for some $q \in \pi(G) - \pi(U)$. If $N \lhd G$ and $\bar{G} = G/N$, then $\bar{U}$ is permutable with all Sylow q-subgroups of $\bar{G}$ for the same q.

This is obvious.

In what follows we use the following

Lemma A.71.2 (= Lemma A.28.16 (Wielandt)). *Suppose that proper subgroups A, B of a group G satisfy $AB^g = B^gA$ for all $g \in G$. If $G = A^GB = AB^G$, then $G = AB^g$ for some $g \in G$.*

Proof of Theorem A.71.1. Assume that G is a counterexample of minimal order; then P^G is nonsolvable so is G. In particular,

(∗) $|\pi(G)| \geq |\pi(P^G)| \geq 3$ (Burnside).

Let $q \in \pi(G)-\{p\}$ and $Q \in \mathrm{Syl}_q(G)$; then $PQ^g = Q^gP$ for any $g \in G$. As $|\pi(PQ^g)| = 2 < 3$, we get, by $(*)$, $PQ^g < G$ for all $g \in G$.

(i) Assume that $P^G < G$. Then, by Exercise 1, P is permutable with all Sylow q-subgroups of P^G, where a prime q runs over the set $\pi(P^G) - \{p\}$. By induction, $P \le S = \mathrm{S}(P^G)$. As S is characteristic in $P^G \lhd G$, it follows that the solvable subgroup $S \lhd G$ so that $P^G \le S \le \mathrm{S}(G)$, i.e., G is not a counterexample. Thus,

(ii) $P^G = G$.

Let $q \in \pi(G) - \{p\}$ and $Q \in \mathrm{Syl}_q(G)$. Assume that $PQ^G = G$. Then, by (ii), $P^GQ = G = PQ^G$. Therefore, by Lemma A.71.2, $G = PQ^g$ for some $g \in G$ whence $|\pi(G)| = 2 < 3$, contrary to $(*)$. Thus,

(iii) $PQ^G < G$, where $Q \in \mathrm{Syl}_q(G)$ ($q \in \pi(G) - \{p\}$. In particular, G is not simple.

Let R be a minimal normal subgroup of G. Then $P \not\le R$, by (ii). Next, R has no non-trivial characteristic subgroups so $R = Q^G$ for any nonidentity Sylow subgroup Q of R.

Assume that $PR = G$; then R is nonsolvable since G is and G/R is a p-group. Let $q \in \pi(R) - \{p\}$ and $Q \in \mathrm{Syl}_q(R) = \mathrm{Syl}_q(G)$. In that case, $Q^G = R$ (see the previous paragraph) so that, by (ii), $P^GQ = G = PR = PQ^G$, hence, by Lemma A.71.2, $PQ^g = G$ for some $g \in G$, contrary to $(*)$. Thus,

(iv) $PR < G$.

Write $\bar{G} = G/R$. By Exercise 2, $\bar{P}$ is permutable with any Sylow r-subgroup of $\bar{G}$, where r runs over the set $\pi(\bar{G}) - \{p\}$. By induction, $\bar{P}^{\bar{G}} \le \bar{H} = \mathrm{S}(\bar{G})$ so that $\bar{H} = H/R$ is solvable and $P^G \le H$. It follows from (ii) that $H = G$ so that $G/R = \bar{G}$ is solvable. In particular, as R is an arbitrary minimal normal subgroup of G, we get

(v) R is nonsolvable and G/R is solvable and R is a unique minimal normal subgroup of G, i.e., G is a monolith.

By Exercise 1, P is permutable with any Sylow r-subgroup of PR for all $r \in \pi(PR) - \{p\} = \pi(R) - \{p\}$. As, by (iv), $PR < G$, it follows, by induction, that $P \le \mathrm{S}(PR)$. As R is nonsolvable characteristically simple, we get $P = \mathrm{S}(PR)$ and $P \cap R = \{1\}$ so that

(vi) $PR = P \times R$.

Write $C = C_G(R)$. As $C \cap R = \mathrm{Z}(R) = \{1\}$ and G/R is solvable, by (v), it follows that C, being isomorphic to a subgroup of G/R, is solvable. As $P \le C$, by (vi), and $C \lhd G$, we conclude that P^G is solvable, i.e., $P^G \le \mathrm{S}(G)$ and G is not a counterexample. The proof is complete. □

The proof of the following theorem is a small modification of the proof of Theorem A.71.1.

Theorem A.71.3 (Isaacs [Isa21], Corollary C). *Let U be a nilpotent Hall subgroup of a group G. If, for all $q \in \pi(G) - \pi(U)$, $UQ = QU$ for all $Q \in \mathrm{Syl}_q(G)$, then $U^G \le \mathrm{S}(G)$.*

Proof. One may assume that $\pi(U) \neq \pi(G)$ (otherwise, $U = G$ and it is nothing to prove further). Assume that G is a counterexample of minimal order; then U^G is nonsolvable so is G. Let $q \in \pi(G) - \pi(U)$ and $Q \in \mathrm{Syl}_q(G)$. Then $UQ = QU$, by hypothesis. As UQ is solvable, by Wielandt's theorem on factorization (see, for example, Theorem A.28.14), one has $UQ < G$.

Assume that $U^G < G$. Then, by Exercise 1, U is permutable with all r-Sylow subgroups of U^G for any $r \in \pi(U^G) - \pi(U)$ (as U is a Hall subgroup of G, the last set is nonempty). By induction, $U \leq S = \mathrm{S}(U^G)$. As S is characteristic in $U^G \lhd G$, it follows that $S \lhd G$, and inclusion $U^G \leq S$ shows that G is not a counterexample. Thus,

(i) $U^G = G$.

Assume that $UQ^G = G$, where $q \in \pi(G) - \pi(U)$ and $Q \in \mathrm{Syl}_q(G)$. Then, by (i), $U^G Q = G = UQ^G$. It follows from Lemma A.71.2 that $G = UQ^g$ for some $g \in G$ so that G is solvable, by Wielandt's Theorem A.28.14, i.e., G is not a counterexample. Thus,

(ii) $UQ^G < G$, where $Q \in \mathrm{Syl}_q(G)$ for any $q \in \pi(G) - \pi(U)$. In particular, G is not simple and $|\pi(G) - \pi(U)| > 1$.

Let R be a minimal normal subgroup of G. Then $U \nleq R$, by (i), and R is characteristically simple.

Assume that $UR = G$. Let $q \in \pi(G) - \pi(U)$ and $Q \in \mathrm{Syl}_q(R)(= \mathrm{Syl}_q(G))$. In that case, $Q^G = R$ so that, by (i), $U^G Q = G = UR = UQ^G$. Then, by Lemma A.71.2, $UQ^g = G$ for some $g \in G$. By Wielandt's Theorem A.28.14, G is solvable, i.e., G is not a counterexample. Thus,

(iii) $UR < G$.

Write $\bar{G} = G/R$; then $\bar{U} > \{1\}$ (see (i)) is a nilpotent $\pi(\bar{U})$-Hall subgroup of $\bar{G}$. By Exercise 2, $\bar{U}$ is permutable with any Sylow r-subgroup of $\bar{G}$, where r runs over the set $\pi(\bar{G}) - \pi(\bar{U})$. By induction, applied to $\bar{G}$, $\bar{U}^{\bar{G}} \leq \bar{H} = \mathrm{S}(\bar{G})$; then $U^G \leq H$. It follows from (i) that $H = G$ so that $G/R = \bar{G}$ is solvable. In particular,

(iv) R is nonsolvable and, as G/R is solvable for any choice of R, it follows that R is a unique minimal normal subgroup of G, i.e., G is a monolith.

By Exercise 1, U is permutable with any Sylow r-subgroup of UR for all $r \in \pi(UR) - \pi(U)$. As, by (iii), $UR < G$, and U is a nilpotent $\pi(U)$-Hall subgroup of UR, it follows, by induction, that $U \leq \mathrm{S}(UR)$. As R is a direct product of isomorphic nonabelian simple groups, we get $U = \mathrm{S}(UR) \lhd UR$ and $U \cap R = \{1\}$ so that

(v) $UR = U \times R$.

Write $C = C_G(R)$. Then $C \lhd G$. As $C \cap R = \{1\}$ and G/R is solvable, by (iv), it follows that C, as a group isomorphic to a subgroup of G/R, is solvable. As $U \leq C$, by (v), and $C \lhd G$, we conclude that $U^G \leq C$ is solvable, i.e., G is not a counterexample. □

In [Hup1, Satz VI.3.1], the groups in which any two Sylow subgroups of distinct orders are permutable, are described.

The following result, in some sense, is dual to Theorem A.71.3.

Exercise 3 (Kegel). Let $U > \{1\}$ be a solvable subgroup of a group G. If for each $p \in \pi(U)$ one has $U \cap P \in \mathrm{Syl}_p(U)$ for all $P \in \mathrm{Syl}_p(G)$, then U is subnormal in G.

Solution. Let $p \in \pi(\mathrm{F}(U))$, where $\mathrm{F}(U)$ is the Fitting subgroup of U, and let $P_1 \in \mathrm{Syl}_p(\mathrm{F}(U))$. If $P \in \mathrm{Syl}_p(G)$, then, by hypothesis, $P \cap U = P_2 \in \mathrm{Syl}_p(U)$. Since P_1 is contained in all Sylow p-subgroups of U, it follows that $P_1 \leq P$ for all $P \in \mathrm{Syl}_p(G)$ so that $P_1 \leq P_G$, the core of P in G. Write $\bar{G} = G/P_G$. By induction applied to $\bar{G}$, the subgroup $\bar{U}$ is subnormal in $\bar{G}$ so that U is subnormal in G.

Problem. Study the groups G containing a nilpotent subgroup U such that all $\pi(U)$-subgroups of G are solvable and U is permutable with all Sylow q-subgroups of G, where q runs over the set $\pi(G) - \pi(U)$ (if $\pi(U) = \pi(G)$, then G is solvable, by hypothesis).

Appendix 72 Nonabelian p-groups with an abelian subgroup of index p covered by minimal nonabelian subgroups

Here we prove the following

Theorem A.72.1 (Janko). *Let A be an abelian subgroup of index p in a nonabelian p-group G. If G is covered by minimal nonabelian subgroups, then $G' \le Z(G)$ has exponent p.*

Proof. Let $a \in A$ and $g \in G - A$ be such that $[a, g] \neq 1$. We shall prove that $[a, g] \in Z(G)$ is of order p, and this completes the proof. By hypothesis, there is a minimal nonabelian $M \le G$ containing a. Since $A \in \Gamma_1$ and $M \nleq A$, we get $G = AM = MA$. It follows that $g = a_1 g_1$ for some $a_1 \in A$ and $g_1 \in M$. Obviously, $g_1 \notin A \Rightarrow g_1 \in M - A$. By the choice, $a \notin Z(G)$ so that $C_G(a) = A$ hence $g_1 \notin C_G(a)$, and we conclude that $[a, g_1] \neq 1$. As $a, g_1 \in M$ do not commute, we get $M = \langle a, g_1 \rangle$ and so $M' = \langle [a, g_1] \rangle$. It follows from $C_G([a, g_1]) \ge MA = G$ that $[a, g_1] \le Z(G)$. Next, $[a, g_1]$ has order p (Lemma 65.1). We get

$$[a, g] = [a, a_1 g_1] = [a, g_1][a, a_1]^{g_1} = [a, g_1] .$$

As $o([a, g_1]) = p$ and $[a, g_1] \in Z(G)$, an arbitrary $\neq 1$ commutator $[a, g]$ in G is of order p and contained in $Z(G)$ (see the displayed formula), completing the proof. □

Note that the nonabelian p-groups covered by minimal nonabelian subgroups are not classified yet (see #860).

Problem 1. Study the nonabelian p-groups with an abelian subgroup of index p and covered by $\mathcal{A}_2$-subgroups.

Problem 2. Study the nonabelian p-groups with a normal abelian subgroup of index p^2 and covered by $\mathcal{A}_1$-subgroups.

Problem 3. Study the nonabelian p-groups with subgroup of index p and exponent p covered by $\mathcal{A}_1$-subgroups.

Problem 4. Classify the nonabelian p-groups covered by $M_p(n, n)$-($M_p(n, n, 1)$-) subgroups for a fixed $n > 1$.

Problem 5. Classify the nonabelian p-groups covered by subgroups $\cong M_{p^n}$ for a fixed $n > 3$.

Appendix 73 On metacyclic and modular p-groups

Recall that a p-group is said to be modular if any two of its subgroups are permutable.

Proposition A.73.1. *Let G be a modular minimal nonabelian p-group. Then G is metacyclic and if $p = 2$, then $G \cong Q_8$ or G is ordinary metacyclic.*

Proof. (Janko). Suppose that G is not a quaternion group. Then Theorem 73.15 implies that G has an abelian normal subgroup N and an element $t \in G$ such that $G = \langle N, t\rangle$ and there is a fixed positive integer s which is at least 2 in the case $p = 2$ such that $a^t = a^{1+p^s}$ for all $a \in N$. Since G is nonabelian, there is $n \in N$ such that $n^t = n^{1+p^s} \neq n$. It follows that $\langle n, t\rangle = G$ and so G is metacyclic. Moreover, if $p = 2$ then $n^t = n \cdot n^{2^s}$, where $n^{2^s} \in \mho_2(\langle n\rangle)$ since $s \geq 2$ and so G is ordinary metacyclic, completing the proof. □

Proposition A.73.2. *Let G be a metacyclic p-group. If $p > 2$ or $p = 2$ and G is ordinary metacyclic, then G is modular.*

Proof. (Janko) One may assume that G is nonabelian. Let $G = \langle a, b\rangle$ with $A = \langle a\rangle \trianglelefteq G$ and, in case $p = 2$, let us assume that G centralizes $A/\mho_2(A)$. Then each subgroup of A is normal in G. In case $p = 2$, set $o(a) = 2^n$, where $n \geq 3$; then $a^b = av$ with $v \in \langle a^4\rangle$. This gives

$$(a^{2^{n-2}})^b = (a^b)^{2^{n-2}} = (av)^{2^{n-2}} = a^{2^{n-2}}v^{2^{n-2}} = a^{2^{n-2}},$$

and so $\Omega_2(A) \leq Z(G)$. By Proposition 73.1, G is modular. If $p > 2$, then G is $S(p^3)$-free so modular. □

Proposition A.73.3. *Let G be a nonabelian modular metacyclic p-group. Then either $p > 2$ or $p = 2$ and $G \cong Q_8$, $G \cong M_{2^n}$ $(n \geq 4)$ or G is any ordinary metacyclic 2-group.*

Proof. (Janko) Assume that $p = 2$. Assume in addition that G has a cyclic subgroup of index 2. By Theorem 1.2, $G \cong Q_8$ or $G \cong M_{2^n}$ $(n \geq 4)$. Here the fact is used that if G is a 2-group of maximal class and order $\geq 2^4$, then $G/Z(G)$ is dihedral and so in this case G is nonmodular. From now on, let G have no cyclic subgroup of index 2. Let $G = \langle a, b\rangle$ be a metacyclic 2-group with $\langle a\rangle \trianglelefteq G$, $o(a) \geq 4$, and assume that G is not ordinary metacyclic with respect to $\langle a\rangle$. Hence G does not centralize $\langle a\rangle/\langle a^4\rangle$, where $\langle a^4\rangle = \mho_2(\langle a\rangle)$ and so $a^b = a^{-1}v$ with some $v \in \langle a^4\rangle$. Since G has no cyclic subgroup of index 2, one has also that $\langle a\rangle \cap \langle b\rangle \leq \langle a^4\rangle$. Also note that b^2 centralizes $\langle a\rangle/\langle a^4\rangle$ and so $\langle a^4, b^2\rangle$ is normal in G and $G/\langle a^4, b^2\rangle \cong D_8$. Hence G is nonmodular. □

Proposition A.73.4. *A minimal nonabelian p-group G of order $> p^3$ is powerful (see §26) $\iff$ it is metacyclic nonisomorphic to $\langle a, b \mid a^{2^n} = b^4 = 1,\ b^a = b^3\rangle$.*

Proof (Janko). Let $G = \langle a, b\rangle$. Assume that G is nonmetacyclic. In that case, a and b can be chosen so that $G' \nleq \langle a\rangle$ and $G' \nleq \langle b\rangle$. Then $G/\langle a^p, b^p\rangle \in \{S(p^3), D_8\}$ so that G is not powerful. Thus, G is metacyclic. In that case, $G = \langle a, b \mid a^{p^m} = b^{p^n} = 1,\ b^a =$

$bb^{p^{m-1}}\rangle$. If $p > 2$, then $G/\mho_1(G) \cong E_{p^2}$ so G is powerful. Now let $p = 2$. If $n = 1$, then $G \cong M_{2^{m+1}}$ is powerful $\iff m > 3$. Now let $n > 1$. In that case $G' \le \mho_2(G)$ and this holds $\iff m > 2$. □

Proposition A.73.5. *If G is a metacyclic 2-group and G/G' has no cyclic subgroup of index 2, then G is powerful.*

Proof. Indeed, if the abelian 2-group G/G' has no cyclic subgroup of index 2, then $G' \le \mho_2(G)$, i.e., $G/\mho_2(G)$ is abelian, so G is powerful (see § 26). □

Proposition A.73.6. *Suppose that a nonabelian metacyclic p-group G contains an abelian subgroup A of index p. If G is not minimal nonabelian, then $p = 2$ and $G/Z(G)$ is dihedral. (See Corollary 148.2.)*

Proof. In the case under consideration $|G : Z(G)| > p^2$. By Lemma 1.1, $A/Z(G) \cong G'$ so cyclic. If $A/Z(G), B/Z(G) < G/Z(G)$ are distinct cyclic of index p, then $A, B \in \Gamma_1$ are distinct abelian so that $A \cap B = Z(G)$ is of index p^2, a contradiction. Thus, $A/Z(G)$ is the unique cyclic subgroup of $G/Z(G)$ of index p so that $p = 2$ and $G/Z(G)$ is of maximal class (Theorem 1.2). Assume that $G/Z(G)$ is not dihedral. Then $G/Z(G)$ contains a cyclic subgroup $C/Z(G)$ of order 4 such that $C \nleq A$. Then $A \cap C > Z(G)$, a contradiction. Thus, $G/Z(G)$ is dihedral. □

Proposition A.73.7. *If G is a nonabelian two-generator p-group, then $Z(G) \le \Phi(G)$.*

Proof. (This result was cited in the book in many places.) Assume that this is false. Then $G = HZ(G)$ for some $H \in \Gamma_1$ so that H is nonabelian. Then H/L is noncyclic for $L = H \cap Z(G)$. As $G/L = (H/L) \times (Z(G)/L)$, it follows that

$$d(G) \ge d(H/L) + d(Z(G)/L) \ge 2 + 1 = 3 > 2 = d(G)\,,$$

a contradiction. □

Proposition A.73.8. *Suppose that the Frattini subgroup of a nonabelian metacyclic p-group G, $p > 2$, is abelian. Then G is either an $\mathcal{A}_1$- or an $\mathcal{A}_2$-group.*

Proof. Let $A \in \Gamma_1$. As the abelian $\Phi(G)$ has index p in A, it follows from Proposition A.73.5 that A is either abelian or minimal nonabelian, and hence G is either an $\mathcal{A}_1$- or an $\mathcal{A}_2$-group. □

Proposition A.73.9. *Let L be a central subgroup of a two-generator p-group G. If G/L has two distinct cyclic subgroups of index p, then G is either abelian or minimal nonabelian.*

Proposition A.73.10. *Let a metacyclic p-group G be such that $|G/Z(G)| = p^3$. Then G is an $\mathcal{A}_2$-group.*

Proof. There is an abelian $A \in \Gamma_1$. By Lemma 1.1, $|G'| = \frac{1}{p}|G/Z(G)| = p^2$ so that G is $\mathcal{A}_2$-group, by Corollary 65.3. □

Proposition A.73.11. *Suppose that all minimal nonabelian subgroups of a p-group G are metacyclic with noncyclic center. Then $\Omega_1(G)$ is contained in all maximal normal abelian subgroups of G.*

Proof. Let $A < G$ be maximal normal abelian and $x \in G - A$. By Lemma 57.1, there is $a \in A$ such that $M = \langle a, x \rangle$ is minimal nonabelian so metacyclic. Since $\mathrm{Z}(M) = \Phi(M)$ and all elements of the set $M - \Phi(M)$ have orders $> p$ and $x \notin \Phi(M)$, we get $o(x) > p$. Thus, $\Omega_1(G) < A$. □

Appendix 74 On p-groups, $p > 2$, without subgroups isomorphic to $S(p^3)$

Let $G = \Omega_1(G)$ be a nonabelian 2-group. Then there are noncommuting involutions x, y. In that case, the subgroup $\langle x, y\rangle$ contains a subgroup $D \cong D_8$. If G is not of maximal class, then $C_G(D) \nleq D$ (Proposition 10.17). If $D < H \leq DC_G(D)$ with $|H : D| = 2$, then H contains at least three subgroups isomorphic with D. If G is of maximal class and $D < G$, then G has at least two subgroups $\cong D$, unless $G \cong SG_{16}$.

In this appendix, we consider a similar situation in case p-groups, $p > 2$.

Theorem A.74.1. *Let G be a nonabelian p-group, $p > 2$. Then one of the following holds:*
(a) *The subgroup $\Omega_1(G)$ is elementary abelian.*
(b) *There is in G a subgroup $\cong S(p^3)$.*

In particular, if G has no subgroup $\cong S(p^3)$, then the subgroup $\Omega_1(G)$ is elementary abelian.

Proof. Let E be a maximal elementary abelian subgroup of G. Write $N = N_G(E)$. Suppose that $E \neq \Omega_1(N)$. Then there is in the set $N - E$ an element x of order p. Write $H = \langle x, E\rangle$. The subgroup H is nonabelian (otherwise, $H > E$ is an elementary abelian subgroup of G). By Lemma A.57.1, there is $a \in E$ such that the subgroup $M = \langle a, x\rangle$ is minimal nonabelian. It follows from $\Omega_1(M) = M$ and $p > 2$ that $M \cong S(p^3)$ (see Lemma 65.1), and we have case (b). Now assume that $E = \Omega_1(N)$. Then the subgroup E is characteristic in N, and this implies that $N = G$, so that G is as in (a). □

Corollary A.74.2. *Let E be a maximal elementary abelian subgroup of a p-group G, $p > 2$. Then one of the following holds:*
(a) *$E = \Omega_1(G)$.*
(b) *There is in $N_G(E)$ a subgroup $\cong S(p^3)$.*

Proof. Obviously, E is a maximal elementary abelian subgroup of $N_G(E)$. Assume that $N_G(E)$ has no subgroup $\cong S(p^3)$. Then $E = \Omega_1(N_G(E))$ is characteristic in $N_G(E)$ (see the proof of the previous theorem), and we conclude that $N_G(E) = G$ hence $E = \Omega_1(G)$. □

Corollary A.74.3. *Let $G = \Omega_1(G)$ be a nonabelian p-group, $p > 2$. If $C_G(H) < H$ for any minimal nonabelian subgroup $H < G$ of minimal order, then G is of maximal class.*

Proof. Let E be a maximal normal elementary abelian subgroup of G and let $x \in G - E$ be of order p. Then, as in Theorem A.74.1, there is in G a subgroup $H \cong S(p^3)$ so that H is minimal nonabelian subgroup of G of minimal order. By hypothesis, $C_G(H) < H$, and it follows from Proposition 10.17 that G is of maximal class. □

Exercise 1. Corollary A.74.3 also holds for $p = 2$. where $S(p^3)$ should be replaced by D_8. (**Hint.** See the first paragraph of the appendix.) The same is true for Theorem A.74.1 and Corollary A.74.2.

Corollary A.74.4. *Let an $\mathcal{A}_n$-group G, $n > 2$, be a p-group, $p > 2$. Suppose that $\Omega_1(G)$ is nonabelian. Then there is in G a nonmetacyclic $\mathcal{A}_2$-subgroup of minimal order.*

Proof. By hypothesis, G is nonmetacyclic since $\Omega_1(G)$ is. Assume that all $\mathcal{A}_2$-subgroups of minimal order in G are metacyclic. Let $E \leq G$ be minimal nonmetacyclic. Then either E is of order p^3 and exponent p or a 3-group of maximal class and order 3^4 (Theorem 66.1). In the second case, E is a nonmetacyclic $\mathcal{A}_2$-subgroup of minimal order in G, contrary to the assumption. Thus, E is of order p^3 and exponent p.

If $E \cong S(p^3)$ and $E < M \leq G$ with $|M : E| = p$, then M is a nonmetacyclic $\mathcal{A}_2$-subgroup of minimal order in G, contrary to the assumption.

Let $E \cong \mathrm{E}_{p^3}$ and let $E \leq E_1$, where E_1 is a maximal elementary abelian subgroup of G. By hypothesis, $E_1 \neq \Omega_1(G)$. Therefore, by Theorem A.74.1, there is in G a subgroup $S \cong S(p^3)$, contrary to the previous paragraph.

Thus, there is in G a nonmetacyclic $\mathcal{A}_2$-subgroup of minimal order. □

Exercise 2. Let an $\mathcal{A}_n$-group G, $n > 2$, be a 2-group, and let G be not of maximal class. Suppose that $\Omega_1(G)$ is nonabelian. Then there is in G a nonmetacyclic $\mathcal{A}_2$-subgroup of minimal order. (**Hint.** Use Proposition 10.17.)

Exercise 3. Let G be a nonabelian group of exponent p and order $> p^3$. If G has no subgroup of order p^4 and class 2, then it is of maximal class with abelian subgroup of index p. (**Hint.** Let $A < G$ be minimal nonabelian and let $H = AC_G(A)$. Use Proposition 10.17.)

Exercise 4. Is it true that the group G from Corollary 74.3 has an abelian subgroup of index p?

Appendix 75 Irregular p-groups with $< p$ absolutely regular subgroups of maximal possible order

Recall that a p-group X is *absolutely regular* if $|X/\mho_1(X)| < p^p$. Absolutely regular p-groups are regular (Theorem 9.8 (a)). Absolute regular 2-groups are cyclic. Absolutely regular 3-groups are metacyclic (Corollary 36.6). In this appendix, we show that the title groups are of maximal class. We prove the following

Theorem A.75.1. *Let A be an absolutely regular subgroup of an irregular p-group G such that $|A| \geq |B|$ for any absolutely regular subgroup $B < G$. If the number of absolutely regular subgroups of order $|A|$ in G is $< p$, then G is of maximal class.*

Remark 1. We claim that any irregular p-group G contains an absolutely regular subgroup of order p^p. Indeed, let $C < G$ be cyclic of order p^2 (C exists since $\exp(G) > p$, by Theorem 7.2 (b)). Let $C < R \leq G$, where $|R| = p^p$. As $\exp(R) > p$, it follows that $|R/\mho_1(R)| \leq p^{p-1}$ so that R is absolutely regular. Thus, in Theorem A.75.1, $|A| \geq p^p$.

Proof of Theorem A.75.1. It follows from Theorem 9.6 (e) that any (irregular) p-group G of maximal class and order $> p^{p+1}$ satisfies the hypothesis (indeed, there is in G exactly one absolutely regular subgroup of order $\frac{1}{p}|G|$). Next, all irregular groups of order p^{p+1} are of maximal class (Theorem 7.1 (b)) and such group may contain exactly p absolutely regular subgroups of index p for p odd (according to information of Mann, such groups exist for $p > 2$; the group D_8 has only one absolutely regular subgroup of order 2^2). Therefore, one may assume that $|G| > p^{p+1}$ is not of maximal class.

Let $A < G$ be an absolutely regular subgroup of maximal possible order; then $A \triangleleft G$ (to see this, consider the action of G by conjugation on the set of $< p$ absolutely regular subgroups of G of order $|A|$). Let $A < H \leq G$, where $|H : A| = p$. By Remark 1, $|A| \geq p^p \Rightarrow |H| \geq p^{p+1}$.

(i) Let $|A| = p^p$; then $|H| = p^{p+1}$. Assume that H is regular. Then $\Omega_1(H)$ is of order p^p and exponent p since $\exp(H) > p$ and H is not absolutely regular (Theorem 7.2 (b, d)). In that case, all maximal subgroups of H not containing $\Omega_1(G)$ are absolutely regular and their number is $\geq p$ since $d(G) \geq 2$ (see Theorem 7.1 (b)), and this is a contradiction. Thus, H is irregular so of maximal class (Theorem 7.1 (b)). It follows that any subgroup of G containing A as a subgroup of index p, is of maximal class. Then the group G, by Exercise 10.10, is also of maximal class, and we are done in this case.

(ii) Let $|A| > p^p$; then $|H| > p^{p+1}$. As in (i), one may assume that H is irregular. Indeed, if H is regular, then $\Omega_1(H)$ is of order p^p and exponent p, and therefore $\geq p$ maximal subgroups of H (of order $|A|$) not containing $\Omega_1(H)$ are absolutely regular, contrary to the hypothesis. Assume that H is not of maximal class. Then $H = A\Omega_1(H)$, where $\Omega_1(H)$ is of order p^p and exponent p (Theorem 12.1 (b)). Let $M < H$ be of index

p such that $\Omega_1(H) \nleq M$ (by the given factorization of H, one has $\Omega_1(G) \nleq \Phi(H)$); then $|M| = |A|$. The number of such M is a multiple of p so $\geq p$. Therefore, by hypothesis, it is possible to choose M so that it is not absolutely regular. As $\Omega_1(M) = M \cap \Omega_1(H)$ is of order p^{p-1} and exponent p, it follows that M is of maximal class (Theorem 12.1 (a)). In that case, $H/K_p(H)$ is of order p^{p+1} and exponent p, by Theorem 12.12 (b). Such group H has no absolutely regular subgroup of index p, and this is a contradiction since $A < H$ is absolutely regular of index p. Thus, for any choice, the subgroup H is of maximal class. Then, by Exercise 10.10, G is also of maximal class, completing the proof. □

Remark 2. The proof of Theorem A.75.1 shows that if A, B are maximal subgroups of an irregular p-group G of order $> p^{p+1}$ such that A is absolutely regular and B is of maximal class, then G is of maximal class.

We suggest to the readers to prove, using Hall's enumeration principle, the following generalization of Theorem A.75.1.

Theorem A.75.2. *Let G be an irregular p-group which is not of maximal class. The number of absolutely regular subgroups of order p^n, where $n \geq p$ is fixed, is a multiple of p.*

For $n = p$, this theorem follows immediately from Theorem 13.5. Theorem A.75.1 is also a generalization of Theorem 1.17 (b).

It follows that if $n > p$ is fixed and an irregular p-group G has exactly one absolutely regular subgroup of order p^n, then G is of maximal class and order p^{n+1} (here we use Theorem 9.6 (e) and Exercise 10.10).

Problem 1. Given $n > p$, study the p-groups containing exactly p^2 subgroups of maximal class of order p^n.

Problem 2. Given $n \geq p$, study the p-groups containing exactly p absolutely regular subgroups of order p^n.

Problem 3. Study the 2-groups containing exactly two metacyclic subgroups of index 2. (See § 87 where the 2-groups containing exactly one nonmetacyclic subgroup of index 2 are classified.)

Appendix 76 On a class of p-groups

The second author has classified the nonabelian p-groups $G = \Omega_1(G)$ such that $\langle A, x\rangle$ is minimal nonabelian for any maximal abelian $A < G$ and $x \in G - A$ of order p. In this appendix, we slightly improve this result for $p > 2$.

Theorem A.76.1. *Suppose that a nonabelian p-group $G = \Omega_1(G)$, $p > 2$, is such that the subgroup $\langle A, x\rangle$ is minimal nonabelian for any maximal normal abelian $A < G$ and $x \in G - A$ of order p. Then $G \cong S(p^3)$.*

Lemma A.76.2. *If G is as in Theorem A.76.1, then it has no normal elementary abelian subgroup of order p^3.*

Proof. Assume that $E \triangleleft G$ is elementary abelian of order p^3. Let $E \le A < G$, where A is a maximal normal abelian subgroup of G and let $x \in G - A$ be of order p. Then the nonabelian $B = \langle x, A\rangle$ is not minimal nonabelian since it contains a subgroup $C = \langle x, E\rangle$ satisfying $|\Omega_1(C)| = p^4$, contrary to Lemma 65.1. □

Recall that the fundamental subgroup G_1 of a 3-group G of maximal class is either abelian or minimal nonabelian. Indeed, assume that G_1 is nonabelian. Then $|G_1| \ge 3^4$ and G_1 is metacyclic (Corollary 36.6). As G_1 has a G-invariant subgroup of type $(3, 3)$ (Lemma 1.4), it follows that the cyclic derived subgroup of G_1 has order 3 (see Exercise 9.1 (b)). Then, by Lemma 65.2 (a), the two-generator subgroup G_1 is minimal nonabelian, as asserted.

Proof of Theorem A.76.1. If $|G| = p^3$, then $G \cong S(p^3)$. Next we assume that $|G| > p^3$. By Lemma A.76.2, G has no normal subgroup $\cong E_{p^3}$. Then, by Theorem 13.7, the group G, being nonmetacyclic and satisfying $\Omega_1(G) = G$, is a 3-group of maximal class.

Let G_1 be the fundamental subgroup of G and let $x \in G - G_1$ be of order 3. As $G = \langle x, G_1\rangle$ is not minimal nonabelian (indeed, $\mathrm{cl}(G) > 2$; see Lemma 65.1), it follows that G_1 is nonabelian. Then $|G| > 3^4$. As we know, G_1 is metacyclic minimal nonabelian and $\Phi(G)$ is noncyclic abelian. As $|G : \Phi(G)| = 3^2$, it follows that $\Phi(G)$ is a maximal abelian subgroup of G (see Theorem 9.6 (e)). By hypothesis, $M = \langle x, \Phi(G)\rangle \in \Gamma_1$ is minimal nonabelian so that $\mathrm{cl}(M) = 2$. This is a contradiction since, by Theorem 9.6 (e), M, being nonmetacyclic, is of maximal class and order $\ge 3^4$ so $\mathrm{cl}(M) > 2$. Now assume that $|G| = 3^4$. In that case, there is in G an abelian subgroup B of index 3. If $y \in G - B$ is of order 3, then $G = \langle y, B\rangle$, by hypothesis, is minimal nonabelian, a contradiction since $\mathrm{cl}(G) = 3 > 2$ (see Lemma 65.1). Thus, in the case under consideration, $|G| = p^3$ so that $G \cong S(3^3)$. □

Exercise 1. Suppose that a nonabelian 2-group $G = \Omega_1(G)$ is such that the subgroup $\langle A, x\rangle$ is minimal nonabelian for any maximal normal abelian $A < G$ and $x \in G - A$ of order 2. If $Z(G)$ is cyclic, then $G \cong D_8$.

Solution. One may assume that $|G| > 8$. If G is of maximal class, then $G \cong D_{2^n}$, $n > 3$. However $G = \langle C, x\rangle$, where C is cyclic and $x \in G - C$ is of order 2. As $\mathrm{cl}(G) > 2$, G is not minimal nonabelian, a contradiction. Now let $E < G$ be a G-invariant elementary abelian subgroup of maximal order and let $E \leq A \lhd G$, where A is maximal abelian; then $C_G(A) = A$. Obviously, $E = \Omega_1(A)$ and $|A| > 4$ (Proposition 1.8). Let $x \in G - A$ be an involution. Then $H = \langle A, x\rangle$ is minimal nonabelian, by hypothesis, so $|H| > 8$ and $|E| = 4$. Set $F = \langle E, x\rangle$; then $|F| = 8$ so that F, being a proper subgroup of H, is elementary abelian for any choice of x. It follows that $C_G(E) \geq \Omega_1(G) = G$; then $Z(G) \geq E$ is noncyclic, contrary to the hypothesis.

Problem 1. Study the nonabelian p-groups $G = \Omega_1(G)$ such that for any noncyclic maximal normal abelian subgroup A and $x \in G - A$ of order p the subgroup $\langle x, A\rangle$ is minimal nonabelian.

Problem 2. Study the p-groups $G = \Omega_1(G)$, $p > 2$, such that for any maximal normal regular subgroup R of G and any $x \in G - R$ of order p the subgroup $\langle x, R\rangle$ is minimal irregular.

Appendix 77 The p-groups, $p > 2$, containing only one subgroup $\cong S(p^3)$

In this appendix, we classify the p-groups, $p > 2$, containing exactly one subgroup isomorphic to $S(p^3)$.[1]

Theorem A.77.1 ([ZG2], Theorem 3.4). *Suppose that a p-group G of order $> p^3$, $p > 2$, contains only one subgroup, say D, which is $\cong S(p^3)$. Then one of the following holds:*
(a) *$G = DC$, where $\Omega_1(G) = D$ and C is cyclic of order $> p$.*
(b) *$p = 3$, $G \cong \Sigma_9$, a Sylow subgroup of the symmetric group S_9 of degree 9.*
(c) *$p = 3$, G is a 3-group of maximal class and order 3^4 such that $\Omega_1(G) = D$.*

Proof. By hypothesis, $D \lhd G$. Suppose that $D < \Omega_1(G)$. Then there is $x \in G - D$ of order p. Set $H = \langle x, D\rangle$; then $|H| = p^4$. If $\exp(H) = p$, then there are in H at least p subgroups isomorphic to $S(p^3)$ (Theorem 5.8 (b)), contrary to the hypothesis. Thus, $\exp(H) > p$ so that H is irregular (Theorem 7.2 (b)), and we conclude that $p = 3$ and H is of maximal class (Theorems 7.1 (b) and 7.2 (b)). Let $R < D$ be G-invariant of order 3^2. Then $K = \langle x, R\rangle \cong E_{27}$, by hypothesis, since the subgroup K of order 3^3 and exponent 3 (Theorems 7.1 (b) and 7.2 (b) again) is not equal to D, and we conclude that $H \cong \Sigma_9$ (Exercise 9.13). It is known that Σ_9 satisfies the hypothesis. Let $e_3(G)$ be the number of subgroups of G of order 3^3 and exponent 3 and $e_3'(G)$ be the number of elementary abelian subgroups of G of order 3^3. Assume that G is not of maximal class. Then $e_3(G) \equiv 1 \pmod 3$ (Theorem 18.2 (a)) and $e_3'(G) \equiv 1 \pmod p$ (Theorem 10.4). Therefore, the number $t = e_3(G) - e_3'(G) \equiv 0 \pmod 3$, where t is the number of nonabelian subgroups of order 3^3 and exponent e in G. Since, by hypothesis, $t = 1$, we get a contradiction. Thus, G is a 3-group of maximal class. Assume that $|G| = 3^m > 3^4$. Since the fundamental subgroup G_1 is either abelian or minimal nonabelian, G has no normal subgroup $\cong S(3^3)$, a contradiction. Thus, $|G| = 3^4$. It follows that either $\Omega_1(G) \cong \Sigma_9$ or $\Omega_1(G) = D$.

It remains that $\Omega_1(G) = D$ and $G = DC$, where C is cyclic of order $> p$ (Theorem 13.7). The proof is complete. □

A classification of 2-groups with only one subgroup, say Q, $Q \cong Q_8$, is essentially more difficult task (about such 2-groups see [ZG2, Theorem 4.1]).

Exercise 1 ([ZG2], Theorem 4.6). Suppose that a 2-group G of order > 8 contains only one subgroup $\cong D_8$. Then $G \cong SD_{16}$.

Solution. Let D be the unique subgroup of G which is $\cong D_8$. If $C_G(D) \nleq D$, then $C_G(D)$ contains at least three subgroups $\cong D_8$ (see Appendix 16), a contradiction. If $C_G(D) <$

1 Such groups were classified in [ZG2, Theorem 3.4]. The p-groups, $p > 2$, without subgroups $\cong S(p^3)$ are classified in Appendix 74. The first author wrote this section before acquaintance with [ZG2].

D, then G is of maximal class (Proposition 10.17). It is easily seen that in this case $G \cong \mathrm{SD}_{16}$ (Proposition 1.6 and Theorem 1.2).

Exercise 2 ([ZG2] Theorem 4.6). Classify the p-groups, $p > 2$, containing exactly one proper subgroup, say D, which is isomorphic to M_{p^3}.

Solution. By Proposition 10.19, G is nonmetacyclic. If $\mathrm{C}_G(D) < D$, then G is of maximal class (Proposition 10.17). One has $D \triangleleft G$. Then $|G| = p^4$ (note that $|\mathrm{Aut}(D)|_p = p^3$). As $\exp(G) > p$, the set Γ_1 contains at least two nonabelian members of exponent p, G is irregular (Theorem 7.2 (b)), hence, $p = 3$. Now let $\mathrm{C}_G(D) \nleq D$. Clearly, $\mathrm{C}_G(D)$ is cyclic of order $> p$. Set $S = D\mathrm{C}_G(D)$ and $F = \Omega_2(S)$. Then $F' = \Phi(F)$. By Theorem 7.2 (d), $|\Omega_1(F)| = p^3$. All maximal subgroups of F not containing $\Omega_1(F)$, are metacyclic. By Exercise 1.6, G has at most $p + 1$ abelian subgroups of index p. Therefore, there is in F at least $(p^2 + p + 1) - (p + 1) - 1 > 1$ subgroups $\cong \mathrm{M}_{p^3}$, a contradiction.

Problem 1. Classify the p-groups of order $> p^4$, $p > 2$, containing exactly one nonabelian metacyclic subgroup of order p^4 and exponent p^2.

Appendix 78 Further criterion of p-nilpotence and π-nilpotence

Recall that a group G is p-nilpotent if it has a normal p-complement.

We begin with the following

Definition 1. A p-group P is said to be a $(*)$-group if, whenever $p \in \pi(G)$, $P \in \mathrm{Syl}_p(G)$ and $\mathrm{N}_G(P)$ is p-nilpotent, then G is also p-nilpotent.

Regular p-groups are $(*)$-groups, by Wielandt's Theorem A.52.5 (a).

Let G be a metacyclic p-group. Set

$$w = w(G) = \max\{n \mid |\Omega_n(G)| = p^{2n}\}, \quad R(G) = \Omega_{w(G)}(G).$$

Then $G/R(G)$ is either cyclic or a 2-group of maximal class (Lemma 1.4).

Proposition A.78.1. *Let p be the minimal prime divisor of the order of a group G and let $P \in \mathrm{Syl}_p(G)$ be metacyclic. If $P/R(P) > \{1\}$ is cyclic, then P is a $(*)$-group.*

Proof. Assume that P is a counterexample of minimal order. Then there is a group G satisfying the following condition: $\mathrm{N}_G(P)$ is p-nilpotent, where $P \in \mathrm{Syl}_p(G)$ but G is not p-nilpotent. The subgroup P is nonabelian (Burnside; see Theorem A.52.1). Therefore, since $P/R(P)$ is cyclic, it follows that $R(P) > \{1\}$. By hypothesis, P is not a 2-group of maximal class. We proceed by induction on $w = w(G)$. By induction, if $P \le T < G$, then T is p-nilpotent (indeed, $\mathrm{N}_T(P) = \mathrm{N}_G(P) \cap T$ is p-nilpotent). By Theorem A.52.2, in G there is a minimal nonnilpotent subgroup $H = Q{\cdot}H'$, where $H' \in \mathrm{Syl}_p(H)$ is metacyclic and $Q \in \mathrm{Syl}_q(H)$ is cyclic, $q > p$ is a prime. Then $H' \cong \mathrm{E}_{p^2}$ (Proposition 10.19 and Lemma A.22.1)). Since $q \mid p^2 - 1$ and $q > p$, one has $p = 2$, $q = 3$. (Equality $p = 2$ follows from Wielandt's Theorem A.52.5 (a) since the metacyclic p-groups, $p > 2$, are regular.) One may assume that $H' \le P$. As P is nonabelian, one has $H' < P$. Moreover, $H' = \Omega_1(P) \lhd P$ since $\Omega_1(P) \cong \mathrm{E}_4$. Write $N = \mathrm{N}_G(H')$; then $P < N$ since $H < N$.

Let $w(P) = 1$. Then $P \cong \mathrm{M}_{2^n}$ (Theorem 1.2) and G is 2-nilpotent, by Proposition A.52.4. Now let $w > 1$. Then N/H' is 2-nilpotent, by induction (note that $(P/H')/R(P/H')$ is nonidentity cyclic), but N is not 2-nilpotent since $H < N$. By induction, $N = G$ so that $H' \lhd G$. By induction, G/H' is 2-nilpotent. Let U/H' be the normal $2'$-Hall subgroup of G/H' and let $V(> \{1\})$ be a $2'$-Hall subgroup of U (V exists, by Schur–Zassenhaus). One has $U = V \cdot H'$ (semidirect product with kernel H'). Then V is a π'-Hall subgroup of G. By assumption, V is not normal in G. Again, by the theorem of Schur–Zassenhaus and Frattini's argument, one has

$$G = \mathrm{N}_G(V)U = \mathrm{N}_G(V)VH' = \mathrm{N}_G(V)H'.$$

By the modular law, $P = H'(\mathrm{N}_G(V) \cap P)$, which is a contradiction since, in view of $w > 1$, one has $H' \le \Phi(P)$. Thus, H does not exist so G is 2-nilpotent. Therefore, P is a $(*)$-group, as was to be shown. □

If $p > 2$ is the minimal prime divisor of the order of a group G and $P \in \mathrm{Syl}_p(G)$ is metacyclic, then G is p-nilpotent (here we do not assume that $\mathrm{N}_G(P)$ is p-nilpotent). Indeed, G has no minimal nonnilpotent subgroup, say H, whose Sylow p-subgroup coincides with H' (see the proof of Theorem A.78.1).

For many related results, see Appendix 52.

Exercise 1. Suppose that $Q \cong \mathrm{Q}_8$ is a Sylow 2-group of a group G. If $\mathrm{N}_G(Q)$ is 2-nilpotent, so is G. What will be if we replace Q by $Q \times A$, where A is abelian?

Exercise 2. Suppose that a Sylow 2-subgroup of a group G is a unique nonabelian metacyclic group P of order 2^4 and exponent 4. Is it true that if $\mathrm{N}_G(P)$ is 2-nilpotent so is G?

Problem. Is it true that if $P = R(P) \in \mathrm{Syl}_2(G)$ is metacyclic, then G is 2-nilpotent if $\mathrm{N}_G(P)$ is 2-nilpotent?

Appendix 79 2-groups G containing a nonabelian metacyclic subgroup H of order 2^{2e} and exponent 2^e such that $N_G(H)$ is metacyclic

Let H be a proper nonabelian metacyclic subgroup of order p^{2e} and exponent p^e in a p-group G such that $N_G(H)$ is metacyclic. If $p > 2$, then $H = \Omega_e(N_G(H))$ is characteristic in $N_G(H)$ so $G = N_G(H)$ is metacyclic.[1] In Theorem A.79.3, we consider a more complicated situation that arises in the case when H is nonnormal in a nonmetacyclic 2-group G.

Let G be a metacyclic p-group and let $R < G$ be abelian of type (p, p). Suppose that $R \neq \Omega_1(G)$. In that case, $p = 2$ and there is $x \in G - R$ of order 2. Then $\langle x, R\rangle \cong D_8$ so that G is a 2-group of maximal class (Proposition 10.19).

Remark 1. Let H be a proper normal subgroup of order p^{2e} and exponent p^e in a metacyclic p-group G. If $p > 2$, then $\Omega_e(G) = H$ since G is regular. Next assume that $p = 2$ and $e > 1$. Then $\Omega_1(H) \cong E_4 \cong \Omega_1(G)$ (here we use Lemma 1.4 since H is not of maximal class) so that $\Omega_1(H) = \Omega_1(G)$ (Proposition 10.19) is characteristic in G. It is easy to prove by induction that then $\Omega_{e-1}(H) = \Omega_{e-1}(G)$ is characteristic in G. Assume that H is not characteristic in G; then $H/\Omega_1(G)$ is not characteristic in $G/\Omega_1(G)$. But $\Omega_{e-1}(H) = \mho_1(H)$. Write $\bar{G} = G/\mho_1(H)$. Then $\bar{H}$ is not characteristic in $\bar{G}$. It follows that $\Omega_1(\bar{G}) > \bar{H}$. Therefore, there is an involution $\bar{x} \in \bar{G} - \bar{H}$. Then $\langle \bar{x}, \bar{H}\rangle \cong D_8$. By Proposition 10.19, $\bar{G}$ is a 2-group of maximal class, dihedral, or semidihedral.

Lemma A.79.1. *Let H be a proper subgroup of a p-group G, $R = \Omega_1(H)$ is of exponent p. If $\Omega_1(N_G(H)) = R$, then all G-invariant subgroups of exponent p are contained in H.*

Proof. Assume that $S \not\leq H$ is a normal subgroup of G of exponent p. Let $S_0 \leq S$ be G-invariant such that $S_0 \not\leq H$ and $|S_0|$ is as small as possible. Then $|HS_0 : H| = p$ so that $S_0 < N_G(H)$. Since $\Omega_1(HS_0) > R = \Omega_1(N)$, we get a contradiction. Thus, S does not exist. □

Exercise 1. Suppose that H is a proper absolutely regular subgroup of a p-group G and let $\Omega_1(N_G(H)) = \Omega_1(H)$. Then G is either absolutely regular or of maximal class.

Solution. Set $S = \Omega_1(H)$; then $|S| = p^k \leq p^{p-1}$. Write $N = N_G(H)$. If N is of maximal class so is G (Remark 10.5). Next we assume that N is not of maximal class. Then N is

1 Here we have used the following obvious fact which holds for arbitrary p-groups. If H is a proper subgroup of a p-group G such that H is characteristic in $N = N_G(H)$, then $N = G$.

absolutely regular (Theorem 12.1 (a)). By Lemma A.79.1, G has no normal subgroup of order p^p and exponent p so it is of maximal class (Theorem 12.1 (a)).

Let G be a metacyclic p-group. Let (see § 124)

$$p^{w(G)} = p^w = \max\{p^i \mid |\Omega_i(G)| = p^{2i}\}, \quad R(G) = \Omega_i(G).$$

Now we are ready to prove the main result of this appendix.

Theorem A.79.2. *Let H be a nonnormal metacyclic subgroup of order 2^{2e} and exponent 2^e of a 2-group G. Then, if $\mathrm{N}_G(H)$ is metacyclic, then $\mho_1(H) \lhd G$ and $G/\mho_1(H)$ is of maximal class.*

Proof. Let $e = 1$; then $H \cong \mathrm{E}_4$. If H is characteristic in $N = \mathrm{N}_G(H)$, then $N = G$. Now assume that H is not characteristic in N. Then $H^\phi \neq H$ for some $\phi \in \mathrm{Aut}(N)$. In that case, HH^ϕ, being metacyclic of order 2^3 and generated by elements of order 2, is $\cong \mathrm{D}_8$. Then N is of maximal class (Proposition 10.19). By Remark 10.5, G is of maximal class, and we are done in this case.

Now let $e > 1$. Then $R = \Omega_1(H)$ is the unique normal abelian subgroup of type $(2, 2)$ in G (see Lemma A.79.1). By hypothesis, $\mathrm{N}_{G/R}(H/R) = N/R$ is metacyclic. By induction, $F/R = \Omega_{e-2}(H/R) \lhd G/R$ so that $F = \Omega_{e-1}(H) \lhd G$. Next, $\mathrm{N}_{G/F}(H/F)$ is metacyclic, by hypothesis, so that G/F, by case $e = 1$, is of maximal class. As $F = \mho_1(H)$, we are done. □

Exercise 2. Study the metacyclic p-groups G containing a proper nonabelian subgroup H of order p^4 and exponent p^2.

Hint. One has $\mathrm{Z}(H) = \Omega_1(H) = \Omega_1(G) \lhd G$. Set $C = \mathrm{C}_G(\mathrm{Z}(H))$; then $|G : C| \leq p$. Let $L < \mathrm{Z}(H)$ of order p be such that $L \neq H'$. Then H/L is a nonabelian subgroup of C/L of order p^3. If $p > 2$, then $C/L = H/L$ (Proposition 10.19) so that $|G : H| \leq p \Rightarrow |G| = p^5$. Now let $p = 2$. Then C/L is of maximal class (Proposition 10.19).

Problem 1. Let H be a proper metacyclic subgroup of a p-group G. Study the structure of G, provided $\mathrm{N}_G(H)$ is metacyclic.

Problem 2. Let S be a proper special subgroup of a p-group G such that $\mathrm{N}_G(S)$ is special. Study the structure of G. Consider in detail the case when S and $\mathrm{N}_G(S)$ are extraspecial.

Problem 3. Given $n > 2$, study the primary $\mathcal{A}_n$-groups G such that $\mathrm{N}_G(M)$ is an $\mathcal{A}_2$-group for all minimal nonabelian $M < G$.

Problem 4. Let H be a proper metacyclic subgroup of a nonmetacyclic 2-group G, H is noncyclic. Study the structure of G provided any subgroup of G containing H as a subgroup of index 2 is metacyclic.

Appendix 80 On minimal nonabelian groups of order 2^{2n} and exponent 2^n

In this appendix we prove the following

Proposition A.80.1. *Let* $G = \langle a, b \mid o(a) = o(b) = 2^n,\ n > 2,\ a^b = a^{1+2^{n-1}}\rangle$ *be a minimal nonabelian group of order* 2^{2n} *and exponent* 2^n*. Then*

(a) $c_n(G) = 3 \cdot 2^{n-1}$.

(b) *Let D be generated by all normal cyclic subgroups of G of order* 2^n*. Then* $D \in \Gamma_1$ *is characteristic in G and all cyclic subgroups of order* 2^n *in D are G-invariant. If* $G/G' = (U/G') \times (V/G')$*. where* U/G'*,* V/G' *are cyclic of orders* $2^n, 2^{n-1}$*, respectively, then V is cyclic of order* 2^n*.*

(c) *The number of G-invariant cyclic subgroups of G of order* 2^n *is equal to* 2^{n-1} *so that all such subgroups are contained in D. The subgroup D is characteristic in G.*

(d) *Every subgroup of G of order 2 is contained in exactly* 2^{n-1} *cyclic subgroups of order* 2^n*. (This is not true for* $n = 2$*.)*

(e) *Let* $\Gamma_1 = \{D = D_1, D_2, D_3\}$ *and let* $C_i < D_i$ *be cyclic of order* 2^n*,* $i = 1, 2, 3$*. Then* $\mho_{n-1}(D_i) \neq \mho_{n-1}(D_j)$ *for* $i \neq j$*. Therefore,* $C_i \cap C_j = \{1\}$ *so that* $C_iC_j = G$*.*

(f) *If* $C \triangleleft G$ *is cyclic of order* 2^{n-1} *such that* $G' \not\leq C$*, then* $G/C \cong M_{2^{n+1}}$*. Every such C is contained in exactly two cyclic subgroups of order* 2^n*, and these subgroups are not G-invariant.*

Proof. The group G is metacyclic. Write $A = \langle a\rangle$, $B = \langle b\rangle$; then $G = B \cdot A$ is a semidirect product with kernel A.

(a) Since $\exp(\Omega_{e-1}(G)) = 2^{e-1}$ and $|\Omega_{e-1}(G)| = 2^{2e-2}$ for $e \leq n$, one has

$$c_n(G) = \frac{|G - \Omega_{n-1}(G)|}{2^{n-1}} = \frac{2^{2n} - 2^{2n-2}}{2^{n-1}} = 3 \cdot 2^{n-1}$$

(here 2^{n-1} is the number of generators of a cyclic group of order 2^n). Similarly, for $e < n$, one has $c_e(G) = 3 \cdot 2^{e-1}$.

(b,c) The subgroup $\Omega_{n-1}(G) = \langle a^2\rangle \times \langle b^2\rangle = \mho_1(G) = Z(G) = \Phi(G)$ is abelian of type $(2^{n-1}, 2^{n-1})$. If $L \triangleleft G$ is cyclic of order 2^n, then $L \not\leq \Phi(G)$ since $\exp(\Phi(G)) = 2^{n-1} < 2^n$, and hence the quotient group G/L is cyclic. Set $T = \langle ab^2\rangle$. Then $|T| = 2^n$, $A \cap T = \Omega_1(A) = G'$ so that $T \triangleleft G$ and $D = AT \in \Gamma_1$, by the product formula. Let $V < D$ be cyclic of order 2^n, $V \neq A$. Then $A \cap V > \{1|$, by the product formula, so that $G' = \Omega_1(A) < V$. Thus, all cyclic subgroups of D of order 2^n are G-invariant. If L is as in the second sentence of this paragraph, then $A \cap L \geq G'$ (otherwise, $AL = G$ is abelian) and hence $AL \leq D$ (since D is the unique maximal subgroup of G containing A) so that D contains all G-invariant cyclic subgroups of order 2^n of G. It follows that D is characteristic in G. Next,

$$c_n(D) = \frac{|D - \Omega_{n-1}(D)|}{2^{n-1}} = \frac{2^{2n-1} - 2^{2n-2}}{2^{n-1}} = 2^{n-1},$$

so that G has exactly 2^{n-1} normal cyclic subgroups of order 2^n and all these subgroups contain G', completing the proof of (c). Also, G has exactly $3 \cdot 2^{n-1} - 2^{n-1} = 2^n$ nonnormal cyclic subgroups of order 2^n, by (a). If L is as above and $M < G$ is nonnormal cyclic of order 2^n, then $L \cap M = \{1\}$ since $\Omega_1(L) = G'$, and so $M \cdot L = G$. Let U, V be such as in the statement of (b) and assume that V is noncyclic; then $\Omega_1(G) < V$. In this case, $|U| = 2^{n+1}$ and $R = \Omega_1(G) < U$. As U is noncyclic, we have $U = U_1 \times G' = U_1 R$, where $U_1 \cong C_{2^n}$. Then $G = UV = U_1RV = U_1V$ and so $U_1 \cap V = \{1\}$, by the product formula. This is a contradiction since $U_1 \cap V \geq U_1 \cap R > \{1\}$. Thus, V is cyclic, completing the proof of (b).

(d) Let

$$L_1 = \Omega_1(A), \quad L_2 = \Omega_1(B), \quad L_3 = \langle x \rangle, \quad \text{where } x \in \Omega_1^\#(G),\ x \notin A,\ x \notin B\},$$
$$\Gamma_1 = \{D = D_1, D_2, D_3\}, \quad A < D = D_1, \quad B < D_2 .$$

All members of the set Γ_1 are abelian of type $(2^n, 2^{n-1})$. One has $c_n(D_i) = 2^{n-1}$ and $D_i \cap D_j = \Phi(G) = \Omega_{n-1}(G)$ $(i \neq j)$ so that, $\sum_{i=1}^{3} c_n(D_i) = 3 \cdot 2^{n-1} = c_n(G)$. It follows from (a) and (c) that all cyclic subgroups of order 2^n from D_2 and D_3 are nonnormal in G. Note, that $c_n(D_i)$ cyclic subgroups and only these ones contain L_i, $i = 1, 2, 3$. The proof of (d) is complete. (Note that $(ab)^{2n-1} = a^{2n-1}b^{2n-1}$ generates L_3. However, if $n = 2$, then$(ab)^2 = b^2$ generates L_2.)

(e) Follows from (d).

(f) Let $C < G$ be cyclic of order 2^{n-1} such that $G' \nleq C$ (C exists since $\Phi(G)$ is abelian of type $(2^{n-1}, 2^{n-1})$ and $|G'| = p$). One has $C < \Omega_{n-1}(G) = Z(G)$. As G/C is minimal nonabelian with cyclic subgroup $\langle a \rangle C/C = AC/C$ of index 2 and $n > 2$, it follows that $G/C \cong M_{2^{n+1}}$ (Theorem 1.2). Clearly, C is contained in three subgroups of G of order $2|C| = 2^n$. One of these subgroups, namely CR, is noncyclic. Let U/C, $V/C < G/C$ be nonnormal of order 2. Then $G' \nleq U, V \Rightarrow R \nleq U, V \Rightarrow \Omega_1(G) \nleq U, Y$. It follows that U, V are cyclic, completing the proof of (f) and thereby the theorem. □

It follows from (e) that the groups of Theorem A.80.1 are modular (this is not true for $n = 2$).

Exercise 1. Prove an analog of Theorem A.80.1 for metacyclic minimal nonabelian groups of order p^{2n} and exponent p^n, $p > 2$.

Exercise 2. Classify the characteristic subgroups of groups of Theorem A.80.1 and Exercise 1.

Exercise 3. Prove analogs of Theorem A.80.1 and Exercise 2 for arbitrary minimal nonabelian p-groups. (See Lemma 65.1.)

Appendix 81 On p-groups with a cyclic subgroup of index p^3

In this appendix, we prove one property of p-groups with a cyclic subgroup of index p^3. Such groups were classified in [Nei] and [ZL2] (see also [Tit]); our exposition is independent of those papers.

Proposition A.81.1. *Suppose that a p-group G of order $> p^5$, $p > 2$, has a cyclic subgroup Z of index p^3. Then either G is metacyclic or possesses a normal abelian subgroup of index p^2.*

(1) *If G is metacyclic, it possesses a normal abelian subgroup of index p^3.*

(2) *If G is a 3-group of maximal class, then $|G| = 3^6$.*[1]

Proof. One may assume that G is nonabelian. Assume that G has no normal subgroup $\cong \mathrm{E}_{p^3}$. Then, by Theorem 13.7, one of the following holds:

(i) G is metacyclic.

(ii) $G = ZR$, where $R = \Omega_1(G) \cong S(p^3)$.

(iii) G is a 3-group of maximal class.

(a) Let G be as in (ii). Let $L < R$ be G-invariant of index p. Then $\mathrm{C}_G(L)$ is abelian of index p in G. Note that $\mathrm{C}_G(L)$ has a cyclic subgroup of index p; in this case G has a cyclic subgroup of index p^2. Next, $\Phi(\mathrm{C}_G(L))$ is G-invariant cyclic of index p^3.

(b) Now let G be as in (iii), i.e., G is a 3-group of maximal class and order $> 3^5$. Then G_1, the metacyclic fundamental subgroup of G, is either abelian or minimal nonabelian.[2] In that case, $\Phi(G)$ is a G-invariant metacyclic abelian subgroup of index 3^2 in G. By Theorem 3.19, $Z < G_1$. Next, $|Z_G| \le 3$ (see footnote). Let $w = w(G_1)$ (see § 124). Then $3^2 = |G_1 : Z| = 3^w$ so that $w = 2$. In that case, by Theorem 9.6, $|G_1| \le 3^{2w+1} = 3^5$ so that $|G| = 3|G_1| = 3^6$, proving (2).

(c) Now let G have a normal subgroup $E \cong \mathrm{E}_{p^3}$. Then $G/\mathrm{C}_G(E)$ is of order at most p^3 and exponent p. Write $H = EZ$, where Z is cyclic of index p^3 in G. Then $|G : H| \le p$. If $H = G$, then $\mathrm{C}_G(E)$ is abelian of index p in G. Now let $|G : H| = p$. Then $E\mho_1(Z)$ is the abelian subgroup of index $\le p$ in H so index p^2 in G. In that case, by Exercise 1.7, G has a normal abelian subgroup of index p^2.

(d) Now let G be metacyclic with cyclic subgroup Z of index p^3. Assume that $Z < A < G$, where A is abelian of order $p|Z|$; then $|G : A| = p^2$. In that case, G has a normal abelian subgroup of index p^2 (Exercise 1.7). Now assume that A does not exist; then

---

1 Any 3-group of maximal class possesses a normal abelian subgroup of index 3^2. A 3-group of maximal class and order 3^6 has a cyclic subgroup of index 3^3.

2 As G has no normal cyclic subgroup of order 3^2 (Lemma 1.4 and Exercise 9.1 (b)), one has $|G_1'| \le 3$ so, if metacyclic subgroup G_1 is nonabelian, it is minimal nonabelian, by Lemma 65.2 (a).

$C_G(Z) = Z$. Let $Z < M < G$, where $|M : Z| = p$. Then $M \cong \mathrm{M}_{p|Z|}$ (Theorem 1.2) has index p^2 in G. In that case, G/M_G is isomorphic to a subgroup of $\Sigma_{p^2} \in \mathrm{Syl}_p(\mathrm{S}_{p^2})$. Since G/M_G is metacyclic, it follows from the structure of Σ_{p^2} that $|G/M_G| \le p^3$. If $M \lhd G$, then the noncyclic subgroup of index p in M is characteristic in M so normal in G and has there index p^3. Now let M be nonnormal in G. Since $M \cong \mathrm{M}_{p|Z|}$ is minimal nonabelian, M_G is abelian of index p in M so of index p^3 in G.[3] □

Proposition A.81.2. *Suppose that a 2-group G has a cyclic subgroup Z of index 2^3. Then G has a normal abelian subgroup of index 2^3.*

Proof. One may assume that G has no cyclic subgroup of index 2. Let $Z < M < G$, where $|M : Z| = 2$. As in the proof of Theorem A.81.1, one may assume that M is nonabelian. By Exercise 10.10, one can choose M so that it is not of maximal class. Then, by Theorem 1.2, $M \cong \mathrm{M}_{2|Z|}$. As M contains an abelian characteristic subgroup of index 2 so of index 2^3 in G, one may assume that M is not normal in G. Since $|G : M| = 4$, G/M_G is isomorphic to a subgroup of $\mathrm{S}_{|G:M|} = \mathrm{S}_4$. Therefore, M_G is abelian of index 2^3 in G. □

Problem. Study the p-groups containing a cyclic subgroup of index p^4.

3 Since our metacyclic p-group is metabelian, it contains a normal abelian subgroup of index $= |G : Z| = p^3$; see Mann's Theorem 39.7. However, our solution is independent of that result. The general result also follows from Theorem 39.1.

Appendix 82 On nonmodular p-groups all of whose subgroups of order $> p$ are quasinormal

In Passman's Theorem 1.25, the non-Dedekindian p-groups all of whose subgroups of order $> p$ are normal, are classified. Below we initiate the study of the nonmodular p-groups all of whose subgroups of order $> p$ are quasinormal. We did not attain the full classification of such p-groups (this is a fairly difficult problem; difficulty is due to the fact that the intersection of two quasinormal subgroups need not quasinormal).

Exercise 1. Suppose that all subgroups of a p-group G having order $> p^k$, are quasinormal. If $L \triangleleft G$ is of order p^k, then G/L is modular.

Exercise 2. Suppose that all minimal nonabelian subgroups of a p-group G are quasinormal. If A is a maximal abelian normal subgroup of G, then G/A is modular.

Proposition A.82.1. *Let G be a nonmodular p-group all of whose subgroups of order $> p$ are quasinormal. Then*

(a) *G contains a subgroup $R \cong D_8$ if $p = 2$ or $S(p^3)$ if $p > 2$, $Z(R) = \Omega_1(Z(G))$ so that $Z(G)$ is cyclic.*
(b) *$C_G(R)$ is either cyclic or $\cong Q_8$.*
(c) *All subgroups of G of order p that are $\neq Z(R)$ are not quasinormal.*
(d) *If $\exp(G) = p$, then $|G| = p^3$.*

Proof. By Theorem A.59.1, G has a section R/L isomorphic either with D_8 or $S(p^3)$ (both these groups are nonmodular and have a nonquasinormal subgroup of prime order). If $U/L < R/L$ is not quasinormal in R/L, then $|U| = p$, by hypothesis, and this implies that $L = \{1\}$. Thus, $|R| = p^3$ and $R \in \{S(p^3), D_8\}$ so that $\Omega_1(R) = R$. Next, we suppose that $R < G$. Assume that $V \neq Z(R)$ is a subgroup of order p in $C_G(R)$. Then $F = RV = R \times V$ contains a nonquasinormal subgroup (in F so in G) of order p^2, a contradiction. Thus, $C_G(R)$ contains only one subgroup of order p, namely $Z(R)$. If $C_G(R) < R$, then G is of maximal class (Proposition 10.17). In that case, G has a nonquasinormal subgroup of order p^2 (consider $G/Z(G)$), a contradiction. Since $R < G$, we get $|C_G(R)| > p$ so that $C_G(R)$ is quasinormal. By Proposition 1.3, $C_G(R)$ is either cyclic or isomorphic to Q_8, completing the proof of (a) and (b).

Assume that $T < G$ is quasinormal of order p such that $T \nleq R$. Write $H = RT$ (by hypothesis, R is quasinormal in G). If $U, V < R$ are not permutable, then $\langle U, V\rangle = R$, $UT = U \times T$ and $VT = V \times T$ so that $RT = R \times T$, contrary to (b). Thus, T does not exist. As all noncental subgroups of R of order p are not quasinormal in R, it follows that $Z(R)$ is the unique quasinormal subgroup of order p in G, completing the proof of (c).

Now assume that $\exp(G) = p$ and $|G| > p^3$. One has $p > 2$. If $R < G$ is isomorphic to $S(p^3)$, then, by (b), $C_G(R) < R$ so Proposition 10.17 implies that G is of maximal class.

Let $L \lhd G$ be of order p. By (c), $L = Z(R)$. Then G/L is modular so elementary abelian, and it follows from Lemma 4.3 that $|G| = p^3$, proving (d) and thereby the theorem. □

Problem 1. Study the nonmodular p-groups of exponent $> p$ all of whose cyclic subgroups of order $> p$ are quasinormal.

Problem 2. Study the irregular nonmodular p-groups all of whose maximal regular subgroups are quasinormal.

Problem 3. Study the p-groups all of whose subgroups of order $\geq p^k$, $k \in \{2, 3\}$ are quasinormal.

Problem 4. Study the p-groups all of whose minimal nonabelian subgroups (and so all nonabelian subgroups) are quasinormal.

Problem 5. Study the p-groups all of whose non-quasinormal subgroups are cyclic. (See § 16.)

Problem 6. Study the nonmodular p-groups all of whose nonquasinormal cyclic subgroups are conjugate.

Problem 7. Study the nonmodular p-groups not generated by nonquasinormal subgroups.

Appendix 83 Nonabelian regular p-groups of exponent p^e are generated by minimal nonabelian subgroups of exponent p^e

By Theorem 10.28, a nonabelian p-group G is generated by minimal nonabelian subgroups. Let G be a nonabelian group of exponent p^e. It is possible that G has no minimal nonabelian subgroup of exponent p^e. However, if, in addition, G is regular, it contains such subgroup. Indeed, $\Omega_{e-1}(G) < G$. By Theorem 10.28, there is in G a minimal nonabelian subgroup M not contained in $\Omega_{e-1}(G)$; then $\exp(M) = p^e$. Moreover, as Proposition A.83.1 shows, a nonabelian regular group of exponent p^e is generated by minimal nonabelian subgroups of exponent p^e.

Proposition A.83.1. *A nonabelian regular group of exponent p^e is generated by minimal nonabelian subgroups of exponent p^e.*

Proof. As regular 2-groups are abelian, one has $p > 2$. One may assume that G is not minimal nonabelian (otherwise, it is nothing to prove further). In view of Theorem 10.28, one may assume that $e > 1$. Assume that the theorem has proved for all proper nonabelian subgroups of G of exponent p^e. All members of the set Γ_1 have exponent $\geq p^{p-1}$. By Exercise 1.6, there are in G at least $p(> 2)$ nonabelian maximal subgroups. Let $M_1, \dots, M_p \in \Gamma_1$ are pairwise distinct nonabelian. As $\Omega_{e-1}(G) < G$, there is among M_i's at most one that has exponent p^{e-1}. Therefore, one may assume, without loss of generality, that $M_1, \dots, M_{p-1}$ have exponent p^e. By induction, M_i is generated by minimal nonabelian subgroups of exponent p^e for $i = 1, \dots, p-1$ (note, that $p - 1 \geq 2$). Then $G = M_1 M_2$ is generated by minimal nonabelian subgroups of exponent p^e, completing the proof. □

Exercise 1. Prove that the (irregular) group $G = \Sigma_{p^2}$, $p > 2$, is generated by minimal nonabelian subgroups of exponent p^2.

Hint. One has $\exp(G) = p^2$ and G is minimal irregular. The set Γ_1 has exactly $p-1 \geq 2$ nonabelian (regular) members of exponent p^2. Use Theorem A.83.1.

Exercise 2. Given $e > 2$, describe the subgroup of $G = \Sigma_{p^e}$ that is generated by all minimal nonabelian subgroups of G of exponent p^e. Is it true that this subgroup coincides with G?

Exercise 3. Let G be a p-group of maximal class and exponent $p^e \geq p^3$. Prove that G is not generated by minimal nonabelian subgroups of exponent p^e.

Hint. If G_1 is abelian, G has no minimal nonabelian subgroups of exponent p^e (Theorems A.83.1 and 13.19.) If G_1 is nonabelian, it is generated by minimal nonabelian subgroups of exponent p^e (the same theorems). Since all minimal nonabelian subgroups of G not contained in G_1 have exponent $\leq p^2 < p^3$ (Theorem 13.19), we are done.

Exercise 4. Let $G = \Omega_e^*(G)$ be an irregular group of exponent p^e and suppose that there is in G a maximal abelian normal subgroup A of exponent $< p^e$. Is it true that, provided $p > 2$, G is generated by minimal nonabelian subgroups of exponent p^e?

Exercise 5. Is it true that the group $G = \Sigma_{p^e}$, $e > 2$, is generated by minimal nonabelian subgroups $\cong S(p^3)$ for $p > 2$ and D_8 for $p = 2$?

Appendix 84 Noncyclic 2-groups in which all cyclic subgroups of any equal order > 2 are conjugate

Here we prove the following

Theorem A.84.1. *Let G be a non-Dedekindian 2-group of exponent $2^e > 2$. If all cyclic subgroups of G of order 2^k for $k \in \{2, \dots, e\}$ are conjugate, then $G \cong D_{2^{e+2}}$.*

Proof. Let $C < G$ be cyclic of order 2^e. Set $H = C^G$; then $H < G$. By hypothesis, $H = H_2(G)$ since C contains a member of each conjugate class of cyclic subgroups of order 2^k, $k \in \{2, \dots, e\}$. In that case, by Burnside, H is abelian, $|G : H| = 2$ and all elements of the set $G - H$ are involutions. Let A be a minimal nonabelian subgroup of G. Then all elements of the set $A - (A \cap H)$ are involutions so that $A = \Omega_1(A)$. Since $|\Omega_1(A)| \leq 2^3$ (Lemma 65.1), it follows that $A \cong D_8$. In that case, the subgroup $A \cap H \cong C_4$ is G-invariant. It follows that our group G contains only one cyclic subgroup of order 4, and so, by Theorem 1.17 (b), G is of maximal class. It follows from Theorem 1.2 that G is a dihedral group. □

Problem 1. Classify the p-groups in which all cyclic subgroups of order $> p$ are normal (quasinormal).

Problem 2. Classify the p-groups in which any two nonnormal (non-quasinormal) cyclic subgroups of equal order are conjugate.

Appendix 85 On p-groups with given epimorphic images

In Propositions A.85.1–A.85.4 some ideas of Theorem 12.9 are developed.

Proposition A.85.1. *Suppose that a noncyclic p-group G of order $p^n > p^3$ has only one normal subgroup, say R, of order p, and the quotient group G/R is metacyclic. Then G is metacyclic. If, in addition, G/R is abelian, then $G \cong \mathrm{M}_{p^n}$.*

Proof. The group G is nonabelian, $\mathrm{Z}(G)$ is cyclic, G/R is noncyclic, $\mathrm{d}(G) = 2$ since $R \le \Phi(G)$ (indeed, $\Phi(G) > \{1\}$ possesses a G-invariant subgroup of order p which coincides with R, by hypothesis).

(a) Suppose that the quotient group G/R is abelian. Then $G' = R$ and so, by Lemma 65.2 (a), G is minimal nonabelian. In that case, $\Phi(G) = \mathrm{Z}(G)$ is cyclic, and we conclude, using Lemma 65.1 that $G \cong \mathrm{M}_{p^n}$, $n > 3$.

(b) Now suppose that the quotient group G/R is nonabelian; then $R < G'$ (indeed, minimal normal subgroup of G, contained in G', must coincide with R since G is a monolith). Let L/R be a subgroup of index p in $(G/R)' = G'/R$. Since G/L, being minimal nonabelian (Lemma 65.2 (a)), is metacyclic as an epimorphic image of the metacyclic group G/R, then G is also metacyclic (Theorem 36.1). □

Proposition A.85.2. *Suppose that a p-group G of order $p^n \ge p^{p+1}$ has only one normal subgroup, say R, of order p, and the quotient group G/R is absolutely regular. Then G is also absolutely regular.*

Proof. By Remark 7.2, the group G is regular. Assume, by way of contradiction, that G is not absolutely regular. In that case, $|\Omega_1(G)| = p^p$ (Theorem 12.1 (a) since G, in addition, is not of maximal class, by Theorems 9.5, 9.6, and 9.8 (a)). By Theorem 7.2 (c), $|G/\mho_1(G)| = |\Omega_1(G)| = p^p$. Since $|G| \ge p^{p+1}$, one has $\mho_1(G) > \{1\}$, and therefore $R \le \mho_1(G)$. In that case, G/R is not absolutely regular since its proper epimorphic image $G/\mho_1(G)$ is not absolutely regular, a final contradiction. □

Theorem A.85.3. *Suppose that a p-group G is such that the quotient group $G/\mathrm{Z}(G)$ is absolutely regular of order $\ge p^p$. If $\mathrm{Z}(G)$ is a unique normal subgroup of its order in G, then G is also absolutely regular.*

Proof. Let $L \triangleleft G$ be of order $\frac{1}{p}|\mathrm{Z}(G)|$ and let $K/L \triangleleft G/L$ be of order p. As $|K| = |\mathrm{Z}(G)|$, then $K = \mathrm{Z}(G)$, by hypothesis. We see that all G-invariant subgroups of order $\frac{1}{p}|\mathrm{Z}(G)|$ are contained in $\mathrm{Z}(G)$. Therefore, by Lemma A.85.2, the quotient group G/L is absolutely regular since $\mathrm{Z}(G)/L$ is the unique normal subgroup of order p in G/L, by hypothesis. The number of choices of L is $\equiv 1 \pmod{p}$, by Sylow's theorem. By the above, if L_1 is one of such subgroups, then G/L_1 is absolutely regular. Therefore, by Supplement to Theorem 47.1, G is absolutely regular, as was to be shown. □

Exercise 1. Suppose that a p-group G contains a normal subgroup T such that G/T is absolutely regular of order $\geq p^p$. If T is the unique G-invariant subgroup of its order in G, then G is also absolutely regular. (**Hint.** Mimic the proof of Theorem A.85.3. This is a generalization of that theorem.)

Proposition A.85.4. *Suppose that a 2-group G contains a normal subgroup $T > \{1\}$ such that G/T is of maximal class and order $\geq 2^3$. If T is the unique G-invariant subgroup of its order in G, then G is also of maximal class.*

Proof. If $T \leq G'$, then $|G : G'| = |(G/T) : (G/T)'| = 4$ so G is of maximal class, by Taussky (see Proposition 1.6).

Now assume that $T \nleq G'$. Then $|T| > |G'|$ so $|G/T| < |G/G'|$. However, the noncyclic abelian group G/G' contains two distinct (normal) subgroups T_1/G' and T_2/G' of index $|G : T|$. Thus, the different G-invariant subgroups T_1 and T_2 have order $|T|$, contrary to the hypothesis. □

Exercise 2. Try to prove analog of Proposition A.85.4 for p-groups, p is odd. **Hint.** We consider the case $|T| = p$ only (here $|G/T| \geq p^{p+1}$). In that case, G is a monolith and so $Z(G)$ is cyclic. Assume that G is not of maximal class. Then $Z(G) \cong C_{p^2}$. By Lemma 1.4, there is in G a normal subgroup $R \cong E_{p^2}$; then $T < R$. As $RZ(G)/T \cong E_{p^2}$ is a central subgroup of G/T, the last group is not of maximal class, contrary to the hypothesis. If $|T| > p$ and $L < T$ is a G-invariant of index p, then G/L is of maximal class by what has just been proved. The number of such L is $\equiv 1 \pmod p$. Note that any G-invariant subgroup of order $\frac{1}{p}|T|$ is contained in T. Thus, it suffices to prove that if $n \geq p + 2$ and the number of G-invariant subgroups L in a p-group G of order $> p^n$ such that G/L is of maximal class and order p^n is $\equiv 1 \pmod p$, then G is of maximal class. (See Appendix 87, for $p = 2$.)

Appendix 86 p-groups with nonabelian derived subgroup of order p^4

If G is a p-group with $|G'| = p^3$, then G' is abelian (Burnside; see also Lemma 1.4). Here we prove the following

Theorem A.86.1. *If G is a p-group such that G' is nonabelian of order p^4, then one of the following holds:*
(a) *$G' = D \times C$, where D is nonabelian of order p^3, $|C| = p$.*
(b) *$p > 2$ and G' is metacyclic of exponent p^2, $G' \cap Z(G) = G''$.*

Proof. By Lemma 1.4, $|Z(G')| = p^2$. Clearly, $d(G') \le 3$.

(i) Assume that $d(G') = 2$. Then G' is metacyclic (Theorem 44.12) and $\exp(G') = p^2$ since, if $\exp(G') = p^3$, then $\mho_1(G') \le Z(G')$ (any G-invariant subgroup of G' of order p^2 is contained in $Z(G')$) and $\Omega_1(G') \le Z(G')$ (Lemma 1.4) and then G' is abelian, a contradiction. In the case under consideration, G' is the unique nonabelian metacyclic group of order p^4 and exponent p^2.

However, if, in addition, $p = 2$, then $G' \cong \mathcal{H}_{2,2}$. In that case, all subgroups of G' of order 2 are characteristic so normal in G. If $L < G'$ is of order 2 such that G'/L is nonabelian, we get a contradiction (Lemma 1.4).

Thus, $p > 2$. Assume that $K = G' \cap Z(G) > G''$. Then $K \cong E_{p^2}$ since $Z(G') \cong E_{p^2}$. Let $L < K$ be of order p such that $L \ne G''$. Then $G'/L = (G/L)'$ is nonabelian of order p^3, contrary to Lemma 1.4. Thus, $G' \cap Z(G) = G''$ so that G' is as in (b).

(ii) Now let $d(G') = 3$. Then there is in G' a minimal nonabelian subgroup A (of order p^3). By Lemma 1.4, $|Z(G')| = p^2$. Therefore, $G' = AZ(G)$. If $Z(G) \cong E_{p^2}$, then $G' = A \times C$, where $|C| = p$.

Assume that $Z(G') \cong C_{p^2}$. If $p > 2$, then $G' = AZ(G') = \Omega_1(G')Z(G')$ is nonabelian (Theorem 7.2 (c)), and so $\Omega_1(G') \cong S(p^3)$ is nonabelian G-invariant of order p^3 (Theorem 7.2 (d)), contrary to Lemma 1.4. If $p = 2$ and $G' \cong Q_8 * C_4 (\cong D_8 * C_4)$ (central product), then Q_8 is characteristic in G' (Appendix 16); in that case, $Q_8 \triangleleft G$, and this is impossible. □

We suggest to readers to prove the following

Theorem A.86.2. *Let G be a p-group and $N \le \Phi(G)$ be a nonabelian G-invariant subgroup of order p^4. Then one of the following holds:*
(a) *$G' = D \times C$, where D is nonabelian of order p^3, $|C| = p$.*
(b) *$p > 2$ and G' is metacyclic of exponent p^2, $G' \cap Z(G) = N'$.*

Theorem A.86.1 follows from Theorem A.86.2.

Problem 1. Let G be a p-group with nonabelian metacyclic derived subgroup G'. Study the structure of G'.

Problem 2. Study the p-groups with special (i) derived subgroup, (ii) Frattini subgroup (see § 60; see also Problem 2628).

Appendix 87 On the number of epimorphic images of maximal class and a given order of a 2-group

By Theorem 47.1, if G is a nonmetacyclic 2-group of order $2^m > 2^n > 2^3$, then the number of $D \lhd G$ such that the quotient group G/D is metacyclic of order 2^n is even. In this appendix, we prove a similar result on the number of $D \lhd G$ such that G/D is of maximal class and order 2^n, $n \geq 3$.

We introduce some notation (see the paragraph containing formulas (1) and (2) in § 47).

Let $\mathrm{Sc}(G) = \Omega_1(Z(G))$ be the socle of a p-group G (= the subgroup generated by all minimal normal subgroups of G). Let a positive integer i be such that $p^i \leq p^s = |\mathrm{Sc}(G)|$ and let Δ_i denote the set of all subgroups of order p^i in $\mathrm{Sc}(G)$; then $|\Delta_i| \equiv 1 \pmod p$, by Sylow's theorem. Let $\mathcal{M}$ be a set of nonidentity normal subgroups of G. Given $H \in \Delta_1 \cup \cdots \cup \Delta_s$, let $\mu_n(H)$ be the number of $D \in \mathcal{M}$ such that $|G/D| = p^n$ and $H \leq D$. We claim that the following identity holds:

$$|\mathcal{M}| = \sum_{i=1}^{s} (-1)^{i-1} p^{i(i-1)/2} \sum_{H \in \Delta_i} \mu_i(H). \tag{1}$$

Indeed, let $D \in \mathcal{M}$ and $|D \cap \mathrm{Sc}(G)| = p^k$; then $k \geq 1$ since $D > \{1\}$. For natural numbers $u \geq v$, let $\phi_{u,v}$ denote the number of subgroups of order p^v in E_{p^u} (one has $\phi_{u,u} = 1$ and $\phi_{u,v} \equiv 1 \pmod p$, by Sylow's theorem). Then the contribution of D in the right-hand side of (1) is equal to

$$\sum_{j=1}^{k} (-1)^{j-1} p^{j(j-1)/2} \phi_{j,k}\,,$$

and this number is equal to 1, by formula (2) in § 5. Since the contribution of D in left-hand side of (1) is also equal to 1, formula (1) is true. In particular,

$$|\mathcal{M}| \equiv \sum_{H \in \Delta_1} \mu_n(H) \pmod p. \tag{2}$$

Exercise 1. Let G be a noncyclic p-group of order p^m and let $n > 1$ and $n < m$. Prove that the number of $D \lhd G$ such that G/D is cyclic of order p^n is a multiple of p.

Hint. First consider the trivial case $n = m - 1$ and use the basic theorem on abelian p-groups. For $n < m - 1$, use (2).

Lemma A.87.1. *Let $D \lhd G$ be the unique minimal normal subgroup of a p-group G (in that case, $|\mathrm{Sc}(G)| = p$). If G/D is of maximal class and order $> p^{p+1}$, then G is also of maximal class.*

Proof. If $D = Z(G)$, then G is of maximal class (to see this, let us consider the upper central series of G). Now assume that $D < Z(G)$. Then $Z(G) \cong C_{p^2}$ since G is a monolith. If G has no normal abelian subgroup of type (p, p), then G is a 2-group of maximal class (Lemma 1.4). Assume that there is in G a normal subgroup $R \cong E_{p^2}$. Then $R \cap Z(G) = D$ since D is the unique minimal normal subgroup of G. In that case, $RZ(G)/D = (R/D) \times (Z(G)/D)) \cong E_{p^2}$ is a central subgroup of the group G/D, hence G/D is not of maximal class, contrary to the hypothesis. □

It is possible to give an alternate proof of Lemma A.87.1 by means of the argument from the proof of Proposition A.87.4, below.

Lemma A.87.2. *Suppose that a nonabelian 2-group G of order 2^{n+1} is not of maximal class and $n > 2$. Then the number $\mu_n(G)$ of subgroups $D \le Z(G)$ of order 2 such that G/D is of maximal class and order 2^n, is even.*

Proof. Assume that $\mu_n(G) > 0$ (otherwise, it is nothing to prove further). In view of Lemma A.87.1, one has $|\mathrm{Sc}(G)| > 2$. Therefore, G has no cyclic subgroup of index 2. Let $D \lhd G$ be such that G/D is of maximal class and order $2^n (= \frac{1}{2}|G|)$. As G is not of maximal class, $|Z(G)| = 4$ (moreover, $Z(G) \cong E_4$ since G is not a monolith).

(i) Let $n = 3$; then $|G| = 2^4$. If $D < G'$, then $|G : G'| = 4$ so that G is of maximal class, by Proposition 1.6 (Taussky), contrary to the hypothesis. Thus, $D \nleq G'$; then $|G'| = 2$. If $D \nleq \Phi(G)$, then D is a direct factor of G; then $\mu_n(G) = 4$ and we are done in this case. Next, we assume that $D < \Phi(G)$; then $d(G) = d(G/D) = 2$. Then $Z(G) = \Phi(G)$, and we conclude that G is minimal nonabelian. In that case, $Z(G)$ is noncyclic (otherwise, $Z(G) = \Phi(G) = \mho_1(G)$ so G has a cyclic subgroup of index 2 and G/D is not of maximal class). In the case under consideration, $Z(G)$ contains two subgroups of order 2, say D_1 and D_2, that are $\neq G'$ (one of these subgroups coincides with D). Then G/D_1 and G/D_2 are nonabelian so of maximal class so that $\mu_2(G) = 2$ is even.

(ii) Next, we assume that $n > 3$. If D is a direct factor of G, then $\mu_n(G) = 4$ is even. Therefore, one may assume that $D < \Phi(G)$; then $d(G) = d(G/D) = 2$. Let $T/D < G/D$ be cyclic of index 2; then the noncyclic subgroup T (see Theorem 1.2) is abelian of type $(2^{n-1}, 2)$, $T = D \times C$, where $C \cong C_{2^{n-1}}$. In that case, $\Phi(T) = \Phi(C)$. Let $D_1 = \Omega_1(\Phi(T))$; then $Z(G) = D \times D_1 \cong E_4$. Let D, D_1, D_2 be all subgroups of order 2 in $Z(G)$. The quotient group G/D_1, whose center $\cong E_4$ (in view of $n > 3$), is not of maximal class. Therefore, to complete the proof, it remains to show that G/D_2 is of maximal class. By Lemma 1.1, $|G : G'| = 2|Z(G)| = 8$. By Proposition 1.6 (Taussky), $D \nleq G'$, and this implies that G' is cyclic of order 2^{n-2} (indeed, $G' < T$). Being two-generator of order 8, G/G' is abelian of type (4, 2). As $D_2 \nleq G'$ (otherwise, $Z(G) = D_1 \times D_2 < G'$ so that $|G : G'| = |G/Z(G) : G'/Z(G)| = 4$ and G is of maximal class, contrary to the hypothesis), it follows that $|(G/D_2) : (G/D_2)'| = 4$, hence G/D_2 is of maximal class (Proposition 1.6), $\mu_n(G) = 2$, completing the proof.

□

Theorem A.87.3. *Let a nonabelian 2-group G of order $2^m > 2^n > 2^2$ be not of maximal class. Then the number of $D \lhd G$ such that G/D is of maximal class and order 2^n is even.*

Proof. Let $\mathcal{M}$ be the set of $D \lhd G$ such that G/D is of maximal class and order 2^n. One may assume that $\mathcal{M} \neq \emptyset$. Next, let $\mu_n(H)$ be the number of those members of the set $\mathcal{M}$ that contain $H \in \Delta_1$ (see paragraph preceding Exercise 1). In view of Lemmas A.87.1 and A.87.2, one may assume that $|\mathrm{Sc}(G)| > 2$ and $n < m - 1$. We proceed by induction on m.

If $H \in \Delta_1$ is such that G/H is of maximal class, then $\mu_n(H) = 1$ (Exercise 9.1 (b)). Let Δ_1' be the set of $H \in \Delta_1$ such that G/H is of maximal class. Then, by Lemma A.87.2, $|\Delta_1'| \equiv 0 \pmod 2$. Therefore, the contribution of the members of the set Δ_1' in the left-hand side of (2) is equal to $|\Delta_1'| \equiv 0 \pmod 2$.

Let $H \in \Delta_1$ be such that G/H is not of maximal class or, what is the same, $H \in \Delta_1 - \Delta_1'$. Then, by induction, $\mu_n(H) \equiv 0 \pmod 2$. It follows from (2) that $2 \mid \mu_n(G)$, as was to be shown. □

Theorem A.87.3 may be considered as dual to Theorem 5.4.

It is interesting to prove an analog of Theorem A.87.3 for $p > 2$. One step in this direction was done in Lemma A.87.1. Let G be a p-group, $p > 2$, which is neither absolutely regular nor of maximal class and $|G| = p^m > p^{p+2}$, $p + 1 < n < m$. If it is known that the number of $D \lhd G$ such that G/D is of maximal class and order p^{m-1} is a multiple of p, then it is easy to prove, using (2), that p divides the number of $H \lhd G$ such that G/H is of maximal class and order p^n (see the proof of Theorem A.87.3).

Proposition A.87.4. *Let G be a p-group of order $p^m > p^{p+2}$. If all epimorphic images of G of order p^{m-1} are of maximal class (in that case, all nonabelian epimorphic images of G are of maximal class, by Exercise 9.1 (b)), then G is also of maximal class.*

Proof. Let $D \lhd G$ be of order p. Then G/D is of maximal class and order $\geq p^{p+2}$, by hypothesis. It follows that $\mathrm{cl}(G) \geq \mathrm{cl}(G/D) \geq p + 1$ so that $\mathrm{K}_{p+1}(G) > \{1\}$. Therefore, by hypothesis, the quotient group $G/\mathrm{K}_{p+1}(G)$, being a nonabelian epimorphic image of G, is of maximal class so its order equals p^{p+1}. Obviously, $\mathrm{K}_{p+1}(G)$ is the unique normal subgroup of G of index $p^{p+1} < |G|$. Then, by Theorem 12.9, G is of maximal class. □

The following exercise asks to check validity of Supplement to Theorem 47.1.

Exercise 2. Let G be a p-group of order p^m, $m > n \geq p$. Is it true that if G is neither absolutely regular nor of maximal class, then the number of $D \lhd G$ such that G/D is absolutely regular of order p^n is a multiple of p?

Exercise 3. Let G be a p-group of order p^{n+1} without cyclic subgroup of index p, $n > 2$ and $n > 3$ if $p = 2$. Find the number $\mu_n(G)$ of those $D \lhd G$ that $G/D \cong \mathrm{M}_{p^n}$.

Hint. Let D be such that $G/D \cong \mathrm{M}_{p^n}$. If $D \nleq \Phi(G)$, then $\mu_n(G) = p^2$. Now let $D < \Phi(G)$. Then G is minimal nonabelian.

Exercise 4. Let G be a 2-group without cyclic subgroup of index 2. Find the number $\mu_n(G)$ of $D \lhd G$ such that $G/D \in \{D_{2^n}, Q_{2^n}, SD_{2^n}\}$ provided $|G| = 2^{n+1}$ (three problems). Consider in detail the case $|G| = 2^{n+1}$.

Exercise 5. Let G be a 2-group without cyclic subgroup of index 2. Find the number $\mu_n(G)$ of $D \lhd G$ such that G/D has a cyclic subgroup of index 2. (See Theorem A.87.3 and Exercise 3.)

Problem 1. Let G be a p-group which is neither absolutely regular nor of maximal class. Find the number of $D \lhd G$ such that G/D of given order $> p^p$ is either absolutely regular or of maximal class.

Problem 2. Let G be a p-group. Find the number of $D \lhd G$ such that G/D of given order $> p^{p+1}$ is an L_p-group (see §§ 17–19).

Problem 3. Let G be a 2-group. Find the number of $D \lhd G$ such that G/D is an U_2-group (see § 67.).

Problem 4. Let G be a p-group of order $p^{n+1} \geq p^5$. Find the number of $D \lhd G$ such that G/D is an $\mathcal{A}_i$-group, $i = 1, 2$, of order p^n.

Problem 5. Let G be a p-group of order p^m without abelian subgroup of index p. Find the number $\mu_n(G)$ of $D \lhd G$ such that G/D has order p^{m-1} and abelian subgroup of index p.

Appendix 88 Minimal nonabelian p-groups $\not\cong Q_8$ in which the intersection of all their nonnormal subgroups is $> \{1\}$

Blackburn [Bla7] has classified the non-Dedekindian p-groups in which the intersection of all their nonnormal subgroups is $> \{1\}$. Here we prove [Bla7, Lemma 2], a partial case of that result.

Let G be a minimal nonabelian p-group. Then $G = \mathrm{M}_p(m, n, 1)$ if G is nonmetacyclic (in that case, G/G' is abelian of type (p^m, p^n)). Next, $G = \mathrm{M}_p(m, n)$ if $G = B \cdot A$ is the semidirect product with kernel $A \cong \mathrm{C}_{p^m}$ and $B \cong \mathrm{C}_{p^n}$ (in that case, $m > 1$).

Exercise 1. Let $G = \langle a, b\rangle$ be a minimal nonabelian p-group $\not\cong Q_8$, $A = \langle a\rangle$, $B = \langle b\rangle$. If the intersection of any two distinct conjugate cyclic subgroups of G is $> \{1\}$, then one and only one of the following holds:

(a) If $G = M_p(m, n, 1)$, then $m > 1$, $n > 1$.

(b) If $G = M_p(m, n)$, then $n > 1$.

Solution. If $C < G$ is nonnormal cyclic of order $> p$, then $C_G = \mho_1(C) > \{1\}$ (indeed, $\mho_1(C) \leq \Phi(G) = \mathrm{Z}(G)$). If G is nonmetacyclic, then $m > 1$ and $n > 1$. In that case, $\Omega_1(G) \leq \mathrm{Z}(G)$ so that any nonnormal cyclic subgroup has non-trivial core. Thus, minimal nonabelian group $G = M_p(m, n, 1)$ satisfies the hypothesis $\iff$ $m > 1$ and $n > 1$. A metacyclic minimal nonabelian group $M_p(m, n)$ satisfies the hypothesis $\iff$ $n > 1$: then $\Omega_1(G) \leq \mathrm{Z}(G)$.

If G is a minimal nonabelian p-group of order $> p^3$ and exponent $p^e (> p)$, then $\exp(\Omega_k(G)) = p^k$ for all $k \leq e$ (see Lemma 65.1).

Proposition A.88.1. *Let $G = \langle a, b\rangle$ be a minimal nonabelian p-group $\not\cong Q_8$, Suppose that any two distinct nonnormal cyclic subgroups of G have intersection $> \{1\}$. Then $G \cong \mathcal{H}_{2,2}$, the unique nonabelian metacyclic group of order 16 and exponent 4.*

Proof. If the group G is nonmetacyclic, then, by Lemma 65.1,

$$G = \langle a, b \mid o(a) = p^m,\ o(b) = p^n,\ [a, b] = c,\ o(c) = p,\ [a, c] = [b, c] = 1\rangle .$$

In that case, the subgroups $A = \langle a\rangle$, $B = \langle b\rangle$ are nonnormal in G and $A \cap B = \{1\}$, i.e., this G does not satisfy the hypothesis.

Thus, the group G is metacyclic; then, by Lemma 65.1,

$$G = \langle a, b \mid o(a) = p^m,\ o(b) = p^n,\ a^b = a^{1+p^{m-1}}\rangle .$$

In that case, $A = \langle a\rangle \lhd G$ is of order p^m, $B = \langle b\rangle$ is nonnormal of order p^n, $A \cap B = \{1\}$ and $|G| = p^{m+n}$. As B is not normal in G, it follows that $B \cap B^x > \{1\}$ for any $x \in G - \mathrm{N}_G(B)$ so that $n > 1$.

(i) Let $m = n = 2$. All G-invariant cyclic subgroups of order p^2 contain $G' = \Omega_1(A)$. Since G/G' is abelian of type (p^2, p) and contains exactly $p + 1$ subgroups of order p, it follows that G has exactly p normal cyclic subgroups of order p^2. As $c_2(G) = \frac{|G|-|\Omega_1(G)|}{p(p-1)} = p^2 + p$, it follows that G contains exactly $(p^2 + p) - p = p^2$ nonnormal cyclic subgroups. Let $p > 2$. The cyclic subgroup $L = \langle ab \rangle$ is not b-invariant (indeed, $(ab)^b = a^{1+p}b \notin \langle ab \rangle$) and $L \cap B = \{1\}$ so that G does not satisfy the hypothesis. Thus, $p = 2$. As $G/\Omega_1(B) \cong D_8$ has exactly 2^2 nonnormal subgroups of order 2 (their inverse images are cyclic of order 4, all nonnormal cyclic subgroups of G contain $\Omega_1(B)$, i.e., $G \cong \mathcal{H}_{2,2}$ satisfies the hypothesis.

(ii) Let $m = n > 2$. In that case, all nonnormal cyclic subgroups of G have order p^m since $\Omega_{m-1}(G) = Z(G)$. Then the subgroup $C = \langle ab \rangle$ is not b-invariant so nonnormal in G, $|C| = p^m$ and $C \cap B = \{1\}$, i.e, G does not satisfy the hypothesis.

(iii) Let $m > n$. Then the subgroup $D = \langle a^{p^{m-n}} b \rangle$ of order p^n is not a-invariant so nonnormal in G and $D \cap B = \{1\}$, i.e, G does not satisfy the hypothesis.

(iv) Let $m < n$. Then the subgroup $E = \langle ab^{p^{n-m}} \rangle$ of order p^m is not b-invariant so nonnormal in G and $E \cap B = \{1\}$, i.e, G does not satisfy the hypothesis.

Thus, $G = \mathcal{H}_{2,2}$, as was to be shown. □

Exercise 2. Classify all minimal nonabelian p-groups G such that any two non-G-invariant cyclic subgroups of equal order have intersection > {1}. (**Hint.** The proof of Theorem A.88.1 shows that if G is metacyclic, then it is isomorphic to $\mathcal{H}_{2.2}$.)

Corollary A.88.2. *Suppose that G is a non-Dedekindian p-group in which the intersection of any two non-G-invariant cyclic subgroups is $> \{1\}$. If $A \leq G$ is minimal nonabelian, then $A \in \{Q_8, \mathcal{H}_{2,2}\}$; in particular, $p = 2$.*

Proof. Let $A \leq G$ be minimal nonabelian; then $A \in \{Q_8, \mathcal{H}_{2,2}\}$ so that $p = 2$ (Theorem A.88.1). □

Problem 1. Classify the nonabelian 2-groups all of whose minimal nonabelian subgroups are isomorphic to Q_8 or $\mathcal{H}_{2,2}$.

Problem 2. Study a nonabelian p-group G such that the intersection of any two distinct conjugate maximal abelian subgroups is $> Z(G)$.

Problem 3. Study the p-groups G such that, whenever $A, B < G$ are distinct nonabelian subgroups, then $A \cap B > \{1\}$.

Problem 4. Study the 2-groups G such that, whenever $A, B < G$ are distinct minimal nonabelian subgroups, then $A \cap B > \{1\}$.

Appendix 89 Metacyclic 2-groups containing an abelian subgroup of order 2^{2n} and exponent 2^n

In this appendix we prove the following

Theorem A.89.1 (Sambale [Sam1], Lemma 4.13). *If a metacyclic* 2*-group G contains a proper abelian subgroup H of order 2^{2n} and exponent $2^n > 2$, then $H = \Omega_n(G)$.*

Proof. Assume that G is a counterexample of minimal order. If $H \leq M \in \Gamma_1$, then $H = \Omega_n(M)$, by induction, hence H is characteristic in $M \triangleleft G$, so that $H \triangleleft G$.

(i) Let $n = 2$. We proceed by induction on $|G|$. Since H is not characteristic in $N = \mathrm{N}_G(H)$, we get $H < \Omega_2(N) \Rightarrow N = G$, by induction. Assume that there is $x \in G - H$ of order 4. Set $X = \langle x \rangle$. As $H < \Omega_2(XH)$, we get $XH = G$, by induction. By Proposition 10.19, $\Omega_1(G) = \Omega_1(H)$. Now $x_2 \in \Omega_1(G) \rightarrow |G| = 2^5$ so that $H \in \Gamma_1$ and $\bar{G} = G/\Omega_1(G) = \bar{X} \cdot \bar{H} \cong \mathrm{D}_8$. By Proposition 10.19, $X\Omega_1(G)$ of order 8 is abelian so that $\mathrm{C}_G(\Omega_1(G)) \geq XH = G$; then $\Omega_1(G) \leq \mathrm{Z}(G)$. If $\bar{S} < \bar{G}$ is cyclic of order 4, then $S \in \Gamma_1$ is abelian hence $S \cap H = \mathrm{Z}(G)$ has index 4 is G. In that case G is minimal nonabelian so that $G' < \Omega_1(G)$ (Lemma 65.1) and therefore $G/\Omega_1(G) = \bar{G}$ is abelian, a contradiction: by the above, $\bar{G} \cong \mathrm{D}_8$.

(ii) Now let $n > 2$. Then $\Omega_1(H) = \Omega_1(G)$ and $H/\Omega_1(H)$ is abelian of order $2^{2(n-1)}$ and exponent 2^{n-1}. Then, by induction, $H/\Omega_1(G) = \Omega_{n-1}(G/\Omega_1(G))$ and this implies that $H = \Omega_n(G)$. The proof is complete. □

Exercise 1. Let H be an abelian subgroup of order p^{2n} and exponent p^n of a metacyclic p-group G, $p > 2$. Then $H = \Omega_n(G)$. (**Hint.** The group G, by Theorem 7.1 (c), is regular. Use Theorem 7.2 (b).)

Exercise 2. Let H be an absolutely regular subgroup of order p^{kn} and exponent p^n in an absolutely regular p-group G such that $|\Omega_1(G)| = p^k$, $k \leq p - 1$. Then $H = \Omega_n(G)$. State a similar assertion for a p-group G of maximal class.

Clearly, Exercise 1 follows from Exercise 2.

Problem 1. Study the metacyclic 2-groups G containing a nonabelian subgroup H of order 2^{2n} and exponent 2^n. Consider in detail the case when H is minimal nonabelian.

Appendix 90 Two alternate proofs of G. A. Miller's theorem on minimal non-Dedekindian groups and a corollary

Let Θ be a group-theoretical property inherited by subgroups. A group is said to be a Θ^1-group if it is not a Θ-group, but all its proper subgroups are Θ-groups. If not all (finite) groups are Θ-groups, then Θ^1-groups exist (indeed, if G is a non-Θ-group of minimal order, then it is a Θ^1-group). In some sense, for many properties Θ, in contrast to class of Θ-groups, the class of Θ^1-groups is fairly narrow (e.g., the number of solvable groups of order ≤ 1000 is essentially bigger than the number of minimal nonsolvable groups of order ≤ 1000). Examples of Θ^1-groups for appropriate Θ: minimal nonabelian, minimal nonnilpotent, minimal nonsolvable, minimal nonsupersolvable groups. In most cases, classification of Θ^1-groups allows us to obtain nontrivial characterizations of Θ-groups. Let us give one example. We claim that a finite group G is nilpotent if its any two elements of equal prime power order generate a nilpotent subgroup. Assume, by way of contradiction, that G is not nilpotent. Then it contains a Θ^1-subgroup (= minimal nonnilpotent subgroup), say S. We have $S = P \cdot Q$, where $Q = S' \in \mathrm{Syl}_q(S)$ and $P = \langle x \rangle \in \mathrm{Syl}_p(S)$ is cyclic (Appendix 22). If $P_1 = \langle y \rangle \neq P$ is another Sylow p-subgroup of S, then $S = \langle x, y \rangle$, a contradiction since S is nonnilpotent. There are many examples of this phenomenon in the papers and books devoted to finite groups.

A group is said to be Dedekindian if all its subgroups are normal. All Dedekindian groups are nilpotent. Let us show that if G is a nonabelian Dedekindian p-group (sometimes, such groups are termed *hamiltonian*), then $p = 2$. Let $A \leq G$ be minimal nonabelian. It follows from Lemma 65.1 that $A \cong Q_8$, the ordinary quaternion group, and we conclude that $p = 2$. This argument shows importance of Θ^1-groups (in the considered case Θ = commutativity). If G is a nonabelian Dedekindian 2-group, then $G = Q \times E$, where $Q \cong Q_8$ and $\exp(E) \leq 2$ (Theorem 1.20). Is is easy to show that all minimal nonabelian subgroups of a nonabelian Dedekindian 2-group are $\cong Q_8$.

Redei [Red1] in 1947 has classified the minimal nonabelian p-groups (see Lemma 65.1). His result has numerous applications in our book. The nonnilpotent minimal nonabelian groups are classified in [MilM]. Miller [Mil9] many years before Redei, in 1907, has proved the following stronger result for 2-groups.

Theorem A.90.1. *If all proper subgroups of a group G are Dedekindian, then one of the following holds:*
(a) *G is Dedekindian.*
(b) *G is minimal nonabelian.*
(c) *$G \cong Q_{16}$, the generalized quaternion group of order 16.*
(d) *$G = P \cdot Q$ is minimal nonnilpotent, where $Q = G' \cong Q_8$, $P \in Syl_3(G)$ is cyclic.*[1]

1 Miller [Mil9] asserted that here $G \cong \mathrm{SL}(2, 3)$, i.e., $|P| = 3$, which is not correct.

There are a few applications of Theorem A.90.1 in our book. The presented proof of Theorem A.90.1 is the third one in the book (see Appendix A.17 and § 245).

A non-Dedekindian group all of whose proper subgroups are Dedekindian is said to be *minimal non-Dedekindian*. It follows from Theorem A.90.1 that a minimal non-Dedekindian p-group is either minimal nonabelian or $\cong Q_{16}$ (by the above, if $p > 2$, then Dedekindian p-group is abelian).

We collected some known results used in this section, in the following

Lemma A.90.2. *Suppose that G is a nonabelian p-group.*

(a) *(Lemma 65.1) Let G be a minimal nonabelian group. Then* $d(G) = 2$, $|G'| = p$, $|\Omega_1(G)| \le p^3$, $d(K) \le 3$ *for any $K \le G$ and G has no two distinct maximal elementary abelian subgroups, unless* $G \in \{S(p^3), D_8\}$.

(b) *(Appendix 16) Let a 2-group $G = Q * C$ (central product) be of order* 16, *where $Q \cong Q_8$ and $C \cong C_4$; then Q is characteristic in G and G contains exactly three dihedral subgroups. If $G = Q_8 \times C_4$, then there is $L < G$ of order 2 (namely, $Z(Q) \ne L \ne \Omega_1(C_4)$) such that G/L is as in the previous sentence.*

(c) *(Lemma 1.4) If G is not a 2-group of maximal class, then it has a normal subgroup $\cong E_{p^2}$. In any p-group the Frattini subgroup $\Phi(G)$ has no nonabelian G-invariant subgroup of order p^3.*

(d) *(Theorem 1.20) If a Dedekindian p-group G is nonabelian, then $p = 2$ and $G = Q \times E$, where $Q \cong Q_8$ and* $\exp(E) \le 2$.

(e) *(Lemma 4.3) If $|G'| = p$ and $S < G$ is minimal nonabelian, then $G = S * C_G(S)$.*

(f) *(Theorem 10.28) The group G is generated by minimal nonabelian subgroups.*

The proofs of parts (a)–(f) of Lemma A.90.2 are elementary and easy.

We need the following

Remark 1. Let $Q_8 \cong Q$ has index 2 in a 2-group G and assume that all minimal nonabelian subgroups of G are $\cong Q_8$. If $C_G(Q) < Q$, then G is of maximal class (Proposition 10.17) so, as easily follows from Theorem 1.2 and Proposition 1.6, $G \cong Q_{16}$. Now let $C_G(Q) \not\le Q$; then $C_G(Q) = Z(G)$. If $Z(G) \cong E_4$, then $G = Q \times C$, where $|C| = 2$, so that G is Dedekindian. If $Z(G) \cong C_4$, then G has a subgroup $\cong D_8 \not\cong Q_8$ (Lemma A.90.2 (b)), a contradiction. Thus, either G is Dedekindian or $\cong Q_{16}$.

Proof of Theorem A.90.1. All groups (a)–(d) satisfy the hypothesis. Next we suppose that all proper subgroups of a group G are Dedekindian.

We proceed by induction on $|G|$. Assuming that G is neither Dedekindian nor minimal nonabelian, we have to prove that G is as in (c) or (d).

(A) Assume that G is nilpotent. If $|G|$ is not a power of a prime, then all Sylow subgroups of G are Dedekindian; then, by Lemma A.90.2 (d), G is also Dedekindian. In what follows, we assume that G is a p-group, p is prime. As G contains a minimal nonabelian subgroup, say A, it follows that $A \cong Q_8$ (see the first paragraph of this section) so that $p = 2$ (Lemma A.90.2 (d)). By Remark 1, one may assume that $|G| > 16$. In that case, we have to obtain a contradiction.

By assumption, there is in the set Γ_1 a nonabelian Dedekindian subgroup $H = Q{\times}E$, where $Q \cong Q_8$ and $\exp(E) = 2$ (since $|G| > 16 \Rightarrow |H| > 8$). Then $H' = Q' \cong C_2$ is G-invariant since H' is characteristic in $H \lhd G$. As all sections of Dedekindian groups are Dedekindian, all proper subgroups of the quotient group $\bar{G} = G/H'$ are Dedekindian (but $\bar{G}$ need not necessarily Dedekindian). Besides, $\bar{H} = H/H'$ is elementary abelian maximal subgroup of $\bar{G} = G/H' = G/Q'$. By induction, one of the following holds:

(i) $\bar{G}$ is abelian.

(ii) $\bar{G}$ is minimal nonabelian.

(iii) $\bar{G}$ is nonabelian Dedekindian.

(iv) $\bar{G} \cong Q_{16}$.

In what follows, we consider these four possibilities separately. As in the second paragraph of (A), $H = Q \times E \in \Gamma_1$. Recall that $|G| > 16$ so that $E > \{1\}$.

(i) Assume that the quotient group $\bar{G}(= G/H' = G/Q')$ is abelian.

In that case, $G' = H' = Q'$. Then, by Lemma A.90.2 (e), $G = Q * C_G(Q)$ (central product) since Q is minimal nonabelian. Assume that $C_G(Q)$ has a cyclic subgroup C of order 4. Then $Q * C$ is non-Dedekindian (Lemma A.90.2 (b)) whence $QC = G$. If $Q \cap C = Z(Q)$, then the group $G = Q * C$ of order 16 has a subgroup $\cong D_8$ (see Remark 1), a contradiction. Thus, $G = Q \times C$ is of order 2^5. In that case, if $L < G$ is of order 2 and $L \neq G'$, $L \nleq C$, then $G^o = G/L = Q^o * C^o$, where $Q^o \cong Q$ and $C^o \cong C$. By Lemma A.90.2 (b), there is in G^o a (non-Dedekindian) subgroup $D^o \cong D_8$; the inverse image of D^o in G is a non-Dedekindian proper subgroup of G, a contradiction. Thus, $C_G(Q)$ is elementary abelian, and, obviously, $C_G(Q) = Z(G)$. Then $Z(G) = Z(Q) \times E$, where $\exp(E) = 2$, and in this case $G = Q \times E$ is Dedekindian, contrary to the assumption.

(ii) Let the quotient group $\bar{G}$ be minimal nonabelian; then $|G'| = 4$.

In that case, $H' \le \Phi(H) \le \Phi(G)$ so that $d(G) = d(\bar{G}) = 2$ (Lemma A.90.2 (a)). If the set Γ_1 of maximal subgroups of G has two distinct abelian members, say A and B, then $\Phi(G) = A \cap B = Z(G)$ has index 4 in G, so that G is minimal nonabelian, contrary to the assumption. Thus, there is in the set Γ_1 at most one abelian member so the set Γ_1 has two distinct nonabelian members. Now, $d(G) = 2 \Rightarrow |\Gamma_1| = 3$. Let $\Gamma_1 = \{D_1 = H, D_2, A\}$, where D_1, D_2 are nonabelian. Let $D_i = Q_i \times E_i$, where $Q_i \cong Q_8$, $\exp(E_i) = 2$, $i = 1, 2$ (Lemma A.90.2 (d)). One has $D_i' = Q_i'$, $i = 1, 2$.

Assume that $Q_2 < D_1$. Then $N_G(Q_2) \ge D_1D_2 = G$ (indeed, in the case under consideration, $Q_2' = D_1' = D_2'$) so that $Q_2 \lhd G$ and, since $d(G) = 2$, we get $Q_2 \le D_1 \cap D_2 = \Phi(G)$, contrary to Lemma A.90.2 (c). Thus, $Q_2 \nleq D_1$. Then $G = D_1Q_2$ so that $D_1 \cap Q_2$ is cyclic of order 4, by the product formula. It follows that $Q_1 \cap Q_2 > \{1\}$ so $Q_1' = Q_2'$. In that case, the minimal nonabelian 2-group G/Q_1' has two distinct elementary abelian subgroups D_i/Q_1' ($i = 1, 2$) of index 2, contrary to Lemma A.90.2 (a). Thus, case (ii) is impossible.

(iii) The quotient group $\bar{G}$ is nonabelian Dedekindian.

The nonabelian Dedekindian group $\bar{G} = \bar{T} \times \bar{U}$, where $\bar{T} \cong Q_8$ and $\exp(\bar{U}) = 2$ (Lemma A.90.2 (d)). Since T is nonabelian Dedekindian, one has $T = Q_1 \times H' = Q_1 \times Q'$, where $Q_1 \cong Q_8$ (Lemma A.90.2 (d)); then $Q' \nleq Q_1$. As $H/H' = H/Q'$ is elementary abelian, we get

$$T/H' \cong Q_8 \Rightarrow T \nleq H \Rightarrow Q_1 \nleq H \Rightarrow HQ_1 = G$$

(the last implication holds since $H \in \Gamma_1$). Then $H \cap Q_1 \cong C_4$ so that $Q' = \mho_1(H) < Q_1$ contrary to the above proved relation. Thus, case (iii) is also impossible.

(iv) The quotient group $\bar{G} \cong Q_{16}$.

Since $\bar{H}$, a maximal subgroup of $\bar{G}$, is elementary abelian of order ≥ 8 and index 2 in $\bar{G}$, this is a contradiction since any maximal subgroup of $\bar{G} \cong Q_{16}$ has exponent > 2.

(B) Let G be nonnilpotent.

As Dedekindian groups are nilpotent, it follows that $G = P \cdot Q$ is minimal nonnilpotent, where $Q = G' \in \mathrm{Syl}_q(G)$ and $P \in \mathrm{Syl}_p(G)$ is cyclic (Appendix 22). If Q is nonabelian, then, being Dedekindian and special, it is $\cong Q_8$; in that case $p = 3$ since $\mathrm{Aut}(Q_8)$ is isomorphic to a subgroup of the symmetric group of degree 4. If Q is abelian, then G is minimal nonabelian, contrary to the hypothesis. □

The second proof of Theorem A.90.1 for 2-groups[2]. We use induction on $|G|$. Assuming that G is neither Dedekindian nor minimal nonabelian, we have to prove that $G \cong Q_{16}$. In view of Remark 1, one may assume that $|G| > 16$.

By assumption, there is in the set Γ_1 a nonabelian Dedekindian subgroup $H = Q \times E$, where $Q \cong Q_8$ and $\exp(E) = 2$. All minimal nonabelian subgroups of a nonabelian Dedekindian 2-group are $\cong Q_8$. It follows that any minimal nonabelian subgroup of G is $\cong Q_8$ since it is contained in a nonabelian Dedekindian maximal subgroup of G. By Lemma A.90.2 (f), there is in G a minimal nonabelian subgroup $M (\cong Q_8)$ such that $M \nleq H$. Then $L = H \cap M \cong C_4$. Therefore, $\mho_1(L) = \mho_1(H) = H' \cong C_2$. It follows that H' is contained in all minimal nonabelian subgroups of G since H' is contained in all minimal nonabelian subgroups of H. Next, $H' \lhd G$ since H' is characteristic in $H \lhd G$. Write $\bar{G} = G/H'$; then $\bar{G} = \bar{M} \cdot \bar{H}$ (semidirect product with elementary abelian kernel $\bar{H}$ and complement $\bar{M} \cong C_2$). To prove that $\bar{G}$ is abelian, it suffices to show that it has no minimal nonabelian subgroup. Assume, however, that $\bar{F} \leq \bar{G}$ is minimal nonabelian. Then $\bar{U} = \bar{H} \cap \bar{F}$ is $\bar{G}$-invariant elementary abelian subgroup of index 2 in $\bar{F}$ so that $|\bar{F}| \leq 16$ (Lemma A.90.2 (a)).

Assume that $\bar{F} = 16$. Then $\bar{F}$ is non-Dedekindian so F is also non-Dedekindian. It follows that $F = G$. By part A(ii) of the first proof of Theorem A.90.1, this is impossible.

2 For another proof, see § 245.

Now let $|\bar{F}| = 8$; then $\bar{F} \cong D_8$ since $\bar{F} \cap \bar{H} \cong E_4$. In that case, F in non-Dedekindian since it has a section $\bar{F} \cong D_8$. As $|G| > 16$, $F < G$, a contradiction. Thus, $\bar{G}$ has no minimal nonabelian subgroup so it is abelian. By the above, $\Omega_1(\bar{G}) = \bar{G}$, and whence $\bar{G}$ is elementary abelian. Thus, all minimal nonabelian subgroups of G are G-invariant and their centers coincide with $H' = Z(Q) = G' \cong C_2$. By Lemma A.90.2 (e), $G = Q * C_G(Q)$. As in the first proof, $C_G(Q)$ is of exponent 2 so that $C_G(Q) = Z(G)$ is elementary abelian. Then $Z(G) = Z(Q) \times E$, where $E < Z(G)$ is elementary abelian. Then $G = Q \times E$ is Dedekindian, a contradiction. □

Remark 2. Since all minimal nonabelian subgroups of the group G from Theorem A.90.1 are $\cong Q_8$, one may apply Theorem 90.1 to produce the third proof of our theorem. However, Theorem 90.1 is essentially deeper, and therefore we prefer to avoid its use.

Remark 3. Let G be a minimal nonabelian p-group of exponent p^e. We claim that $G = \langle x, y \rangle$, where x, y have equal order. Since any nonabelian group of order p^3 is generated by two elements of equal order, one may assume that $|G| > p^3$. Set $\bar{G} = G/\Omega_1(G)$; then $\exp(\bar{G}) = p^{e-1}$ and $\bar{G} = \langle \bar{x}, \bar{y} \rangle$, where $o(x) = o(y) = p^{e-1}$ since $\bar{G}$ is abelian. As $G = \langle x, y \rangle$ and $o(x) = o(y) = p^e$, we are done. We see that G is generated by two elements of order $\exp(G)$, unless $G \cong D_8$. (Another proof: the result also follows from $\Omega_{e-1}(G) < G$.)

Corollary A.90.3. *A group G is Dedekindian $\iff$ any its subgroup $\langle x, y \rangle$, where $x, y \in G$ are elements of equal prime power order, is Dedekindian.*

Proof. Assume that G is a counterexample of minimal order. Then, by Theorem A.90.1, G is minimal non-Dedekindian so it is either minimal nonabelian or $\cong Q_{16}$ or as in Theorem A.90.1 (d). If G is as in Theorem A.90.1 (d) and P, P_1 are distinct Sylow 3-subgroups of G, then $\langle x, y \rangle = G$, where x, y are generators of P, P_1, respectively. If G is a minimal nonabelian p-group, the result follows from Remark 3. The group $G \cong Q_{16}$ is generated by two elements of order 4. In all these cases, we get a contradiction. □

Problem 1. Classify the non-Dedekindian 2-groups all of whose nonnormal subgroups are Dedekindian. (See [FA], where a partial case is treated.)

Problem 2. A 2-group is called generalized Dedekindian if it is either abelian or $= Q \times A$, where $Q \cong Q_8$ and A is abelian. Classify the 2-groups all of whose nonabelian maximal subgroups are generalized Dedekindian.

Appendix 91 On a p-group whose proper Hughes subgroup has the Frattini subgroup of order p

If G is a group and $p \mid |G|$, then its Hughes subgroup (= H_p-subgroup of G)

$$\mathrm{H}_p(G) = \langle x \in H \mid o(x) \neq p \rangle.$$

Clearly, $G/\mathrm{H}_p(G)$ is of exponent dividing p. Assume that $\mathrm{H}_p(G) < G$. Let $x \in G - \mathrm{H}_p(G)$ and $y \in \mathrm{H}_p(G)^{\#}$. Then $o(x) = o(yx^{-1}) = p$ so that

$$(1) \qquad 1 = (yx^{-1})^p = yy^x \cdots y^{x^{p-1}}.$$

By Kegel's theorem (see Appendix 49) if $H \lhd G$ and for any $y \in H$ and $x \in G - H$ of order p equality (1) holds, then H is nilpotent.

It is interesting to study the structure of the proper H_p-subgroup of a p-group G. The following result is proved in [GMS1]:

Theorem A.91.1 (Gavioli–Mann–Scoppola). *Let E be an extraspecial p-group, $p > 2$. Then there is no p-group G with the proper H_p-subgroup $\cong E$.*

As we know this is the first non-trivial example of a non-Hughes subgroup, i.e., a p-group which is not the proper H_p-subgroup of any p-group.

It follows from Theorem 7.2 (b) that if a p-group G of exponent $> p$ has a proper H_p-subgroup, then G is irregular.

The following assertions are obvious.

(∗) Let G be a p-group of exponent $> p$ and $\mathrm{H}_p(G) < G$. (i) If $\mathrm{H}_p(G) < M \leq G$, then $\mathrm{H}_p(M) = \mathrm{H}_p(G)$. (ii) If $x \in G - \mathrm{H}_p(G)$ and $N < \mathrm{H}_p(G)$ is G-invariant of exponent $> p$, then $\mathrm{H}_p(\langle x, N\rangle) = N$. (iii) If $M < \mathrm{H}_p(G)$ is G-invariant and $\exp(G/M) > p$, then $\mathrm{H}_p(G/M) = \mathrm{H}_p(G)/M$.

Lemma A.91.2. *Suppose that a nonabelian p-group G has an abelian subgroup A of index p. If $|G : G'| = p^2$, then G is of maximal class.*

Proof. Assume that G is a counterexample of minimal order; then $|G| > p^3$. By Lemma 1.1, $|Z(G)| = \frac{1}{p}|G : G'| = p$. By induction, the quotient group $G/Z(G)$ is of maximal class, and we are done since $|Z(G)| = p$. □

Our aim is to prove the following supplement of Theorem A.91.1:

Theorem A.91.3. *Suppose that E be a p-group such that $|\Phi(E)| = p$. If G is a p-group with the proper H_p-subgroup $\cong E$, then G contains a subgroup U of maximal class and order p^{p+1} such that $E \cap U$ is the H_p-subgroup of U.*

Remark 1. Let G be a p-group of maximal class and order $> p^{p+1}$ possessing an abelian subgroup A of index p. Write $\bar{G} = G/K_{p+1}(G)$. Then $\bar{A} = H_p(\bar{G})$ of order p^p is abelian with $|\Phi(\bar{A})| = p$ (Exercise 9.13). Thus, there is a p-group, satisfying the hypothesis of Theorem A.91.3.

Proof of Theorem A.91.3. Suppose that G is a p-group such that $E = H_p(G) < G$ and $|\Phi(E)| = p$. Then $\Phi(E) \lhd G$, $\exp(E) = p^2$ since $E/\Phi(P)$ is elementary abelian and $\exp(E) > p$. It follows from $E < G$ that $\Omega_1(G) = \langle G - E\rangle = G$ is of exponent $> p$, and Theorem 7.2 (b) implies that G is irregular. Let $y \in E$ be of order p^2 and $x \in G - E$; then $o(x) = p$ and $Y = \langle y\rangle \lhd E$. Bearing in mind our aim, one may assume, without loss of generality that $G = \langle x, E\rangle$; then $|G : E| = p$ so that $E \in \Gamma_1$. Write $U = \langle x, y\rangle$ and prove that U is of maximal class and order p^{p+1} with $H_p(U) = U \cap E$.

Let $p = 2$. Then x inverts E so E is abelian (Burnside) of type $(4, 2, \dots, 2)$. In that case, $U = \langle x, y\rangle \cong D_8$, and we are done in this case.

Next we assume that $p > 2$.

We claim that x does not normalize Y. Assume that this is false. Then, by $(*)$, $Y = H_p(U)$ (indeed, all elements in the set $U - Y$ have order p) which is impossible (Theorem 1.2 or Theorem 1.17 (b)).

As we know, all cyclic subgroups of order p^2 are normal in $H_p(G)$. Therefore, $Y, Y^x \lhd H_p(G)$. Then $H_p(U) = \langle y, y^x, \dots, y^{x^{p-1}}\rangle =: V$. Next, $|U : V| = p$ since $V \lhd \langle E, x\rangle = G$ and $\langle V, x\rangle \geq \langle y, x\rangle = U$, and our claim follows (see $(*)$). As $V/\Omega_1(Y)$, as a subgroup of $E/\Omega_1(Y) = E/\Phi(E)$, is elementary abelian of order $\leq p^p$, it follows that $|V| \leq p^{p+1}$. As we have noted, $U = X \cdot V$, semidirect product with kernel V and complement $X = \langle x\rangle$. We claim that $|U| = |XV| \leq p^{p+1}$. Assume that this is false. Then $|V| = p^{p+1}$. Write $\bar{U} = U/\mho_1(Y)$; then $\bar{V} \cong E_{p^p}$. One has $d(U) = 2 = d(\bar{U})$ since $\mho_1(Y) = \Phi(E) \lhd U$ and $\bar{U} = \langle \bar{x}, \bar{y}\rangle$, $o(\bar{y}) = p$ so that $|U/U'| = p^2$. By Lemma A.91.2, $\bar{U}$ is of maximal class. Then, by Exercise 9.13, $\bar{U} \cong \Sigma_{p^2} (\in \mathrm{Syl}_p(S_{p^2})$. However, $H_p(\Sigma_{p^2}) = \Sigma_{p^2}$, contrary to $(*)$. Thus, $|U| = p^{p+1}$ so that $|V| = p^p$. Now $H_p(U) < U$ implies that the subgroup U is irregular[1] so it follows from Theorem 7.1 (b) that U is of maximal class (indeed, $|U| = p^{p+1}$ and, by Theorem 7.1 (b), $\mathrm{cl}(U) > p - 1$). Clearly, $H_p(U) = U \cap E$. □

It follows from Theorem A.91.1 that the subgroup $E \cap U$ from Theorem A.91.3 is not extraspecial (see also Exercise 3).

Recall that a p-group X is absolutely regular if $|X/\mho_1(X)| < p^p$.

Exercise 1. If the proper Hughes subgroup $H_p(G)$ of a p-group G is absolutely regular, then G is of maximal class.

Solution. As $H_p(G) < G$, it follows that G is irregular (see the proof of Theorem A.91.3). Let $H_p(G) < B \leq G$, where $|B : H_p(G)| = p$. Then $H_p(G) = H_p(B)$, by $(*)$. As $H_p(B) <$

---

1 If X is a regular group of exponent $p^e > p$, then $\Omega_{e-1}(X) < X$ so that $H_p(X) \leq \langle X - \Omega_{e-1}(X)\rangle = X$ (here we have used Theorem 7.2 (b)).

B, B is irregular. Assume that B is not of maximal class. Then, by Theorem 12.1 (b), $B = \mathrm{H}_p(G)\Omega_1(B)$, where $\Omega_1(B)$ is of order p^p and exponent p. Since $\mathrm{H}_p(G) \cup \Omega_1(B) \neq B$, we get a contradiction. Thus, B is of maximal class for any its choice. Then, by Exercise 10.10, G is of maximal class as well.

Remark 2. Let G be a p-group of maximal class and exponent $> p^2$. Then the Hughes subgroup of the quotient group $G/\Omega_1(G_1)$ is a proper absolutely regular subgroup. (**Hint.** Use Theorem 13.19.)

Exercise 2. Prove that there does not exist a p-group whose proper Hughes subgroup is irregular of maximal class.

Solution. Let G be a p-group such that $\mathrm{H}_p(G) < G$ is irregular of maximal class. Then $p > 2$ (Burnside). Bearing in mind our aim, one may assume that $|G : \mathrm{H}_p(G)| = p$ (see $(*)$). Let H_1 be the fundamental subgroup of $\mathrm{H}_p(G)$; then H_1 is G-invariant (since H_1 is characteristic in $H \triangleleft G$) of order $\geq p^p$. Let $x \in G - \mathrm{H}_p(G)$. Write $B = \langle x, H_1\rangle$; then $\mathrm{H}_p(B) = H_1$ is absolutely regular (indeed, $\mathrm{H}_p(B) \leq \mathrm{H}_p(G) = H$ and $x \notin \mathrm{H}_p(G)$ so that $\mathrm{H}_p(B) < H$). Then, by Exercise 1, B is of maximal class. Thus, all subgroups of G containing H_1 as a subgroup of index p, are of maximal class. Therefore, by Exercise 10.10, G is of maximal class. In that case, $\mathrm{H}_p(G) = G_1$ (fundamental subgroup) is absolutely regular of order $\frac{1}{p}|G| > p^p$ so it is not of maximal class (Theorem 9.5), contrary to the hypothesis.

Exercise 3. If G is a 2-group of exponent > 2 and $\mathrm{H}_2(G) < G$, then $|G : \mathrm{H}_2(G)| = 2$.

Solution. Let $x, y \in G - \mathrm{H}_2(G)$ are distinct. Then x and y invert $\mathrm{H}_2(G)$ so that xy centralizes $\mathrm{H}_2(G)$. It follows that $xy \in \mathrm{H}_2(G)$, and we conclude that $|G : \mathrm{H}_2(G)| = 2$.

Exercise 4. Let $\{1\} < \mathrm{H}_p(G) < G$, where G is a p-group, $p > 2$. Then $\mathrm{H}_p(G)$ has no characteristic cyclic subgroup of order $> p$. (**Hint.** Use part (ii) of $(*)$.)

Exercise 5. Suppose that a nonabelian p-group G of exponent $> p > 2$ has the abelian H_p-subgroup. Then all $\mathcal{A}_1$-subgroups of G are $\cong \mathrm{S}_{p^3}$. (Use Lemma 57.1 and [HogK]. See Mann's commentary to #115.)

Exercise 6. Let E be a proper nonabelian metacyclic subgroup of order 3^3 of a 3-group G. Then $E \neq \mathrm{H}_3(G)$.

Solution. By [SS], $|G| = 3^4$. By Exercise 1, G is of maximal class. Next, G possesses an abelian subgroup A of index 3 which is $\neq \mathrm{H}_3(G)$. It follows that $A \cong \mathrm{E}_{3^3}$. By Exercise, 9.13, $G \cong \Sigma_{3^3}$. In that case, $\mathrm{H}_3(G) = G$, a contradiction.

Exercise 7. Let $E < G$ be as in Theorem A.91.3. Then
(a) $\exp(G/\Phi(E)) = p$.
(b) If $x \in G - E$, then $\exp(\mathrm{C}_G(x)) = p$.
(c) If $y \in E$ is of order p^2, then $|\langle y\rangle^G| = p^p$.

Exercise 8. Let E be a proper Hughes subgroup of a p-group G, $Y = \langle y \rangle \cong C_{p^2} \lhd E$, $\Omega_1(Y) \lhd G$ and $x \in G - E$. Is it true that $U = \langle x, y \rangle$ is of maximal class and order p^{p+1}?

Problem 1. Study the p-groups, $p > 2$, whose proper Hughes subgroups are special.

Problem 2. Study the p-groups, $p > 2$, whose proper Hughes subgroups are of class 2.

Problem 3. Study the p-groups G such that, whenever $M \leq G$ is nonabelian of exponent $> p$, then $\mathrm{H}_p(M) = M$.

Problem 4. Is it possible to replace, in Theorem A.91.3, the condition $|\Phi(E)| = p$ by $|E'| = p$?

Problem 5. Study the p-groups G of exponent $> p$ such that $\mathrm{H}_p(G) < G$ and $G = U \cdot \mathrm{H}_p(G)$, semidirect product with kernel $\mathrm{H}_p(G)$ and complement U. (Example: any p-group G with $|G : \mathrm{H}_p(G)| = p$.) Is it possible such factorization in the case $|G : \mathrm{H}_p(G)| > p$?

Problem 6. Let a p-group $G = U \cdot H$, semidirect product with kernel H of exponent $> p$ and complement U of order $> p$ and exponent p. Study the structures of H and G provided $\exp(\mathrm{C}_H(x)) = p$ for all $x \in U^\#$.

Appendix 92 p-groups all of whose subgroups of order p^p and exponent p are abelian

In this appendix, we prove the following

Theorem A.92.1. *Suppose that a p-group $G, p > 2$, contains a subgroup of order p^p and exponent p. If all such subgroups are abelian, then $\Omega_1(G)$ is (elementary) abelian.*

Proof. Suppose that G is regular. Then $H = \Omega_1(G)$ has exponent p. Assume that H is nonabelian. Then there is in H a minimal nonabelian subgroup E. By Lemma 65.1, $|E| = p^3$. Let $E \leq M \leq H$, where $|M| = p^p$. As, by hypothesis, M is abelian, we get a contradiction. Thus, G is irregular.

Assume that G has no normal subgroup of order p^p and exponent p. Then G, by Theorem 12.1 (a), is of maximal class. As there is in G a subgroup of order p^p and exponent p, it is elementary abelian. Therefore, $G \cong \Sigma_{p^2}$, a Sylow p-subgroup of the symmetric group of degree p^2 (Exercise 9.13). But Σ_{p^2} contains two distinct subgroups, say E and E_1 of order p^p and exponent p. In that case, $E \cap E_1 = Z(G)$ has index p^2 in G so that $\mathrm{cl}(G) = 2$. As $\mathrm{cl}(\Sigma_{p^2}) = p > 2$, we get a contradiction.

Thus, G contains a normal subgroup E of order p^p and exponent p; by hypothesis, $E \cong \mathrm{E}_{p^p}$. Let $E \leq E_1 < G$, where E_1 is a maximal normal elementary abelian subgroup of G. Assume that $E_1 < \Omega_1(G)$. Then there is in $\Omega_1(G) - E$ an element x of order p. Write $H = \langle x, E_1\rangle$. By Theorem 10.1, H is nonabelian. Then there is in E_1 an element a of order p such that $U = \langle a, x\rangle$ is minimal nonabelian (Lemma 57.1). It follows from $\Omega_1(U) = U$ that $U \cong S(p^3)$, a nonabelian group of order p^3 (Lemma 65.1). Let $U \leq V < H$, where $|V| = p^p$. It follows from $V = \langle x\rangle(V \cap E_1)$ (by modular law) that $\Omega_1(V) = p$, and we conclude that $\exp(V) = p$ (Theorems 7.1 (b) and 7.2 (b)). By hypothesis, V is abelian, a contradiction since $U \leq V$ is nonabelian. Thus, x does not exist so that $\Omega_1(G) = E_1$ is elementary abelian. □

Corollary A.92.2. *Suppose that a p-group G, $p > 2$, contains a subgroup of order p^3 and exponent p and all such subgroups are abelian. Then $\Omega_1(G)$ is abelian.*

Proof. As in the proof of Theorem A.92.1, one may assume that G is irregular. If there is in G a subgroup of order p^p and exponent p, it is abelian (indeed, it has no minimal nonabelian subgroups, by Lemma 65,1), and so the result follows from Theorem A.92.1. Otherwise, G is of maximal class (Theorem 12.1 (a)). By Theorem 10.4, G has a normal subgroup $E \cong \mathrm{E}_{p^3}$. Then $p > 3$ (Exercise 9.13). As above, using Lemma 57.1, one obtains that $\Omega_1(G)$ is abelian. □

Suppose that a p-group G, $p > 2$, contains two noncommuting elements, say x and y, of order p. We claim that the subgroup $H = \langle x, y\rangle$ contains a subgroup $\cong S(p^3)$. Indeed, otherwise, all subgroups of H of order p^3 and exponent p, existing by Lemmas 1.4 and

57.1, are abelian. Then, by Corollary A.92.2, $\Omega_1(H)$ is elementary abelian, contrary to the hypothesis.

Exercise 1. Study the p-groups of order $> p^p$ and exponent $> p$, $p > 3$, all of whose subgroups of order p^p are abelian.

Hint. If G has a subgroup of order p^p and exponent p, then $\Omega_1(G)$ is abelian. Let $x \in G - \Omega_1(G)$ be of order p^2. Then $x^p \in \Omega_1(G)$. If $\langle x, \Omega_1(G)\rangle$ is nonabelian, there is $a \in \Omega_1(G)$ such that $M = \langle a, x\rangle$ is minimal nonabelian (Lemma 57.1). By Lemma 65.1, $|M| \le p^4 < p^p$. If $M < H \le G$ and $|H| = p^p$, then H is nonabelian, contrary to the hypothesis. It follows that $\Omega_2(G) \le C_G(\Omega_1(G))$. If G has no subgroup of order p^p and exponent p, apply Theorem 12.1 (a).

Exercise 2. If all minimal nonabelian subgroups of a p-group G, $p > 2$, are metacyclic, then $\Omega_1(G)$ is abelian.

Solution. Let $E \triangleleft G$ be elementary abelian of maximal possible order. Assume that there is $x \in G - E$ of order p and set $H = \langle x, E\rangle$. Then H is nonabelian, by Theorem 10.1. Then, by Lemma 57.1, there is $a \in E$ such that $A = \langle a, x\rangle$ is minimal nonabelian. As $\Omega_1(A) = A$. It follows that $A \cong S(p^3)$ is nonmetacyclic (Lemma 65.1), a contradiction. Thus, x does not exist so that $E = \Omega_1(G)$.

If a regular p-group G is of exponent p^{e+1}, then $\Omega^*_{e+1}(G) = \langle x \in G \mid o(x) = p^{e+1}\} = G$ since $\Omega^*_{e+1}(G) = \langle G - \Omega_e(G)\rangle = G$ (see Theorem 7.2 (b)).

Exercise 3. Let G be a nonabelian p-group of exponent $> p^e \ge p^2$, $p > 2$. If all maximal regular subgroups of G of exponent p^e are abelian and normalized by all elements of G of order p^e, then $\Omega^*_e(G)$ is abelian.

Solution. Let H be a maximal subgroup among regular subgroups of exponent p^e in G; then H is abelian, by hypothesis and, clearly, $H = \Omega^*_e(H)$ (here we use the remark preceding the exercise). Assume that $H < \Omega^*_e(G)$. Then there is $x \in G - H$ of order p^e. Set $M = \langle x, H\rangle$; by hypothesis, $H \triangleleft M$. Then M is nonabelian (otherwise, $\exp(M) = p^e$ and $M > H$, contrary to the choice of H). By Lemma 57.1, there is $h \in H$ such that $A = \langle h, x\rangle$ is minimal nonabelian of exponent p^e. As $\mathrm{cl}(A) = 2 < p$ and $\Omega^*_e(A) = A$ in view of regularity A, it follows that $\exp(A) = p^e$ (Theorem 7.2 (b)). Let $A \le T < G$, where $\exp(T) = p^e$ and T is regular as large as possible; then $\Omega^*_e(T) = T$. Then, by hypothesis, T is abelian, a contradiction since $T \ge A$. Thus, $\Omega^*_e(G) = H$ is abelian.

The above theorem and corollary have also been proved in § 30; however, here we offered alternate proofs.

Appendix 93 Nonabelian p-groups G with $|G : HC_G(H)| \leq p$ for all nonabelian subgroups H

The following theorem contains some useful information on p-groups satisfying Problem 3269.

Theorem A.93.1 (Janko). *Let G be a nonabelian p-group with $|G : HC_G(H)| \leq p$ for all nonabelian subgroups H in G. Then each maximal abelian subgroup is of index $\leq p^2$ in G and $|G'| \leq p^3$. Moreover, if $|G'| = p^3$, then $|G : Z(G)| = p^3$.*

Proof. Let A be a maximal abelian subgroup of G. Hence $A < G$ and let $H > A$ be a subgroup in G with $|H : A| = p$. Since $C_G(A) = A$, H is nonabelian. By our hypothesis, $|G : HC_G(H)| \leq p$ and so $|G : H| \leq p$ because $C_G(H) < A < H$. We get $|G : A| \leq p^2$ and so each maximal abelian subgroup of G is of index $\leq p^2$ in G.

In particular, we see that the element breadth $b(G)$ of G is ≤ 2. If $b(G) = 1$, then Proposition 121.9 implies that $|G'| = p$. If $b(G) = 2$, then Theorem 121.1 gives that either $|G'| = p^2$ or $|G'| = p^3$ and $|G : Z(G)| = p^3$. Our theorem is proved. □

Problem 1. Classify the nonabelian p-groups G such that $|G : HC_G(H)| \leq p$ for any (i) minimal nonabelian $H < G$, (ii) minimal nonmetacyclic $H < G$, (iii) maximal metacyclic $H < G$, (iv) maximal regular subgroup H of an irregular p-group G.

Problem 2. Classify the non-metahamiltonian p-groups G such that $|G : HC_G(H)| = p$ for any nonnormal nonabelian $H < G$. (See [FA].)

Appendix 94 Non-Dedekindian p-groups with exactly one conjugate class of nonnormal maximal cyclic subgroups

We solve here problem 3746 and prove the following result.

Theorem A.94.1 (Janko). *Let G be a non-Dedekindian p-group with exactly one conjugate class of nonnormal maximal cyclic subgroups. Then G is of class 2 and G' has a cyclic subgroup of index p. Moreover, if $p > 2$, then both G' and G/G_0 are cyclic, where G_0 is the subgroup generated by all nonnormal subgroups in G.*

Proof. Let G be a non-Dedekindian p-group with exactly one conjugate class of nonnormal maximal cyclic subgroups and let Z be a nonnormal maximal cyclic subgroup in G. Set $G_0 = Z^G$ so that $G_0 < G$ and G_0 is the subgroup generated by all nonnormal subgroups of G. Indeed, let X be a nonnormal subgroup in G and assume $X \nleq G_0$. Let $x \in X - G_0$. If $\langle x \rangle$ is nonnormal in G, then let $\langle a \rangle$ be a maximal cyclic subgroup of G containing $\langle x \rangle$ so that $\langle a \rangle$ is nonnormal in G. But $\langle a \rangle \nleq G_0$, contrary to our hypothesis. Thus, for each $x \in X - G_0$, we have $\langle x \rangle \triangleleft G$. Since $X = \langle X - G_0 \rangle$, we get $X \triangleleft G$, a contradiction.

We are in a position to use Theorem 231.1 so that G is of class 2 and G' has a cyclic subgroup of index p. Moreover, if $p > 2$, then both G' and G/G_0 are cyclic, where G_0 is the subgroup generated by all nonnormal subgroups in G. □

Another approach to this problem see in § 240.

Appendix 95 The centralizer of any element from $G - \Phi(G)$ cannot be a nonabelian two-generator subgroup

Let $x \in G - \Phi(G)$ and assume that $C = C_G(x)$ is minimal nonabelian. Then $x \in Z(C) = \Phi(C) \le \Phi(G)$, contrary to the choice of x. It appears that this argument (taken from the letter of the second author) allows us to prove more general result.

Let G be a nonabelian two-generator p-group. We assert that then $Z(G) \le \Phi(G)$. Indeed, $G/Z(G)$ is noncyclic so that $d(G/Z(G) \ge 2 = d(G)$, and this implies that $Z(G) \le \Phi(G)$.

Theorem A.95.1. *Suppose that G is a nonabelian p-group and $x \in G - \Phi(G)$. Then $C_G(x)$ is not a two-generator nonabelian subgroup.*

Proof. Assume that the assertion is false. Then $C = C_G(x)$ is a two-generator nonabelian subgroup for some $x \in G - \Phi(G)$. As $C/Z(C)$ is noncyclic and $d(C) = 2$, it follows that $d(C/Z(C)) = 2$ so that $Z(C) \le \Phi(C)$. In that case, $x \in Z(C) \le \Phi(C) \le \Phi(G)$, contrary to the choice of x. □

Corollary A.95.2. *Let G be a metacyclic p-group. Then $C_G(x)$ is abelian for any $x \in G - \Phi(G)$.*

Proof. Assume that $C_G(x)$ is nonabelian. As $d(C_G(x)) = 2$ since C is metacyclic, we get a contradiction with Theorem A.95.1. □

Corollary A.95.3. *Let A be an abelian subgroup of a p-group G not contained in $\Phi(G)$. Then $C_G(A)$ is not a two-generator nonabelian subgroup.*

Definition. A p-group $G > \{1\}$ is said to be $\mathcal{A}_0$-group if $Z(G) \le \Phi(G)$.

Obviously, $\mathcal{A}_0$-groups are nonabelian.

Examples. Here we present examples of $\mathcal{A}_0$-groups G.

(i) G is nonabelian and $d(G) = 2$. In particular, p-groups of maximal class and minimal irregular p-groups are $\mathcal{A}_0$-groups.
(ii) Special p-groups.
(iii) $G = A \times B$, where A and B are as in (i) or in (ii).

We see that the set of $\mathcal{A}_0$-groups is fairly wide.

Proposition A.95.4. *Suppose that G is a p-group and $x \in G - \Phi(G)$. Then $C_G(x)$ is not an $\mathcal{A}_0$-group. In particular, $C_G(x)$ is not special.*

See the proof of Theorem A.95.1.

Problem 1. Study the primary $\mathcal{A}_n$-groups G, $n > 2$, such that $C_G(x)$ is an $\mathcal{A}_k$-subgroup for $k \leq 2$ and any $x \in G - \Phi(G)$.

Problem 2. Study the p-groups G such that $C_G(x)$ is abelian for any $x \in G - G'$ (for any $x \in G - \Phi(G)$).

Problem 3. Study the nonabelian p-groups G all of whose proper nonabelian subgroups (nonabelian maximal subgroups) are $\mathcal{A}_0$-groups.

Appendix 96 Nonabelian 2-groups in which any two noncommuting elements generate a subgroup of maximal class

We prove the following result which solves Problem Nr.788 for $p = 2$. This result also follows from Theorem 90.1. The corresponding problem for $p > 2$ is very difficult since the p-groups of maximal class are not classified for $p > 2$ (see [GMS1]).

Theorem A.96.1 (Janko). *Let G be a nonabelian 2-group in which any two noncommuting elements generate a group of maximal class. Then one of the following holds:*
(a) $|G : H_2(G)| = 2$ *and* $H_2(G)$ *is noncyclic (i.e., G is quasidihedral but not dihedral);*
(b) $G = HZ(G)$, *where H is of maximal class and* $\mho_1(Z(G)) \leq Z(H)$;
(c) $G = HZ(G)$, *where H is extraspecial of order* $\geq 2^5$ *and* $\mho_1(Z(G)) \leq Z(H)$.
Conversely, each group in (a), (b) and (c) satisfies the assumption of the theorem.

Proof. We may assume that G is not of maximal class.

(i) First we assume that $exp(G) > 4$. Suppose for a moment that each element of order ≥ 8 lies in $Z(G)$. Let $x, y \in G$ with $[x, y] \neq 1$. Since $\langle x, y\rangle$ is of maximal class, we have in our case $\langle x, y\rangle \cong D_8$ or $\langle x, y\rangle \cong Q_8$. Let k be an element of order 8 so that here $k \in Z(G)$. But then kx and ky are elements of order 8 with $[kx, ky] = [x, y] \neq 1$ and therefore $\langle kx, ky\rangle$ is of maximal class, a contradiction. We have proved that G possesses a cyclic subgroup A of order ≥ 8 such that $A \nleq Z(G)$.

It is easy to see that any cyclic subgroup X of order ≥ 8 is normal in G. Indeed, let $g \in G$ so that g either centralizes X or $\langle X, g\rangle$ is of maximal class in which case g normalizes X.

Let $y \in G$ be such that $[A, y] \neq 1$ and so $\langle A, y\rangle$ is of maximal class. Then $\langle A, y\rangle$ contains a subgroup of maximal class $\langle B, y\rangle$ of order 2^4, where $B = \langle b\rangle \cong C_8$, $B \leq A$, and $y^2 \in \Omega_1(B) = \langle z\rangle$. We know that B is normal in G. Set $M = C_G(B)$ so that $G/M \neq \{1\}$ is elementary abelian of order ≤ 4. If $G/M \cong E_4$, then there is $l \in G-M$ such that $l^2 \in M$, l^2 centralizes B and $b^l = bz$. But then $\langle b, l\rangle' = \langle z\rangle$ and so $\langle b, l\rangle$ is not of maximal class, a contradiction. Thus $|G : M| = 2$.

For each $x \in G - M$, $x^2 \in \langle z\rangle$. Indeed, $[b, x] \neq 1$ and so $\langle b, x\rangle$ is of maximal class and therefore $x^2 \in \Omega_1(B) = \langle z\rangle$. Consider $\bar{G} = G/\langle z\rangle$. Then all elements in $\bar{G} - \bar{M}$ are involutions which implies that $M/\langle z\rangle$ is abelian and for each $m \in M$, $m^y = m^{-1}z^{\epsilon}$, $\epsilon = 0, 1$.

Suppose that M is nonabelian. Then $M' = \langle z\rangle$ and let $m, n \in M$ with $[m, n] = z$. In that case (since $\langle m, n\rangle$ is of maximal class), $\langle m, n\rangle \cong D_8$ *or* $\cong Q_8$. But then bm and bn are elements of order 8 with $[bm, bn] = [m, n] = z$ and so $\langle bm, bn\rangle$ is of maximal class, a contradiction. Hence M is abelian. If M is cyclic, then $\langle M, y\rangle = G$ is of maximal class, a contradiction. Thus, M is noncyclic abelian.

If all elements in $G - M$ are involutions, then $H_2(G) = M$ and we have obtained a group in part (a) of our theorem.

We may assume that not all elements in $G - M$ are involutions and so we may suppose $y^2 = z$. Let t be any involution in $M - \langle z \rangle$ and assume that t is a square in M, i.e., there is $k \in M$ such that $k^2 = t$. Since $k^y = k^{-1}z^\epsilon$ ($\epsilon = 0, 1$), $\langle y, k \rangle$ is nonabelian. In that case $\langle y, k \rangle$ is of maximal class containing two distinct involutions z and t which are squares in $\langle y, k \rangle$, a contradiction. We have proved that M is abelian of type $(2^s, 2, \ldots, 2)$, $s \geq 3$. Setting $E = \Omega_1(M)$, we get $M = \langle b' \rangle E$, where $o(b') = 2^s$, $|E| \geq 4$, and $\langle b' \rangle \cap E = \Omega_1(\langle b' \rangle) = \langle z \rangle$ since z is the unique involution in M which is a square in M. Since $(b')^y = (b')^{-1}z^\eta$ ($\eta = 0, 1$), $H = \langle b', y \rangle$ is of maximal class and $G = HE$. For each $t \in E$, we have either $t^y = t$ or $t^y = tz$. If y centralizes E, then $E = Z(G)$. If y does not centralize E, then $E_0 = C_E(y)$ is of index 2 in E. Let v be an element of order 4 in $\langle b' \rangle$ and let $u \in E - E_0$. In that case

$$(vu)^y = v^{-1}(uz) = (vz)(uz) = vu \textit{ and } (vu)^2 = z,$$

and so $Z(G) = E_0\langle vu \rangle$ with $\mho_1(Z(G)) = \langle z \rangle$. In any case we get $G = HZ(G)$, $Z(G) > Z(H) = \langle z \rangle$, and $\mho_1(Z(G)) \leq \langle z \rangle$. We have obtained a group in part (b) of our theorem.

(ii) We examine now the case $exp(G) = 4$. Let $\langle x \rangle$ be a cyclic subgroup of order 4 and $y \in G$. Then either $[x, y] = 1$ or $\langle x, y \rangle \cong D_8$ *or* Q_8. In any case y normalizes $\langle x \rangle$ and so each cyclic subgroup of order 4 is normal in G. We may use Theorem [Kaz1] from § 58. It follows that either $|G : H_2(G)| = 2$ (and we get a group of part (a) of our theorem) or $|G'| = 2$ and $\Phi(G)$ is cyclic.

Suppose that we are in the second case. Since G does not possess elements of order 8, we have $|\Phi(G)| = 2$ and then $\Phi(G) = G'$. The fact $|G'| = 2$ implies that $G = H_1 * H_2 * \ldots * H_nZ(G)$, where H_i ($i = 1, \ldots, n$) is minimal nonabelian (Lemma 4.2). In our case $H_i \cong D_8$ *or* Q_8 and so $H = H_1 * H_2 * \ldots * H_n$ is extraspecial. Also, $\Phi(G) = G' = H' = Z(H)$ implies that $\mho_1(Z(G)) \leq Z(H)$. If $n = 1$, we have obtained a group of part (b) of our theorem and so we may assume that $n > 1$. In that case $|H| \geq 2^5$ and we have obtained a group of part (c) of our theorem.

It is necessary to prove only for groups (b) and (c) of our theorem that any two noncommuting elements generate a group of maximal class. Indeed, let h_1z_1 and h_2z_2 be any noncommuting elements in G, where $h_1, h_2 \in H$ and $z_1, z_2 \in Z(G)$. Then $[h_1z_1, h_2z_2] = [h_1, h_2] \neq 1$ and so $H_0 = \langle h_1, h_2 \rangle \leq H$ is a group of maximal class with $H_0' \geq Z(H)$. On the other hand, a 2-group $\langle h_1, h_2 \rangle$ is of maximal class if and only if $[h_1, h_2] \neq 1$, $\langle [h_1, h_2] \rangle$ is normal in H_0 and $h_1^2, h_2^2 \in \langle [h_1, h_2] \rangle$. Hence $H_1 = \langle h_1z_1, h_2z_2 \rangle$ is of maximal class since $[h_1z_1, h_2z_2] = [h_1, h_2] \neq 1$, h_1z_1 and h_2z_2 normalize $\langle [h_1z_1, h_2z_2] \rangle = \langle [h_1, h_2] \rangle$ and $(h_1z_1)^2, (h_2z_2)^2$ are contained in $\langle [h_1, h_2] \rangle$ (noting that $z_1^2, z_2^2 \in Z(H) \leq H_0' = H_1'$). □

Thus, the 2-groups all of whose nonabelian two-generator subgroups are of maximal class, are classified.

Note that if A is a minimal nonabelian subgroup of a group from Theorem A.96.1, then $|A| = 8$; such 2-groups are classified in Theorem 90.1.

For related results, see § 225.

Problem 1. Classify the p-groups G such that any two non-commuting elements generate either a minimal nonabelian subgroup or a p-group of maximal class.

Problem 2. Classify the p-groups G, $p > 2$, such that any two non-commuting elements generate either an absolutely regular subgroup or a p-group of maximal class.

Appendix 97 p-groups all of whose subgroups of order p^p and exponent p are of maximal class

Let G be a p-group. If G has no subgroup of order p^p and exponent p, then it is either absolutely regular or irregular of maximal class (Theorem 12.1 (a)). Assume that G has a subgroup of order p^p and exponent p and all such subgroups are of maximal class; then $p > 2$. Here we classify such p-groups.

Theorem A.97.1. *Suppose that a p-group G contains a proper subgroup of order p^p and exponent p and all normal subgroups of G of order p^p and exponent p are of maximal class so that $p > 2$. Then one of the following holds:*
(a) G is of maximal class of order p^{p+1}.
(b) $G = H\Omega_1(G)$, where $\Omega_1(G)$ is of order p^p and exponent p (hence $\Omega_1(G)$ is of maximal class) and H is absolutely regular.

Proof. By hypothesis, G is not absolutely regular. Assume that G is not of maximal class. Then, by Theorem 12.1 (a), there is in G a normal subgroup R of order p^p and exponent p. Take in R a G-invariant subgroup L of order p^2; then $L \cong \mathrm{E}_{p^2}$. Write $C = \mathrm{C}_G(L)$. Since $R \nleq C$, it follows that $|G : C| = p$. Assume that there is in C a G-invariant subgroup S of order p^p and exponent p. As S is of maximal class and $L \le \mathrm{Z}(C)$, we get $L \nleq S$. Then $L * S$ is G-invariant of order $> p^p$ and exponent p. Let $L < T \le LS$, where T is G-invariant of order p^p. As, in addition, $\exp(T) = p$ and T is not of maximal class, this contradicts the hypothesis. Thus, S does not exist. If, in addition, C is not absolutely regular, the number $\mathrm{e}_p(C)$ of subgroups in C of order p^p and exponent p is $\equiv 1 \pmod p$ (Theorem 13.5); then C has a G-invariant subgroup of order p^p and exponent p, contrary to what has been proved. Thus, $C \in \Gamma_1$ is absolutely regular, and we get $G = C\Omega_1(G)$, where $\Omega_1(G)$ is of order p^p and exponent p (Theorem 12.1 (b)). By hypothesis, $\Omega_1(G)$ is of maximal class, and so $|G| = p^{p+1}$ (Theorem 9.6(c)). □

Appendix 98 The number of cyclic subgroups of given order in a metacyclic p-group

This is a supplement to § 124 where the number $s_n(G)$ of subgroups of order p^n in a metacyclic p-group G was computed. Here we intend to find the number $c_n(G)$ of cyclic subgroups of order p^n in a metacyclic p-group G in terms of $|R(G)|$ and $G/R(G)$ (see the following paragraph).

Let G be a metacyclic p-group. Write

$$w = w(G) = \max\{i \mid |\Omega_i(G)| = 2^{2i}\}, \quad R(G) = \Omega_w(G).$$

For example, if G is a 2-group of maximal class, then $R(G) = \{1\}$. For any metacyclic p-group G, the quotient group $G/R(G)$ is either cyclic or a 2-group of maximal class (Lemma 1.4).

Let G be a metacyclic p-group. If $G = R(G)$ is of exponent p^e, then, for any $n \leq e$, one has

$$c_n(G) = \frac{|\Omega_n(G) - \Omega_{n-1}(G)|}{(p-1)p^{n-1}} = \frac{p^{2n} - p^{2(n-1)}}{(p-1)p^{n-1}} = (p+1)p^{n-1}$$

(here $|\Omega_n(G) - \Omega_{n-1}(G)|$ is the number of elements of order p^n in G and $(p-1)p^n$ is the number of generators of the cyclic group of order p^n).

If G is a metacyclic p-group and $n \leq w$, then $c_n(G) = c_n(R(G))$, hence it suffices to find $c_n(G)$ for $n > w$.

If a p-group G is such that $\Omega_i(G) = \{x \in G \mid o(x) = p^i\}$ for all $p^i \leq \exp(G)$, then

$$(1) \qquad c_n(G) = \frac{|\Omega_n(G) - \Omega_{n-1}(G)|}{(p-1)p^{n-1}}.$$

Note that

$$c_1(D_{2^{m+1}}) = 2^m + 1,\ c_k(D_{2^{m+1}}) = 1 \text{ for } k \in \{2, \dots, m\},$$
$$c_1(Q_{2^{m+1}}) = 1,\ c_2(Q_{2^{m+1}}) = 2^{m-1} + 1,\ c_k(Q_{2^{m+1}}) = 1 \text{ for } k \in \{3, \dots, m\},$$
$$c_1(SD_{2^{m+1}}) = 2^{m-1} + 1,\ c_2(SD_{2^{m+1}}) = 2^{m-2} + 1,\ c_k(SD_{2^{m+1}}) = 1 \text{ for } k \in \{3, \dots, m\}.$$

Proposition A.98.1. *Suppose that G is a metacyclic p-group such that $G/R(G)) \cong C_{p^t}$ and $n = w + k$, where $k \leq t$. Then $c_{w+k}(G) = p^w$.*

Proof. Our group G satisfies the condition stated in the paragraph preceding the proposition. As $|\Omega_{w+k}(G)| = p^{2w+k}$ and $|\Omega_{w+k-1}(G)| = p^{2w+k-1}$, application of formula (1) yields the desired result. □

Let G be a metacyclic p-group and $M_1/R(G), \dots, M_s/(G)$ be all subgroups of order p in $G/R(G)$, then, by $\Omega_w(M_i) = \Omega_w(M_i)$, one has

$$(2) \qquad c_{w+1}(G) = \sum_{j=1}^{s} c_{w+1}(M_j) = s \cdot c_{w+1}(M_1) = s \cdot p^w.$$

Indeed, for $i \neq j$, the subgroup $M_i \cap M_j = R(G) = \Omega_w(M_i) = \Omega_w(M_j)$, and (2) follows from Proposition A.98.1.

Proposition A.98.2. *Suppose that G is a metacyclic 2-group such that $G/R(G) \cong \mathrm{D}_{2^{m+1}}$.*
(a) $\mathrm{c}_{w+1}(G) = (2^m + 1)2^w$.
(b) If $1 < k \leq m$, then $\mathrm{c}_{w+k}(G) = 2^w$.

Proof. (a) follows from (2) since $G/R(G)$ contains exactly $2^m + 1$ subgroups of order 2. Since $G/R(G)$ contains exactly one cyclic subgroup say, H_k, of order 2^{w+k} for $k \in \{2, \dots, m\}$, one has $\mathrm{c}_{w+k}(G) = \mathrm{c}_{w+k}(H_k)$, (b) follows from Proposition A.98.1 applied to H_k. □

Proposition A.98.3. *Suppose that G is a metacyclic 2-group such that $G/R(G) \cong \mathrm{Q}_{2^{m+1}}$.*
(a) $\mathrm{c}_{w+1}(G) = 2^w$.
(b) $\mathrm{c}_{w+2}(G) = (2^{m-1} + 1)2^w$.
(c) If $2 < k \leq m$, then $\mathrm{c}_{w+k}(G) = 2^w$.

Proof. As $\mathrm{c}_1(G/R(G)) = 1$, (a) follows from Lemma A.98.1. If $M_1/R(G), \dots, M_t/R(G)$, where $t = 2^{m-1}+1$, be all cyclic subgroups of order 4 in $G/R(G)$ and $\Omega_{w+1}(G) = M_i \cap M_j$ for $i \neq j$, we get, by Lemma A.98.1,

$$\mathrm{c}_{w+2}(G) = \sum_{i=1}^{2^{m-1}+1} \mathrm{c}_{w+2}(M_i) = (2^{m-1} + 1)2^w ,$$

completing the proof of (b). As $\mathrm{c}_{w+k}(G/R(G)) = 1$ for $k \in \{3, \dots, m\}$, we get $\mathrm{c}_{w+k}(G) = 2^w$, by Lemma A.98.1. □

Proposition A.98.4. *Suppose that G is a metacyclic 2-group such that $G/R \cong \mathrm{SD}_{2^{m+1}}$.*
(a) $\mathrm{c}_{w+1}(G) = (2^{m-1} + 1)2^w$.
(b) $\mathrm{c}_{w+2}(G) = (2^{m-2} + 1)2^w$.
(c) If $2 < k \leq m$, then $\mathrm{c}_{w+k}(G) = 2^w$.

Proof. Let $D/R(G) \in \Gamma_1(G/R(G))$ be dihedral and let $Q/R(G) \in \Gamma_1(G/R(G))$ be the generalized quaternion group (here $\Gamma_1(G/R(G))$ is the set of all maximal subgroups of $G/R(G)$). As $\Omega_1(G/R(G)) = D/R(G)$ and so $\mathrm{c}_1(G/R(G)) = \mathrm{c}_1(D/R(G) = 2^{m-1} + 1$, it follows from Lemma A.98.2 (a) that $\mathrm{c}_{w+1}(G) = \mathrm{c}_{w+1}(D) = (2^{m-1} + 1)2^w$. As $\Omega_2^*(G/R(G)) = Q/R(G)$ and so $\mathrm{c}_2(G/R(G)) = \mathrm{c}_2(Q/R(G)) = 2^{m-2} + 1$, it follows from Lemma A.98.3 (b) that $\mathrm{c}_{w+2}(G) = (2^{m-2} + 1)2^w$. As $\mathrm{c}_k(G/R(G)) = 1$ for $k \in \{3, \dots, m\}$, we get $\mathrm{c}_{w+k}(G) = 2^w$, by Lemma A.98.1. □

Exercise 1. Let $G = R(G)$ be a metacyclic p-group of order p^{2w} and exponent p^w. Find the number $\mathrm{s}_w(G)$ of subgroups of order p^w in G.

Hint. Let $H < G$ be noncyclic of order p^w; then H contains $\Omega_1(G)$. The number of such H is equal to $\mathrm{s}_{w-2}(G/\Omega_1(G))$. It follows that $\mathrm{s}_w(G) = \mathrm{c}_w(G) + \mathrm{s}_{w-2}(G/\Omega_1(G))$. Use induction on w.

Exercise 2. Let $G = R(G)$ be a metacyclic p-group of order p^{2w} and exponent p^w. Find the number $s_{w+1}(G)$.

Exercise 3. Given $k \leq n$, find the number $c_k(\Sigma_n)$.

See § 124, where the number $s_k(G)$ is computed for an arbitrary metacyclic p-group G. That computation is based on entirely other ideas.

Appendix 99 On existence of L_p-subgroups in a p-group

Given $n > 1$, a p-group G is said to be an L_n-*group*, if $\Omega_1(G)$ is of order p^n and exponent p and $G/\Omega_1(G)$ is cyclic of order $> p$. If G is an L_2-group, then it is either abelian of type (p^e, p) or $\cong M_{p^{e+1}}$, $e > 2$. If $n < p$, then an L_n-group is absolutely regular (Theorems 12.1 (a) and 9.8 (a)). However, there exist irregular L_p-groups. In this appendix we obtain some information on the structure of L_p-groups and prove that a p-group of exponent $p^e > p^2$ having a sufficiently complicated structure, possesses an L_p-subgroup of exponent p^e. If G is an L_n-group of exponent $p^e > p^3$ and $G_0/\Omega_1(G)$ is a subgroup of index p in $G/\Omega_1(G)$, then G_0 is an L_n-group.

Lemma A.99.1. *Let G be an L_p-group of exponent $p^e > p^2$. Then $G = CR$, where $R = \Omega_1(G)$ (of order p^p and exponent p) and $C \cong C_{p^e}$; then $|G| = p^{p+e-1}$.*

(a) $\mho_1(G) = \mho_1(C)$.

(b) *All proper subgroups of G are regular so that, if G is irregular, then $d(G) = 2$. Moreover, $cl(R\mho_1(G)) < p$.*

Proof. By what we said in the first paragraph, one may assume that $p > 2$.

(a) If G is irregular, then $|G/\mho_1(G)| \geq p^p$ (Theorem 9.8 (a)) so that $|\mho_1(G)| \leq p^{e-1}$. However, $|\mho_1(C)| = p^{e-1}$, and we conclude that $|\mho_1(G)| = p^{e-1}$, hence $\mho_1(G) = \mho_1(C)$. If G is regular, then $|G/\mho_1(G)| = |\Omega_1(G)| = p^p$ (Theorem 7.2 (c)) so, as above, $\mho_1(G) = \mho_1(C)$, completing the proof of (a).

(b) Let $C < H \in \Gamma_1$. Then $\Omega_1(H) < \Omega_1(G)$ since $C\Omega_1(H) = H < G$ so that H is absolutely regular (Theorem 12.1 (a)). If $F \in \Gamma_1$ and $\Omega_1(G) = R \nleq F$, then again $\Omega_1(F) < \Omega_1(G)$ so F is absolutely regular (Theorem 12.1 (a)). Let $R < T \in \Gamma_1$. Then $T = R\mho_1(C)$, by the modular law. Since $C \cap R = \mho_1(C) \cap R = \Omega_1(C)$, since $\mho_1(C) = \mho_1(G) \lhd G$, one has

$$T/(R \cap \mho_1(C)) = T/\Omega_1(C) = R\mho_1(C)/\Omega_1(C) \cong (R/\Omega_1(C)) \times (\mho_1(C)/\Omega_1(C))$$

so that $cl(T/\Omega_1(C)) < p - 1$, and we conclude that $cl(T) < p$ so that T is regular (Theorem 7.1 (b)), completing the proof of (b) and thereby the proposition. □

Lemma A.99.2. *Let a p-group G be neither absolutely regular nor of maximal class and let $A < G$ be of order $< p^p$ and exponent p. Then $A < R \leq G$, where R is of order p^p and exponent p. If, in addition, $A \lhd G$, then it is possible to choose R so that it is G-invariant.*

Proof. As G has a normal subgroup of order p^p and exponent p (Theorem 12.1 (a)), there exists $S \lhd G$ of exponent p minimal such that $S \nleq A$. Then AS is of order $p|A|$ and exponent p (Theorem 7.1 (b)). If $|SA| < p^p$, the same argument allows us to find in G a subgroup of exponent p containing AS as a subgroup of index p. Continuing so, we at last find the subgroup of order p^p and exponent p containing A. If $A \lhd G$, then $SA \lhd G$.

So the above argument allows us to find in G a G-invariant subgroup of order p^p and exponent p containing A. □

Theorem A.99.3. *Let a p-group G of exponent $p^e > p^2$ be neither absolutely regular nor of maximal class. Then G contains an L_p-subgroup of exponent p^e.*

Proof. We proceed by induction on $|G|$.

Let $C < G$ be cyclic of order p^e and let $C < H \in \Gamma_1$; then $\exp(H) = p^e$. If H is neither absolutely regular nor of maximal class, then there is in H an L_p-subgroup of exponent p^e, by induction. Next, we assume that H is either absolutely regular or of maximal class.

(i) Assume that H is absolutely regular. Then $G = HR$, where R is of order p^p and exponent p (Theorem 12.1 (b)). In that case, CR is an L_p-subgroup of exponent p^e.

(ii) Now let H be of maximal class. Then $K = \Omega_1(\Phi(H))$ is a G-invariant subgroup (as $\Phi(H)$ is characteristic in $H \triangleleft G$) of order p^{p-1} and exponent p (Theorem 9.6 (e)). Write $M = CK$. As $\mho_1(C) < \Phi(H)$, we get $C \cap K = \Omega_1(C)$, and we conclude that M is an L_{p-1}-subgroup of exponent p^e. By Lemma A.99.2, K is contained in a G-invariant subgroup R of order p^p and exponent p. Set $Q = MR = CKR = CR$. We claim that Q is an L_p-subgroup. It suffices to show that $\Omega_1(Q) = R$ (indeed, then Q/R is cyclic of order $p^{e-1} > p$ since $e > 2$). Assume that this is false. Since Q/R is cyclic, we get $|\Omega_1(Q)| = p^{p+1}$. By the product formula, $C \cap \Omega_1(Q)$ is cyclic of order p^2 so that $\exp(\Omega_1(Q)) = p^2$, and we conclude that $\Omega_1(Q)$ is irregular (Theorem 7.2 (b). Then, by Theorem 7.1 (b), $\Omega_1(Q)$ is of maximal class. As $\Omega_1(Q)$ is a unique subgroup of Q containing R as a subgroup of index p, it follows from Exercise 10.10 that the subgroup Q is of maximal class. This is a contradiction since Q/R is cyclic of order $p^{e-1} \geq p^2$. Thus, $\Omega_1(Q) = R$ so that Q is an L_p-subgroup, as desired. □

If G is as in Theorem A.99.3 and C is a cyclic subgroup of G of order $p^e = \exp(G)$, it follows from the proof of the theorem that there exists in G an L_p-subgroup containing C.

Corollary A.99.4. *Let G be as in Theorem A.99.3. Then there is in G a regular subgroup T of order p^{p+1} such that $\Omega_1(T)$ is of order p^p.*

Proof. By Theorem A.99.3, there is in G an L_p-subgroup H of order p^{p+e-1}. Set $R = \Omega_1(H)$ and let $T/R < H/R$ be of order p. Then T is regular since all proper subgroups of H are regular, by Lemma A.99.1. □

Exercise 1. Let G be an L_p-subgroup. Present a proof independent of Lemma A.99.1 that if $T/\Omega_1(G) < G/\Omega_1(G)$ is of index p, then T is regular. (**Hint.** Let $D < \Omega_1(G)$ be G-invariant of index p^2. Consider $\mathrm{C}_G(\Omega_1(G)/D)$.)

Exercise 2. Is it true that if G is as in Theorem A.99.3, then there exists in G an L_p-subgroup H of exponent p^e such that $\Omega_1(H) \triangleleft G$? (I think that the answer is negative.)

Appendix 100 Nonabelian p-groups with minimal number of conjugate classes of maximal abelian subgroups

Any nonabelian p-group G is covered by its maximal abelian subgroups. By Lemma 116.1 (a), if a p-group G is covered by n proper subgroups, then $n \geq p + 1$. The set $\mathcal{A}(G)$ of maximal abelian subgroups of G is G-invariant. That set is a join of $\mathrm{ka}(G)$ conjugate classes of subgroups. For example, if G is a 2-group of maximal class, then $\mathrm{ka}(G) = 3$. Moreover, if a p-group G of maximal class contains an abelian subgroup of index p, then $\mathrm{ka}(G) = p + 1$. It will be shown (see Lemma A.100.1) that if G is nonabelian, then $\mathrm{ka}(G) \geq p+1$. Therefore, is is natural to classify the nonabelian p-groups with exactly $p+1$ conjugate classes of maximal abelian subgroups. Such p-groups will be considered in Theorem A.100.2. It appears that the condition $\mathrm{ka}(G) = p + 1$ is fairly restrictive.

We begin with the following easy lemma.

Lemma A.100.1. *A nonabelian p-group G contains $p+1$ pairwise nonconjugate maximal abelian subgroups or, what is the same,* $\mathrm{ka}(G) \geq p + 1$.

Proof. Let $A_1, \dots, A_s$ be all pairwise nonconjugate maximal abelian subgroups of G (then $s = \mathrm{ka}(G)$) and let $\mathcal{C}(A_i) = \bigcup_{x \in G} A_i^x$ be the union of all G-conjugates with A_i, $i = 1, \dots, s$. As all maximal abelian subgroups cover G, we get $G = \bigcup_{i=1}^s \mathcal{C}(A_i)$. Let A_i^G be the normal closure of A_i in G for all i. Obviously, $\mathcal{C}(A_i) = \bigcup_{x \in G} A_i^x \subseteq A_i^G$ so that $G = \bigcup_{i=1}^s A_i^G$. One has $A_i^G < G$ for all i. By Lemma 116.3 (a), $s \geq p + 1$, as required. □

Theorem A.100.2. *Suppose that a nonabelian p-group G has exactly $p + 1$ conjugate classes of maximal abelian subgroups, i.e.,* $\mathrm{ka}(G) = p + 1$. *and let $A_1, \dots, A_{p+1}$ be representatives of these classes. Write $A_i^G = M_i$ for $i = 1, \dots, s$. Then*

(a) $D = \bigcap_{i-1}^{p+1} M_i$ *has index* p^2 *in G.*

(b) $M_i = A_i^G \in \Gamma_1, i = 1, \dots, p + 1$.

(c) *There is in the set Γ_1 an abelian member which we denote by A.*

(d) *If $B \in \mathcal{A}(G) - \{A\}$, then $|B| = p|\mathrm{Z}(G)|$, so all members of the set $\mathcal{A}(G) - \{A\}$ have the same order $p|\mathrm{Z}(G)|$.*

(e) *Write $\bar{G} = G/\mathrm{Z}(G)$. Then $\bar{G} = \bigcup_{B \in \mathcal{A}(G)} \bar{B}$ is a partition.*

Proof. (a) Follows from the equality $\bigcup_{i=1}^{p+1} A_i^G = G$ and Lemma 116.3.

(b) Follows from (a) and Lemma 116.3 (b).

(c) There is in G a normal maximal abelian subgroup, say A_1. Then $M_1 = A_1^G = A_1$, so that $A_1 = A = M_1 \in \Gamma_1$ is abelian.

(d) Let $B \in \mathcal{A}(G) - \{A\}$. Clearly, $AB = G$ so that $A \cap B = \mathrm{Z}(G)$ and $|B : \mathrm{Z}(G)| = p$, by the product formula, proving (d).

(e) Write $\bar{G} = G/Z(G)$. It follows from $G = \bigcup_{B \in \mathcal{A}(G)} B$ that $\bar{G} = \bigcup_{B \in \mathcal{A}(G)} \bar{B}$, and this is a partition, by (d). Note that $|\bar{B}| = p$ for all $B \in \mathcal{A}(G) - \{A\}$. □

Problem 1. Classify the nonabelian p-groups G with $\mathrm{ka}(G) = p + 2$. (See § 116.)

Problem 2. Study the nonabelian p-groups satisfying $\mathrm{ka}(G) \leq 2p$. (See § 116.)

Problem 3. Let G be a p-group containing a nonabelian maximal subgroup H. Assume that H has exactly one conjugate class of maximal abelian subgroups of G. Study the structure of H.

Problem 4. Suppose that a p-group G is an $\mathcal{A}_n$-group, $n > 1$, with minimal possible $\mathrm{ka}(G)$ for all $\mathcal{A}_n$-groups. Study the structure of G.

Problem 5. Study the irregular p-groups having exactly $p+1$ conjugate classes of maximal regular subgroups.

Appendix 101 Finite p-groups saturated by isolated subgroups

This appendix is written by the first author after acquaintance with § 251.

1°. A subgroup H of a p-group G is said to be *isolated* if for a cyclic $C < G$, one has $H \cap C > \{1\} \Rightarrow C \leq H$. If $H < M \leq G$ and H is an isolated subgroup of a p-group G with $\exp(H) > p$ and $|M : H| = p$, then all elements of the set $M - H$ have order p so that H is an H_p-subgroup of M. If $H < G$ has index p and isolated in a p-group G, then all maximal cyclic subgroups of H are maximal cyclic in G. Of course, G and $\{1\}$ are isolated. Next, the Hughes subgroup $\mathrm{H}_p(G)$ is isolated in a p-group G. If $C \cong \mathrm{C}_{2^n}$, $n > 3$, then C is not isolated in its holomorph (see Exercise 1, below); in fact, the above assertion also holds for $n = 3$. All subgroups of order p in a p-group $G > \{1\}$ are isolated $\iff$ $\exp(G) = p$.

Exercise 1. If a p-group G has a proper isolated cyclic subgroup L of order $> p$, then $p = 2$, G is dihedral and $L \in \Gamma_1$.

Solution. Let $L < M \leq G$, where $|M : L| = p$. Then L is a unique cyclic subgroup of index p in M so $p = 2$ and M is of maximal class (Theorem 1.2). By Exercise 10.10, G is of maximal class. As $L < G$ is maximal cyclic of order > 2, it follows that $L \in \Gamma_1$ and G is dihedral.

Exercise 2. If and only if all proper subgroups of order p^2 in a p-group G are isolated when $\exp(G) = p$.

Hint. Assuming that $\exp(G) > p$, use Exercise 1. Obviously, all dihedral 2-groups do not satisfy the hypothesis.

Exercise 3. Suppose that a p-group G possesses an isolated metacyclic subgroup L of order p^3, $p > 2$. Then $p = 3$ and G is a 3-group of maximal class and order 3^4 such that all members of the set $\Gamma_1 - \{L\}$ have exponent 3.

Solution. By Exercise 1, L is noncyclic so $\exp(L) = p^2$. Let $L < M \leq G$, where $|M : L| = p$. Then $\Omega_1(M) = M$ so M is irregular (Theorem 7.2 (b)). By Theorem 7.1 (b), $p = 3$ and M is of maximal class. Thus, any subgroup of G, containing L as a subgroup of index 3, is of maximal class; therefore, G is of maximal class (Exercise 10.10). Assume that $|G| > 3^4$. Then G_1, the fundamental subgroup of G, is metacyclic. Consideration of $G_1 \cap L$ shows that $L < G_1$. Taking $M \leq G_1$, we get a contradiction since G_1 is not of maximal class (Theorems 9.6). Thus, $|G| = 3^4$. Now the last assertion is obvious since $L = \mathrm{H}_3(G)$.

Exercise 4. If $M \triangleleft G$ is isolated and $\exp(G/M) = p$, then $\mathrm{H}_p(G) \leq M$. (**Hint.** If $x \in G - M$ has order $> p$, then $\langle x \rangle \cap M > \{1\}$, a contradiction.)

Exercise 5. All maximal subgroups of a p-group G are isolated $\iff$ $\Phi(G)$ is isolated.

Solution. $\Rightarrow$: If $M \in \Gamma_1$, then $\mathrm{H}_p(G) \leq M$. It follows that $\mathrm{H}_p(G) \leq \Phi(G)$ so that all elements of the set $G - \Phi(G)$ have order p; then $\Phi(G)$ is isolated. The reverse implication follows from Exercise 4.

Exercise 6. Let G be a 3-group of maximal class and order $> 3^3$. Then there is in G a nonisolated $\mathcal{A}_1$-subgroup.

Solution. Assume that $|G| = 3^4$. Then it contains a subgroup $C \cong \mathrm{C}_9$. If a nonabelian $M < G$ does not contain C, then $C \cap M > \{1\}$ so the minimal nonabelian M is not isolated. Thus, $C = \Phi(G)$ since the set Γ_1 has three nonabelian members, a contradiction since $\Phi(G)$ must be noncyclic (Lemmas 1.4 and 9.1(b)). If $|G| > 3^4$, it contains a subgroup R of maximal class and order 3^4. By the above, R contains a nonisolated minimal nonabelian subgroup.

Exercise 7. If a subgroup H is isolated in a p-group G and $L < G$, then $H \cap L$ isolated in L.

Solution. Indeed, let $C \leq L$ be cyclic and $C \cap (H \cap L) > \{1\}$. Then $C \cap H > \{1\}$ so that $C \leq H$. It follows that $C \leq L \cap H$ so that $L \cap H$ is isolated in L.

Exercise 8. Let G be a minimal nonabelian group of exponent $p^m > p$, $p > 2$. Then $\Omega^*_m(G) = G$. (**Solution.** As G is regular, $\Omega_{m-1}(G) < G$. Therefore, $\Omega^*_m(G) = \langle G - \Omega_{m-1}(G)\rangle = G$.)

2°. In this subsection, we study the nonabelian p-groups all of whose minimal nonabelian subgroups are isolated (for detailed description of such groups, see § 251).

The following proposition was proved by Mann for $p > 2$ (see his comment to Problem 115); for $p = 2$ this is known.

Proposition A.101.1. *Suppose that G is a nonabelian p-group of exponent $> p$. Then the following conditions are equivalent:*

(a) *All nonabelian subgroups of G are generated by elements of order p. In particular, if $M < G$ is minimal nonabelian, then $M \in \{\mathrm{D}_8, S(p^3)\}$.*
(b) *There is in the set Γ_1 an abelian member A and all $\mathcal{A}_1$-subgroups of G are $\cong \mathrm{D}_8$ or $S(p^3)$. In that case, $A = \mathrm{H}_p(G)$.*

Proof. Lemma holds for $p = 2$ (see Theorems 10.32 and 10.33). Next, we assume that $p > 2$.

(a) $\Rightarrow$ (b): If $A \leq G$ is minimal nonabelian, then $A \cong S(p^3)$ since $\Omega_1(A) = A$ has order p^3 (Lemma 65.1). We proceed by induction on $|G|$. Let $M \in \Gamma_1$ be of exponent $> p$. If M is abelian, we are done. Now let M be nonabelian. In that case, by induction, there is in M an abelian subgroup $A = \mathrm{H}_p(M)$ of index p; then $|G : A| = p^2$. Assume that A is not normal in G. In that case, $A^x \neq A$ for $x \in G - M$. But $A^x < M$ so one may assume that $A \cap A^x = \mathrm{Z}(M)$ and hence $\mathrm{cl}(M) = 2$ and M is regular. Then M is generated by $\mathcal{A}_1$-subgroups which are of exponent p so that $\exp(M) = p$ (Theorem 7.2 (b)), a contradiction. Thus, $A \triangleleft G$ and $|G/A| = p^2$ so that G is metabelian. By [HogK], $|G : \mathrm{H}_p(G)| = p$ so that $M = \mathrm{H}_p(G)$. Next, we use Mann's commentary to Problem 115. Let $x \in M$ be of order $> p$ and $y \in \mathrm{Z}_2(M) - \mathrm{Z}(M)$. Then $\mathrm{cl}(\langle x, y\rangle) \le 2$. If $\langle x, y\rangle$ is nonabelian, it is generated by elements of order p (Theorem 10.28) so has exponent p (Theorem 7.2 (b)), a contradiction. Therefore, x centralizes $\mathrm{Z}_2(M)$. Thus, all elements of M of order $> p$ generate a proper subgroup $H = \mathrm{H}_p(M)$ of M since $\mathrm{Z}_2(M) > \mathrm{Z}(M)$. As we have proved above, $M = \mathrm{H}_p(M)(= H)$, and this is a contradiction. Thus, M must be abelian, proving (b).

(b) $\Rightarrow$ (a): Assume that $A = \mathrm{H}_p(G)$ is abelian. If $K \le G$ is nonabelian, then $K = \langle K - A\rangle$ and all elements of the set $K - A$ have order p so that $\Omega_1(K) = K$, proving (a). □

If $\mathrm{H}_p(G) < G$ is abelian, then $|G : \mathrm{H}_p(G)| = p$. Assume that this is false. Let $\mathrm{H}_p(G) < F \le G$, where $|F : \mathrm{H}_p(G)| = p^2$. As F is metabelian, we get $|F : \mathrm{H}_p(F)| = p$ [HogK] so that $|\mathrm{H}_p(G)| < |\mathrm{H}_p(F)|$, a contradiction.

3°. In what follows, a p-group G of exponent $> p$ is neither abelian nor minimal nonabelian and all $\mathcal{A}_1$-subgroups of G are isolated.

Take a minimal nonabelian subgroup $H < G$ and let $H < M \le G$, where $|M : H| = p$. As all elements of the set $M - H$ have order p, then, by Burnside, $p > 2$ since H is nonabelian. Assume, in addition, that M is a regular subgroup. It follows from $\langle M - H\rangle = M$ that $\exp(M) = p$ (Theorem 7.2 (b)), and then $H \cong S(p^3)$ (Lemma 65.1). Thus, if G is regular, then all minimal nonabelian subgroups of G are $\cong S(p^3)$. By Theorem 10.28, G is generated by minimal nonabelian subgroups so $\exp(G) = p$ (Theorem 7.2 (b)), contrary to the assumption. Therefore, if a regular p-group G of exponent $> p$ and $H < G$ is minimal nonabelian of exponent $> p$, then H is not isolated in G. It follows that our group G is irregular.

Let $H < G$ be an $\mathcal{A}_1$-subgroup of exponent $> p$. Let $H < M \le G$, where $|M : H| = p$. By the previous paragraph and Exercise 8, $H = \mathrm{H}_p(M)$ so that M is irregular (Theorem 7.2 (b)). Let $F \ne H$ be minimal nonabelian in M. Since $F = \langle F - H\rangle$, F is regular and all elements of the set $F - H$ have order p, we get $\exp(F) = p$ (Theorem 7.2 (b)) so that $F \cong S(p^3)$ (Lemma 6.5.1). Thus, all $\ne H$ minimal nonabelian subgroups of M are isomorphic with $S(p^3)$, and we conclude that H is characteristic in M. Considering the intersection $H \cap F \cong \mathrm{E}_{p^2}$ which is isolated in H (Exercise 7), we see that $H = \mathrm{M}_p(m, 1, 1)$ is nonmetacyclic, $m > 1$.

Assume that $p > 3$. As $|H/\mho_1(H)| = p^3 < p^{p-1}$ (Lemma 65.1), then H is absolutely regular. Then M, which is not of maximal class (Theorems 9.5 and 9.6), is equal to $H\Omega_1(M)$, where $\Omega_1(M)$ is of order p^p and exponent p (Theorem 12.1 (b)). In that case, $|\Omega_1(H)| = |H \cap \Omega_1(M)| = p^{p-1} > p^3$, contrary to Lemma 65.1. Thus, $p = 3$. In that case, by Exercise 6, M is not of maximal class.

Thus, G is an irregular 3-group; then $\exp(G) > 3$ (Theorem 7.1 (b)). As above, let $H < G$ be minimal nonabelian of exponent > 3; then $|M| = 3|H|$ (here $H < M$ is of index 3). Then, as we know, M is irregular. If $|H| = 3^3$, then M is of maximal class; then G is also of maximal class (Exercise 10.10). By Exercise 6, $M = G$ is of order 3^4; by Theorem 9.5, $\exp(H) = 3^2$. In that case, as it is easy to see, G has a nonisolated minimal nonabelian subgroup. Thus, if $\exp(H) > 3$, then $|H| > 3^3$.

Next, we assume that the set Γ_1 has no abelian member. We sum up the obtained results. The subgroup $M > H$ of order $3|H| \geq 3^5$ is irregular. Let $F < M$ be minimal nonabelian, $F \neq H$; then $F \cong S(3^3)$. Let $F \cap H < T < H$, where $|T : (F \cap H)| = 3$. As $F \cap H$ is isolated in the abelian group T of order 3^3 (Exercise 7), it follows that $T \cong \mathrm{E}_{3^3}$ so that H is nonmetacyclic, and we conclude that $T = \Omega_1(H)$ (Lemma 65.1) is determined uniquely. Let $F < S < M$, where $|S : F| = 3$; then $\exp(S) = 3$ since $\exp(F) = 3$ and all elements of the set $S - F$ have order 3: F is isolated in S. It follows from the structure of H that $H \cap S = \Omega_1(H)(\cong \mathrm{E}_{3^3})$. As $\exp(S) = 3$ and $|S| = 3^{3+1}$ then, by Theorem 9.5. S is not of maximal class, so that $S = F \times Z$, where $Z \cong \mathrm{C}_3$ (Proposition 10.17).

Let $M < U < G$, where $|U : M| = 3$. By [SS] (see also § 256), $H < \mathrm{H}_3(U)$. It follows that there is $x \in U - H$ of order 3^2. As H is characteristic in M, it follows that $H \lhd U$.

Let $A < H$ be M-invariant of index 3 in H; then $|M/A| = 3^2$. Assume that $M/A \cong \mathrm{C}_{3^2}$. Then $M = AV$, where V is cyclic of order > 3. As, by the product formula, $H \cap V > \{1\}$, we get $V < H$, a contradiction. Thus, $M/A \cong \mathrm{E}_{3^2}$. Let $T/A \neq H/A$ be a subgroup of order 3 in M/A. Assume that $\exp(T) > 3$. All cyclic subgroups of T of order > 3 are contained in $H = \mathrm{H}_3(M)$ so that $\mathrm{H}_3(T) < T$, and we conclude that T is irregular (Theorem 7.2 (b)).

Exercise 9. Let G be a p-groups whose H_p-subgroup is proper minimal nonabelian. Then $|G : \mathrm{H}_p(G)| = p$.

Solution. In view of Exercise 1 one may assume that A is noncyclic; then $p > 2$. Let $A < B \leq G$, where $|B : A| = p$. As all elements of the set $B - A$ have order p, we get $\Omega_1(B) = B$ so B (of exponent $> p$) is irregular (Theorem 7.2 (b)). It follows that $|\Omega_1(A)| = p^{p-1}$ (see Theorem 12.1 (b)) and $|A| \geq p^p$ since $\exp(A) > p$). By Theorem 12.1 (b), B is of maximal class. Then, by Exercise 10.10, G is also of maximal class. As the fundamental subgroup G_1 of G is absolutely regular, A is not a proper subgroup of G_1 (otherwise, one can take B in G_1). Therefore, if $A = G_1$ then $A = \mathrm{H}_p(G)$. Now assume that $A \nleq G_1$ and $|G_1| > p^p$. Then $A \cap G_1$ is a proper isolated subgroup of G_1. By the product formula, $A \cap G_1 \geq \Omega_1(G_1)$, which is impossible (then all cyclic subgroups of G_1 are contained in $A \cap G_1$). Thus, $|G_1| = p^p$. In that case, $A = \mathrm{H}_p(G) \in \Gamma_1$ since $\exp(A) > p$ and all elements of the set $G - A_1$ have order p. In that case, $\exp(G_1) = p$.

Exercise 10. If a minimal nonabelian p-group G contains a nontrivial isolated subgroup, say C, then one of the following holds:

(a) G is nonmetacyclic. Next, let G be metacyclic.

(b) $G \cong \mathrm{D}_8$, $C \cong \mathrm{C}_4$,

(c) $G \cong M_{p^n}$, $|C| = p$,
(d) G is metacyclic of order 2^4 and exponent 4, $|C| = 2$.

Hint. One may assume that $|G| > p^3$. If G is nonmetacyclic, then G' is isolated. Now let $G \ncong Q_8$ be metacyclic. Let a cyclic $C > \{1\}$ be isolated in G. If G has a cyclic subgroup of index p, then $|C| = p$. If G has a subgroup of order p^4 and exponent p^2, it coincides with that subgroup; then $p = 2 = |C|$.

Exercise 11. Classify the irregular p-groups of maximal class containing a proper isolated minimal nonabelian subgroup of exponent $> p$.

Exercise 12. Study the p-groups containing an isolated proper abelian subgroup of type (p^n, p) (subgroup $\cong M_{p^{n+1}}$).

Exercise 13. If a nonabelian metacyclic p-group G contains a nontrivial isolated subgroup C, then one of the following holds:
(a) G is minimal nonabelian (see Exercise 10).
(b) $G \in \{D_{2^n}, SD_{2^n}\}$.
(c) G is a group of Lemma 42.1 (c), C is nonnormal of order 2.

Hint. If G has a cyclic subgroup of index p, it is not isomorphic with Q_{2^n}. Next, we assume that G has no cyclic subgroup of index p. Set $R = \Omega_1(G)$; then $R \cong E_{p^2}$ (Lemma 1.4). Let $C < G$ be nontrivial isolated. Clearly, $R \nleq C$ so that C is cyclic. Let $C < B \leq G$, where $|B : C| = p$. Then $\Omega_1(B) = B$. Assume that $|C| > p$. Then B is a 2-group of maximal class so is G (Proposition 10.17), contrary to the assumption. Thus, $|C| = p$. Suppose that $H/R \leq G/R$ is abelian of type (p, p). Then H is of exponent p^2 and $|C| = p$; in that case $p = 2$ and H is nonabelian. If H/R does not exist, then G/R is either cyclic or generalized quaternion. In the first case, $G \cong M_{p^{n+1}}$. In the second case, G is a 2-group from Lemma 42.1 (c), and then $|C| = p$.

Exercise 14. Suppose that a nonabelian p-group G contains an abelian subgroup of index p. Study the structure of G provided all its minimal nonabelian subgroups are isolated.

Exercise 15. Let G be a nonabelian group of order p^{p+1} and let $A \in \Gamma_1$ be isolated absolutely regular. Check that $B \in \Gamma_1 - \{A\}$.

Exercise 16. Let $A < G$ be an isolated minimal nonabelian subgroup of a p-group G. If $\exp(A) > p$, then $p = 3$.

Solution. Bearing in mind our aim, one may assume that $|G : A| = p$. As we know, G is irregular. By Lemma 65.1 and Theorem 12.1 (b), $p^{p-1} \leq \Omega_1(A) \leq p^3$, and this implies that $p = 3$ since $p > 2$.

Exercise 17. A nonabelian p-group of order p^4 has only one abelian subgroup of index $p \iff$ it is of maximal class.

Solution. If G has two distinct abelian subgroups of index p, it is not of maximal class. Now let G has only one abelian subgroup of index p. Then $|Z(G)| = p$. In that case $\bar{G} = G/Z(G) \ncong E_{p^3}$ (otherwise, since $\bar{G}$ is not extraspecial, $|Z(G)| = p^2$). If $\bar{G}$ is abelian of type (p^2, p), then $|Z(G)| = p^2$ and G has two distinct abelian subgroups of index p, a contradiction. Thus, $G/Z(G)$ is nonabelian. In that case, G is of maximal class.

Exercise 18. If the Hughes subgroup $H_p(G)$ of a nonabelian p-group is a proper abelian subgroup, then $|G : H_p(G)| = p$. (**Hint.** Use [HogK].)

4^0. In this subsection we use some results obtained in Subsection 2^0. In particular, we have proved there that if G has no abelian subgroup of index p, then $p = 3$ and G is an irregular 3-group.

Below a 3-group G is an $\mathcal{A}_n$-group, $n > 1$, of exponent > 3, all of whose minimal nonabelian subgroups are isolated; in particular, $|G| \geq 3^{n+3} \geq 3^5$. We claim that there is in the set Γ_1 an abelian member and all minimal nonabelian subgroups of G are $\cong S(3^3)$. (See also § 251 where the same main results are obtained, but our proof is entirely different.)

Lemma A.101.2. *Let a p-group G contain an isolated minimal nonabelian subgroup L of exponent $> p$ and index p, let $S \neq L$ be a minimal nonabelian subgroup of G. Then $p = 3$, $S \cong S(3^3)$ and G contains a nonisolated subgroup $\cong S(3^3)$.*

Proof. Assume, by way of contradiction, that all minimal nonabelian subgroups of G are isolated. All elements of the set $G - L$ have order p so that $L = H_p(G)$, $G = \Omega_1(G)$, and Theorem 7.2 (b) implies that G is irregular. As $|\Omega_1(L)| \leq p^3$, we get $p = 3$. Exercise 6 implies that G is not of maximal class. Since $\Omega_1(G) = G$, it follows that L is nonmetacyclic so that $L \cong M_3(m, .n.1)$, $m > n$ hence $\exp(L) = 3^m > 3$. Let $S \neq L$ be a minimal nonabelian subgroup of G. As $\langle S - L\rangle = S$, we get $\Omega_1(S) = S$ so that $S \cong S(3^3)$. As $LS = G$, we get $L \cap S \cong E_{p^2}$, and this subgroup is isolated in L (Exercise 7).

Assume that $n > 1$. Then $\mho_1(L) (\lhd G)$ is abelian of rank 2. As $\Omega_1(L) \cong E_{3^3}$, it follows, by the product formula, that there is an element $x \in \mho_1(L) \cap (S \cap L)$ of order 3. By Theorem 7.2 (c), there is in L a cyclic subgroup Y of order p^2 containing x. Then $Y \cap S > \{1\}$ and $Y \nleq S$, i.e., S is not isolated in G. Thus, $n = 1$; then $L \cong M_3(m, 1, 1)$, $m > 1$. The subgroup $\mho_1(L) \lhd G$ is cyclic of order 3^{m-1}.

Assume that $m > 2$. Let $x \in G - L$ and write $T = \langle x\rangle \cdot \mho_1(L)$. Since $\exp(C_L(x)) = 3$, then x does not centralizes the cyclic subgroup $\mho_1(L)$ of order 3^{m-1}, and we conclude that $T \cong M_{p|\mho_1(L)|}$ is minimal nonabelian. All elements of the set $T-(T\cap L) = T-\mho_1(L) \subset G - L$ have order 3 so that $\Omega_1(T) = T$, and we conclude that $3^{m-1} = \exp(T) = 3$, which is a contradiction. Thus, $m = 2$. The subgroup $\mho_1(L)$ is G-invariant of order 3.

Let $C < L$ be cyclic of order 3^2. Then $|\mho_1(C)| = |\mho_1(L)| = 3 \Rightarrow |G| = 3^4|\mho_1(L)| = 3^5$. The quotient group $\bar{G} = G/\mho_1(C)$ is of exponent 3 since $\bar{L} \cong S(3^3)$ is of exponent 3 and all elements in the set $G-L$ have order 3, and we conclude that $\mho_1(L) = \mho_1(C) = \mho_1(G)$ is of order 3 and $L \cong M_3(2, 1, 1)$. In that case, there is in $\bar{G}$ a subgroup $\bar{E} \cong E_{3^3}$. One has $\bar{E} \neq \bar{L}(\cong S(3^3))$ since these groups are nonisomorphic so that $E \neq L$. Assume that $\exp(E) = 3^2$. Being regular, E is generated by elements of order 3^2 (Theorems 7.1 (c) and 7.2 (b)). It follows from $|E| = |L|$ and $L = H_3(G)$ that $E = L$, contrary to what we have said above. Thus, $\exp(E) = 3$. Assume that E is nonabelian. Then it contains a minimal nonabelian subgroup $R \cong S(3^3)$. It follows from $E' = R' = \mho_1(G)$ that $R \cap C = \mho_1(G) > \{1\}$, and we conclude that R is not isolated in G, a contradiction. Thus, E is abelian so that $E \cong E_{3^4}$. If $\bar{E}_1 < \bar{G}$ is another subgroup $\cong E_{3^3}$, then $E_1 \cong E_{3^4}$, by the above. Then $E \cap E_1 = Z(G) \cong E_{3^3}$ and $cl(G) = 2$ so G is regular, a contradiction. Thus, $\bar{E}$ is the unique abelian subgroup of index 3 in $\bar{G}$. Then, by Exercise 17, $\bar{G}$ is of maximal class and order 3^{3+1}. By Theorem 9.5, $\exp(\bar{G}) = 3^2$, contrary to what that have said above. Thus, G has a nonisolated minimal nonabelian subgroup. □

Now we are ready to prove the main result of this appendix.

Theorem A.101.3 (= Theorem 251.1 (Janko)). *Suppose that all minimal nonabelian subgroups of an $\mathcal{A}_n$-group G, $n > 1$, of exponent $p^e > p$ are isolated. Then G has an abelian subgroup of index p and all minimal nonabelian subgroups of G are $\cong S(p^3)$.*

Proof. We proceed by induction on $|G|$. Next, we assume that there is no abelian member in the set Γ_1.

If $K \in \Gamma_1$ is not minimal nonabelian, then all (proper) minimal nonabelian subgroups of K are isolated so that, by induction, K has an abelian subgroup of index p and all minimal nonabelian subgroups of K are $\cong S(p^3)$. If $K \in \Gamma_1$ is minimal nonabelian, it has an abelian subgroup of index p as well. Now assume that the set Γ_1 has no minimal nonabelian members; then all minimal nonabelian subgroups of G are $\cong S(p^3)$, by induction. In that case, by Lemma A.101.1, G has an abelian subgroup of index p. Therefore, in what follows we assume that there is a minimal nonabelian $L \in \Gamma_1$; then, since $|L| = \frac{1}{p}|G| \geq p^4$, then $\exp(L) = p^m > p$, by Lemma 65.1. By Lemma A.101.2, $p = 3$. Since $L < G$ is isolated of index 3, all elements of the set $G - L$ have order 3 so that $L = H_3(G)$. By Exercise 8, $\Omega_m^*(L) = L$. Now the theorem follows from Lemma A.101.2. Let $A \in \Gamma_1$ be abelian.

It remains to prove that all minimal nonabelian subgroups of G are $\cong S(p^3)$. We proceed by induction on $|G|$. Let $M < G$ be minimal nonabelian of exponent $> p$. Let $M < F \leq G$, where $|F : M| = p$. Then $H_p(F) = M$ (see Exercise 8) and F is irregular since $\Omega_1(F) = F$ (Theorem 7.2 (b)). By induction, $F = G$ (indeed, not all minimal nonabelian subgroups of F have order p^3) so that $M \in \Gamma_1$. By Exercise 7, $A \cap M$ is isolated in A, and $|A : (A \cap M)| = p$, by the product formula. It follows that $\Omega_1(A) = A$ so that $\exp(A) = p$ since A is abelian. By Lemma 65.1, $|A \cap M| = p^3$ so that $|G| = p^5$. Let $H \neq L$ be minimal nonabelian subgroup of G. As above, $H \cong S(3^3)$. One has $MH = G$ and $M \cap H \cong E_{3^2}$ is isolated in M (Exercise 7). One has $M \cong M_3(2, 1, 1)$ (see the proof of Lemma A.101.2). Let $C = \mho_1(M)$. Then, as above, G/C is of order 3^4 and exponent 3. One has $A/C \cong E_{3^3}$. Assume that $B/C \neq A/C$ is an abelian subgroup of index 3 in G/C. As $cl(G) > 2$, the subgroup B is nonabelian, and regular (Theorem 7.1 (c)). Let S be minimal nonabelian subgroup of B. As $M = \mathrm{H}_3(G)$, one has $\exp(B) = 3$ so that $S \cong S(3^3)$. If $C \nleq S$, then $\mathrm{E}_{3^3} \cong B/C = SC/C \cong S \cong S(3^3)$, a contradiction. Thus $(\mho_1(M) =)C < S$ so that $M \cap S$ is not isolated in M, contrary to Exercise 7 (indeed, $C < C_1 < M$. where $C_1 \cong \mathrm{C}_9$). Thus, M does not exist so that all minimal nonabelian subgroups are isomorphic with $S(p^3)$. □

Solution. Set $H = \mathrm{H}_p(G)$; then $\exp(H) > p$. Assume that $|H| = p^3$; then $H \cong \mathrm{M}_{p^3}$. As above, G is of maximal class. Assume that $|G : H| > p$. Then $H \nleq G_1$. In that case, all elements of the set $G_1 - (G_1 \cap H)$ have order p, a contradiction. Thus, $|H| > p^3$. Then $\exp(H/H') > p$ and H/H' is the abelian H_p-subgroup of G/H'. By Exercise 18, $|G : H| = |(G/H') : (H/H')| = p$. (Moreover, if $H = \mathrm{H}_p(G) < G$ is nonabelian and $\exp(H/H') > p$, then $|G : H| = p$.)

5⁰ Problems.

Problem 1. Study the irregular 3-groups $G = FL$ of order $> 3^4$, where $F \cap L = \{1\}$, $F \cong S(3^3)$, L is cyclic.

Problem 2. Study the primary $\mathcal{A}_n$-groups, $n > 2$, all of whose $\mathcal{A}_2$-subgroups are isolated.

Problem 3. Classify the nonmetacyclic p-groups all of whose minimal nonmetacyclic subgroups are isolated.

Problem 4. Study the irregular p-groups all of whose maximal subgroups of class 2 are isolated.

Problem 5. Study the nonmetacyclic p-groups all of whose maximal metacyclic subgroups are isolated. (This is solved by the second author.)

Problem 6. Classify the nonabelian p-groups all of whose maximal abelian subgroups are isolated. (This is solved by the second author; see § 252.)

Problem 7. Classify the $\mathcal{A}_n$-groups that are p-groups and contain exactly one minimal nonabelian subgroup of order $> p^3$.

Problem 8. Does there exist an irregular p-group, $p > 2$, all of whose minimal irregular subgroups are isolated?

Problem 9. A subgroup H of a p-group G is said to be *quasi-isolated* if, for any cyclic $C \leq G$, $C \cap H > \{1\} \Rightarrow \Omega_1(C) \leq H$. Study the $\mathcal{A}_n$-groups, $n > 1$, all of whose minimal nonabelian subgroups (maximal abelian subgroups) are quasi-isolated.

Appendix 102 A characterization of minimal nonabelian p-groups

There are many challenging problems concerning the p-groups with an abelian subgroup of index p (one of such problems is a classification of their representation groups).

A nonabelian p-group G is minimal nonabelian $\iff$ $d(G) = 2$ and $|G : Z(G)| = p^2$. What will be if $d(G) = 2 = cl(G)$ and G has an abelian subgroup of index p? As Theorem A.102.1 shows, in that case G is also minimal nonabelian.

Theorem A.102.1. *If a two-generator p-group G of class 2 possesses an abelian subgroup of index p, then it is minimal nonabelian.*

Proof. It suffices to prove that $G/Z(G) \cong E_{p^2}$ or, what is the same, there is in the set Γ_1 two distinct abelian members. Let $A \in \Gamma_1$ be abelian. Write $\bar{G} = G/Z(G)$. By hypothesis, $\bar{G}$ is abelian of type (p^m, p^n).

Assume that G is not minimal nonabelian; then $m+n > 2$ so one may assume that $m > 1$ (if $\bar{G} \cong E_{p^2}$, then G is minimal nonabelian as follows from the first sentence of the proof).

Assume that $n = 1$. Then $\bar{G}$, being abelian of type (p^m, p), contains two distinct cyclic subgroups $\bar{U}$ and $\bar{V}$ of index p. In that case, $U, V \in \Gamma_1$ are distinct abelian, and we conclude that $U \cap V = Z(G)$; then $G/Z(G) \cong E_{p^2}$ so that $m = 1$, contrary to the assumption. Thus, $m > 1$, $n > 1$. Any maximal subgroup of the abelian group $\bar{G}$ is either of type (p^{m-1}, p^n) or of type (p^m, p^{n-1}). To fix ideas, suppose that $\bar{A}$ is abelian of type (p^m, p^{n-1}), where $A \in \Gamma_1$ is abelian as in the first paragraph of the proof. Let $\bar{G} = \bar{U} \times \bar{V}$, where $C_{p^m} \cong \bar{U} < \bar{A}$ and $\bar{V} \cong C_{p^n}$ (see Introduction, Exercise 4 (a)). Then V is abelian, $\bar{A}\bar{V} = \bar{G}$ and so $\bar{A} \cap \bar{V} > \{\bar{1}\}$, by the product formula (moreover, that intersection has order p^{n-1}). Therefore, $Z(G) \geq A \cap V$. Since $|A \cap V| = |\bar{A} \cap \bar{V}|\,|Z(G)| > |Z(G)|$, we get a contradiction.

Thus, $m = n = 1$ so that G is a minimal nonabelian group, as required. □

Second proof of Theorem A.102.1. (Janko). Let $\langle g, h\rangle = G$. We want to prove that $o([g, h]) = p$. One may assume that $g \in G - A$; then $g^p \in Z(G)$ (indeed, $C_G(g^p) \geq \langle g, A\rangle = G$ since $|G : A| = p$). As $cl(G) = 2$, one has $[g, h]^p = [g^p, h] = 1$ so that $o([g, h]) = p$ since $[g, h] \neq 1$. Since $cl(G) = 2$, one has $[g, h] \in Z(G)$. Write $\bar{G} = G/\langle[g, h]\rangle$; then $[\bar{g}, \bar{h}] = 1$ and so $\bar{G} = \langle\bar{g}, \bar{h}\rangle$ is abelian. Now, by Lemma 65.2 (a), the group G is minimal nonabelian. □

For an essentially stronger result, see Theorem A.53.1.

Let a nonabelian two-generator p-group G possesses an abelian subgroup of index p. Then $G/K_3(G)$ is minimal nonabelian (Theorem A.102.1).

Exercise 1. Classify the metacyclic p-groups having an abelian subgroup of index p. (See Corollary A.102.3.)

Exercise 2. Suppose that a two-generator p-group G has an abelian subgroup of index p. Is it true that if $\bar{G} = G/Z(G)$ is minimal nonabelian, then $|\bar{G}| = p^3$? (**Hint.** Use Theorem A.102.1.)

Lemma A.102.2. *Suppose that a nonabelian two-generator p-group G possesses an abelian subgroup of index p and $H \in \Gamma_1$ is nonabelian. Then $Z(H) = Z(G)$.*

Proof. Assume that the lemma is false. Then, by Theorem A.102.1, $\mathrm{cl}(G) > 2$. Let $A \in \Gamma_1$ be abelian. Since G is not minimal nonabelian (Lemma 65.1), it follows that A is the unique abelian member of the set Γ_1; therefore, if $H \in \Gamma_1 - \{A\}$, then H is nonabelian. The subgroup $H \cap A$ is maximal abelian in H which implies $Z(H) < H \cap A < A$ so that $C_G(Z(H)) \geq AH = G$, and we conclude that $Z(H) \leq Z(G)$. As $d(G) = 2$, one has $Z(G) \leq \Phi(G) < H$ (in fact, $Z(G) < \Phi(G)$ since G is not minimal nonabelian). It follows that $Z(G) \leq Z(H)$ hence $Z(G) = Z(H)$, and we are done. □

Corollary A.102.3. *If a nonabelian metacyclic p-group G has an abelian subgroup of index p, then either G is minimal nonabelian or $p = 2$ and $G/Z(G)$ is of maximal class.*

Proof. Suppose that G is not minimal nonabelian; then $\mathrm{cl}(G) > 2$, by Theorem A.102.1. Let $A \in \Gamma_1$ be abelian; then A is the unique abelian member of the set Γ_1. Let $H \in \Gamma_1 - \{A\}$; then H is nonabelian. By Lemma A.102.2, $Z(H) = Z(G)$.

(i) Suppose that $\mathrm{cl}(H) = 2$. Then, H is minimal nonabelian since $A \cap H$ is abelian of index p (Theorem A.102.1). In that case, $H/Z(G) = H/Z(H) \cong E_{p^2}$ (recall that $d(H) = 2$ since it is metacyclic as a subgroup of G). Then $|G/Z(G)| = p^3$, and it is easy to see that in that case $p = 2$ and $G/Z(G) \cong D_8$ is of maximal class.

(ii) Now let $\mathrm{cl}(H) > 2$ so that $|H| \geq p^4$. Then, by induction, $H/Z(G)$ is of maximal class. By Proposition 10.19, $G/Z(G)$ is of maximal class of order $\geq p^4$. As $G/Z(G)$ is metacyclic, it follows that $p = 2$. □

Corollary A.102.4. *Suppose that a nonabelian two-generator p-group G possesses an abelian subgroup A of index p and let $H \in \Gamma_1$ be nonabelian metacyclic. Then $G/Z(G)$ is a group of maximal class.*

Proof. By Corollary A.102.3, either H is minimal nonabelian or $H/Z(H)$ is a 2-group of maximal class. By Lemma A.102.2, $Z(H) = Z(G)$.

By Theorem A.102.1 and Corollary A.102.3, either $H/Z(G)$ is $\cong E_{p^2}$ or a 2-group of maximal class. If H is minimal nonabelian, then $|G : Z(G)| = |H : Z(G)|\,|G : H| = p^3$.

(i) Let $H/Z(G) \cong E_{p^2}$. In that case, $G/Z(G)$ is a two-generator group of order p^3. If $G/Z(G)$ is abelian, it, being noncyclic and two-generator, is of type (p^2, p). In that case, G is minimal nonabelian, a contradiction since $H \in \Gamma_1$ is nonabelian. Thus, $G/Z(G)$ is nonabelian (of order p^3) so of maximal class.

(ii) Now let $H/Z(G)$ be a 2-group of maximal class. Then $G/Z(G)$ is a 2-group of maximal class, by Theorem 12.12 (a). □

Appendix 103 p-groups all of whose subgroups of order p^3 are isomorphic

Let G be a noncyclic p-group. All subgroups of G of order p are isomorphic. If all subgroups of G, $|G| > p^2$, of order p^2 are isomorphic, then either $\exp(G) = p$ or G is a generalized quaternion group. In this appendix, we consider the noncyclic p-groups of order $> p^3$ all of whose subgroups of order p^3 are isomorphic. In that case, as follows easily from Proposition 1.3, $\exp(G) \leq p^2$.

Exercise 1. Let G be a noncyclic group of order p^4. If all members of the set Γ_1 are isomorphic, then G is either metacyclic of exponent p^2 or elementary abelian.

Solution. As G has an abelian subgroup of index p, all maximal subgroups of order p^3 are abelian. If G is abelian, it is one of groups from the conclusion. Now let G be nonabelian. Then it is minimal nonabelian. By Lemma 65.1, $\exp(G) > p$. As G has no subgroup $\cong E_{p^3}$, it is metacyclic (Lemma 65.1) without cyclic subgroup of index p, and we are done.

Exercise 2. Study the nonabelian groups or order p^5 all of whose maximal subgroups are isomorphic.

Theorem A.103.1. *Suppose that G is a noncyclic p-group of order $> p^3$ all of whose subgroups of order p^3 are isomorphic. Then* $\exp(G) \leq p^2$ *and one of the following holds:*
(a) *G is elementary abelian.*
(b) *G is metacyclic of order p^4 and exponent p^2.*
(c) *G is minimal nonmetacyclic of order 2^5.*
(d) *Now let $p = 2$ and $|G| > 2^5$. Then*
   (d1) $|G| = 2^6$.
   (d2) *G is special with $|G'| = 4$, $\Omega_1(G) = G'$.*
   (d3) *All maximal subgroups of G are minimal nonmetacyclic so isomorphic.*
   (d4) *All subgroups of index 4 are metacyclic, exactly five of them are abelian.*
   (d5) *G has no normal cyclic subgroup of order 4.*

Sketch of the proof. As G has an abelian subgroup, say A, of order p^3, all subgroups of G of order p^3 are abelian. As G is noncyclic so is A (Lemma 1.4); then $\exp(G) \leq p^2$. If $\exp(A) = p$, then G is elementary abelian (Lemma 65.1). Now let $\exp(A) = p^2$; then $\exp(G) = \exp(A)$.

(i) All subgroups of G of order p^3 are abelian of type (p^2, p).

(ii) All subgroups of G of order p^4 are metacyclic of exponent p^2. This follows from Exercise 1.

(iii) All subgroups of G of order p^5 are minimal nonmetacyclic of order 2^5 so isomorphic. This follows from (ii) and Theorem 66.1.

(iv) All minimal nonabelian subgroups of G are metacyclic of order p^4 and exponent p^2. This follows from (i) and (ii) and Lemma 65.1.

(v) If $p > 2$, then G of order p^4 is as in (ii). This follows from (iii) immediately.

Therefore, it remains to consider the 2-groups. Assume that $|G| > 2^5$ (see (iii)).

(vi) $E_4 \cong \Omega_1(G) = Z(G)$. Indeed, let $H < G$ be of order 2^5 and $R \lhd G$ be abelian of type $(2, 2)$. Considering $R\Omega_1(H)$, we see that $R = \Omega_1(H) = Z(H)$. Let $H_1 \neq H$ be of order 2^5. Then again $R = Z(H_1)$. It follows easily that $R = Z(M)$ for any minimal nonabelian $M < G$. By Theorem 10.28, $R = Z(G)$.

(vii) Let $T \lhd G$ be of order 2^5. Then T contains a unique abelian subgroup B of type $(4, 4)$ and index 2 (see (iii) and Theorem 66.1) so that $B \lhd G$. Let $L < B$ be G-invariant of index 2. Write $C = C_G(L)$. Then $C = B$ (this follows from (iii)). In that case, G/B is isomorphic to a subgroup of $\mathrm{Aut}(L) \cong D_8$. It follows that $|G| = |B|\,|G/B| \leq 2^4 \cdot 2^3 = 2^7$. If $H \in \Gamma_1$, then $Z(H) = \Omega_1(H) \cong E_4$ (see (vi)).

(viii) Assume that $|G| = 2^7$; then $G/B \cong D_8$. Let $H/B < G/B$ be cyclic of order 4. As $\exp(G) = 4$, it follows that $H = \langle x\rangle \cdot B$ (semidirect product with kernel B), where $o(x) = 4$ and $\langle x\rangle \cap B = \{1\}$. Then $\langle x^2, \Omega_1(B)\rangle \leq \Omega_1(G)$ so $|\Omega_1(G)| \geq 8$, contrary to (vi). Thus, $G/B \cong E_4$ so that $|G| = 2^6$.

(ix) By (vi), all elements in the set $G - B$ have order 4. As $K = \langle x, B\rangle$ is minimal nonmetacyclic of order 2^5 for any $x \in G - B$, it follows that $C_K(x)$ is abelian of type $(4, 2)$. Indeed, assuming that this is false, one obtains that $C_K(x)$ is abelian of type $(4, 4)$. In that case, $C_K(x) \cap B = Z(K)$ is abelian of type $(4, 2)$, contrary to (vi).[1]

(x) If $H \in \Gamma_1$, then $\Phi(H) = \Omega_1(G) = Z(G)$ since H is minimal nonmetacyclic group of order 2^5 (see Theorem 66.1). Thus, all maximal subgroups of $G/Z(G)$ are elementary abelian of order 8, as $|G/Z(G)| = 2^4$, so it follows that $G/Z(G) \cong E_{16}$, hence, $d(G) = 4$. One has $Z(G) = \Phi(G) = G'$ (note that $\Phi(H) = H' = Z(H) = Z(G)$). It follows that G and all its maximal subgroups are special with the same center. (See [Cos], where p-groups all of whose maximal subgroups are special, are described.)

(xi) Assume that $L \lhd G$ is cyclic of order 4. By (vi), $|G : C_G(L)| = 2$, so $C_G(L)$ is minimal nonmetacyclic of order 2^5. This is a contradiction since the center of minimal nonmetacyclic group of order 2^5 is abelian of type $(2, 2)$. Thus, G has no normal cyclic subgroup of order 4.

(xii) Let $S < G$ be maximal abelian. Then $G' = Z(G) < S$ so that $S \lhd G$ and $C_G(S) = S$. Assume that $|S| = 8$. It follows that G/S of order 8 is isomorphic with $\mathrm{Aut}(S)$ has a Sylow 2-subgroup $\cong D_8$, a contradiction since $G/Z(G) \cong E_{2^4}$, by (x), and $Z(G) < S$. Thus, all maximal abelian subgroups of G are of type $(4, 4)$.

(xiii) If $M < G$ is minimal nonabelian, then M is metacyclic of order 2^4. Indeed, if $M \leq T < G$, where T is of order 2^5, the result follows from Theorem 66.1. □

Problem 1. Study the p-groups all of whose abelian subgroups of order p^3 are isomorphic.

Problem 2. Study the p-groups all of whose subgroups of order p^4 are isomorphic.

---

1 See § 77, where the 2-groups with self-centralizing abelian subgroups of type $(4, 2)$ are described.

Appendix 104 Alternate proof of the theorem of Janko on nonabelian p-groups all of whose maximal abelian subgroups are isolated

Recall (see §§ 251, 252 and Appendix 101) that a subgroup A of a group G is isolated if, whenever a cyclic $C \leq G$ satisfies $C \cap A > \{1\}$, then $C \leq A$. See above sections and appendix on properties of isolated subgroups of a p-group. In particular, if $A < B \leq G$, where A is isolated and $|B : A| = p$, then all elements of the set $B - A$ have order p so that $\Omega_1(B) = B$.

Exercise 1. Let C be a subgroup of order p in an abelian p-group G. If C is maximal cyclic in G, then it is a direct factor of G.

Solution. Assume that $C = \langle c \rangle \leq \Phi(G)(= \mho_1(G))$. Then there if $x \in G$ such that $x^p = c$, and this is a contradiction since C is a proper subgroup of $\langle x \rangle$. Thus, there is $B \in \Gamma_1$ not containing C. Then $G = C \times B$.

Exercise 2. Let $H > \{1\}$ be a proper isolated subgroup of an abelian p-group G. Then H of exponent p is a direct factor of G.

Solution. If $H < S \leq G$, where $|S : H| = p$, then $\Omega_1(S) = \langle S - H \rangle = S$ so that $\exp(S) = p \Rightarrow \exp(H) = p$. If $C \leq H$ is of order p, then C is isolated in G so it is a maximal cyclic subgroup of G. It follows that $C \not\leq \Phi(G)$ is a direct factor of G (Exercise 1) so that $H \cap \Phi(G) = \{1\}$. Let $A < G$ be minimal such that $AH = G$. Then

$$A \cap H \leq \Phi(A) \cap H \leq \Phi(G) \cap H = \{1\}$$

so that $G = A \times H$.

The second author has proved the following

Theorem A.104.1 (= Theorem 252.1). *Let G be a nonabelian p-group all of whose maximal abelian subgroups are isolated. Then $p > 2$ and G is of exponent p.*

Below we offer another proof of that theorem.

We need the following

Lemma A.104.2. *Let A be an isolated subgroup of a p-group G and $H < G$. Then the subgroup $A \cap H$ is isolated in H.*

Proof. One may assume that $A \cap H > \{1\}$ and $\exp(H) > p$. Let $C \leq H$ be cyclic of order $> p$. Assuming that $(A \cap H) \cap C > \{1\}$, we have to show that $C \leq A \cap H$. One has $C \cap A > \{1\}$ so that $C \leq A$ since A is isolated. In that case, $C \leq A \cap H$ so $A \cap H$ is isolated in H (we do not assert that $A \cap H$ is isolated in G). □

Proof of Theorem A.104.1. Assume that G is a counterexample of minimal order; than $\exp(G) = p^e > p$. Assume that G has a minimal nonabelian subgroup S of exponent $> p$. Then there is in S a nonisolated maximal subgroup B (this is clear if $|S| = p^3$; if $|S| > p^3$, then taking $B \geq \Omega_1(S)$, we see that B is non-isolated in S). Let $B \leq A$, where A is a maximal abelian subgroup of G; then A is isolated in G, by hypothesis and $A \cap S = B$. However, by Lemma A.104.2, $B = A \cap S$ is isolated in S, a contradiction. Thus, all minimal nonabelian subgroups of G have exponent p, and we conclude that $p > 2$. In that case, there is in G an abelian subgroup A of index p (Proposition 101.1). As A is isolated, all elements of the set $G - A$ have order p, and we conclude that $\exp(A) > p$. However, by Theorem 254.1, A (of exponent $p^e > p$) is not isolated in G, a final contradiction. □

Problem 1. Study the nonabelian p-groups of exponent $> p$ all of whose minimal nonabelian subgroups of exponent $> p$ are isolated.

Problem 2. Classify the nonmetacyclic p-groups all of whose maximal metacyclic subgroups are isolated.

Problem 3. Study the nonabsolutely regular p-groups of exponent $> p$ all of whose maximal absolutely regular subgroups are isolated.

Problem 4. Study the p-groups containing a proper isolated special subgroup.

Appendix 105 Nonabelian 2-groups generated by an element of order 4 and an involution

It is well known that a nonabelian 2-group generated by two involutions is dihedral. However the next case, where a nonabelian 2-group is generated by an element of order 4 and an involution is already hopeless. Nevertheless, we are able to prove here the following essential initial result.

Theorem A.105.1. *Let $G = \langle a, c\rangle$ be a nonabelian 2-group, where $o(a) = 4$ and $o(c) = 2$. Set $b = a^2$ so that $\langle b, c\rangle$ is either a four-group or is isomorphic to $D_{2^{n+1}}$ with $n \geq 2$.*

(a) *If $\langle b, c\rangle$ is a four-group, then we have the following two possibilities:*
   (a1) *If G is metacyclic, then $G \cong D_8$ or $G \cong SD_{2^m}$, $m \geq 4$.*
   (a2) *If G is nonmetacyclic, then G has an abelian subgroup $D = \langle d\rangle \times \langle b\rangle$ of index 2, where $o(d) \geq 4$, $o(b) = 2$ and there is an involution $c \in G - D$ such that $d^c = d^{-1}b$ and $b^c = b$.*

(b) *If $\langle b, c\rangle \cong D_{2^{n+1}}$ with $n \geq 2$, then we set $v = bc$ (of order 2^n) and $w = v^a$. In this case G is nonmetacyclic, and G is a splitting and faithful extension of $H = \langle v, w\rangle$ by $\langle a\rangle \cong C_4$, where b inverts $\langle v\rangle$ and $\langle w\rangle$, $w = v^a$, $w^a = v^{-1}$ and $|\langle v\rangle \cap \langle w\rangle| \leq 2$. Also,*

$$G' = \langle v^2, w^2\rangle\langle vw\rangle \text{ and } G/G' \cong C_4 \times C_2 .$$

If $o(ca) \leq 4$, then $H = \langle v, w\rangle$ is abelian and so in this case the structure of G is uniquely determined.

Proof. Let G be a nonabelian 2-group with $G = \langle a, c\rangle$, where $o(a) = 4$ and $o(c) = 2$. Set $b = a^2$.

(i) Suppose that $[b, c] = 1$ so that $b \in Z(G)$ and $G/\langle b\rangle$ is a four-group or dihedral. If $G/\langle b\rangle \cong E_4$, then $G \cong D_8$ and so assume that $G/\langle b\rangle$ is dihedral. Then $G/\langle b\rangle$ has a cyclic subgroup $D/\langle b\rangle$ of index 2, where $|D/\langle b\rangle| \geq 4$, $a, c \in G - D$, and c inverts $D/\langle b\rangle$. Also, D is abelian. If D is cyclic (of order ≥ 8), then $G' \geq \langle b\rangle$ and so the fact that $G/\langle b\rangle$ is dihedral gives $|G : G'| = 4$. By a result of O. Taussky, G is of maximal class and order ≥ 16. Since G is generated by an involution and an element of order 4, G is semidihedral. In this case G is metacyclic and so we have obtained the groups stated in part (a1) of our theorem.

Suppose that D is noncyclic so that D splits over $\langle b\rangle$ and we have $D = \langle b\rangle \times \langle d\rangle$ with $o(d) \geq 4$. Here $\Omega_1(\langle d\rangle) \in Z(G)$ and so $Z(G) = \Omega_1(\langle d\rangle) \times \langle b\rangle \cong E_4$. Since c inverts $D/\langle b\rangle$, we have either $d^c = d^{-1}$ or $d^c = d^{-1}b$. If $d^c = d^{-1}$, then $G = \langle d, c\rangle \times \langle b\rangle$ and $d(G) = 3$, contrary to $G = \langle a, c\rangle$. Hence, we have $d^c = d^{-1}b$. Note that in this case,

$$(dc)^2 = d(cdc) = dd^{-1}b = b \text{ and so } G = \langle dc, c\rangle \text{ with } o(dc) = 4 \text{ and } o(c) = 2 .$$

Since c centralizes $\Omega_1(\langle d\rangle)$ and b, we have $\langle \Omega_1(\langle d\rangle), b, c\rangle \cong E_8$ and so G is nonmetacyclic. We have obtained the groups from part (a2) of our theorem.

(ii) From now on we assume that $[b,c] \neq 1$ so that $v = bc$ is of order 2^n, $n \geq 2$, and $\langle b,c\rangle \cong D_{2^{n+1}}$. Set $w = v^a$ and we note that b inverts $\langle v\rangle$ and $\langle w\rangle$ (since $w = v^a$ and a centralizes b) and we get $w^a = v^{a^2} = v^b = v^{-1}$. Thus $H = \langle v,w\rangle$ is normalized by $\langle a\rangle$ implying that $G = H\langle a\rangle$ (noting that $c = bv$). We have $|\langle v\rangle \cap \langle w\rangle| \leq 2$. Indeed, if $M = \langle v\rangle \cap \langle w\rangle$ is of order ≥ 4, then $M^a = M$ and $a^2 = b$ inverts M, a contradiction. Note that an automorphism α of a cyclic 2-group Z of order ≥ 4 which inverts Z is not a square in $\mathrm{Aut}(Z)$.

If $a \in H$, then $G = H$ and G is generated by two distinct conjugates $\langle v\rangle$ and $\langle w\rangle$ in G, a contradiction. Hence, $a \notin H$.

Suppose $b \in H$. Then $|G : H| = 2$. Because b inverts $\langle v\rangle \cong C_{2^n}$ and $\langle w\rangle \cong C_{2^n}$, $n \geq 2$, we get $v^2, w^2 \in H'$ and so H/H' is generated by two distinct involutions (noting that $|\langle v\rangle \cap \langle w\rangle| \leq 2$). Thus $|H/H'| = 4$ implies (by a result of O. Taussky) that H is of maximal class and order $\geq 2^3$. But then H' is cyclic and so $\langle v^2\rangle = \langle w^2\rangle \cong C_2$ and H is generated by two distinct cyclic subgroups $\langle v\rangle$ and $\langle w\rangle$ of order 4. It follows that $H \cong Q_{2^m}$, $m \geq 3$, contrary to the fact that $b \in H$ and $b \notin Z(H)$. We have proved that $b \notin H$ and so G is a splitting and faithful extension of $H = \langle v,w\rangle$ by $\langle a\rangle \cong C_4$.

Set $K = H\langle b\rangle$. It is easy to see that G is nonmetacyclic. Note that $|H| \geq 2^3$ and so $|K| \geq 2^4$. Also, $K' \geq \langle v^2, w^2\rangle < H$. If $|H : K'| = 2$, then $|K/K'| = 4$ and so (by a result of O. Taussky) K is of maximal class. In that case K' is cyclic and we get

$$|\langle v\rangle| = |\langle w\rangle| = 4 \text{ and } K' \cap \langle v\rangle = K' \cap \langle w\rangle = \langle v\rangle \cap \langle w\rangle \cong C_2\,.$$

It follows that

$$H \cong Q_{2^m}\,, \quad m \geq 3, \text{ and } K \cong SD_{2^{m+1}}\,.$$

But then $\langle v\rangle$ and $\langle w\rangle$ must be fused in K. This is not the case since b inverts $\langle v\rangle$ and $\langle w\rangle$. This contradiction proves that $|H/K'| \geq 4$ and since H/K' is abelian and is generated by two involutions, it follows $H/K' \cong E_4$ and $K' = \langle v^2, w^2\rangle$. Since b centralizes H/K', we have $K/K' \cong E_8$. In particular, K is nonmetacyclic and so G is also nonmetacyclic. Since $v^a = w$, we have $wv^{-1} \in G'$ and $(wv^{-1})^{-1} = vw^{-1} \in G'$ and $vw = (vw^{-1})w^2 \in G'$. Hence, $G' = \langle v^2, w^2\rangle\langle vw\rangle$ and G/G' is abelian of type $(4,2)$.

We compute

$$(ca)^2 = caca = (bvab)va = v^{-1}ava = v^{-1}a^2(a^{-1}va) = v^{-1}bw\,,$$

and so

$$(ca)^4 = v^{-1}(bwv^{-1}b)w = v^{-1}w^{-1}vw = [v,w]\,.$$

Hence, if $o(ca) \leq 4$, then $[v,w] = 1$ and so $H = \langle v,w\rangle$ is abelian (of order 2^{2n} or 2^{2n-1}). Because the action of $\langle a\rangle$ on H is known, the structure of G in this case is uniquely determined. Our theorem is proved. □

Appendix 106 Nonabelian 2-groups not covered by proper nonabelian subgroups

Let G be a p-group with the title property. One may assume that, in addition, G is not minimal nonabelian. This will be if and only if $G \neq \bigcup_{H \in \Gamma_1'} H$, where Γ_1' is the set of all nonabelian maximal subgroups of G. Let $S \in \Gamma_2$, there Γ_2 be the set of all subgroups of index p^2 in G containing $\Phi(G)$. Then all members of the set Γ_1 containing S cover G. It follows that S is contained in an abelian $A(S) \in \Gamma_1$. Therefore if $\mathrm{d}(G) = 2$, then G has the title property if and only if it has an abelian subgroup of index p. It is natural to ask if there exists a nonabelian not two-generator p-group that is not covered by their nonabelian maximal subgroups. The following theorem answers on this question.

Theorem A.106.1. *Suppose the p-group G possesses a nonabelian maximal subgroup. Then G is not covered by its nonabelian maximal subgroups if and only if one of the following holds:*

(a) $\mathrm{d}(G) = 2$ *and there is in the set Γ_1 an abelian member.*

(b) $\mathrm{d}(G) = 3$, $\Phi(G) < \mathrm{Z}(G)$.

Proof. Any group from (a) satisfies the hypothesis. Below we shall show the any group from (b) also satisfies the hypothesis. Note that if G is a nonabelian p-group and $|G : \mathrm{Z}(G)| = p^2$, then $\Phi(G) \leq \mathrm{Z}(G)$.

Now let G has a nonabelian subgroup of index p and suppose that G is not covered by its nonabelian maximal subgroups. If $\mathrm{d}(G) = 2$, then $G = \bigcup_{H \in \Gamma_1} H$ is the irredundant covering so that there is in the set Γ_1 an abelian member.

Therefore, one may assume in what follows that $\mathrm{d}(G) > 2$. Let $S \in \Gamma_2$. Since all maximal subgroups of G containing S, cover G, one has $S < A(S) \in \Gamma_1$, where $A(S)$ is abelian. The subgroup $A(S)$ contains at least $p + 1$ members of the set Γ_2. As $|\Gamma_2| \geq 1 + p + p^2 > 1 + p$, there is $T \in \Gamma_2$ such that $T \nleq A(S)$. As above, $T < A(T) \in \Gamma_1$, where $A(T)$ is abelian. As $A(S) \neq A(T)$, by the choice of T, the subgroup $\mathrm{Z}(G) = A(S) \cap A(T)$ has index p^2 in G. By the first paragraph of the proof, $\Phi(G) < \mathrm{Z}(G)$. Assume that $\mathrm{d}(G) > 3$. Let $N \in \Gamma_1$ be nonabelian; then $\mathrm{d}(N) \geq 3$ so that N contains $\geq 1 + p + p^2 > 1 + p$ members of the set Γ_2. Then, by Exercise 1.6, there is in N a nonabelian $T \in \Gamma_2$, a contradiction: all members of the set Γ_2 are abelian. Thus, $\mathrm{d}(G) = 3$.

Let $H \in \Gamma_1$ be nonabelian; then $\mathrm{Z}(G) \cap H = \Phi(G)$ (indeed, $|\mathrm{Z}(G) \cap H| = |\Phi(G)|$, by the product formula, and $\Phi(G) \leq \mathrm{Z}(G) \cap H$ since $\Phi(G) = S \cap T = \mathrm{Z}(G)$ and $\Phi(G) < H \in \Gamma_1$). We have $G = H\mathrm{Z}(G)$ since $\mathrm{Z}(G) \nleq H \in \Gamma_1'$, the set in nonabelian members of the set Γ_1. Let $x \in \mathrm{Z}(G) - \Phi(G)$; then $\mathrm{Z}(G) = \langle x, \Phi(G) \rangle$ since $|\mathrm{Z}(G) : \Phi(G)| = p$. We claim that $x \notin \bigcup_{K \in \Gamma_1'} M$. Assume that this is false. Then there is a nonabelian $N \in \Gamma_1'$ such that $x \in N$. In that case, $\mathrm{Z}(G) = \langle x, \Phi(G) \rangle < N$. As $|N : \langle x, \Phi(G) \rangle| = p$, the subgroup N is abelian, a contradiction. Thus, G is not covered by its nonabelian maximal subgroups. □

Problem 1. Study the irregular p-groups that are not covered by their nonabelian maximal regular subgroups.

Problem 2. Study the nonabelian nonmetacyclic p-groups that are not covered by their nonabelian maximal metacyclic subgroups.

Problem 3. Classify the non-Dedekindian p-groups that are not generated by nonnormal subgroups (cyclic subgroups).

Appendix 107 Nonabelian p-groups all of whose minimal nonabelian subgroups have the same center

Here, we solve Problem 1728 by proving the following result.

Theorem A.107.1. *Let G be a nonabelian p-group all of whose minimal nonabelian subgroups have the same center Z. Then $Z \leq Z(G)$ and either G has an abelian subgroup of index p or $\mho_1(G) \leq Z$.*

Proof. Let G be a nonabelian p-group all of whose minimal nonabelian subgroups have the same center Z. In that case, $C_G(Z)$ contains all minimal nonabelian subgroups of G. Since, by Theorem 10.28, G is generated by its minimal nonabelian subgroups, we get $Z \leq Z(G)$. Let $A < G$ be a maximal normal abelian subgroup of G. Let $h \in G-A$ be such that $h^p \in A$. By Lemma 57.1, there is $a \in A$ so that $\langle h, a\rangle$ is minimal nonabelian. Then

$$Z = Z(\langle h, a\rangle) = \langle h^p, a^p, [h, a]\rangle < A .$$

Suppose that $\exp(G/A) > p$. Then there is $g \in G - A$ with $g^p \notin A$. But by Lemma 57.1, there is a minimal nonabelian subgroup M in G with $g \in M$. In this case $g^p \in \Phi(M) = Z(M) = Z \leq Z(G) < A$, contrary to a choice of g. We have proved that $\exp(G/A) = p$. If, in addition, $g \in G - A$ such that $g^p \in A$ and M is as above, then $g^p \in \Phi(M) = Z$.

Suppose that $|G/A| > p$. Assume, in addition, that $\exp(A/Z) > p$. If $p = 2$, then all elements in $(G/Z) - (A/Z)$, by the previous paragraph, are involutions and, therefore, each such involution inverts A/Z, a contradiction (noting that G/A is elementary abelian of order ≥ 4). Now let $p > 2$ and let H/A be a subgroup of order p^2 in G/A; then $H' \leq A$ is abelian and therefore H and H/Z are metabelian. We have $H_p(H/Z) = A/Z$ and this subgroup has index p^2 in H/Z. This contradicts a result of Hogan–Kappe (HogK) about H_p-subgroups of metabelian p-groups (see also the introduction to § 256). We have proved that, provided $|G/A| > p$, we must have $\exp(A/Z) = p$ and so $\mho_1(G) \leq Z$. Our theorem is proved. □

Problem. Study the nonabelian p-groups G such that, whenever M is a minimal nonabelian subgroup of G, then $C_G(M)$ is minimal nonabelian.

Appendix 108 Nonabelian p-groups G in which the center of each nonabelian subgroup is contained in $Z(G)$

Here we solve Problem 1794 by proving the following result.

Theorem A.108.1. *Let G be a nonabelian p-group in which the center of each nonabelian subgroup is contained in $Z(G)$. Then either G has an abelian subgroup of index p or $\mho_1(G) \le Z(G)$.*

Proof. Let G be a nonabelian p-group in which the center of each nonabelian subgroup is contained in $Z(G)$. Then for any two distinct maximal abelian subgroups A, B of G, the subgroup $\langle A, B\rangle$ is nonabelian so, by hypothesis, we have

$$Z(\langle A, B\rangle) \le Z(G) \le A \cap B. \tag{1}$$

Next,

$$C_G(A \cap B) \ge \langle A, B\rangle \Rightarrow A \cap B \le Z(\langle A, B\rangle). \tag{2}$$

It follows from (1) and (2) that $A \cap B = Z(G)$. By Theorem 255.1, either G has an abelian subgroup of index p or $\mho_1(G) \le Z(G)$ and we are done. □

Problem 1. Study the nonabelian p-groups G such that, whenever $M \le G$ is minimal nonabelian, then $Z(M) \le Z(G)$.

Problem 2. Study the primary $\mathcal{A}_n$-groups G, $n > 1$, such that, whenever $M \le G$ is an $\mathcal{A}_2$-subgroup, then $Z(M) \le Z(G)$.

Problem 3. Study the primary $\mathcal{A}_n$-groups G, $n > 1$, all of whose $\mathcal{A}_2$-subgroups have the same center.

Appendix 109 O. Schmidt's theorem on groups all of whose nonnormal subgroups are conjugate

In this appendix we prove the following

Theorem A.109.1 (O. Schmidt [Sch3]). *If all nonnormal subgroups of a non-Dedekindian p-group G are conjugate, then $G \cong \mathrm{M}_{p^n}$.*

Proof. Let $H < G$ be nonnormal. As all maximal subgroups of H, being G-invariant, do not generate H, it follows that H is cyclic.

If $|H| = p$, then $G \cong \mathrm{M}_{p^n}$ (Theorem 1.25).

Next assume that $|H| > p$. We proceed by induction on $|G|$. Let $R = \Omega_1(H)$; then $R \triangleleft G$ and all nonnormal subgroups of a non-Dedekindian group G/R are conjugate. By induction, $G/R \cong \mathrm{M}_{p^m}$ for some m. The group G/R has two distinct cyclic subgroups A/R and B/R of index p. Then $A, B \in \Gamma_1$ are distinct abelian so that $A \cap B = \mathrm{Z}(G)$. It follows from $R \le \Phi(H) \le \Phi(G)$ that $\mathrm{d}(G) = \mathrm{d}(G/R) = 2$, and we conclude that G is minimal nonabelian. If G is nonmetacyclic, it has two non-conjugate nonnormal cyclic subgroups (Lemma 65.1), a contradiction. Thus, the group G is metacyclic. In that case,

$$G = \langle x, y \mid o(x) = p^n, n > 1, \ o(y) = p^2, \ x^y = x^{1+p^{n-1}} \rangle .$$

As $\mathrm{c}_2(G) = p^2 + p$ and there are in G exactly p normal cyclic subgroups of order p, there are in G exactly $p (> 1)$ conjugate classes of nonnormal cyclic subgroups of order p^2, a contradiction. Thus, $|H| = p$. □

Schmidt [Sch3] also has proved the following

Theorem A.109.2. *If all nonnormal subgroups of a nonnilpotent group G are conjugate, then $G = P \cdot Q$ is minimal nonabelian, where $P \in \mathrm{Syl}_p(G)$ is cyclic and $Q = G' \in \mathrm{Syl}_q(G)$ is of order q, $q \equiv 1 \pmod p$.*

Proof. Let $H = PQ \le G$ be minimal nonnilpotent, where $P \in \mathrm{Syl}_p(G)$ is cyclic and $Q = H' \in \mathrm{Syl}_q(G)$. As all subgroup of Q is normal, it follows that $|Q| = q$. The subgroup $H = \langle P^x \mid x \in G \rangle \triangleleft G$. By Frattini's lemma, $G = H\mathrm{N}_G(P)$. Since $H \cap \mathrm{N}_G(P) = P$, it follows that the subgroup $\mathrm{N}_G(P)$ is nonnormal in G, and we conclude that $\mathrm{N}_G(P) = P$ so that $G = H\mathrm{N}_G(P) = HP = H$, completing the proof. □

This is a partial case of the main result of [Bra].

In [Sch4] Schmidt classified the groups G with exactly two classes of nonnormal conjugate subgroups. If $A, B < G$ are representatives of those classes, then A, B are cyclic.

Problem 1. Classify the non-Dedekindian p-groups all of whose nonnormal maximal cyclic subgroups are conjugate.[1]

Problem 2. Classify the non-Dedekindian p-groups all of whose maximal nonnormal subgroups are conjugate.

Problem 3. Classify the non-Dedekindian p-groups with exactly two conjugate classes of nonnormal maximal cyclic subgroups.

1 This was solved in Theorem A.94.1.

Research problems and themes V

I am convinced that the special problems in all their complexity constitute the stock and core of mathematics; and to master their difficulties requires on the whole the harder labor.
– H. Weyl, The Classical Groups, Princeton University Press, 1973.

The heart of mathematics is its problems.
– P. Halmos

Theories change, but the groups remain.
– M. Hall and J. K. Senior, The Groups of Order 2^n, $n \leq 6$.

A well-chosen problem can isolate an essential difficulty in a particular area serving as a particular benchmark against which progress in this area can be measured.
– P. Erdos

A scientific person will never understand why he should believe opinions only because they are written in a certain book. Furthermore, he will never believe that the results of his own attempts are final.
– A. Einstein

Science will stagnate if it is made to serve practical goals.
– A. Einstein

The gradation of all science is to cover the greatest number of empirical facts by logical deduction from the smallest number of hypotheses or axioms.
– A. Einstein

You are never sure whether or not a problem is good unless you actually solve it.
– M. Gromov

This is the fifth list of research problems and themes written by the first author. As in the previous lists, I have used numerous constructive remarks and comments by Zvonimir Janko and Avinoam Mann. Qinhai Zhang sent me a great number of remarks. This allowed me to improve this list considerably. As in the previous lists, a short information about solved problems is presented after their statements. Only few problems from this list were solved since it was started in very beginning of 2011.

In many cases, we suggest to find a solution of a problem for p-groups of class two or groups of exponent p. These partial cases may appear fairly difficult. In many cases will be useful at first to solve some problems for some partial cases of p-groups ($\mathcal{A}_n$-groups with small n, or metacyclic or regular p-groups).

Most items in this list as in the previous ones are devoted to subgroup structure of p-groups. Many problems order to investigate impact of minimal nonabelian subgroups on the structure of a group. Any item beginning with “Study,” we consider as a theme for research. Many items from this and the previous lists make sense for non-nilpotent groups.

These lists show how little we know (or how much we do not know) about p-groups.

2801 Let a p-group possess an abelian subgroup of index p. Present an elementary approach to classification of its representation groups.

2802 Describe the representation groups of the two-generator p-groups with a metacyclic maximal subgroup.

2803 Classify the nonabelian p-groups G containing a maximal subgroup H such that all abelian subgroups of G not contained in H are cyclic. (Example: $G = \mathrm{SD}_{2^n}$, $H = \mathrm{D}_{2^{n-1}}$.)

2804 Given $e > 1$, does there exist a group G of exponent p^e such that the subgroups $\Omega_k^*(G)$ are (i) pairwise nonincident, (ii) abelian for $k = 1, 2, \dots, e$.

2805 Let Δ_1 and Δ_2 be representation groups of a p-group G. Compare their Schur multipliers. Consider in detail the case when $\mathrm{M}(\Delta_1)$ is trivial.

2806 Study the nonabelian p-groups G containing a noncentral element a such that $\langle a, x\rangle$ is minimal nonabelian for all $x \in G - \mathrm{C}_G(a)$.

2807 Let R be a subgroup generated by all real elements of a 2-group G. Study the quotient group G/R.

2808 Study the p-groups G such that for every $H < G$ there exists $x \in G$ with $H^G = \langle H, H^x\rangle$.

2809 Classify the nonmetacyclic p-groups all of whose metacyclic subgroups are either abelian or minimal nonabelian.

2810 Study the nonabelian p-groups G such that, whenever a nonabelian $H < G$ satisfies $H_G > \{1\}$, then $H \lhd G$. Consider in detail the case when all subgroups of G satisfy the above condition.

2811 Study the p-groups G such that for every noncyclic $H < G$ one has $\mathrm{d}(H^G) = \mathrm{d}(H)$.

2812 Find the maximal order of a group of exponent 4 generated by n involutions. (See § 61.)

2813 Study the p-groups G of exponent $> p$ in which any subgroup of order p is contained in exactly p maximal cyclic subgroups of G.

2814 Study the p-groups G such that for every cyclic $H < G$ of order p^2 the subgroup $\mathrm{C}_G(H)$ is abelian of type (p^n, p). (See § 77 and Appendix 50.)

2815 Suppose that an irregular p-group $G = H\Omega_1(G) > \Omega_1(G)$, where $p > 3$, $H < G$ is absolutely regular and $|\Omega_1(G)| = p^p$ (see Theorem 12.1 (b).) Treat the cases: (i) $\Omega_1(G)$ is of maximal class, (ii) all subgroups of G containing $\Omega_1(G)$ as a subgroup of index p are regular.

2816 Given a p-group G, does there exist an overgroup W of G such that all maximal subgroups (normal subgroups) of W are characteristic?

2817 Study the two-generator p-groups G such that $G/\mathrm{Z}(G) \cong \mathrm{M}_p(m, m)$.

2818 Classify the p-groups with ≤ 2 conjugate classes non-quasinormal subgroups. (See § 59 and § 96.)

2819 Does there exist n such that, if $G/\mathrm{K}_n(G)$ is metabelian so is G?

2820 Classify the p-groups all of whose maximal nonnormal abelian subgroups are conjugate (this was solved by the second author; see § 234).

2821 Given $n > 2$, does there exist a p-group containing n self centralizing cyclic subgroups of pairwise distinct orders?

2822 Present an elementary method for computing the Schur multiplier of extraspecial (metacyclic, minimal nonmetacyclic) p-groups.

2823 Classify the p-groups G such that $\beta_1(G, H)(= \alpha_1(G) - \alpha_1(H)) \leq p$ for all $H \in \Gamma_1$. (See § 76.)

2824 Study the p-groups G such that, whenever $A \lhd G$ and $x \in G - A$, then $\mathrm{cl}(\langle x, A\rangle) \leq \mathrm{cl}(A) + 1$. (A partial case of this is solved by the second author; see Corollary 224.2.)

2825 Study the p-groups all of whose nonabelian two-generator subgroups are either metacyclic, or minimal nonabelian, or of maximal class.

2826 (Old problem) Study the p-groups G such that all elements of G' are commutators.

2827 Study the special p-groups G such that $H \cap \mathrm{Z}(G) = \mathrm{Z}(H)$ for all nonabelian $H \leq G$.

2828 Does there exist a p-group G satisfying $A^\# \subseteq \bigcup_{\chi\in\mathrm{Irr}_1(G)} \mathrm{T}_\chi$ for all minimal nonabelian $A < G$? Here $\mathrm{T}_\chi = \{x \in G \mid \chi(x) = 0\}$.

2829 Estimate the number $|\mathrm{Aut}(G) : \mathrm{Inn}(G)|_p$, where G is a p-group of maximal class and order p^n.

2830 Let G, G_0 be p-groups such that $\mathrm{s}_n(G) = \mathrm{s}_n(G_0)$ for all n. Compare (i) $\exp(G)$ and $\exp(G_0)$, (ii) $|G/G'|$ and $|G_0/G_0'|$, (iii) $|\mathrm{Z}(G)|$ and $|\mathrm{Z}(G_0)|$, (iv) $\mathrm{d}(G)$ and $\mathrm{d}(G_0)$.

2831 Study the nonmetacyclic p-groups all of whose maximal metacyclic subgroups have index $\leq p^2$.

2832 Classify the p-groups all of whose outer p-automorphisms have $\leq p^2$ fixed points. (See #2008.)

2833 Study the p-groups with at most two conjugate classes of nonnormal cyclic (abelian) subgroups.

2834 Classify the 2-groups with exactly one minimal nonmetacyclic subgroup (i) of order 2^4, (ii) of order 2^5.

2835 Classify the p-groups G, $p > 2$. All of whose maximal (i) absolutely regular subgroups, (ii) regular subgroups have index p^2. Is it true that in case (ii) there is in G a normal regular subgroup of index p^2?

2836 Study the p-groups G all of whose maximal subgroups of class 2 contain $\Phi(G)$.

2837 Classify the p-groups G such that, whenever $A < B \leq G$ are nonabelian and $|B : A| = p$, then $\beta_1(B, A)(= \alpha_1(B) - \alpha_1(A)) \in \{p - 1, p\}$. (See § 76.)

2838 Find the minimal order of a p-group G such that all maximal subgroups of $\Phi(G)$ are irregular. Is it possible that $\mathrm{cl}(G) = p$ in that case?

2839 Classify the 2-groups containing a normal subgroup $R \cong C_2$ such that G/R is nonabelian Dedekindian.

2840 Study the primary $\mathcal{A}_n$-groups, $n > 2$, in which every $\mathcal{A}_k$-subgroup is contained in an appropriate $\mathcal{A}_{k+1}$-subgroup for all $k \leq n - 1$.

2841 Study the p-groups G satisfying $\bigcup_{\chi \in \mathrm{Irr}_1(G)} \mathrm{T}_\chi = G - \mathrm{Z}(G)$.

2842 (Ito) A special p-group G is said to be a *Golfand group* (see [Gol]) if for any $N < G'$ the quotient group G/N is special; such G we also call a $\mathcal{G}_p(2d)$-group, where $2d = \mathrm{d}(G)$. Given d, find the maximal order of a $\mathcal{G}_p(2d)$-group.

2843 Study the p-groups G such that $\chi(1)^2 = |G : \mathrm{Z}(\chi)|$ for each $\chi \in \mathrm{Irr}(G)$.

2844 Study the groups G of order p^n such that $\Omega_1(G) \leq \mathrm{Z}(G)$ and all epimorphic images of G of order p^{n-1} are pairwise nonisomorphic. Is it true that then $|\Omega_1(G)|$ is bounded?

2845 Classify the p-groups G such that, whenever $A < G$ is non-quasinormal and $A < B < G$, where B is quasinormal in G, then $B \in \Gamma_1$; see § 62.

2846 Study the p-groups all of whose metacyclic sections are abelian. (See § 247.)

2847 Study the p-groups such that, whenever $A < H \in \Gamma_1$ and A is non-G-invariant, then $\mathrm{N}_G(A) \leq H$.

2848 Study the p-groups $G = \Omega_n^*(G)$, $n > 1$, all of whose cyclic subgroups of order p^n are normal. (See § 230 where more general problem is treated.)

2849 Study the p-groups G such that $\Phi(G) \leq \mathcal{N}(G)$, where $\mathcal{N}(G)$ is the norm of G (see § 143).

2850 Describe the representation groups of p-groups G in which G' is a unique minimal normal subgroup.

2851 Study the p-groups in which every maximal cyclic subgroup ($\mathcal{A}_1$-subgroup) is a direct factor in its normalizer.

2852 Study the p-groups G of exponent $> p^2$ such that for any two distinct maximal conjugate cyclic $A, B < G$, one has $|A \cap B| \leq p$.

2853 Does there exist a nonabelian p-group all of whose nonabelian subgroups with cyclic derived subgroup are nonnormal?

2854 Study the nonabelian p-groups G such that the set $G - \mathrm{Z}(G)$ is covered by $\mathcal{A}_1$-subgroups. (See § 860.)

2855 Study the p-groups G such that, whenever $A < G$, there is $B \triangleleft G$ with $B \cong A$.

2856 Study the p-groups G such that $\mathrm{dl}(G) = \mathrm{dl}(H)$ for all $H \in \Gamma_1$.

2857 Given n, estimate the number of groups of order p^n that are metabelian of maximal class.

2858 Does there exist a p-group G such that $|\mathrm{Aut}(G)| < |\mathrm{Aut}(H)|$ for all $H \in \Gamma_1$?

2859 Given $n > 1$, does there exist a p-group in which every maximal subgroup is a direct product of n nonabelian subgroups of order p^3?

2860 Classify the p-groups G such that $H^G \leq N_G(H)$ for all $H < G$.[2]
2861 Study the p-groups G covered by homocyclic subgroups.
2862 Study the p-groups all of whose maximal regular subgroups have orders $\leq p^{p+2}$.
2863 Is it possible to estimate $d(G)$, where G is a p-group with cyclic Schur multiplier?
2864 Classify the p-groups all of whose maximal subgroups have (i) derived subgroups of order $\leq p$, (ii) cyclic derived subgroups. (These problems were solved by the second author; see § 137 and § 139. See also [Leo].)
2865 Study the p-groups G such that, whenever $A < B$ are nonnormal subgroups of G, then $A_G < B_G$.
2866 Classify the p-groups that are lattice (N-lattice) isomorphic with p-groups having an abelian subgroup of index p.
2867 Classify the nonmodular p-groups G such that (i) $|G : H^G| = p$, (ii) $|N_G(H) : H| = p$, (iii) $|H : H_G| = p$ for each non-quasinormal $H < G$.
2868 Study the irregular p-groups in which any two distinct subgroups of order p^p and exponent p have intersection of order p^{p-1}.
2869 Classify the p-groups G of order p^n, $p > 2$, with cyclic G' for which $p^{[n/2]+1} = |G/G'|$ (see § 204.).
2870 Classify the nonabelian p-groups G with cyclic G' all of whose $\mathcal{A}_1$-subgroups have a cyclic subgroup of index p.
2871 Is it true that $\mathrm{Aut}(G)$ is not a p-group for any group G of exponent p?
2872 Study the p-groups G such that, whenever $A \leq G$, there is $i = i(A)$ such that $A = \Omega_i^*(A)$.
2873 Study the p-groups that are lattice isomorphic to irregular p-groups containing a subgroup of exponent p and index p. (See Appendix 42.)
2874 Study the p-groups that are lattice isomorphic to a Sylow p-subgroup of the affine general linear group $\mathrm{AGL}(n, p)$.
2875 Study the p-groups all of whose subgroups are quasinormal in their normal (quasinormal) closures.
2876 Does there exist a nonabelian p-group G satisfying $\mathrm{Aut}(G) \cong A$ for a given abelian group A? (See #1906.)
2877 Classify the p-groups all of whose non-quasinormal subgroups are abelian. (See [FA].)
2878 Let E be a normal subgroup of order p^p and exponent p of an irregular p-group $G = \Omega_1(G)$. Study the structure of G if, whenever $x \in G - E$ is of order p, then the subgroup $\langle x, E \rangle$ is of maximal class. (See Exercise 10.10.)

2 As Mann has noted, it is known that $\mathrm{cl}(G) \leq 3$, but the converse is not true (examples: D_{16} and SD_{16} but not Q_{16}; in the last group any subgroup is normalized by its conjugates. The p-groups with normal normalizers of all subgroups also have this property.

2879 Study the p-groups covered by non-quasinormal abelian subgroups.
2880 Study the p-groups G such that $\mathrm{H}_p(G) < G$ is (i) minimal nonabelian, (ii) $\mathcal{A}_2$-subgroup.
2881 Study the p-groups G such that, whenever nonabelian $A < B \le G$, then $|B' : A'| = |B : A|$. (See § 148.)
2882 Study the p-groups G with irregular G' ($\Phi(G)$) such that H' ($\Phi(H)$) is absolutely regular for all $H < G$.
2883 Study the p-groups all of whose normal (nonnormal) abelian subgroups have exponent p.
2884 Classify the p-groups G with $R \lhd G$ of order p^k, $k \in \{1, 2\}$, such that all nonnormal subgroups of G/R have order p (have equal order). (See Theorem 1.25.)
2885 Study the p-groups all of whose subgroups of index p^2, except one, are either abelian or minimal nonabelian.
2886 Study the p-groups G such that, for all nonabelian $A < G$, one has (i) $\mathrm{cl}(A^G) = \mathrm{cl}(A)$, (ii) $\mathrm{dl}(A^G) = \mathrm{dl}(A)$.
2887 Classify the p-groups G such that $\mathcal{L}_N(G) \cong \mathcal{L}(G_0)$ for some minimal nonabelian p-group G_0.
2888 Study the p-groups in which any two normal abelian subgroups are incident.
2889 Find all possible minimal degrees of faithful representations by permutations of a p-group of maximal class and order p^n.
2890 Describe subgroup, normal and power structures of the 2-groups generated by three involutions. (According to J. P. Serre, the derived length of such groups is not bounded.)
2891 Study the 2-groups G all of whose $\mathcal{A}_1$-subgroups have order ≤ 16.
2892 (Inspired by [BlaDM]) Classify the p-groups G such that, whenever $H = AB$, where $H, A, B \le G$, then one of the subgroups A, B is quasinormal in H.
2893 Study the p-groups of exponent $> p$ containing the same number of subgroups of any order $> p$ as a group of exponent p.
2894 Study the primary $\mathcal{A}_n$-groups, $n > 2$, without three pairwise nonisomorphic $\mathcal{A}_1$-subgroups. (See § 206 and [BerZhang].)
2895 Classify the 2-groups G such that, whenever $H \in \Gamma_1$, then H is either abelian or has the form $D \times E$, where D is nonabelian of order 8 and E is elementary abelian (abelian). (See Appendix 90 and § 245.)
2896 Study the irregular p-groups (i) all of whose maximal absolutely regular subgroups are nonnormal, (ii) containing a maximal absolutely regular subgroup of order p^p.
2897 Study the nonabelian p-groups G, $p > 2$, such that, whenever $A < G$ is maximal abelian, then $A/\mathrm{Z}(G)$ is cyclic. (See Theorem 20.4. This is solved for $p = 2$ by the second author; see Theorem 91.4.)
2898 (Inspired by one Wielandt's example) Given $p > 5$, construct an irregular p-group $G = M_1 \times \cdots \times M_k$, where $k = (p + 1)/2$ and all M_i are metacyclic with $|M_i'| > p$ for all i.

2899 Classify the p-groups G containing a maximal cyclic subgroup A such that $\langle A, B\rangle$ is minimal nonabelian for any maximal cyclic subgroup $B \neq A$ of G. (This is solved by the second author; see Theorem 182.1.)

2900 Classify the 2-groups G such that $\mathrm{Aut}(G) \cong \mathrm{Aut}(M)$, where M is a 2-group of maximal class.

2901 Study the p-groups G such that, whenever $H \in \Gamma_1$ then $\mathrm{C}_H(x) \leq \mathrm{Z}(H)$ for each $x \in G - H$.

2902 Classify the p-groups in which the normal closure of any elementary abelian subgroup is abelian.

2903 Study the 3-groups G such that all subgroups of G containing A as a subgroup of index 3 are irregular for each $\mathcal{A}_1$-subgroup $A < G$.

2904 Suppose that p-groups G and W are lattice isomorphic and G is covered by minimal nonabelian subgroups. Classify all those W that are not covered by minimal nonabelian subgroups.

2905 Classify the special 2-groups G satisfying $|G/\Omega_1(G)| = 2$.

2906 (Old problem) Study the minimal irregular p-groups, $p > 2$. (See Theorem 7.4.)

2907 Describe the subgroup, power and normal structures of the p-groups G with a special subgroup of index p.

2908 Study the p-groups all of whose subgroups of index p^2 have class ≤ 2.

2909 Study the p-groups G of maximal class such that all elements of the set $G - G_1$ have order p (see § 9).

2910 Classify the special p-groups G such that, whenever $A, B \leq \mathrm{Z}(G)$ are different of equal order, then $G/A \ncong G/B$.

2911 Study the p-groups G with $H \in \Gamma_1$ such that each maximal cyclic subgroup of G of order $> p$ which is not contained in H, is self centralizing. (See #237.)

2912 Study the p-groups G all of whose maximal subgroups, except one, are generated by elements of order p.

2913 Study the p-groups G of exponent $> p^2$ satisfying $|\mathrm{N}_G(Z) : \mathrm{C}_G(Z)| \leq p$ for any cyclic $Z < G$.

2914 Study the p-groups G such that $|G : \mathrm{C}_G(x)| = p$ for any $x \in \mathrm{Z}_2(G) - \mathrm{Z}(G)$.

2915 Find norms of all $\mathcal{A}_2$-groups. (See § 71.)

2916 Study the p-groups G of exponent $> p$ such that, whenever $A, B < G$ are cyclic with $A \cap B = \{1\}$, then $[A, B] = \{1\}$. (See § 159 and § 160.)

2917 Classify the noncyclic 2-groups G such that, whenever $L < B \leq G$, where L is maximal cyclic in G and $|B : L| = 2$, then either B is minimal nonabelian or of maximal class. (This was solved by the second author; see § 237.)

2918 Study the irregular p-groups G such that, whenever $H < G$ is maximal absolutely regular, then all subgroups of G containing H as a subgroup of index p are irregular.

2919 Classify the p-groups all of whose nonlinear irreducible characters, except one, have degree p.

2920 Classify the 2-groups G such that $N_G(A)$ is metacyclic for any abelian $A < G$ of type $(4, 2)$.

2921 Study the p-groups all of whose maximal subgroups of class 2 are either $\mathcal{A}_1$- or $\mathcal{A}_2$-groups.

2922 Classify the p-groups G such that, for any minimal nonabelian $A < G$, (i) $\mathrm{cl}(N_G(A)) = 2$, (ii) $|N_G(A) : A| = p$.

2923 Let H be a subgroup of index p^3 in a p-group G and $\mu \in \mathrm{Lin}(H)$. List all possible decompositions of μ^G in irreducible constituents.

2924 Study the nonmetacyclic p-groups G such that, whenever $H < G$ is maximal metacyclic (minimal nonabelian), then all subgroups of G containing H as a subgroup of index p are two-generator.

2925 Study the p-groups containing the same number of subgroups of any order as a special p-group.

2926 Study the minimal irregular p-groups of order p^{p+2}.

2927 Study the p-groups containing exactly two non-trivial characteristic subgroups.

2928 Find the minimal order of a p-group G with the irregular subgroup $\Phi(\Phi(G))$.

2929 Study the p-groups G generated by any two distinct maximal abelian subgroups (this was solved by the second author; see Theorem 183.1).

2930 Study the nonabelian p-groups G such that the intersection of any two distinct maximal abelian subgroups of G is cyclic. (This was solved for $p = 2$ by the second author; see Theorem 91.1.)

2931 Study the p-groups G all of whose maximal subgroups admit a nontrivial partition but G does not admit a nontrivial partition.

2932 Given n, is it possible to estimate the derived length of the p-groups having at most n pairwise noncommuting elements? (See § 116.)

2933 Study the nonabelian p-groups in which the centralizers of all nonabelian subgroups are elementary abelian.

2934 Study the p-groups containing a maximal set S of pairwise noncommuting elements such that any two distinct elements of S generate (i) $\mathcal{A}_1$-subgroup, (ii) metacyclic subgroup, (iii) subgroup of maximal class, (iv) regular subgroup.

2935 Does there exist a nonabelian p-group G containing a maximal set S of pairwise noncommuting elements such that $S \subset G'$?

2936 Given a p-group G, let $\mathcal{M}$ be the set of those subgroups of G that are connected with G by only one maximal chain. Study the structure of G provided that all minimal members of the set $\mathcal{M}$ are (i) cyclic, (ii) abelian, (iii) minimal nonabelian, (iv) maximal metacyclic, (v) minimal nonmetacyclic, (vi) of maximal class, (vii) elementary abelian, (viii) absolutely regular.

2937 Classify the p-groups G satisfying $|N_G(A) : C_G(A)| = p^2$ for any minimal nonabelian $A < G$.

2938 For all "interesting" p-groups G, compare the numbers

$$\exp(G) \text{ and } \exp(G/\Omega_1(G)) \cdot \exp(\Omega_1(G)) .$$

2939 Study the p-groups having at most $2(p-1)$ nonlinear irreducible characters of each degree. (See Exercise A.51.150.)

2940 Classify the p-groups, having a representation group with abelian subgroup of index p.

2941 Classify the nonabelian p-groups with exactly one noncyclic maximal abelian subgroup. (This is solved by the second author for $p = 2$ in § 173.)

2942 Study the p-groups G, $p > 2$, such that, whenever $C < B \leq G$, where C is cyclic of order $> p$ and $|B : C| = p$, then any two cyclic subgroups of index p in B are nonconjugate in G. (See § 63.)

2943 Study the p-groups all of whose maximal elementary abelian subgroups are soft. (See § 130.)

2944 Classify the nonabelian p-groups G containing a maximal abelian subgroup A such that, whenever $a \in A - Z(G)$, there is $b \in G - A$ such that $\langle a, b \rangle$ is an $\mathcal{A}_1$-subgroup. (See Lemma 57.1.)

2945 Study the p-groups G such that $\langle [A, B] \rangle \lhd G$ for all distinct $A, B < G$.

2946 Study the p-groups G such that $N_G(H) \leq H^G$ for all nonnormal $H < G$.

2947 Classify the p-groups with exactly $p+1$ conjugate classes of $\mathcal{A}_1$-subgroups. (See §§ 76, 116.)

2948 Study the p-groups G containing a normal homocyclic subgroup of index $\leq p^2$.

2949 Let $\nu_k(G)$ be the number of normal subgroups of order p^k in a p-group G. Study the p-groups G of order p^m satisfying $\nu_k(G) = \nu_{m-k}(G)$ for all $k \leq m$.

2950 Study the structure of the intersection of normalizers of all noncyclic subgroups of a p-group G.

2951 Classify the 2-groups containing only one subgroup (i) $\cong \mathcal{H}_{2,2}$ (= the nonabelian metacyclic group of order 16 and exponent 4), (ii) nonmetacyclic minimal nonabelian subgroup of order 2^4.

2952 Let $G = A \times B$, where $A, B < G$, $|A| = p^m > p$ and $|B| = p^n > p$. Study the p-groups G_0 such that $s_m(G_0) = s_m(G)$ and $s_n(G_0) = s_n(G)$.

2953 Given $n > p$, find the maximal possible nilpotence class of a group of order p^n and exponent p. Study the groups G of exponent p and order p^n such that $\text{cl}(G) = n - 2$.

2954 Let x be a fixed noncentral element of a p-group G. Study the structure of G if $\langle x, y \rangle$ is metacyclic (of maximal class) for all $y \in G$.

2955 (Isaacs) Does there exist a p-group admitting a nontrivial partition by nonabelian subgroups?

2956 Let $G = \Sigma_{p^n} = \Sigma_{p^{n-1}} \text{ wr } C_p$ and B is the base subgroup of G. Estimate $\beta_1(G, B) = \alpha_1(G) - \alpha_1(B)$.

2957 Study the p-groups in which the normal closures of cyclic subgroups are abelian. (This is solved by the second author; see § 224.)

2958 Study the p-groups G such that for all $\mathcal{A}_1$-subgroups $B < G$ one has $C_G(A) = A$ for all maximal $A < B$. (This was solved by the second author; see Theorems 92.1 and 92.2.)

2959 Study the p-groups in which any maximal set of pairwise noncommuting elements has cardinality $p + 2$. (See #116.)

2960 Study the irregular 3-groups G such that $|M'| \leq p$ for all regular $M < G$.

2961 Given a group G_0 of exponent p, classify the groups G of exponent p satisfying $s_k(G) = s_k(G_0)$ for all k.

2962 Classify the special 2-groups G such that 4 does not divide $c_2(G)$.

2963 Suppose that a p-group $G = Z_1 \dots Z_n$, where $p > 2$, $Z_i \cong C_{p^e}$ for $i = 1, \dots, n$, $Z_1, \dots, Z_n$ are pairwise permutable. Is it true that if $|G| = p^{ne}$, then all maximal cyclic subgroups of G have equal order?[3]

2964 Study the 2-groups G containing a normal nonabelian Dedekindian subgroup of index 4.

2965 Study the two-generator p-groups with cyclic derived subgroup, find their Schur multipliers and representation groups.

2966 Classify the maximal abelian subgroups of the p-group $A \operatorname{wr} B$ (of order $|A|^{|B|}|B|$), where A and B are elementary abelian (cyclic).

2967 Study the p-groups all of whose nonnormal maximal abelian subgroups are cyclic. (See § 16.)

2968 Study the metacyclic capable p-groups.

2969 (A. Zalesskii) Study the nonabelian p-groups G containing a maximal normal abelian subgroup A such that $\exp(G) = \exp(A)\exp(G/A)$.

2970 Given k, estimate the derived length of a p-group G provided all its nonnormal subgroups have derived length $\leq k$. (For $k = 1$, see [FA].)

2971 Study the nonabelian p-groups G such that, for any $\chi \in \operatorname{Irr}_1(G)$, one has $\chi\bar{\chi} = \rho_{G/\ker(\chi)}$, the regular character of $G/\ker(\chi)$. (See [HKS1].)

2972 Study the p-groups in which the centralizers of all noncentral elements (i) are abelian, (ii) have class ≤ 2.

2973 Does there exist a nonabelian p-group G such that the minimal sublattice $\mathcal{L}'(G)$ containing all minimal nonabelian subgroups of G coincides with the lattice $\mathcal{L}(G)$ of all subgroups of G?

2974 Study the p-groups G satisfying $|\Omega_1(H)| = |H/\mho_1(H)|$ for all $H \leq G$.[4]

2975 Study the p-groups whose lattices of normal subgroups are isomorphic to the lattices of subgroups of appropriate special p-groups.

2976 Classify the special p-groups G all of whose normal subgroups are incident with $Z(G)$.

---

3 For $p = 2$ this is not true as the group $\mathcal{H}_{2,2}$ shows.

4 As Mann noted, there is an irregular p-group G of order p^{p+1}, $p > 2$, satisfying this condition.

2977 Study the p-groups G of exponent $> p$ such that $|G : C_G(x)| \leq p$ for all $x \in G$ of order $> p$.

2978 Classify the p-groups G such that, whenever $A \triangleleft H \in \Gamma_1$, then $A \triangleleft G$.

2979 Study the p-groups all of whose nonnormal subgroups are two-generator. (See § 16.)

2980 (Mann) Let a p-group $G = AB$, where A is abelian and $\mathrm{cl}(B) = 2$. Is it true that the derived length $\mathrm{dl}(G)$ is bounded by function of B?

2981 Study the 2-groups $G = AB$, where $A, B < G$ are of maximal class and $A \cap B = \{1\} = A_G = B_G$.

2982 Classify the p-groups all of whose nonnormal subgroups have class at most 2.

2983 Study the minimal irregular p-groups G, $p > 2$, such that $|G/Z(G)| = p^p$.

2984 Classify the special p-groups $G = AB$, where $A, B < G$ are abelian.

2985 Study the p-groups G such that, whenever $A/Z(G) < G/Z(G)$ is abelian, then A is abelian. (This is solved by the second author; see Exercise 200.1.)

2986 Study the nonabelian p-groups G such that for any $x \in G - Z(G)$, the group $C_G(x)/Z(G)$ has a cyclic subgroup of index p.

2987 Classify the p-groups G such that for any noncharacteristic $H < G$ its characteristic closure coincides with G.

2988 Study the p-groups G such that $Z(G)$ is covered by $\mathcal{A}_1$-subgroups of G.

2989 Describe the normal and power structures of the p-groups containing an irregular normal subgroup of maximal class and index p^2. (See Theorem 12.12.)

2990 Classify the primary $\mathcal{A}_n$-groups G, $n > 2$, such that $\alpha_1(A) = \alpha_1(B)$ for any nonabelian $A, B < G$ of equal order.

2991 Study the p-groups all of whose nonabelian subgroups have trivial Schur multipliers.

2992 Study the p-groups all of whose nontrivial characteristic subgroups are contained in $\Phi(G)$.

2993 Classify the maximal subgroups of all $\mathcal{A}_3$-groups. (See [ZZLS] and § 205.)

2994 Study the power structure of an irregular p-group containing a normal cyclic subgroup of index p^p.

2995 Find the maximal n such that any p-group containing a two-generator abelian subgroup of index p^n has a normal abelian subgroup of index p^n.

2996 Suppose that a p-group G is covered by $\mathcal{A}_1$-subgroups of equal order. (See #860.) Is it true that all $\mathcal{A}_1$-subgroups of G have equal order?

2997 Classify the p-groups G containing an abelian subgroup of index p and covered by $\mathcal{A}_1$-subgroups. (This is solved by the second author; see Appendix 72.)

2998 Study the p-groups G such that, whenever $A < G$ is minimal nonabelian, then the degrees of all members of the set $\mathrm{Irr}(\lambda^G)$ are equal for any $\lambda \in \mathrm{Irr}^{\#}(A)$.

2999 Given $s \geq 3$, suppose that a p-group G has a maximal abelian subgroups of orders $p^s, p^{s+1}, \dots, p^{s+k}$. Is it true that k is bounded?

3000 Study the p-groups G all of whose subgroups of index p^2 are two-generator.

3001 Study the p-groups of maximal class and order $> p^{p+1}$, all of whose irregular members of the set Γ_1 are isomorphic.

3002 Study the special p-groups G satisfying $c_1(G) = |\Gamma_1|$.

3003 Study the p-groups G such that H/H_G is cyclic for all $H < G$. (See §§ 16, 63, 223.)

3004 Study the p-groups G such that p^2 does not divide $\exp(\mathrm{Aut}(G))$ and estimate $|G|$. (See § 150 and #1906.)

3005 (Inspired by [GMS1]) Study the p-groups G such that for all $H \in \Gamma_1$ with $\exp(H) > p$, one has $N_G(C) \leq H$ for all cyclic $C \leq H$ of order p^2.

3006 Describe the automorphism groups of 3-groups of maximal class.

3007 Given $m > 6$, study the p-groups G of order p^m such that for every $k \in \{3, \dots, m-2\}$ there is in G an $\mathcal{A}_1$-subgroup of order p^k. (See [QYXA].)

3008 Study the p-groups G such that, whenever $A, B \leq G$ are nonabelian of equal order, then (i) $|A'| = |B'|$, (ii) $|\Omega_1(A)| = |\Omega_1(B)|$, (iii) $|\mho_1(A)| = |\mho_1(B)|$, (iv) $|Z(A)| = |Z(B)|$, (v) $c_1(A) = c_1(B)$, (vi) $\mathrm{cd}(A) = \mathrm{cd}(B)$, (vii) $|\mathrm{Irr}_1(A)| = |\mathrm{Irr}_1(B)|$, (viii) $k(A) = k(B)$.

3009 Study the p-groups G such that, whenever $\chi, \psi \in \mathrm{Irr}_1(G)$ with $\ker(\chi) = \ker(\psi)$, then $\chi(1) = \psi(1)$. (See § 148 and [Man30, Proposition 11].)

3010 (Inspired by [GMS1]) Study the p-groups G with $H \in \Gamma_1$ such that all maximal abelian subgroups of G are not contained in H.

3011 Classify the p-groups G such that $|N_G(A)| = |N_G(B)|$ for any nonnormal $A, B < G$.

3012 (Mann) Study the irregular p-groups G such that, whenever $H < G$, then, for all $p^k \leq \exp(H)$, one has (i) $\exp(\Omega_k(H)) = p^k$, (ii) $\mho_k(H) = \{x^{p^k} \mid x \in H\}$, (iii) $|\Omega_k(H)| = |H : \mho_k(H)|$. (See § 11.)

3013 Suppose that a p-group G has a cyclic subgroup of index p^4. Find the minimal index of a normal abelian subgroup in G.

3014 Classify the nonabelian p-groups G such that whenever $A \leq G$ is minimal nonabelian, then all subgroups of A are quasinormal in G.

3015 Estimate the number of groups of maximal class of order p^p and exponent $p > 5$.

3016 Study the Schur multipliers, representation groups and the automorphism groups of the nonabelian Dedekindian 2-groups.

3017 Let H be a proper noncyclic metacyclic subgroup of a p-group G. Study the structure of G provided $N_G(H)$ is metacyclic.

3018 Study the nonabelian two-generator p-groups all of whose factors of the upper (lower) central series, apart of the last one (apart of the first one), are cyclic. (See § 218.)

3019 Let $\beta(G)$ be the number of normal subgroups D of a p-group G such that G/D is minimal nonabelian. Find all residues of $\beta(G)$ modulo p.

3020 Does there exist an irregular p-group G such that all elements of the set $G - A$ have order p for any maximal abelian subgroup A of G?

3021 Given $k > 1$, classify the metacyclic p-groups covered by $\mathcal{A}_k$-subgroups.

3022 Classify the 2-groups all of whose maximal subgroups are either minimal nonabelian or Dedekindian. (See § 245.)

3023 Study the nonabelian p-groups of exponent $p^e > p$ all of whose maximal abelian subgroups, except one, have (i) exponent p^e, (ii) exponent $< p^e$.

3024 Given a p-group $H = \Omega_1(H)$, does the exist a p-group G such that $\Omega_1(G) \cong H$ and $|G/\mho_1(G)| = p^p$?

3025 Estimate the number of primary $\mathcal{A}_4$-groups of given order.

3026 Suppose that the Frattini subgroup of a p-group G has order p^n. Find the maximal possible derived length (class) of $\Phi(G)$.

3027 Classify the p-groups G of order $p^m > p^3$ satisfying the following condition: there is only one $k \in \{1, \dots, m-2\}$ such that G contain more than one normal subgroup of order p^k. (See Exercise 9.1 (b) and Lemma 9.1.)

3028 Given $e > 1$, study the p-groups of exponent p^e covered by abelian subgroups of rank two and exponent p^e.

3029 Let G be a p-group and let $H \in \Gamma_1$ be such that the set $G - H$ is covered by minimal nonabelian subgroups. Describe the structure of H. (See #860.)

3030 Let A be an abelian p-group of rank two, let Γ be a representation group of A. Study the structures of the representation groups of Γ.

3031 Study the p-groups G such that $\Psi = \langle \Phi(H) \mid H \in \Gamma_1 \rangle < \Phi(G)$. (In that case, $p > 2$ and $G/\Psi \cong S(p^3)$; see § 245.)

3032 Study the nonabelian p-groups G of exponent $> p$ such that, whenever $L < G$ is maximal cyclic of order $> p$, then the cyclic subgroups of G containing $\Phi(L)$ generate G.

3033 Study the p-groups G such that, whenever $C < G$ is nonnormal cyclic and $x \in G - \mathrm{N}_G(C)$, then $|\langle C, C^x \rangle : C| \le p^2$. (See #1762.)

3034 Does there exist n such that, for any p-group G, $p > 2$, it is regular if (i) $G/\mathrm{K}_n(G)$ is regular, (ii) $G/G^{(n)}$ is regular?

3035 Describe $\mathrm{Aut}(\mathrm{UT}(n, p))$, $\mathrm{Aut}(\Sigma_{p^n})$.

3036 Let Γ be a representation group of the group E_{p^d}. Find the number of orders of characteristic subgroups of Γ.

3037 Given a p-group G, does there exist a p-overgroup W of G such that $G \lhd W$ and $\mathrm{Aut}(W)$ is a p-group?

3038 Find the norm of a (i) special p-group, (ii) metacyclic p-group, (iii) p-group with cyclic derived subgroup, (iv) p-group with abelian subgroup of index p. (See §§ 140, 143, 171, 175.)

3039 Study the nonabelian p-groups G satisfying $|\mathrm{Z}(G/\ker(\chi))| = p$ for any $\chi \in \mathrm{Irr}^{\#}(G)$.

3040 Study the subgroup and normal structure of the direct product of two modular p-groups.

3041 Study the p-groups G, such that $\mathrm{Z}_p(G)$ is minimal irregular.

3042 Estimate $\alpha_1(G)$, where G is a special p-group with $|\mathrm{Z}(G)| = p^z$ and $\mathrm{d}(G) = d$. (See Appendix 87.)

3043 Study a nonabelian p-group G containing an abelian subgroup A of index p such that any two noncommuting elements of the set $G-A$ generate a subgroup of maximal class. (See [GMS1].)

3044 Study the p-groups G such that $x \in Z_2(G) - Z(G) \Rightarrow |G : C_G(x)| = p$.

3045 Study the p-groups G all of whose representation groups Γ satisfy $\mathcal{N}(\Gamma) = Z(\Gamma)$, where $\mathcal{N}(\Gamma)$ is the norm of Γ.

3046 Given a p-group G, describe the set of all nonabelian subgroups $H \triangleleft G$ such that $\bigcup_{\lambda \in \mathrm{Irr}_1(H)} \mathrm{Irr}(\lambda^G) \neq \mathrm{Irr}_1(G)$.

3047 Study the p-groups all of whose TI-subgroups are quasinormal.

3048 Given $n > 1$, study the p-groups G such that $Z_n(G)$ is minimal nonabelian.

3049 Study the p-groups G such that, whenever $A < G$, then $|B : A| \leq p$, where B is the quasinormal closure of A in G.

3050 Find the norm of the standard wreath product $A \operatorname{wr} B$ of p-groups A and B.

3051 Let S be a Sylow p-subgroup of the holomorph of an abelian p-group G. Estimate $\mathrm{cl}(S)$ and $\mathrm{dl}(S)$ in the terms of G.

3052 Study the nonabelian p-groups $G = \Omega_1(G)$ of order $> p^3$ all of whose proper two-generator subgroups are metacyclic. (In that case, $p = 2$.)

3053 Describe $\mathrm{Aut}(\Gamma)$, where Γ is a representation group of E_{p^n}.

3054 Let $\Phi_k(G)$ be the intersection of all subgroups of index p^k in a p-group G. Describe the quotient group $G/\Phi_k(G)$.

3055 Classify the nonabelian p-groups of order $> p^3$ containing exactly one minimal nonabelian subgroup of order $> p^3$.

3056 Given $e > 1$, study the irregular p-groups G of exponent p^e such that $\Omega_k^*(G)$ is regular for all $k \leq e$. (See Exercise A.45.97.)

3057 Estimate the class of a p-group G all of whose two-generator subgroups are of class ≤ 2. (This was solved by the second author in Theorem 224.1.)

3058 Classify the irregular p-groups G such that, whenever $A, B < G$ are maximal regular, there is $\phi \in \mathrm{Aut}(G)$ with $B = A^\phi$.

3059 Study the irregular p-groups, $p > 2$, all of whose maximal abelian subgroups are maximal regular.

3060 Find $\mathrm{cd}(\Gamma)$, where Γ is a representation group of an abelian p-group G of a given type.

3061 Study the p-groups G such that the set Γ_1 contains only one member of (i) rank 2, (ii) rank > 2.

3062 Classify the p-groups G that have no subgroups A, B with $|A| = p^3$ and $|B| = p^2$ such that $A \cap B = \{1\}$.

3063 Let p be a minimal prime divisor of the order of a group G and let $p^4 \mid |G|$. Study the "p-structure" of G (and, in particular, the structure and embedding of a Sylow p-subgroup of G provided G is not p-nilpotent) if all its subgroups of order p^4 are G-invariant. (See Appendix 56.)

3064 Study the p-groups in which any maximal subgroup is a direct product of two nonidentity subgroups.

3065 Describe the subgroup $\bigcap_A N_G(A)$, where A runs over all nonnormal cyclic (abelian) subgroups of a non-Dedekindian p-group G.

3066 Classify the p-groups containing exactly one maximal subgroup with noncyclic center. (This was solved by the second author; see § 141.)

3067 Given $k > 1$, study the p-groups all of whose $\mathcal{A}_k$-subgroups have derived subgroup of order p.

3068 (Old problem) Study the 2-groups all of whose elements are real.

3069 Let G be a p-group with known $c_n(G)$ for all n. Is it possible to find the number of elements of any order in G?

3070 Does there exist a p-group G such that $\Phi(G)$ contains a subgroup H with $\mathrm{Aut}(H) \cong \mathrm{Aut}(G)$?

3071 (A stronger variant of #1841.) Study the nonabelian p-groups G containing a maximal abelian subgroup A such that $C_A(x)$ is cyclic for any $x \in G - A$. (Original #1841 was solved by the second author; see Theorem 209.2)

3072 Given a metacyclic minimal nonabelian p-group A with noncyclic $Z(A)$, does there exist a p-group G such that $\Phi(G) \cong A$?

3073 Classify the 2-groups G whose nonreal elements are generators of conjugate cyclic subgroups.

3074 Study the 2-groups G satisfying $d(G) > d(H)$ ($|G : G'| > |H : H'|$) for all (i) $H < G$, (ii) $H < G$ of class ≤ 2. (See [Laf1] and §§ 113, 156. See also PhD thesis of E. Crestani and [Cre].)

3075 Study the p-groups G that do not have $p + 1$ isomorphic $\mathcal{A}_1$-subgroups. (See § 206.)

3076 Study the p-groups G such that $C_G(M)$ (i) is metacyclic for any minimal nonabelian $M < G$, (ii) minimal nonabelian for any metacyclic $M < G$.

3077 Study the structure and embedding of the intersection of normalizers of all those subgroups of a p-group whose orders are $> p$.

3078 Study the nonabelian p-groups G such that $|G/\ker(\chi) : Z(G/\ker(\chi))| = \chi(1)^2$ for all $\chi \in \mathrm{Irr}_1(G)$.

3079 Study the p-groups G all of whose subgroups of index p^2 are isomorphic.

3080 Classify the p-groups G satisfying $|T_\chi| = |Z(\chi)|(\chi(1)^2 - 1)$ for all $\chi \in \mathrm{Irr}(G)$.

3081 Let $\mathcal{L}_{\Omega\mho}(G)$ be the lattice of subgroups generated by Ω- and $\mho$-subgroups of a p-group G and let G_0 be a regular p-group. (i) Study the p-groups G satisfying $\mathcal{L}_{\Omega\mho}(G) \cong \mathcal{L}_{\Omega\mho}(G_0)$. (ii) Study the abelian (regular) p-groups G such that any characteristic subgroup of G is a member of the set $\mathcal{L}_{\Omega\mho}(G)$.

3082 Given $e > 1$, study the p-groups G of exponent p^e satisfying $\mathrm{cl}(\Omega_n^*(G)) < p$ for all $n \leq e$.

3083 Study the p-groups G such that, whenever (i) $x, y \in G - Z(G)$ are of equal order, then $|C_G(x)| = |C_G(y)|$, (ii) nonnormal $A, B < G$ are of equal order, then $|N_G(A)| = |N_G(B)|$.

3084 Find the representation groups of a minimal nonabelian p-group.

3085 Let G, G_0 be p-groups, $\exp(G_0) = p$. Suppose that $\mathcal{L}_N(G) = \mathcal{L}_N(G_0)$. Is it true that $\exp(G)$ is bounded?

3086 Study the p-groups G such that $G \neq \langle \mathcal{A}_1(G) - \{M\}\rangle$ for any $M \in \mathcal{A}_1(G)$.

3087 Classify the p-groups that are lattice isomorphic to prime power $\mathcal{A}_3$-groups. (See #1307 and [ZZLS].)

3088 Study the p-groups G of maximal class satisfying $\mathrm{d}(G_1) = p - 1$.

3089 Given $n > 2$, estimate the minimal order of a p-group containing all types of minimal nonabelian subgroups of order $\leq p^n$.

3090 Estimate the minimal order of a p-group containing all types of irregular groups of order p^{p+1}.

3091 Does there exist a p-group $G > \{1\}$ such that $\mathrm{Aut}(G) \cong \mathrm{Aut}(H)$ for some (i) $H \leq G'$, (ii) $H \leq \Phi(G)$, (iii) $H \leq \mho_1(G)$?

3092 Does there exist a special p-group (group of exponent p) all of whose maximal subgroups are pairwise nonisomorphic?

3093 Study the p-groups G such that $H_G = H \cap G'$ for all nonnormal $H < G$ not contained in G'. (Groups from Theorem 1.25 satisfy the condition.)

3094 Study the p-groups G with (i) $|\Phi(G)| = p^2$, (ii) $\mathrm{d}(G) = 2$ and $G/\mathrm{Z}(G) \cong \Sigma_{p^2}$.

3095 Classify the p-groups G such that $N^G = NG'$ for all nonnormal $N < G$. (This was solved by the second author; see § 191.)

3096 Study the nonmetacyclic p-groups G such that, whenever $M < G$ is maximal metacyclic, then $|M| = 2^{2n}$ and $\exp(M) = p^n$ for some n.

3097 Does there exist a minimal irregular p-group all of whose maximal subgroups are pairwise nonisomorphic?

3098 Classify the 2-groups G generated by subgroups $\cong Q_8$ in which any maximal subgroup of G containing a subgroup $\cong Q_8$ is Dedekindian.

3099 Does there exist a capable p-group containing a normal nonabelian metacyclic subgroup of order p^{2e} and exponent p^e? Consider in detail the case $p = e = 2$.

3100 (Mann) Classify the p-groups G isomorphic to a Sylow p-subgroup of $\mathrm{Aut}(G)$.

3101 Study the p-groups G such that $\mathrm{cl}(\mathrm{N}_G(H)) \leq 2$ for all nonnormal $H < G$ of class ≤ 2.

3102 Classify the nonmetacyclic p-groups all of whose maximal metacyclic subgroups have normal complements.

3103 (Ying [Yin2]) Study the p-groups G all of whose quotient groups are isomorphic to subgroups of G. (See § 192.) Consider the following cases: (i) $\exp(G) = p$, (ii) $\mathrm{cl}(G) = 2$, (iii) G is metacyclic.

3104 Study the irregular p-groups G all of whose regular subgroups are isomorphic to quotient groups of G.

3105 Classify the p-groups G in which the normalizer of any nonnormal cyclic subgroup is minimal nonabelian.

3106 Study the p-groups G such that, whenever $C < G$ is cyclic and $\phi \in \mathrm{Aut}(G)$, then $CC^\phi = C^\phi C$.

3107 Study the p-groups all of whose $\mathcal{A}_1$-subgroups are (i) metacyclic, (ii) metacyclic of orders $\leq p^4$.

3108 Is it true that if a p-group contains a regular subgroup of index p^2, then it contains a normal regular subgroup of index p^2?

3109 Given an abelian p-group A, find $\mathrm{dl}(P)$, where $P \in \mathrm{Syl}_p(\mathrm{Aut}(A))$.

3110 Let a p-group G contain an elementary abelian subgroup H such that all subgroups of H are characteristic in G. Is it true that $|H|$ is bounded?

3111 Study the p-groups G such that G/C is (i) minimal nonabelian, (ii) of maximal class for a cyclic $C \lhd G$.

3112 Does there exist a nonabelian group G of exponent p such that $|\mathrm{Aut}(G)|_p < |G|$?

3113 Given a nonabelian p-group G, let $a(G)$ be the number of maximal abelian subgroups of G of pairwise distinct orders. Find all possible values of $a(G)$ provided G has a maximal abelian subgroup of index p^2.

3114 Present the defining relations for the 2-groups G containing a normal subgroup H of maximal class such that G/H is either cyclic or generalized quaternion.

3115 Study the p-groups G such that $\mathrm{Z}(H) = \mathcal{N}(H)$ for all $H < G$.

3116 Study the nonabelian p-groups G such that $\mathrm{Z}(G)$ is isomorphic to the Schur multiplier of the quotient group $G/\mathrm{Z}(G)$.

3117 (Blackburn [Bla13]) Study the p-groups G, $p > 2$, possessing a subgroup L of order p such that it is contained in exactly one noncyclic subgroup of G of order p^2.

3118 Let G, G_0 be p-groups, $\mathrm{d}(G) = \mathrm{d}(G_0)$, $\Gamma_1 = \{A_1, \dots, A_k\}$ and the set of maximal subgroups of G_0 is $\{B_1, \dots, B_k\}$. Suppose that $A_i \cong B_i$ for $i = 1, \dots, k$. Find $|\mathrm{Z}(G)|$, $|G/G'|$ and other invariants of G if similar invariants for G_0 are known.

3119 Study the p-groups containing an $\mathcal{A}_2$-subgroup of index p.

3120 Study the nonabelian metacyclic p-groups (groups of exponent p) all of whose maximal abelian subgroups are normal.

3121 Study the p-groups G such that $H^G = H\mathrm{Z}(G)$ for any nonnormal $H < G$. (This is solved by the second author; see § 248.)

3122 Does there exist a group G of exponent $p^e > p$ such that $\Omega_e^*(G) \leq \Phi(\Phi(G))$?

3123 Given n, describe all nonabelian groups G of order p^n and exponent p such that $|\mathrm{Aut}(G)|_p$ is minimal possible.

3124 Study the irregular p-groups $G = A \times B$ such that, whenever $A_1 < A$ and $B_1 < B$, then $A_1 \times B$ and $A \times B_1$ are regular.

3125 Study the two-generator 2-groups G such that $G/\mathrm{Z}(G)$ is of maximal class.

3126 Study the irregular p-groups, $p > 2$, all of whose nonnormal regular subgroups are abelian. (See [FA].)

3127 Does there exist an irregular p-group G, $p > 2$, such that $L_G = \{1\}$ for any minimal irregular $L < G$? (For $p = 2$, this is not impossible. See Theorem 200.1.)

3128 Study the p-groups G containing $R \lhd G$ such that $G/R \cong \Phi(G)$ and all proper subgroups of G satisfy the above condition relative to an appropriate normal subgroup.

3129 Study the 2-groups G such that $G/Z(G)$ is minimal nonmetacyclic.

3130 Classify the p-groups G such that for any $H \lhd G$ there is $R \lhd G$ satisfying $G/R \cong H$.

3131 Study the p-groups G of exponent $p^e > p$ satisfying $\Omega_e^*(G) \leq \Omega_1(G) \leq \Phi(G)$.

3132 Study the structure of Aut(S), where S is a special p-group.

3133 Study the p-groups all of whose maximal subgroups are $\cong M_p(n, n) \times C_{p^k}$ for some n and k.

3134 Study the irregular p-groups G with absolutely regular derived subgroup. (See Appendix 8.)

3135 Describe Aut(G), where G is a p-group with cyclic subgroups $\Phi(G)$ and $Z(G)$.

3136 Study the nonabelian p-groups G such that $C_G(Z(A)) = A$ for all (i) minimal nonabelian $A \leq G$, (ii) $\mathcal{A}_2$-subgroups $A \leq G$. (Part (i) was solved by the second author; see Theorem 92.2.)

3137 (i) Classify the groups of order p^{2n+1} having an irreducible character of degree p^n. (ii) (Kazarin) Is it true that if $p = 2$ in (i), then G is extraspecial?

3138 Study the irregular p-groups G such that for any (i) $H < G$ of exponent p there is $N \lhd G$ satisfying $G/N \cong H$, (ii) epimorphic image $\bar{G}$ of exponent p there is $H < G$ satisfying $H \cong \bar{G}$.

3139 Let $M \in \Gamma_1$, where G is a p-group. Study the structure of G if, whenever $L < G$ is such that $L \nleq M$, then $L^G \in \Gamma_1$.

3140 Study the p-groups G of exponent p such that, whenever $H < G$ is of index p^2, there is $N \lhd G$ satisfying $G/N \cong H$.

3141 Study the p-groups G all of whose maximal subgroups have absolutely regular Frattini subgroups but $\Phi(G)$ is irregular.

3142 Study the 2-groups G all of whose proper nonabelian subgroups are of the form $Q \times A$, where Q is a generalized quaternion group and A is abelian. Consider a more general case when Q is of maximal class. (See § 245.)

3143 (Janko) Classify the p-groups (i) all of whose maximal subgroups, except one, are s-self dual, (ii) which are minimal non-s-self dual. (See § 192. Part (ii) was solved by the second author; see § 193.)

3144 Classify the p-groups G of order $> p^{p+2}$ all of whose maximal subgroups, except one, are irregular. (If G is of order p^{p+2}, it is of maximal class, by Theorem 13.6.)

3145 Study the 2-groups $G = AB$, where A is abelian, B is generalized quaternion and $A_G B_G = \{1\}$.

3146 Study the p-groups G such that $N_G(N_G(L)) = G$ for all cyclic $L < G$.

3147 Let H be the subgroup generated by the centers of all nonabelian subgroups of a p-group G. Study the structure of G provided H is (i) abelian, (ii) minimal nonabelian, (iii) metacyclic.

3148 Classify the p-groups G such that $N_G(H)$ is modular for every nonnormal $H < G$.

3149 Classify the irregular p-groups all of whose absolutely regular subgroups have order $\leq p^{p+1}$.

3150 Classify the p-groups, $p > 2$, all of whose $\mathcal{A}_1$-subgroups are isomorphic either to $S(p^3)$ or $M_p(n, n)$.

3151 Study the nonabelian p-groups all of whose proper subgroups have an abelian subgroup of index p.

3152 Study the p-groups G such that $\mho_1(H) = \Phi(H)$ for all $H \leq G$.

3153 Study the nonabelian p-groups G such that, whenever $A < G$ is maximal abelian, then $A/Z(G) \cong E_{p^2}$. (See § 20.)

3154 Given an abelian p-group A, find $\text{cl}(P)$ and $\text{cl}(H)$, where P, H are Sylow p-subgroups of $\text{Aut}(A)$ and holomorph of A, respectively.

3155 Given n, estimate the number of pairwise nonisomorphic extensions of E_{p^n} by a group of order p.

3156 Classify the p-groups in which the normalizer of each nonnormal abelian subgroup is either abelian or minimal nonabelian.

3157 Study the p-groups G such that $\Omega_2(G)$ is special.

3158 Study the p-group all of whose metacyclic subgroup are abelian.

3159 Study the nonabelian p-groups in which the centralizer of each minimal nonabelian subgroup is cyclic. (For a solution, given by the second author, see § 227.)

3160 Classify the q-self dual p-groups G (i) with cyclic G', (ii) of exponent p. (See § 192.)

3161 Classify the p-groups all of whose nonnormal subgroups are either abelian or minimal nonabelian. (See [FA].)

3162 Study the irregular p-groups $G = AB$, where $|A/\mho_1(A)| = p^u$, $|B/\mho_1(B)| = p^v$ and $u + v = p$.

3163 Study the 2-groups $G = \langle x, E \rangle$, where $o(x) = 2$, $E \cong E_4$.

3164 Given $d > 2$, study the d-generator special p-groups G such that $|Z(G)| = p^k$ where $k \in \{\frac{1}{2}d(d-1), \frac{1}{2}d(d-1) - 1\}$.

3165 Let Γ be a representation group of E_{p^d}, $d > 2$. Classify all epimorphic images of Γ of order (i) p^{d+1}, (ii) p^{d+2}, (iii) $\frac{1}{p}|\Gamma|$.

3166 Let Γ be as in #3165. Classify all normal subgroups of Γ nonincident with $Z(G)$.

3167 Study the s-self dual (q-self-dual) special p-groups.

3168 Study the structure of the quotient group $G/\Omega_e^*(G)$ for a group G of exponent $p^e > p$. Is it true that any p-group of exponent $< p^e$ is isomorphic to a subgroup of $G/\Omega_e^*(G)$ for an appropriate G?

3169 Study the 2-groups $G = \langle x, y \rangle$, where $o(x) = 2$, $o(y) = 2^n$, $n > 1$.

3170 Study the p-groups all of whose non-quasinormal subgroups have equal order.

3171 Study the irregular p-groups all of whose maximal regular subgroups have index p.

3172 Study the p-groups of exponent $> p$ all whose subgroups of order p, except one, are maximal cyclic. (Examples: M_{p^n}, any irregular group of order p^{p+1}.)

3173 Classify the p-groups all of whose metacyclic subgroups have cyclic subgroups of index p.

3174 Let G be a nonabelian p-group containing a metacyclic and a nonmetacyclic $\mathcal{A}_1$-subgroup. Study the structure of the quotient group G/K, where K is generated by all metacyclic (nonmetacyclic) $\mathcal{A}_1$-subgroups.

3175 Study the p-groups all of whose nonabelian maximal subgroups are two-generator.

3176 Study the p-groups G possessing an $\mathcal{A}_1$-subgroup A such that all maximal subgroups of A are 2-uniserial in G.

3177 Classify the p-groups all of whose maximal abelian subgroups have index p^2.

3178 Study the p-groups G such that $A \le \mathrm{Z}(A^G)$ for any abelian $A < G$. (This is solved by the second author; see Exercise 200.2.)

3179 Classify the nonabelian two-generator p-groups in which the center of any proper nonabelian epimorphic image is cyclic. (See § 218.)

3180 Suppose that $E < G$ is a special subgroup of a p-group G. Study the structure of G if any subgroup of G of order $p|E|$ containing E is special.

3181 Given n, classify the abelian p-groups G such that p^n does not divide $\exp(\mathrm{Aut}(G))$. (Obviously, $\exp(G) \le p^n$ if $p > 2$ and $\le p^{n+1}$ if $p = 2$.)

3182 Study the p-groups G such that $|\mathrm{Z}_2(H)| \le p^3$ for all nonabelian $H \in \Gamma_1$.

3183 Study the p-groups G such that $\mathrm{Z}(M) \le \mathrm{Z}_2(G)$ for any nonabelian (i) $M < G$. (ii) $M \in \Gamma_1$.

3184 Study the nonabelian p-groups G such that $\mathrm{C}_G(A) \cong A$ for any minimal nonabelian $A < G$.

3185 Study the connection between the Schur multiplier of a p-group G and the Schur multiplier of a representation group of G.

3186 Classify the q-self dual (i) p-groups of maximal class (examples: D_{2^n} and SD_{2^n} but not Q_{2^n}), (ii) p-groups with cyclic derived subgroup, (iii) groups of exponent p, (iv) $\mathcal{A}_2$-groups.

3187 Classify the p-groups all of whose nonnormal metacyclic subgroups are cyclic. (See § 16.)

3188 Study the p-groups G all of whose subgroups of order p^p and exponent p are normal but G has no subgroup of order p^{p+1} and exponent p.

3189 Study the p-groups G all of whose subgroups $H < G$ satisfying $|H|^3 \le p|G|$ are normal. (See § 210.)

3190 Study the p-groups $G = \Omega_1(G)$ such that $\mathrm{C}_G(x)$ is abelian for any $x \in G - \mathrm{Z}(G)$ of order p.

3191 Classify the 2-groups G such that $\mathrm{Aut}(G)$ is metacyclic.

3192 Study the p-groups containing exactly two conjugacy classes of subgroups of order p^p and exponent p.

3193 Study the p-groups all of whose normal subgroups are incident with (i) $\Phi(G)$, (ii) G', (iii) $\mho_1(G)$, (iv) $\mathrm{Z}(G)$ (four problems).

3194 Let $p^{\nu(X)}$ be the maximal order of an abelian subgroup of a nonabelian p-group X. Classify the nonabelian p-groups G all of whose abelian subgroups of order $< p^{\nu(G)-1}$ are G-invariant. (This is solved in § 222.)

3195 Study the p-groups all of whose maximal subgroups, except one, are $\mathcal{A}_2$-groups. (See §§ 100–102.)

3196 Suppose that a p-group G has a non-trivial Schur multiplier and Γ is a representation group of G. Is it true that $\alpha_1(\Gamma) > \alpha_1(G)$?

3197 Study the nonabelian p-groups G such that $A' \leq A_G$ for any $A < G$.

3198 Study the p-groups G containing a normal metacyclic subgroup M such that G/M is (i) cyclic (ii) $\cong \mathrm{E}_{p^2}$.

3199 Study the structure of G', where a p-group G is an $\mathcal{A}_n$-group, $n \in \{4, 5\}$. (See § 76.)

3200 Study the irregular p-groups in which all minimal irregular subgroups are $\cong \Sigma_{p^2}$. (For $p = 2$, this is solved in Proposition 10.28 and § 90.)

3201 Classify the nonabelian p-groups containing a maximal abelian subgroup which is of (i) type (p^2, p), (ii) type (p^2, p^2).

3202 Study the p-groups of class 2 all of whose $\mathcal{A}_1$-subgroups are nonnormal.

3203 Study the nonabelian p-groups G such that $[A, B] \leq A \cap B$ for any distinct maximal abelian $A, B < G$.

3204 Classify the p-groups one of whose representation groups is an $\mathcal{A}_2$-group.

3205 Given a noncentral $H < G$, study the p-groups G such that, for any $x \in G - H$ there is $a \in H$ such that $\langle a, x\rangle$ is minimal nonabelian. (Compare with Lemma 57.1.)

3206 Study the p-groups G such that, whenever $A, B < G$ are distinct conjugate subgroups, then $A \cap B = A_G$. (See #2740.)

3207 Study the p-groups G such that $\mathcal{N}(H) = \mathrm{Z}(H)$ for any $H \leq G$.

3208 Study the p-groups, $p > 2$, in which all factors of the lower (upper) central series have order $\leq p^2$.

3209 Find the minimal order of a 2-group containing minimal nonmetacyclic subgroups of all possible types.

3210 Classify the p-groups G containing a 2-uniserial subgroup (i) of order p^2, (ii) of maximal class.

3211 Classify the p-groups G such that, whenever $H < G$ is nonnormal, then $|H_G| \leq p^2$.

3212 Classify the p-groups all of whose nonnormal abelian subgroups are (i) cyclic, (ii) of rank ≤ 2. (See § 16.)

3213 Study the p-groups G such that $|C^G : C| \leq p^2$ for any cyclic $C < G$.

3214 Let A be a regular p-group with $\exp(A') > p$ and let $\mathcal{M}$ be the set of those regular p-groups B such that $A \times B$ is irregular (by Grün, $\exp(B') > p$). Describe the minimal members of the set $\mathcal{M}$.

3215 Given $k > 1$, classify the p-groups G such that $|G'| = p^{2k+1}$ and $|H'| = p^k$ for all $H \in \Gamma_1$. (See Exercise 1.69.)

3216 Study the p-groups G such that $|H'| \leq p$ for all nonnormal $H < G$. (See §§ 137, 139.)

3217 Study the p-groups G such that if $N_1, N_2 \trianglelefteq G$, then $G/N_1 \cong G/N_2 \iff N_1 = N_2$.

3218 Study the p-groups G all of whose commutators have orders $\leq p$ but $\exp(G') > p$.

3219 Study the p-groups in which the intersection of any two distinct abelian subgroups is normal.

3220 Let $H \in \Gamma_1$. Describe the structure of a p-group G, provided $A \cap H \triangleleft G$ for any $A < G$ nonincident with H.

3221 Study the p-groups G all of whose subgroups $\nleq \Phi(G)$ are complemented.

3222 Classify the nonabelian p-groups G such that $A \cap B$ is maximal either in A or in B for any two nonincident $A, B < G$ (This was solved by the second author; see Theorem 213.1.)

3223 Study the nonabelian p-groups G such that for any minimal nonabelian $A < G$ there is an abelian $B < G$ such that $\langle A, B \rangle = G$.

3224 Does there exist a primary $\mathcal{A}_n$-group G, $n > 2$, such that $M \cap Z(G) = \{1\}$ for any $\mathcal{A}_2$-subgroup $M < G$?

3225 Study the p-groups G containing only one normal maximal abelian subgroup and all nonnormal maximal abelian subgroups of G are soft.

3226 Study the 2-groups G such that, for any nonabelian $H \in \Gamma_1$, one has $H = Q_8 \times A$, where A is abelian.

3227 Classify the minimal nonabelian p-groups (p-groups of maximal class) with trivial Schur multiplier.

3228 Study the irregular p-groups G, $p > 2$, all of whose subgroups H of class > 2 satisfy $|H : H'| = p^2$.

3229 Classify the minimal nonabelian p-groups which can be normal subgroups of capable p-groups.

3230 Classify the p-groups G such that, whenever $H < G$ is of order $> p$ and $|H|^2 < |G|$, then $H \triangleleft G$. (Compare with Theorems 1.25 and 210.1.)

3231 Study the irregular p-groups all of whose minimal irregular subgroups have order p^{p+1}.

3232 Given $n > k > 3$, study the p-groups of order p^n all of whose subgroups of order p^k have equal minimal number of generators.

3233 Study the two-generator p-groups G such that $G/Z(G)$ is minimal nonabelian. (See [QZ2, QH2, QZ].)

3234 Let P be a noncyclic p-group and $\alpha \in \mathrm{Aut}(P)$ be of order p. Study the structure of P provided all subgroups of P are stabilized by α.

3235 Study the p-groups G such that, whenever $H < G$ is nonnormal, there is a cyclic $C \leq H$ satisfying $H^G = C^G$.

3236 Study the p-groups all of whose nonnormal nonabelian subgroups (i) have equal order, (ii) are conjugate.

3237 (Correction of #1021.) Study the p-groups G such that, whenever cyclic $A, B < G$ are noncentral and not conjugate, then $C_G(A) \neq C_G(B)$.

3238 Classify the p-groups all of whose nonnormal abelian subgroups have abelian normalizers. (See § 224.)

3239 Study the p-groups G such that, whenever $A, B \triangleleft G$ have equal index $> p$, then $A \cong B$.

3240 Given an abelian p-group G, construct $\phi \in \mathrm{Aut}(G)$ fixing the minimal possible number of elements of G.

3241 Classify the p-groups all of whose maximal s-self dual subgroups have index p. (A partial case is solved by the second author; see § 193.)

3242 Study the nonmetacyclic 2-groups all of whose minimal nonmetacyclic subgroups have order 2^5. (See § 50.)

3243 Study the p-groups G such that $\exp(\mathrm{C}_G(x)) = o(x)$ for all $x \in G - \mathrm{Z}(G)$.

3244 Let G, G_0 be nonabelian p-groups satisfying the following condition: there is $1-1$ correspondence between the sets $G - \mathrm{Z}(G)$ and $G_0 - \mathrm{Z}(G_0)$ such that, whenever $x \in G - \mathrm{Z}(G)$ and $x_0 \in G_0 - \mathrm{Z}(G_0)$ are corresponding elements, then $\mathrm{C}_G(x) \cong \mathrm{C}_{G_0}(x_0)$. Compare the structures of G and G_0.

3245 (Correction of #1018.) Study the nonabelian p-groups G such that, whenever cyclic $F, H < G$ are nonnormal and $F^G = H^G$, then F and H are conjugate in G.

3246 Classify the p-groups in which normal closures of all nonnormal cyclic subgroups (i) are minimal nonabelian, (ii) $\mathcal{A}_2$-subgroups (both these problems were solved by the second author; see § 223.)

3247 Study the p-groups G containing exactly one normal subgroup of any index $\geq p^{\mathrm{d}(G)}$.

3248 Study the p-groups G, $p > 2$, such that $|H : H'| = p^2$ for all, except one, $H \in \Gamma_1$.

3249 Classify the special p-groups with the Schur multiplier of order $\leq p$.

3250 Let G be a p-group with nonabelian $\Phi(G)$ and let $\alpha_1(G \mid \Phi(G))$ be the number of minimal nonabelian subgroups of G not contained in $\Phi(G)$. Find minimal possible value of $\alpha_1(G \mid \Phi(G))$ provided $\mathrm{d}(G)$ is known.

3251 Study the p-groups such that, whenever $H < G$ is non-quasinormal, then all non-quasinormal subgroups of G of order $|H|$ are conjugate.

3252 Does there exist a nonabelian p-group G such that $\mathrm{C}_G(x)$ is minimal nonabelian for any $x \in G - \mathrm{Z}(G)$? (This is solved by the second author; see Exercise 200.3. See also Appendix 60.)

3253 Study the p-groups G such that $|\mathrm{cd}(H)| \leq 2$ for all $H < G$.

3254 (Old problem). Study the irregular p-groups all of whose regular subgroups are metabelian.

3255 Does there exist a p-group G satisfying (i) $|\mathrm{Aut}(G)| \leq |\mathrm{Aut}(G')|$, (ii) $|\mathrm{Aut}(G)| \leq |\mathrm{Aut}(\Phi(G))|$, (iii) $|\mathrm{Aut}(G)| \leq |\mathrm{Aut}(\mathrm{Z}(G))|$, (iv) $|\mathrm{Aut}(G)| \leq |\mathrm{Aut}(\mho_1(G))|$ (four problems).

3256 Study the p-groups G in which any proper subgroup is isomorphic to a subgroup of an appropriate proper epimorphic image of G.

3257 Study the nonabelian p-groups all of whose subgroups of class two are metacyclic.

3258 Classify the capable (i) minimal nonabelian p-groups, (ii) metacyclic p-groups, (iii) p-groups of maximal class, (iv) $\mathcal{A}_2$-groups.

3259 Study the nonabelian p-groups, $p > 2$, all of whose minimal nonabelian subgroups are isomorphic to M_{p^3}. (See Mann's commentary to #115.)

3260 Classify the p-groups, $p > 2$, containing exactly one proper irregular subgroup. (For $p = 2$ such groups do not exist; see § 76.)

3261 Classify the nonabelian p-groups G such that for all nonnormal abelian $A < G$ one has (i) $\mathrm{cl}(A^G) \le 2$, (ii) $\mathrm{dl}(A^G) \le 2$.

3262 Study the p-groups G such that $\mathrm{s}_n(G) = \mathrm{s}_n(E)$ for all n, where E is (i) extraspecial, (ii) special, (iii) an $\mathcal{A}_i$-group. $i = 1, 2, 3$.

3263 Study the p-groups G such that, whenever $H < G$ is nonnormal, there is $M \lhd N_G(H)$ such that $N_G(H)/M \cong H$.

3264 Study the p-groups G such that, whenever $A < G$ is noncyclic, then $A_G > \{1\}$. (Example: G is a 2-group of maximal class.)

3265 (Inspired by one Zhmud's result) Study the p-groups G such that $|C_G(x)| \ge \mathrm{k}(G)$ for all $x \in G$.

3266 Study the p-groups G such that $|\langle H, H^x\rangle : H| \le p$ for all $H < G$ and $x \in G$.

3267 Study the p-groups containing an extraspecial (special) subgroup whose normalizer is (i) extraspecial, (ii) special.

3268 Classify the irregular p-groups all of whose regular subgroups of maximal order are (i) metacyclic, (ii) absolutely regular.

3269 Classify the nonabelian p-groups G such that $|G : HC_G(H)| \le p$ for each nonabelian $H < G$. (This is solved by the second author; see § 238.)

3270 Classify the nonmetacyclic p-groups containing a subgroup of (i) order p with metacyclic centralizer (see § 198), (ii) type (p, p) with metacyclic centralizer.

3271 Study the p-groups G all of whose nonnormal subgroups (proper subgroups) containing $Z(G)$, are Dedekindian.

3273 Study the p-groups G such that whenever $H < G$ with $p|H|^2 < |G|$, then $H \lhd G$. (See § 210. This was solved by Malinowska [Mal7].)

3274 Study the p-groups $G = \langle a_1, \dots, a_d\rangle$ such that $|\langle a_i, a_j\rangle'| \le p$ for all $i \ne j$.

3275 Study the irregular p-groups G such that $C_G(x)$ is absolutely regular for any $x \in G - Z(G)$.

3276 Classify the p-groups in which the normal closure of any nonnormal cyclic subgroup is nonabelian. (This was solved by the second author; see Theorem 223.1.)

3277 Does there exist a p-group whose derived subgroup is minimal nonabelian and coincides with derived subgroups of all its maximal subgroups?

3278 Study the p-groups $G = \Omega_1(G)$ such that for any $x \in G - \Phi(G)$ of order p the subgroup $\langle x, \Phi(G)\rangle$ is of maximal class.

3279 (Alperin [Alp1]) Study the 2-groups all of whose two-generator subgroups have cyclic derived subgroups.

3280 Study the p-groups G such that for any $x \in G$ there is a cyclic $L < G$ with $G = \langle L, C_G(x)\rangle$.

3281 Suppose that an irregular 3-group $G = A \times B$, where A and B are metacyclic. Describe the minimal irregular subgroups of G.

3282 Study a special p-group S such that $S = \Phi(G)$ for an appropriate p-group G.

3283 Study the nonabelian p-groups G such that $Z(G) = A \cap B$ for any distinct maximal abelian $A, B < G$. (This was solved by the second author for $p = 2$; see Theorem 91.3.)

3284 Does there exist a 2-group G of exponent $p^e > p$ such that $H_p(G)$ is contained in (i) G', (ii) $\mho_1(G)$, (iii) $\Phi(G)$?

3285 Study the p-groups covered by normal (i) subgroups nonincident with $\Phi(G)$, (ii) cyclic subgroups, (iii) abelian subgroups, (iv) subgroups of class 2, (v) metabelian subgroups, (vi) minimal nonmetacyclic subgroups, (vii) minimal nonabelian subgroups. (See ## 2638, 2452, 860.)

3286 Study the p-groups G that are not generated by nonnormal noncyclic subgroups.

3287 Study the p-groups G that are not covered by nonnormal noncyclic subgroups. (See #3285.)

3288 Study the non-Dedekindian p-groups G such that the normal closure A^G is characteristic for any nonnormal subgroup A of G.

3289 Given $n > 3$, classify the 2-groups containing exactly one proper subgroup isomorphic with (i) D_{2^n}, (ii) Q_{2^n}, (iii) SD_{2^n}, (iv) M_{2^n}.

3290 Given a minimal nonabelian p-group A, study the p-groups containing exactly one subgroup isomorphic with A.

3291 Given a p-group H, does there exist p-groups G_0 and G such that $G' \cong H \cong \Phi(G_0)$.

3292 Classify the 2-groups containing exactly one subgroup $\cong \mathcal{H}_{2,2}$.

3293 Study the p-groups all whose noncyclic abelian subgroups are normal.

3294 Study the primary $\mathcal{A}_n$-groups G, $n > 2$, such that, whenever $C < M \le G$, where C is a maximal cyclic (abelian) subgroup of an $\mathcal{A}_1$-subgroup ($\mathcal{A}_2$-subgroup) M, then C is a maximal cyclic (abelian) subgroup of G.

3295 Study the p-groups all of whose $\mathcal{A}_2$-subgroups have no abelian subgroup of index p. (See § 71.)

3296 Classify the p-groups in which the normal closures of all nonnormal subgroups coincide. (This is solved by the second author; see § 242.)

3297 Given $2 < k < n$, classify the $\mathcal{A}_n$-groups which are 2-groups all of whose $\mathcal{A}_k$-subgroups are metacyclic. (See § 204.)

3298 Classify the non-Dedekindian p-groups G such that G/H^G is cyclic for any nonnormal cyclic $H < G$.

3299 Study the p-groups in which any two-generator subgroup is metacyclic (minimal nonabelian).

3300 Study the p-groups G such that, $a_1, \dots a_k \in G \Rightarrow |\Omega_1(\langle a_1, \dots, a_k\rangle)| \le p^k$.

3301 Study the p-groups G such that, whenever $H \le G$, then $|\Omega_1(H)| \le p^{d(H)}$. The same problem where H runs over all proper subgroups (all maximal subgroups) of G.

3302 Study the p-groups G satisfying $|H'|^2 < |H|$ for all $H \le G$.

3303 Classify the p-groups having at most $p+2$ pairwise noncommuting elements. (See Theorem 116.4.)

3304 Study the p-groups G such that, whenever $\{a_1, \dots, a_n\}$ is a maximal set of pairwise noncommuting elements of $H \in \Gamma_1$, then $\bigcup_{i=1}^{n} C_G(a_i) = H$. (See § 116.)

3305 A subgroup $H \leq G$ is a *waist* of a p-group G if it is a unique G-invariant subgroup of its order. Classify the p-groups in which all members of their upper central series are waists. (See § 225.)

3306 Study the p-groups G such that $A \lhd H \in \Gamma_1 \Rightarrow A \lhd G$.[5]

3307 Study the primary $\mathcal{A}_n$-groups, $n > 2$, that are not generated by nonnormal minimal nonabelian subgroups.

3308 Classify the non-Dedekindian p-groups G such that $|G : H^G| \leq p^2$ for all nonnormal $H < G$.

3309 Let $c(G)$ be the number of nonidentity cyclic subgroups in a group G. Find $c(\Sigma_{p^n})$. (For example, $C(D_{2^{n+1}}) = 2^n + n$.)

3310 Let G is an extension of C_{p^m} be C_{p^n}. Find $c(G)$ (See #3309.)

3311 Study the p-groups G in which any noncyclic abelian subgroup, say A, is contained in a minimal nonabelian subgroup of order $p|A|$.

3312 A subgroup H of a p-group G is said to be *isolated* if for any cyclic $C < G$, $H \cap C > \{1\} \Rightarrow C \leq H$. (i) Classify the prime power $\mathcal{A}_n$-groups, $n > 1$, all of whose minimal nonabelian subgroups are isolated. (This was solved by the second author; see § 251.) (ii) Study the p-groups, $p > 2$, containing an isolated maximal subgroup. (iii) Study the irregular p-groups containing an isolated maximal among subgroups of exponent p. (iv) Study the nonabelian p-groups containing exactly one nonisolated maximal abelian subgroup.

3313 Study the p-groups G such that $A \cap B$ is an $\mathcal{A}_1$-subgroup for any distinct $A, B \in \Gamma_1$.

3314 Let $G = AB$, where $A, B < G$ are abelian. Estimate $d(H)$ for $H < G$.

3315 Study the p-groups G all of whose abelian subgroups not contained in a fixed maximal subgroup of G, are cyclic.

3316 Study the 2-groups G containing a subgroup $Q \cong Q_8$ such that $Q \leq H \in \Gamma_1 \Rightarrow H$ is Dedekindian.

3317 Study the non-Dedekindian p-groups G such that whenever H is nonnormal in G, there is only one maximal chain connecting H with G.

3318 Let $a(G)$ be the number of orders of maximal abelian subgroups in a p-group G. Find maximal $a(G)$ if G runs through all (i) p-groups, (ii) Sylow p-subgroups of the symmetric groups, (iii) Sylow p-subgroups of general linear groups over Galois fields, (iv) metacyclic groups.

---

5 If H is a subgroup of maximal class and index p in a p-group G which is not of maximal class, then $A \lhd H \Rightarrow A \lhd G$. Indeed, if A is non-G-invariant, when $N_G(A) = H$ is of maximal class so G if of maximal class (Remark 10.5), a contradiction.

3319 (Janko) Classify the non-Dedekindian p-groups G such that $\Phi(G) \le H^G$ ($G' \le H^G$) for all nonnormal $H < G$.

3320 Let D be the subgroup generated by centers of all minimal nonabelian subgroups of a nonabelian p-group G. Study the structure of the quotient group G/D. (If G is regular, then $\exp(G/R) = p$.)

3321 Let G be a nonabelian p-group. Clearly, if any abelian subgroup of G has the trivial Schur multiplier, then $G \cong Q_{2^n}$. Classify the nonabelian p-groups all of whose nonabelian subgroups (minimal nonabelian subgroups) have trivial Schur multiplier.

3322 Let G be a nonabelian p-group. Suppose that $N_G(A)$ is abelian for all nonnormal abelian $A < G$. Describe the structure of G.

3323 Classify the nonabelian p-groups all of whose maximal abelian subgroups are isolated. (This was solved by the second author; see § 252.)

3324 Classify the p-groups all of whose nonnormal subgroups are either absolutely regular or of maximal class. (See Exercise A.51.39.)

3325 Classify the nonabelian p-groups G, containing a nonabelian maximal subgroup H such that all maximal abelian subgroups of H, except one, are maximal abelian in G. (Janko) In particular, consider the case when the above condition holds for all nonabelian $H \in \Gamma_1$. (Example: D_{2^n}, $n > 3$.)

3326 Classify the p-groups G, $p > 2$, such that, whenever $H \in \Gamma_1$, then one of the following holds: H is either abelian, or minimal nonabelian, or $\exp(H) = p$.

3327 Classify the p-groups G, $p > 2$, (i) all of whose maximal subgroups are either abelian or of exponent p. (ii) with an abelian subgroup of index p all of whose minimal nonabelian subgroups have order p^3.

3328 Classify the irregular p-groups all of whose maximal regular subgroup are either isolated or of exponent p. (See #3312. Let G be a n irregular p-group of maximal class. Then the group $G/K_{p+1}(G)$ satisfies the above condition.)

3329 Study the irregular p-groups of exponent $> p > 2$, all of whose nonabelian (minimal nonabelian) subgroups of exponent $> p$ are isolated.

3330 Study the p-groups G, $p > 2$, such that, whenever $A \le H \in \Gamma_1$, where A is a maximal abelian subgroup of G, then A is isolated in H.

3331 Study the primary $\mathcal{A}_n$-groups G, $n > 1$, such that, whenever $A \le H \in \Gamma_1$, where A is an $\mathcal{A}_1$-subgroup, then A is isolated in G.

3332 Study a 3-group G whose H_3-subgroup is minimal nonabelian. (See Appendix 101.)

3333 Study the p-groups H that are isolated in G, where G is a Sylow p-subgroup of the holomorph of H. Consider in detail the case when H is abelian.

3334 Classify the nonabelian p-groups G such that, whenever $A \le G$ is minimal nonabelian, then all normal subgroups of A are normal in G. (A partial case of this was solved in [AQXY].)

3335 Classify the p-groups G such that, whenever $L \le G$ is nonnormal cyclic, then $C_G(L) = N_G(L)$. (Example: G is a minimal nonabelian p-group.)

3336 Classify the 3-groups G containing a nonmetacyclic minimal nonabelian subgroup M of index 3 such that all elements of the set $G - M$ have order 3. (See Appendix 101.)

3337 Study the p-groups in which any cyclic (abelian) subgroup is contained in an appropriate $\mathcal{A}_2$-subgroup.

3338 Study the nonabelian p-groups G admitting a p-automorphism ϕ that has no fixed points on the set $G - Z(G)$ (on the set $G - \Phi(G)$).

3339 Study the nonabelian p-groups G admitting a p-automorphism ϕ such that, whenever H is a noncentral subgroup of G ($H < G$ is nonincident with $Z(G)$), then $H^\phi \neq H$.

3340 Study the 2-group whose nonabelian subgroup (derived subgroup) is nonabelian Dedekindian.

3341 Find all possible residues $s_n(G) \pmod{2^4}$, where G runs through all 2-groups.

3342 Study the 3-groups G containing an isolated $H \in \Gamma_1$. (For a partial case, see #3336.)

3343 Study the noncyclic 2-groups all of whose maximal subgroups have pairwise distinct classes.

3344 Study the primary $\mathcal{A}_n$-groups G such that $N_G(H)/H$ is cyclic for any $\mathcal{A}_1$-subgroup $H < G$. (See Appendix 101.)

3345 Study the p-groups G such that, whenever $A < B < G$, where A is a maximal abelian and $|B : A| = p$, then B is minimal nonabelian.

3346 Classify the nonmetacyclic 2-groups G such that, whenever $M < G$ is metacyclic, then either $|\Omega_1(M)| \leq 4$ or M is of maximal class.

3347 Classify the p-groups without abelian subgroup of type (p^2, p^2) (nonabelian metacyclic subgroup of order p^4 and exponent p^2).

3348 Study the p-groups G such that, whenever $M < G$ is a maximal normal of exponent p, then M is isolated.

3349 Classify the nonabelian primary groups all of whose minimal nonabelian subgroups (maximal abelian subgroups) are isolated in their normalizers.

3350 Study the irregular p-groups all of whose maximal subgroups of exponent p have orders $\leq p^p$. (Example: any irregular p-group of maximal class.)

3351 Classify the nonabelian groups of order $> p^3$ and exponent p such that, whenever $A, B < G$ are distinct minimal nonabelian, then $A \cap B > \{1\}$.

3352 Study the nonabelian p-groups G such that, whenever $A, B < G$, where A is maximal abelian and B is minimal nonabelian, then $A \cap B$ is maximal in B.

3353 Study the nonabelian p-groups G containing exactly one minimal nonabelian subgroup A with $A_G > \{1\}$.

3354 Study the irregular p-groups of order p^{2p} and exponent p^p.

3355 Study the irregular p-groups G such that for any two distinct $A, B \in \Gamma_1$, one has $\mathrm{cl}(A) + \mathrm{cl}(B) \leq p + 1$.

3356 Study the p-groups G such that, whenever distinct $A, B < G$ satisfy $| |A| |B| \geq |G|$, then $AB = BA$.

3357 Study the nonmetacyclic (non-absolutely regular) p-groups, all of whose maximal metacyclic (absolutely regular) subgroups are isolated (see #3312).

3358 Study the irregular p-groups generated by any two distinct maximal regular subgroups.

3359 Let $G = \Omega_1(G)$ be a nonabelian p-group; then G contains a subgroup $\cong S(p^3)$. Study the structure of G provided all its subgroups $\cong S(p^3)$ are isolated.

3360 Study the structure of a p-group G provided it is isolated in a Sylow p-subgroup W of the holomorph of G. Consider in detail the case when $G = \mathrm{H}_p(W)$.

3361 Study a p-Group G whose Schur multiplier is isolated in some representation group of G. (Example: Any abelian group of type (p^n, p).)

3362 Study the irregular p-groups, $p > 2$, all of whose minimal irregular subgroups are isolated.

3363 Study the nonabelian p-groups G of exponent $> p^2$ such that, whenever $A, C \le G$, where A is minimal nonabelian and C is cyclic, then $|A \cap C| > p$, implies $C < A$ (in that case, the subgroup A is said to be *weakly isolated*). (See § 251 and Appendix 101.)

3364 Study the nonabelian p-groups G of exponent $> p^2$ in which all maximal abelian subgroups of G are weakly isolated. (See §§ 252, 254 and Appendix 104.)

3365 Suppose that a nonabelian p-group G of exponent $> p^2$ contains a weakly isolated subgroup of index p. Study the structure of G. (In any p-group of maximal class and exponent $> p^2$, $p > 2$, its fundamental subgroup is weakly isolated.)

3366 Study the $\mathcal{A}_n$-groups, $n > 2$, of exponent $> p > 2$, all of whose $\mathcal{A}_n$-subgroups have exponent p. Is it true that such groups have an abelian subgroup of index p? (See Mann's commentary to #115 or Lemma A.101.1.)

3367 Study the p-groups G of maximal class such that $|G_1'| = p$.

3368 Classify the nonabelian p-groups of exponent $> p$ all of whose maximal abelian subgroups of exponent $> p$ are isolated. (Example: D_{2^n}.)

3369 Study the p-groups whose derived subgroup (Frattini subgroup) is isolated.

3370 Study the p-groups G of exponent $> p$ all of whose normal subgroups of exponent p are isolated.

3371 A subgroup H of a p-group G is said to be *quasi-isolated* if, for any cyclic $C \le G$, $C \cap H > \{1\} \Rightarrow \Omega_2(C) \le H$. Study the $\mathcal{A}_n$-groups, $n > 1$, all of whose minimal nonabelian subgroups (maximal abelian subgroups) are quasi-isolated.

3372 A subgroup H of a p-group G is said to be *partially isolated* if, for any cyclic $C \le G$, $C \cap H > \{1\} \Rightarrow \mho_1(C) \le H$. Study the $\mathcal{A}_n$-groups, $n > 1$, all of whose minimal nonabelian subgroups (maximal abelian subgroups) are partially isolated.

Bibliography

[ADZ] L. J. An, J. F. Ding and Q. H. Zhang, Finite self-dual groups, *J. Algebra* **341** (2011), 35–44.

[AHZ] L. J. An, R. F. Hu and Q. H. Zhang, Finite p-groups with a minimal nonabelian subgroup of index p (IV), *J. Algebra Appl.* **14**(2015), no. 3, 1550020 (54 pages).

[Alle] R. B. J. T. Allenby, Normal subgroups contained in Frattini subgroups are Frattini subgroups, *Proc. Amer. Math. Soc.* **78** (1980), 315–318.

[Alp1] J. L. Alperin, On a special class of regular p-groups, *Trans. Amer. Math. Soc.* **106** (1963), 77–99.

[Alp2] J. L. Alperin, Large abelian subgroups of p-groups, *Trans. Amer. Math. Soc.* **117** (1965), 10–20.

[Alp3] J. L. Alperin, Centralizers of abelian normal subgroups of p-groups, *J. Algebra* **1** (1964), 110–113.

[AK1] B. Amberg and L. Kazarin, On the rank of a finite product of two p-groups, Groups–Korea 94, Kim/Johnson, W. de Gruyter, Berlin (1995), 1–8.

[ALQZ] L. J. An, L. L. Li, H. P. Qu and Q. H. Zhang, Finite p-groups with a minimal nonabelian subgroup of index p (II), *Science China, Math.* **56** (2013), 1–17.

[AQXY] L. J. An, H. P. Qu, M. Y. Xu and C. S. Yang, Quasi-NC groups, *Comm. Algebra* **36**(2008), no. 11, 4011–4019.

[Art] E. Artin, The orders of the linear groups, *Comm. Pure and Appl. Math.* **8** (1955), 355–365.

[AS] M. Aschbacher and S. D. Smith, The Classification of Quasithin Groups 1, Mathematical Surveys and Monographs, Volume 111, AMS, 2004.

[Att] M. S. Attar, A necessary condition of nonabelian p-groups with second centre of order p^2, *Indian J. Pure Appl. Math.* **42**(3) (2011), 183–186.

[AZ] L. J. An and Q. H. Zhang, Finite metahamiltonian p-groups, *J. Algebra* **442** (2015), 23–35.

[BW] A. H. Baartman and J. J. Woeppel, The automorphism group of a p-group of maximal class with abelian maximal subgroup, *Fund. Math.* **93**, 1 (1976), 41–46.

[Bae2] R. Baer, Partitionen endlicher Gruppen, *Math. Z.* **75** (1961), 333–372.

[Bae3] R. Baer, Gruppen mit Hamiltonschen Kern, *Comp. Math.* **2** (1935), 241–246.

[Baer10] R. Baer, Groups with abelian norm quotient group, *Amer. J. Math.* **61**, 3 (1939), 700–708.

[BauBl] G. Baumslag and N. Blackburn, Groups with cyclic upper central factors, *Proc. London. Math. Soc.* **(3) 10** (1960), 531–544.

[Bei2] B. Beisigel, Die Automorphismengruppen homozyklisher p-Gruppen, *Arch. Math.* **29**, 4 (1977), 363–366.

[Ber1] Y. Berkovich, On p-groups of finite order, *Siberian Math. Zh.* **9** (1968), 1284–1306 (Russian).

[Ber3] Y. Berkovich, A generalization of theorems of Ph. Hall and Blackburn and an application to non-regular p-groups, *Math. USSR Izv.* **35** (1971), 815–844.

[Ber25] Y. Berkovich, Finite Groups. Selected Topics, Parts I, II, in preparation.

[Ber26] Y. Berkovich, Maximal abelian and minimal nonabelian subgroups of some finite two-generator p-groups especially metacyclic, *Isr. J. Math.* **194**, 2 (2013) 813–869.

[Ber30] Y. Berkovich, Finite p-groups with few minimal nonabelian subgroups. With an appendix by Z. Janko, *J. Algebra* **297**, 1 (2006), 62–100.

[Ber43] Y. Berkovich, Groups of Prime Power Order, vol. 1, Walter de Gruyter, Berlin, 2008.

[Ber53] Y. Berkovich, Some remarks on subgroups of maximal class of a finite p-group, *J. Algebra and Applications* **14**, 1550043 (2015) (pages 20).

[Ber54] Y. Berkovich, Finite groups all of whose small subgroups are normal, J. Algebra and Applications 13, 6 (2014), 1450006, 10pp.

[Ber55] Y. Berkovich, Sufficient conditions for 2-nilpotence of a finite groups, *J. Algebra and Applications* **12** 4 (2012), pp7.

[Ber56] Y. Berkovich, On the theorem of Mann about p-groups all of whose nonnormal subgroups are elementary abelian, *Isr. J. Math.* **208**, 1 (2015), 451–460.

[Ber57] Y. Berkovich, Alternate proofs of the theorem of G. A. Miller of 2-groups all of whose proper subgroups are Dedekindian, manuscript.

[BCH] Y. Berkovich, D. Chillag, and M. Herzog, Finite groups in which the degrees of the nonlinear irreducible characters are distinct, *Proc. Amer. Math. Soc.* **115** (1992), 955–959.

[BI] Y. Berkovich and I. M. Isaacs, p-supersolvability and actions of p-groups stabilizing certain subgroups, *J. Algebra* **414** (2014), 82–94.

[BJ1] Y. Berkovich and Z. Janko, Groups of Prime Power Order, vol. 2, Walter de Gruyter, Berlin, 2008.

[BJ2] Y. Berkovich, and Z. Janko, Structure of finite p-groups with given subgroups, *Contemporary Mathematics* **402** (2006), 13–93.

[BJ3] Y. Berkovich, and Z. Janko, On subgroups of finite p-groups, *Israel J. Math.* **171** (2009), 29–49.

[BJ4] Y. Berkovich, and Z. Janko, Groups of Prime Power Order, vol. 3, Walter de Gruyter, Berlin, 2011.

[BerZhang] Y. Berkovich and Q. H. Zhang, $\mathcal{A}_1$- and $\mathcal{A}_2$-subgroups of finite p-groups, *J. Algebra Appl.* **13**, 2(2013), 1350095, 26pp.

[BZ] Y. Berkovich and E. Zhmud, Characters of Finite Groups, Parts 1, 2, Translations of Mathematical Monographs 172, 181, American Mathematical Society, Providence, RI, 1998, 1999.

[BT] E. Best and O. Taussky, A class of groups, *Proc. Roy. Irish. Acad. Sect. A.* **47** (1942), 55–62.

[Bla3] N. Blackburn, On a special class of p-groups, *Acta Math.* **100** (1958), 45–92; (1958), 422–427.

[Bla5] N. Blackburn, Generalizations of certain elementary theorems on p-groups, *Proc. London Math. Soc.* **11** (1961), 1–22.

[Bla7] N. Blackburn, Finite groups in which the nonnormal subgroups have nontrivial intersection, *J. Algebra* **3** (1966), 30–37.

[Bla13] N. Blackburn, Groups of prime-power order having an abelian centralizer of type $(r, 1)$, *Monatshefth. Math.* **99** (1985), 1–18.

[BlaDM] N. Blackburn, M. Deaconescu and A. Mann, Equilibrated groups, *Proc. Cambridge Philos. Soc.* **120**, 2 (1996), 579–588.

[BlaH] N. Blackburn and L. Hethelyi, Some further properties of soft subgroups, *Arch. Math.* **65** **(5)** (1997), 365–371.

[Blac2] S. R. Blackburn, Groups of prime power order with derived subgroup of prime order, *J. Algebra* **219** (1999), 625–657.

[Bra] R. Brandl, Groups with few non-normal subgroups, *Comm. Alg.* **23**, 6 (1995), 2091–2098.

[Carl] J. Carlson, Maximal elementary abelian subgroups of rank 2, *J. Group Theory* **10** (2007), 5–13.

[Car1] R. W. Carter, Normal complements of nilpotent self-normalizing subgroups **78** (1962), 149–150.

[Che] Y. Cheng, On finite p-groups with cyclic commutator subgroups, *Arch. Math.* **39** (1982), 295–298.

[ChM] D. Chillag and A. Mann, Nearly odd order and nearly real finite groups, *Comm. Algebra* (1998), 2041–2064.

[Cos] J. Cossey, Special maximal subgroups of p-groups, *Bull. Aust. Math. Soc.* **89**, 3 (2014), 415–419.

[CT] G. Corsi Tani, Automorphisms fixing every normal subgroup of a p-group, *Bull. Un. Mat. Ital. A* **64** (1985), 245–252.

[Cre1] E. Crestani, On 3-generated 2-groups whose maximal subgroups are 2-generated, *Comm. Alg.* **39** (2001), 881–887.

[Cre] E. Crestani, Monotone 2-groups, PhD thesis, Univ. of Padova.

[CrM] E. Crestani and F. Menegazzo, Monotone 2-groups, *J. Group Theory* **15** (2012), 359–383.

[Dar] R. Dark and M. L. Newell, On condition for commutators to form a subgroup, *J. London Math. Soc.* **17** (1978), 251–252.

[Ded] R. Dedekind, Über Gruppen, deren sämtliche Teiler Normalteiler sind, *Math. Ann.* **48** (1897), 548–561.

[Dor] L. Dornhoff, Group Representation Theory, Part A, Marcel Dekker, New York, 1971.

[ERNS] S. Evans-Riley, M. F. Newman and C. Schneider, On the soluble length of groups with prime-power order, *Bull. Aust. Math. Soc.* **59** (1999) 343–346.

[FA] X. G. Fang and L. J. An, The classification of finite metahamiltonian p-groups, submitted.

[Fei1] W. Feit, On large Zsigmondy primes, *Proc. Amer. Math. Soc.* **102** (1988), 29–36.

[FO3] R. Fouladi and R. Orfi, Maximal subsets of pairwise noncommuting elements of some p-groups of maximal class, *Bull. Aust. Math. Soc.* **84** (2011), 447–451.

[FT] W. Feit and J. G. Thompson, Solvability of groups of odd order, *Pacific J. Math.* **13** (1963), 775–1029.

[FZ] W. Feit and G. Seitz, On finite rational groups and related topics, *Illinois J. Math.* **33** (1988), 103–131.

[Fuch] L. Fuchs, Abelian Groups, Oxford, 1960.

[F-A1] G. A. Fernandez-Alcober, An introduction to finite p-groups: regular p-groups and groups of maximal class, *Math. Contemp.* **20** (2001), 155–226.

[GMS1] N. Gavioli, A. Mann, and C. Scoppola, Two applications of the Hughes subgroup of finite group, Ischia Group Theory 2006, World Scientific, Singapore, 2007, 138–146.

[GMS2] N. Gavioli, V. Monti, and C. Scoppola, Pro-p-groups with waists, *J. Algebra* **351** (2012), 130–137.

[Gil] J. D. Gillam, A note on finite metabelian p-groups, *Proc. Amer. Math. Soc.* **25** (1970), 189–190.

[Gla1] G. Glauberman, Large abelian subgroups of finite p-groups, *J. Algebra* **196** (1997), 301–338.

[Gla2] G. Glauberman, Large abelian subgroups of groups of prime exponent, *J. Algebra* **237** (2001), 735–768.

[Gla5] G. Glauberman, Large subgroups of small class in finite p-groups, *J. Algebra* **272** (2004), 128–153.

[Gla6] G. Glauberman, Centrally large subgroups of finite p-groups, *J. Algebra* **300** (2006), 480–508.

[Gla7] G. Glauberman, Existence of normal subgroups in finite p-groups, *J. Algebra* **319** (2008), 800–805.

[Gla8] G. Glauberman, A p-group with no normal large abelian subgroup, *Contemp. Math.* **524**, AMS (2010), 61–65.

[GM] G. Glauberman and N. Mazza, p-groups with maximal elementary abelian subgroup of rank 2, *J. Algebra* **323** (2010), 1729–1737.

[Gol] Y. A. Golfand, On groups all of whose subgroups are nilpotent, *Dokl. Akad. Nauk SSSR* **125** (1948), 1313–1315.

[Golo1] M. I. Golovanov, Certain classes of finite p-groups, Proc. Sem. Algebraic Systems, No. 2, Krasnoyarsk (1970), 5–24.

[Golo2] M. I. Golovanov, Finite groups with condition $|I_m(G)| = p^{m-1}$, Some questions on the theory groups and rings, Krasnoyarsk (1973), 58–82.

[G-S] J. Gonsalez-Sanchez, A p-nilpotency criterion, *Arch. Math.* **94** (2010), 201–205.

[Gor1] D. Gorenstein, Finite Groups, Harper and Row, N. Y., 1968.

[Gor3] D. Gorenstein (Editor), Reviews on Finite Groups, Amer. Math. Soc., Providence, RI, 1974.

[GLS] D. Gorenstein, R. Lyons and R. Solomon, The Classification of the Finite Simple Groups, Part 1, Chapter G: General Group Theory, AMS, Providence, RI, 1995.

[GLS5] D. Gorenstein, R. Lyons and R. Solomon, The Classification of the Finite Simple Groups, Number 5, Providence, RI, 2002.

[Grov1] J. R. J. Groves, On minimal irregular p-groups, *J. Aust. Math. Soc.* **16** (1973), 78–89.

[Grov2] J. R. J. Groves, On direct products of regular p-groups, *Proc. Amer. Math. Soc.* **37** (1973), 377–379.

[Grov3] J. R. J. Groves, Some criteria for the regularity of a direct product of regular p-groups, *J. Aust. Math. Soc. Ser A* **24** (1977), 35–49.

[Grun2] O. Grün, Über das direkte Produkt regulärer p-Gruppen, *Arch. Math.* **5** (1954), 241–243.

[Grun4] O. Grün, Einige Sätze über Automorphismen abelschen p-Gruppen, *Abh. Math. Sem. Univ. Hamburg* **24** (1960), 54–58.

[HalM1] M. Hall, The Theory of Groups, Macmillan, New York, 1959.

[HS] M. Hall and J. K. Senior, On Groups of Order 2^n, ($n \leq 6$), Macmillan, New York, 1964.

[Hal1] P. Hall, A contribution to the theory of groups of prime power order, *Proc. London Math. Soc.* **36** (1933), 29–95.

[Hal2] P. Hall, On a theorem of Frobenius, *Proc. London Math. Soc.* **40** (1936), 468–501.

[Hal3] P. Hall, The classification of prime power groups, *J. reine angew. Math.* **182** (1940), 130–141.

[Hal4] P. Hall, Nilpotent groups, Can. Math. Congr., Alberta, 1957.

[Hal5] P. Hall, On groups of automorphisms, *J. Math.* **182** (1940), 194–204.

[Hal7] P. Hall. Verbal and marginal subgroups, *J. reine angew. Math.* **182** (1940), 156–167.

[Hal8] P. Hall, The Eulerian functions of a group, *Quart. J. Math.* **7** (1936), 134–151.

[Har] K. Harada, On some 2-groups of normal 2-rank 2, *J. Algebra* **20**, 1 (1972), 90–93.

[Haw] T. O. Hawkes, On the automorphism group of a 2-group, *Proc. London Math. Soc.* **26** (1973), 207–225.

[Hei3] H. Heineken, Nilpotente Gruppen, deren sämtliche Normalteiler charakteristisch sind, *Arch. Math.* **33** (1979/80), 497–503.

[HL1] H. Heineken and H. Liebeck, On p-groups with odd order automorphism groups, *Arch. Math.* **24** (1973), 465–471.

[Hel1] G. T. Helleloid, A survey of automorphism groups of finite p-groups, arXiv.mathGR/0610294 v2 25 Oct 2006, 1–20.

[HK] M. Herzog and G. Kaplan, Large cyclic subgroups contain non-trivial normal subgroups, *J. Group Theory* **4** (2001), 247–253.

[Het2] L. Hethelyi, On subgroups of p-groups having soft subgroups, J. London Math. Soc. (2) 41 (1990), 425–427.

[Het3] L. Hethelyi, Some additional properties of subgroups of p-groups having soft subgroups, *Studia Sci. Math. Hungar.* **29** (1994), 317–320.

[Het4] L. Hethelyi, Some remarks on 2-groups having soft subgroups, *Studia Sci. Math. Hungar.* **27** (1992), 295–299.

[Het5] L. Hethelyi, Soft subgroups of p-groups, *Ann. Univ. Sci. Budapest Eötvös Sect. Math.* **27** (1984), 81–85.

[HetK] L. Hethelyi and B. Külshammer, Characters, conjugacy classes and centrally large subgroups of p-groups of small rank, *J. Algebra* **340** (2011), 199–210.

[HKS1] L. Hethelyi and B. Külshammer and B. Sambale, Conjugacy classes and characters of finite p-groups, *Comm. Alg.* **39** (2011), 657–685.

[HP] W. M. Hill and D. B. Parker, The nilpotence class of the Frattini subgroup, *Isr. J. Math.* **15** (1973), 211–215.

[HW] W. M. Hill and C. R. B. Wright, Normal subgroups contained in the Frattini subgroup, *Proc. Amer. Math. Soc.* **35**)1972), 413–415.

[HogK] G. T. Hogan and W. P. Kappe, On the H_p-problem for finite p-groups, *Proc. Amer. Math. Soc.* **20** (1969), 450–454.

[Hop2] C. Hopkins, Nonabelian groups whose groups of automorphisms are abelian, *Ann. Math.* **29** (1927/28), 508–520.

[Hor1] M. V. Horosevskii, The automorphism groups of finite p-groups. (Russian) *Algebra i Logika* **10** (1971), 81–86.

[Hor2] M. V. Horosevskii, The automorphism group of wreath products of finite groups, *Sibirsk. Mat. Z.* **14** (1973), 651–659.

[Hor3] M. V. Horosevskii, Automorphisms of finite groups. (Russian) *Mat. Sb. (N. S.)* **93** (1974), 576–587.

[How] R. B. Howlett, On the exponent of certain factorizable groups, *J. London Math. Soc.* **231** (1985), 265–271.

[HI] R. B. Howlett and I. M. Isaacs, On groups of central type, *Math. Z.* **179** (1982), 555–569.

[HTh] D. R. Hughes and J. G. Thompson, The H_p-problem and the structure of H_p-groups, *Pacif. J. Math.* **9** (1959), 1097–1101.

[Hum] K. Hummel, The order of the automorphism group of a central product, *Proc. Amer. Math. Soc.* **47** (1975), 37–40.

[Hup1] B. Huppert, Endliche Gruppen, Bd. 1, Springer, Berlin, 1967.

[HupB] B. Huppert and N. Blackburn, Finite Groups II, Springer, Berlin, 1982.

[Isa1] I. M. Isaacs, Character Theory of Finite Groups, Acad. Press, N. Y., 1976.

[Isa5] I. M. Isaacs, Algebra: a graduate course, Brooks/Cole, 1994.

[Isa6] I. M. Isaacs, Commutators and commutator subgroup, *Amer. Math. Monthly* **84** (1977), 720–722.

[Isa8] I. M. Isaacs, Equally partitioned groups, *Pacific J. Math.* **49** (1973), 109–116.

[Isa14] I. M. Isaacs, Large orbits in nilpotent action, *Proc. Amer. Math. Soc.* **127** (1999), 45–50.

[Isa19] I. M. Isaacs, Finite Group Theory, AMS, New York, 2008.

[Isa21] I. M. Isaacs, Semipermutable π-subgroups, *Arch. Math (Basel)* **102**, 1 (2014), 1–6.

[IN2] I. M. Isaacs and G. Navarro, Normal p-complements and fixed elements, *Arch. Math.* **95** (2010), 207–211.

[INS] I. M. Isaacs, G. Navarro and J. Sangroniz, p-groups having few almost-rational irreducible characters, *Israel J. Math.* **189** (2012), 65–96.

[Ish] K. Ishikawa, On finite p-groups which have only two conjugacy lengths, *Isr. J. Math.* **129** (2002), 119–123.

[Ito1] N. Ito, On the degrees of irreducible representations of a finite group, *Nagoya Math. J.* **3** (1951), 5–6.

[Ito10] N. Ito, Über eine Frattini-Gruppe duale Bildung, *Nagoya Math. J.* **9** (1955), 123–127.

[IM] N. Ito and A. Mann, Counting classes and characters of groups of prime exponent, *Israel J. Math.* **156** (2006), 205–220.

[Iwa] K. Iwasawa, Über die endlichen Gruppen und die Verbände ihrer Untergruppen, *J. Univ. Tokyo* **4** (1941), 171–199.

[Jan8] Z. Janko, Finite 2-groups with exactly four cyclic subgroups of order 2^n, *J. reine angew. Math.* **566** (2004), 135–181.

[Jan27] Z. Janko, Finite 2-groups all of whose maximal cyclic subgroups of composite order are self-centralizing, *J. Group Theory* **10** (2007), 1–4.

[Jan36] Z. Janko, Finite nonabelian 2-groups all of whose minimal nonabelian subgroups are isomorphic to $M_{2^{n+1}}$, Ichia group theory, 2010, 170–174, World Sci. Publ. Hackensack, NJ, 2012.

[Jan37] Z. Janko, Finite nonabelian 2-groups all of whose minimal nonabelian subgroups are metacyclic and have exponent 4, *J. Algebra*, **321**, 10 (2009), 2890–2897.

[Jan38] Z. Janko, Finite *p*-groups with many minimal nonabelian subgroups, *J. Algebra* **367** (2012), 363–370.

[Jan39] Z. Janko, Finite *p*-groups all of whose nonabelian maximal subgroups have the largest possible centers, manuscript.

[Jan42] Z. Janko, Finite nonabelian *p*-groups having exactly one maximal subgroup with a noncyclic center, *Arch. Math* **96** (2011), 101–103.

[Jan47] Z. Janko, Finite nonabelian *p*-groups all of whose nonabelian maximal subgroups are either metacyclic or minimal nonabelian, *Arch. Math.* **96** (2011), 105–107.

[Jan48] Z. Janko, Finite *p*-groups all of whose maximal subgroups, except one, have its derived subgroup of order $\leq p$, *Glas. Mat.* **48** (67) (2012), 325–332.

[Jan49] Z. Janko, Finite *p*-groups all of whose maximal subgroups, except one, have cyclic derived subgroup, *J. Algebra and its Applications* **14**, 1 (2015), 1450080, 8pp.

[Jan50] Z. Janko, Finite *p*-groups with exactly two sizes of conjugate classes, manuscript.

[Jan51] Z. Janko, Finite *p*-groups with $A \cap B$ being maximal in A or B for any two nonincident subgroups A and B, *Glas. Math.* **49(69)**, 1 (2014), 113–118.

[Jan52] Z. Janko, Finite nonabelian *p*-groups all of whose subgroups are q-self dual, *J. Algebra and Appl.* **6**, 1 (2014), 1450008, 7pp.

[Jan53] Z. Janko, Finite *p*-groups in which the normal closure of each nonnormal cyclic subgroup is nonabelian, Glas. Mat., Ser. III **49(69)**, 1 (2014), 113–118.

[Jan54] Z. Janko, Finite *p*-groups of exponent p^e all of whose cyclic subgroups of order p^e are normal, *J. Algebra* **416** (2014), 274–286.

[Jan55] Z. Janko, Classification of finite *p*-groups with cyclic intersection of any two distinct conjugate subgroups, Glas. Math., Ser III **50(70)**, 1 (2015), 101–161.

[Joh] D. L. Johnson, A property of finite *p*-groups with trivial multiplicator, *Amer. J. Math.* **98** (1976), 105–108.

[Kal1] L. Kaloujnine, La structure des *p*-groupes de Sylow des groupes symetriques finis, *Ann. Sci Ecole Norm. Super.* **65** (1968), 239–276.

[Kar1] G. Karpilovsky, The Schur Multiplier, Clarendon Press, Oxford, 1986.

[Kar2] G. Karpilovsky, Group Representations, vol. 2, North-Holland, Amsterdam, 1993.

[Kar3] G. Karpilovsky, Projective Representations of Finite Groups, Dekker, 1985.

[Kaz1] L. S. Kazarin, Groups with certain conditions for normalizers of subgroups, *Uchen. zapiski Perm Univ.* **218** (1969), 268–279 (Russian).

[Kaz2] L. S. Kazarin, On some classes of finite groups, *Soviet Math. Dokl.* **12, 2** (1971), 549–553.

[Kaz3] L. S. Kazarin, Groups with restrictions on normalizers of subgroups, *Izv. vuzov (mathematics)* **2** (1973), 41–50 (Russian).

[Kaz4] L. S. Kazarin, On product of two nilpotent groups, *Questions of group theory and homological algebra* (1981), 62–66; II, ibid (1982), 47–49 (Russian).

[KazSi] L. S. Kazarin and V. Sidel'nikov, On the automorphism group of the Suzuki p-algebra, *Math. Notes* **80** (2006), 500–508.

[Keg1] O. H. Kegel, Produkte nilpotenter Gruppen, *Arch. Math.* **12** (1961), 90–93.

[Keg2] O. H. Kegel, Die Nilpotenz der H_p-Gruppen, *Math. Z.* **75** (1960), 373–376.

[Keg3] O. H. Kegel, Sylow-Gruppen und Subnormalteiler endlichen Gruppen, *Math. Z.* **78** (1062), 205–221.

[Khu1] E. I. Khukhro, Nilpotent Groups and their Automorphisms, W. de Gruyter, Berlin, 1993.

[Khu2] E. I. Khukhro, p-automorphisms of finite p-groups, CUP, Cambridge, 1998.

[Kim2] I. Kiming, Structure and derived length of finite p-groups possessing an automorphism of p-power order having exactly p fixed points, *Math. Scand.* **62** (1988), 153–172.

[Kin1] B. W. King, Normal subgroups of groups of prime-power order, in "Proc. 2nd Int. Con. Theory Groups," Lecture Notes in Math., **372**, Springer, Berlin (1973), 401–408.

[Kin3] B. W. King, Presentation of metacyclic groups, *Bull. Aust. Math. Soc.* **8** (1973), 101–131.

[Kirt1] J. Kirtland, Finite groups with subgroups not contained in Frattini subgroup are permutable, *Arch. Math.* **97** (2011), 399–406.

[Klu] F. L. Kluempen, The power structure of 2-generator 2-groups of class two, *Algebra Colloq.* **9**, 3 (2002), 287–302.

[Kno] H. G. Knoche, Über den Frobeniusschen Klassenbegriff in nilpotenten Gruppen, I, II, *Math. Z.* **55** (1951), 71–83; ibid **59** (1953), 8–16.

[Kon1] M. Konvisser, Embedding of abelian subgroups in p-groups, *Trans. Amer. Math. Soc.* **153** (1971), 469–481.

[KonJ] M. Konvisser and D. Jonah, Counting abelian subgroups of p-groups. A projective approach, *J. Algebra* **34** (1975), 309–330.

[Kor] I. Korchagina (Capdeboscq), Ranks of p-groups, *Proc. Amer. Math. Soc.* **140**, 11 (2012), 3769–3775.

[Kur] J. Kurdics, Automorphisms of a minimal nonabelian p-group with odd p I, *Glas. Math. Set III* **46**(66), 2 (2011), 367–383.

[KS] H. Kurzweil und B. Stellmacher, Theorie der endlichen Gruppen. Eine Einführung, Springer, 1998.

[Laf1] T. J. Laffey, The minimum number of generators of a finite p-group, *Bull. London Math. Soc.* **5** (1973), 288–290.

[Laf2] T. J. Laffey, Bounding the order of a finite p-group, *Proc. Roy. Ir. Acad.* 80a **2** (1980), 131–134.

[Laf3] T. J. Laffey, A lemma on finite p-groups and some consequences, *Proc. Camb. Philos. Soc.* **75** (1974), 133–137.

[Laf4] T. J. Laffey, Centralizers of elementary abelian subgroups in finite p-groups, *J. Algebra* **51** (1978), 88–96.

[Lang] S. Lang, Algebra, GTM 211, Springer, 2002.

[Leo] A. Leone, Finite minimal non-KC-groups, *Matematiche* **38** (1987), 191–200.

[Leon1] Y. K. Leong, Finite 2-groups of class two with cyclic centre, *J. Aust. Math. Soc. Ser. A* **27** (1979), 125–140.

[Leon2] Y. K. Leong, Odd order nilpotent groups of class two with cyclic centre, *J. Aust. Math. Soc.* **17** (1974), 142–153.

[Lev1] F. W. Levi, Groups in which the commutator operations satisfy certain algebraic conditions, *J. Indian Math. Soc.* **6** (1942), 87–97.

[Lew] M. L. Lewis, On a p-group Camina pairs, *J. Group Theory* **15** (2012), 469–483.

[LP] P. J. Li, A classification of finite p-groups whose proper subgroups are of class ≤ 2 (I), (II), *J. Algebra Appl.* **12**, 3 (2013), 1250171–1–22, 1250171–1–29.

[LW] H. G. Liu and Y. L. Wang, The automorphism group of a generalized extraspecial p-group, *Sci. China Math.* **53** (2010), 315–334.

[LM] P. Longobardi and M. Maj, On p-groups of breadth two, *Algebra Colloq.* **6** (1999), 121–124.

[LQC] L. L. Li, H. P. Qu and G. Y. Chen, Central extension of inner abelian p-groups (I), *Acta Math. Sinica (Chin. Ser.)* **53**, 4 (2010), 675–684.

[LQZ] P. J. Li, Haipeng Qu and Jiwen Zeng, Finite p-groups whose proper subgroups are of class $\leq n$ (manuscript).

[LubM] A. Lubotzky and A. Mann, Powerful p-groups 1, *J. Algebra* **105** (1987), 484–505.

[LZG] Heng Lv, Wei Zhou and Xiuyun Guo, Finite 2-group with index of every cyclic subgroup in its normal closure no greater than 4, *J. Algebra* **342** (2011), 256–264.

[LZY] Heng Lv, Wei Zhou and Dapeng Xu, Some finite p-groups with bounded index of every cyclic subgroup in its normal closure, *J. Algebra* **338** (2011), 169–179.

[LZX] Heng Lv, Wei Zhou and Haijing Xu, Finite p-groups with small subgroups generated by two conjugate elements, manuscript.

[Macd1] I. D. Macdonald, Generalizations of a classical theorem on nilpotent groups, *Illinois J. Math.* **8** (1964), 556–570.

[Macd3] I. D. Macdonald, Commutators and their products, *Amer. Math. Monthly* **93** (1986), 440–444.

[Macd5] I. D. Macdonald, The breadth of finite p-groups, *Proc. Roy. Soc. Edinburgh, Sect. A* **78** (1977/78), 1–39.

[Macd7] I. D. Macdonald, Computer results on Burnside groups, *Bull. Aust. Math. Soc.* **9** (1973), 433–438.

[Macd9] I. D. Macdonald, Some examples in the theory of groups, Mathematical Essays dedicated to A. J. Macintyre, Ohio Univ. Press, Athens, Ohio (1970), 263–269.

[Macd10] I. D. Macdonald, On cyclic commutator subgroups, *J. London Math. Soc.* **38** (1963), 419–422.

[Macd14] I. D. Macdonald, A computer application to finite p-groups, *J. Aust. Math. Soc., Ser. A* **17** (1974), 102–112.

[Mal2] I. Malinowska, p-automorphisms of finite p-groups: problems and questions, *Advances in Group Theory* (2002), 111–127.

[Mal6] I. Malinowska, On the structure of the automorphism group of a minimal nonabelian p-group (metacyclic case), *Glas. Mat.* **47** (67) (2012), 153–164.

[Mal7] I. Malinowska, Finite groups all of whose small subgroups are pronormal, manuscript.

[Man2] A. Mann, Regular p-groups, *Israel J. Math* **10** (1971), 471–477.

[Man3] A. Mann, Regular p-groups, II, *Israel J. Math.* **14** (1973), 294–303.

[Man4] A. Mann, Regular p-groups, III, *J. Algebra* **70** (1981), 89–101.

[Man5] A. Mann, The power structure of p-groups I, *J. Algebra* **42** (1976), 121–135; II, *J. Algebra* **318** (2007), 953–956.

[Man6] A. Mann, Regular p-groups and groups of maximal class, *J. Algebra* **42** (1976), 136–141.

[Man7] A. Mann, Conjugacy classes in finite groups, *Israel J. Math.* **31** (1978), 78–84.

[Man11] A. Mann, The number of generators of finite p-groups, *J. Group Theory* **8** (2005), 317–338.

[Man15] A. Mann, Generators of p-groups, Proc. Groups-St. Andrews 1985, Cambridge (1986), 273–281.

[Man16] A. Mann, Some applications of powerful p-groups, Proc. Groups-St. Andrews 1989, Cambridge (1991), 370–385.

[Man18] A. Mann, Finite groups with maximal normalizers, *Illinois J. Math.* **12** (1968), 67–75.

[Man20] A. Mann, Some questions about p-groups, *J. Aust. Math. Soc. (Series A)* **67** (1999), 356–379.

[Man21] A. Mann, The derived length of p-groups, *J. Algebra* **224** (2000), 263–267.

[Man22] A. Mann, Finite p-Groups, in preparation.

[Man24] A. Mann, On the power structure of some p-groups, *Circ. Mat. Palermo II* **23** (1990), 227–235.

[Man25] A. Mann, Groups with few class sizes and the centralizer equality subgroup, *Isr. J. Math.* **142** (2004), 367–380.

[Man29] A. Mann, On characters-classes duality and orders of centralizers, *Cont. Math.* **402** (2006), 215–217.

[Man30] A. Mann, Normally monomial p-groups, *J. Algebra* **300** (2006), 2–9.

[Man31] A. Mann, On the exponent of the product of two groups, *Rend. Sem. Mat. Univ. Padova* **115** (2006), 205–207.

[Man33] A. Mann, The number of subgroups of metacyclic groups, *Contemporary Mathematics* **524** (2010), 93–95.

[Man34] A. Mann, On skew p-groups, *Publ. Math. Debrecen* **69** (2006), 353–360.

[Man35] A. Mann, The exponents of central factor and commutator subgroups, *J. Group Th.* **10** (2007), 435-436.

[Man36] A. Mann, Corrigendum The number of generators of finite p-groups, *J. Group Theory* **14** (2011), 329–331.

[Man37] A. Mann, More on normally monomial p-groups, manuscript.

[MM] A. Mann and C. Martinez, The exponent of finite groups, *Arch. Math.* **67** (1996), 8–10.

[MP-F] A. Mann and F. Poznick-Fradkin, Subgroups of powerful groups, *Israel I. Math.* **138** (2003), 19–28.

[Mar] Z. Marcinjak, Fixed elements in 2-groups revisited, *Arch. Math.* **97** (2011), 207–208.

[Mat] H. Matsuyama, Solvability of groups of order $2^a p^b$, *Osaka J. Math.* **10** (1973), 375–378.

[Maz2] V. D. Mazurov, 2-groups with an automorphism of odd order fixing all involutions, *Algebra and Logika* **8, 6** (1969), 874–885 (Russian).

[McK1] S. McKay, Finite p-groups, Queen Mary Math. Notes 18, London, 2000.

[McK2] S. McKay, On the structure of a special class of p-groups, *Quart. J. math. (2)* **38** (1987), 489–502.

[Men] F. Menegazzo, Automorphisms of p-groups with cyclic commutator subgroup, *Rend. Sem. Mat. Padova* **90** (1993), 81–101.

[Mie1] R. J. Miech, metabelian p-groups of maximal class, *Trans. Amer. Math. Soc.* **152** (1970), 331–373.

[Mie2] R. J. Miech, Some p-groups of maximal class, *Trans. Amer. Math. Soc.* **189** (1974), 1–47.

[Mie3] R. J. Miech, On p-groups with a cyclic commutator subgroup, *J. Aust. Math. Soc.* **20** (1975), 178–198.

[Mie4] R. J. Miech, The metabelian p-groups of maximal class, *Trans. Amer. Math. Soc.* **236** (1978), 93–119.

[Mie5] R. J. Miech, The metabelian p-groups of maximal class, *II. Trans. Amer. Math. Soc.* **272** (1982), 465–484.

[Mil8] G. A. Miller, On the holomorph of the cyclic group of order p^m, *Trans. Amer. Math. Soc.* **9** (1908), 232–236.

[Mil9] G. A. Miller, The groups in which every subgroup is either abelian or Hamiltonian, *Trans. Amer. Math. Soc.* **8** (1907), 25–29.

[MilM] G. Miller and H. Moreno, Nonabelian groups in which every subgroup is abelian, *Trans. Amer. Math. Soc.* **4** (1903), 398–404.

[Mill] W. H. Mills, The automorphisms of the holomorph of a finite abelian group, *Trans. Amer. Math. Soc.* **85** (1956), 1–34.

[MLC] E. Morgado Morales and M. Lazo Cortis, On the Sylow p-groups of the automorphism group of a finite homocyclic p-group, *Rev. Cienc. Mat.* **6** (1985), 35–44 (Spanish).

[Mori2] M. Morigi, Power automorphisms of finite p-groups, *Comm. Algebra* **70** (1999), 4853–4877.

[Mori4] M. Morigi, On p-groups with abelian automorphism group, *Rend. Sem. Mat. Univ. Padova* **92** (1994), 47–58.

[Mul] O. Müller, On p-automorphisms of finite p-groups, *Arch. Math.* **32** (1979), 533–538.

[NRSB] L. A. Nazarova, A. V. Roiter, V. V. Sergeichuk and V. M. Bondarenko, Application of modules over a dyad to the classification of finite p-groups that have an abelian subgroup of index p and to the classification of pairs of mutually annihilating operators (Russian), Investigations on the theory of representations, *Zap. Nauchn. Sem. LOMI* **28** (1972), 69–92.

[Nei] L. I. Neikirk, Groups of order p^m which contain cyclic subgroups of order p^{m-3}, *Trans. Amer. Math. Soc.* **6** (1905), 316–325.

[Nek2] K. G. Nekrasov, On some 2-groups with a small noncyclic Frattini subgroup, Algebraic and logical constructions, *Tver*, 1994, 53–65 (Russian).

[Nin] Y. Ninomiya, Finite p-groups with cyclic subgroups of index p^2, *Math. J. Okayama Univ.* **36** (1994), 1–21.

[Ore] O. Ore, Contributions to the theory of groups of finite order, *Duke Math. J.* **5** (1938), 431–460.

[Pas] D. S. Passman, Nonnormal subgroups of p-groups, *J. Algebra* **15, 3** (1970), 352–370.

[Pas1] D. S. Passman, Character theory and group rings, *Contemp. Math.* **524** (2010), 139–148.

[Paz] G. Pazdersky, Prime power groups which are cyclic extensions of elementary abelian groups, *Math. Nachr.* **97** (1980), 57–68.

[QW] G. H. Qian, Y. M. Wang, A note on character kernels in finite groups of prime power order, *Arch. Math.* **9**, 3 (2008), 193–199.

[Qu1] H. P. Qu, Finite p-groups all of whose maximal normal abelian subgroups are soft, *Sci. China Math.* **53**, 11 (2010), 3037–3040.

[QZ2] H. P. Qu and Q. H. Zhang, A central extension of minimal nonabelian p-groups (II), *Acta. Math. Sinica* **53:5** (2010), 933–944.

[QH2] H. P. Qu and R. F. Hu, A central extension of minimal nonabelian p-groups (III), *Acta. Math. Sinica* **53:6** (2010), 1051–1064.

[QZ] H. P. Qu and L. F. Zheng, A central extension of minimal nonabelian p-groups (IV), *Acta. Math. Sinica* **54:5** (2011), 739–752 (in Chinese).

[QZGA] H. P. Qu, L. P. Zhao, J. Gao and L. J. An, Finite p-groups with a minimal nonabelian subgroup of index p (V), *J. Algebra Appl.* **13**(2014), no. 7, 1450032 (35 pages).

[QXA] H. P. Qu, M. Y. Xu and L. J. An, Finite p-groups with a minimal nonabelian subgroup of index p (III), *Sci. China Ser. A* **58**(2015), no. 4, 763–780.

[QYXA] H. P. Qu, S. S. Yang, M. Y. Xu and L. J. An, Finite p-groups with a minimal nonabelian subgroup of index p (I), *J. Algebra* **358** (2012), 178–188.

[Rea1] E. W. Read, On the Schur multiplier of a wreath product, *Illinois J. Math.* **20** (1976), 456–466.

[Rea2] E. W. Read, On the centre of a representation group, *J. London Math. Soc. (2)* **16** (1977), 43–50.

[Red1] L. Redei, Das schiefe Produkt in der Gruppentheorie, *Comment. Math. Helvet.* **20** (1947), 225–267.

[Red2] L. Redei, Die endlichen einstufig nichtnilpotenten Gruppen, *Publ. Math. Debrecen* **4** (1956), 303–324.

[Rib] P. Ribenboim, Catalan's conjecture, Academic Press, New York, 1994.

[Rob] D. J. S. Robinson, A course of the theory of groups, NY, Springer, 1996.

[Rock] D. M. Rocke, p-groups with abelian centralizers, *Proc. London Math. Soc.* **(3) 30** (1975), 55–75.

[Roi1] M. Roitman, On Zsigmondy primes, *Proc. Amer. th. Soc.* **125** (1997), 1913–1919.

[Ron] C. Ronse, On centralizers of involutions in 2-groups, *Math. Ser. Cambridge Philos. Soc.* **86** (1979), 199–204.

[Rus2] D. J. Rusin, The 2-groups of rank 2, *J. Algebra* **149** (1992), 1–31.

[Sh] S. Shahriari, On normal subgroups of capable groups, *Arch. Math.* **48** (1987), 193–198.

[Sam1] Sambale B., Fusion systems of bicyclic 2-groups, arXiv: 1206.2962v1 [mat.GR] 123 June 2012.

[Sam2] Sambale B., Exponent and p-rank of finite p-groups and applications, *Arch. Math.* **103** (2014), 11–20.

[ST] J. Sangroniz and J. Tent, 2-groups with few rational conjugate classes, *J. Algebra* **338** (2011), 114–121.

[Schm3] P. Schmid, On the automorphism group of extraspecial 2-groups, *J. Algebra* **234** (2000), 492–506.

[Sch3] O. Schmidt, Groups having only one class of nonnormal subgroups (Russian), *Mat. Sb.* **33** (1926), 161–172.

[Schm5] R. Schmidt, Subgroup Lattices of Groups, Expositions in Mathematics 14, W. de Gruyter, Berlin, 1994.

[Sch4] O. Schmidt, Groups with two classes of nonnormal subgroups (Russian), Proc. Seminar on group theory (1938), 7–26.

[Scot] W. R. Scott, Group theory, Prentice Hall, 1964.

[Sha1] A. Shalev, The structure of finite

[Sha2] A. Shalev, Finite p-groups, in "*Finite and locally finite groups*", Kluwer Acad. Publ., Dordrecht, 1995, 401–450.

[SQG] Q. W. Song, H. P. Qu and X. Y. Guo, Finite p-groups of class 3 all of whose proper subgroups have class at most 2, *Algebra Coll.* **17** (2010), 191–201.

[Son] Q. W. Song, Finite two-generator p-groups with cyclic derived group, manuscript.

[SS] E. G. Straus and G. Szekeres, On a problem of D. R. Hughes, *Proc. Amer. Math. Soc.* **9** (1958), 157–158.

[Suz1] M. Suzuki, Group Theory I, II, Springer, Berlin, 1982, 1986.

[Suz2] M. Suzuki, Structure of a Group and the Structure of its Lattice of Subgroups, Erg. Math. 10, Berlin: Springer 1956.

[Sze] G. Szekeres, On finite metabelian p-groups with two generators, *Acta Sci. Math. Szeged* **21** (1960), 270–291.

[Tah] K. Tahara, On the second cohomology groups of semidirect products, *Math. Z.* **129** (1972), 365–379.

[Tes] L. Teschke, Über die Normalteiler der p-Sylowgruppe der symmetrische Gruppe von Grade p^m, *Math. Nachr.* **87** (1979), 197–212.

[Tho2] J. G. Thompson, Finite groups with fixed-point-free automorphisms of prime order, *Proc. Nat. Acad. Sci. USA* **45** (1959), 578–581.

[Tho3] J. G. Thompson, Nonsolvable finite groups all of whose local subgroups are solvable, 1, *Bull. Amer. Math. Soc.* **74** (1968), 383–437.

[Tho4] J. G. Thompson, Fixed points of p-groups acting on p-groups, *Math. Z.* **86** (1964), 12–13.

[Tho5] J. G. Thompson, Centralizers of elements in p-groups, *Math. Z.* **96** (1967), 292–293.

[Tit] G. N. Titov, Groups containing a cyclic subgroup of index p^3, *Mat. Zametki* **28** (1980), 17–24.

[Tua1] H. F. Tuan, A theorem about p-groups with abelian subgroup of index p, *Acad. Sinica Science Record* **3** (1950), 17–23.

[Tua2] H. F. Tuan, An Anzahl theorem of Kulakoff's type for p-groups, *Sci. Rep. Nat. Tsing-Hua Univ. A* **5** (1948), 182–189.

[T-V] H. P. Tong-Viet, Symmetric groups are determined by their character degrees, *J. Algebra* **334** (2011), 275–284.

[V-LF1] A. Vera-Lopez and G. A. Fernandez-Alcober, Centralizers of small order in p-groups of maximal class, *Comm. Algebra* **20 (40)** (1992), 1051–1059.

[V-LF2] A. Vera-Lopez and G. A. Fernandez-Alcober, The conjugacy vector of a p-group of maximal class, *Israel J. Math.* **86** (1994), 233–255.

[V-LLO] A. Vera-Lopez and L. Ortiz de Elguea, The conjugacy-vectors of all relative holomorphs of an elementary abelian group of order 16, *Portugal Math.* **47**, 3 (1990), 243–257.

[V-LS] A. Vera-Lopez and J. Sangroniz, Classification of all relative holomorphs of an abelian group of type $C_4 \times C_4$, Rev. Acad. Cienc. Zaragoza (2) 43 ('988), 7–12 (Spanish).

[Ver1] L. Verardi, On groups whose non-central elements have the same finite number of conjugates, *Boll. U. M. I. (7)* **2-A** (1988), 391–400.

[Ver2] L. Verardi, A class of finite groups of exponent p in which every normal subgroup is characteristic, *Boll. Un. Mat. Ital. B (6)* **4** (1988), 307–317.

[Waa1] R. W. van der Waall, On finite p-groups whose commutator subgroups are cyclic, *Indag. Math.* **35** (1973), 342–345.

[Waa2] R. W. van der Waall, Certain normal subgroups of the Frattini subgroup of a finite group, *Indag. Math.* **36** (1974), 382–386.

[Waa3] R. W. van der Waall, On modular p-groups, *J. Algebra* **25** (1973), 125–127.

[Waa4] R. W. van der Waall, On modular p-groups II, *Indag. Math.* **37** (1975), 79–82.

[Waa5] R. W. van der Waall, A note on modular p-groups, *Indag. Math.* **35** (1973), 47–48.

[Waa6] R. W. van der Waall, On monomial groups, *J. reine angew. Math.* **264** (1973), 103–134.

[Waa7] R. W. van der Waall, On p-nilpotent forcing groups, *Indag. Math.* **2** (1991), 367–384.

[WaaN] R. W. van der Waall and C. H. W. M. de Nijs, On the embedding of a finite group as a Frattini subgroup, *Bull. Belg. Math. Soc. Simon Stevin* **2** (1995), 519–527.

[Wam] J. W. Wamsley, The deficiency of metacyclic groups, *Proc. Amer. Math. Soc.* **24** (1970), 724–730.

[Wa1] G. E. Wall, On Hughes' H_p-problem, Proc. Internat. Conf. Theory of Groups (Canberra, 1965), Gordon and Breach, New York (1967), 357–362.

[WZ] Lifang Wand and Qinhai Zhang, Finite p-groups whose nonabelian subgroups have the same center, *J. Group Theory* **17** (2014), 689–703.

[Web2] U. H. M. Webb, The number of stem covers of an elementary abelian p-group, *Math. Z.* **182**, 3 (1983), 327–337.

[Web3] U. H. M. Webb, The occurrence of groups as automorphisms of nilpotent p-groups, *Arch. Math.* **37**, 6 (1981), 481–498.

[Web4] U. H. M. Webb, On the rank of a p-group of class 2, *Canad. Math. Bull.* **26** (1983), 101–105.

[Web5] U. H. M. Webb, The Schur multiplier of a nilpotent group, *Trans. Amer. Math. Soc.* **291** (1985), 755–763.

[Web6] U. H. M. Webb, The using graphs to investigate the automorphism groups of nilpotent groups, Graphs and applications, Wiley, New York (1985), 223–247.

[Weir2] A. Weir, The Sylow subgroups of the symmetric groups, *Proc. Amer. Math. Soc.* **6** (1955), 534–541.

[Wie2] J. Wiegold, The Schur multiplier: an elementary approach, Groups — St. Andrews, 1981, *London Math. Soc. Lect. Notes* **71**, 137–154, Cambridge, 1982.

[Wie3] J. Wiegold, The multiplicator of direct product, *Quart. J. Math. Oxford* (2) (1971), 103–105.

[Wiel2] H. Wielandt, Über Produkte von nilpotenten Gruppen, *Illinois J. Math.* **2** (1958), 611–618.

[Wiel5] H. Wielandt, Zum Satz von Sylow. II. *Math. Z.* **71** (1959), 401–402.

[Wil3] B. Wilkens, 2-groups of breadth 3, *J. Algebra* **318** (2007), 202–224.

[Wil5] B. Wilkens, Finite p-groups not characteristic in any p-group in which they are properly contained, *Israel J. Math.* **166** (2008), 97–112.

[Wil6] B. Wilkens, p-groups without noncharacteristic normal subgroups, *Israel J. Math.* **172**, 357–369.

[Wol] B. Wolf, A note on p'-automorphisms of p-groups P of maximal class centralizing the center of P, *J. Algebra* **190**, 1 (1997), 163–171.

[Won2] W. J. Wong, On finite groups whose 2-Sylow subgroups have cyclic subgroups of index 2, *L. Australian Math. Soc.* **4** (1964), 90–112.

[Wri] D. Wright, Degree of minimal permutation representation of covering groups of abelian groups, *Amer. J. Math.* **96** (1974), 578–592.

[XAZ] M. Y. Xu, L. J. An and Q. H. Zhang, Finite p-groups all of whose nonabelian proper subgroups are generated by two elements, *J. Algebra* **319**(2008), 3603–3620.

[Xu1] M. Y. Xu, The power structure of finite p-groups, *Bull. Aust. Math. Soc.* **36**, 1 (1987), 1–10.

[Xu3] M. Y. Xu, A theorem on metabelian p-groups and some consequences, *Chin. Ann. Math., Ser. B5* (1984), 1–6.

[Xu4] M. Y. Xu, Regular p-groups and their generalizations, manuscript.

[XZ1] M. Y. Xu and Q. H. Zhang, A classification of metacyclic 2-groups, *Algebra Colloq.* **13** (2006), 25–34.

[XZA] M. Y. Xu, Q. H. Zhang and L. J. An, Finite p-groups all of whose proper nonabelian subgroups are generated by two elements, *J. Algebra* **319, 9** (2008), 3603–3620.

[Yin1] J. H. Ying, Finite groups with all subgroups isomorphic to quotient groups, *Arch. Math.* **24** (1973), 561–570.

[Yin2] J. H. Ying, Finite groups with restricted quotient structures, *Bull. Inst. Math. Acad. Sinica* **8** (1980), 1–14.

[Zas2] H. Zassenhaus, A group-theoretic proof of a theorem of Maclagan Wedderburn, *Proc. Glasgow Math. Assoc.* **1** (1852), 53–63.

[ZL2] Q. H. Zhang and P. J. Li, Finite p-groups with a cyclic subgroup of index p^3, *J. Math. Res. Appl.* **32**, 5(2012), 505–529.

[Zha2] Q. H. Zhang, A characterization of metacyclic p-groups by counting subgroups, ICA-Proceedings, September 30, 2011, 713–720.

[Zha3] Q. H. Zhang, A characterization of smallest Suzuki 2-group, Acta. Math. Sinica, *English Series* **24**:12 (2008), 2011–2014.

[ZG1] X. H. Zhang and X. Y. Guo, Finite p-groups whose non-normal cyclic subgroups have small index in their normalizers, *J. Group Theory* **15** (2012), 641–659.

[ZG2] L. B. Zhao and X. Y. Guo, Finite p-groups with exactly one $\mathcal{A}_1$-subgroup of given structure of order p^3, *Acta Math. Sin. Engl. Ser.* **29**, No. 11 (2013), 2099–2110.

[ZGQX] Q. H. Zhang, X. Q. Guo, H. P. Qu and M. Y. Xu, Finite groups which have many normal subgroups, *J. Korean Math. Soc.* **46**, 6 (2009), 1165–1278.

[ZZLS] Q. H. Zhang. L. B. Zhao, M. M. Li and Y. Q. Shen, Finite p-groups all of whose subgroups of index p^3 are abelian, Commun. Math. Stat. **3**, 1 (2015), 69–162.

[ZLS] Q. H. Zhang, X. X. Li and M. J. Su, Finite p-groups whose nonnormal subgroups have orders $\leq p^3$, Front. Math. China **17**, 4 (2014), 1169–1194.

[ZS] Q. H. Zhang and M. J. Su, Finite 2-groups whose nonnormal subgroups have orders $\leq 2^3$, *Front. Math. China* **7**, 5 (2012), 971–1003.

[ZW] Q. H. Zhang and J. J. Wei, The intersection of subgroups of finite groups, *Arch. Math.* **96** (2011), 9–17.

[Zhm5] E. M. Zhmud, The isomorphisms of the lattice of normal subgroups of a finite nilpotent group, *Vestnik Kharkov Univ.* **26** (1967), 3–17 (Russian).

Author index

Subject index

O

P

De Gruyter Expositions in Mathematics

Volume 61
Yakov Berkovich, Zvonimir Janko
Groups of Prime Power Order, 2016
ISBN 978-3-11-028145-3, e-ISBN 978-3-11-028147-7, Set-ISBN 978-3-11-028148-4

Volume 60
Benjamin Fine, Anthony Gaglione, Alexei Myasnikov, Gerhard Rosenberger, Dennis Spellman
The Elementary Theory of Groups, 2014
ISBN 978-3-11-034199-7, e-ISBN 978-3-11-034203-1, Set-ISBN 978-3-11-034204-8

Volume 59
Friedrich Haslinger
The d-bar Neumann Problem and Schrödinger Operators, 2014
ISBN 978-3-11-031530-1, e-ISBN 978-3-11-031535-6, Set-ISBN 978-3-11-031536-3

Volume 58
Oleg K. Sheinman
Current Algebras on Riemann Surfaces, 2012
ISBN 978-3-11-026452-4, e-ISBN 978-3-11-026452-4, Set-ISBN 978-3-11-916387-3

Volume 57
Helmut Strade
Simple Lie Algebras, Completion of the Classification, 2012
ISBN 978-3-11-026298-8, e-ISBN 978-3-11-026301-5, Set-ISBN 978-3-11-916682-9

Volume 56
Yakov Berkovich, Zvonimir Janko
Groups of Prime Power Order 3, 2011
ISBN 978-3-11-020717-0, e-ISBN 978-3-11-025448-8, Set-ISBN 978-3-11-218909-2

Volume 55
Rainer Picard, Des McGhee
Partial Differential Equations, 2011
ISBN 978-3-11-025026-8, e-ISBN 978-3-11-025027-5, Set-ISBN 978-3-11-218895-8

Volume 54
Edgar E. Enochs, Overtoun M. G. Jenda
Relative Homological Algebra, 2011
ISBN 978-3-11-021522-9, e-ISBN 978-3-11-021523-6, Set-ISBN 978-3-11-173442-2

Bei Fragen zur Produktsicherheit wenden Sie sich bitte an:
If you have any questions regarding product safety,
please contact:

Walter de Gruyter GmbH
Genthiner Straße 13
10785 Berlin
productsafety@degruyterbrill.com